Clean Energy Production Technologies

Series Editors

Neha Srivastava, Department of Chemical Engineering and Technology, IIT (BHU) Varanasi, Varanasi, India

P. K. Mishra, Department of Chemical Engineering and Technology, IIT (BHU) Varanasi, Varanasi, India

The consumption of fossil fuels has been continuously increasing around the globe and simultaneously becoming the primary cause of global warming as well as environmental pollution. Due to limited life span of fossil fuels and limited alternate energy options, energy crises is important concern faced by the world. Amidst these complex environmental and economic scenarios, renewable energy alternates such as biodiesel, hydrogen, wind, solar and bioenergy sources, which can produce energy with zero carbon residue are emerging as excellent clean energy source. For maximizing the efficiency and productivity of clean fuels via green & renewable methods, it's crucial to understand the configuration, sustainability and techno-economic feasibility of these promising energy alternates. The book series presents a comprehensive coverage combining the domains of exploring clean sources of energy and ensuring its production in an economical as well as ecologically feasible fashion. Series involves renowned experts and academicians as volume-editors and authors, from all the regions of the world. Series brings forth latest research, approaches and perspectives on clean energy production from both developed and developing parts of world under one umbrella. It is curated and developed by authoritative institutions and experts to serves global readership on this theme.

Shiv Prasad • Vikas Chandra Gupta
Meenu Singh • Mohd. Zafar
Editors

Sustainable Bioethanol Production from Lignocellulosic Biomass

Process Engineering and Innovation

Springer

Editors
Shiv Prasad
Division of Environmental Sciences
ICAR-Indian Agriculture Research Institute
New Delhi, Delhi, India

Vikas Chandra Gupta
Department of Biotechnology
IILM University, School of Engineering
Greater Noida, Uttar Pradesh, India

Meenu Singh
Department of Biotechnology
IILM University, School of Engineering
Greater Noida, Uttar Pradesh, India

Mohd. Zafar
Department of Biotechnology, College of Applied Sciences and Pharmacy
University of Technology and Applied Sciences
Sur, Oman

ISSN 2662-6861 ISSN 2662-687X (electronic)
Clean Energy Production Technologies
ISBN 978-981-95-9689-8 ISBN 978-981-95-9690-4 (eBook)
https://doi.org/10.1007/978-981-95-9690-4

This Springer imprint is published by the registered company Springer Nature Singapore Pte Ltd.
The registered company address is: 152 Beach Road, #21-01/04 Gateway East, Singapore 189721, Singapore

Preface

Sustainable Bioethanol Production from Lignocellulosic Biomass is inspired by the global imperative to accelerate the transition to clean, renewable energy in response to the realities of climate change, resource scarcity, and the urgent need for environmental stewardship. This book is envisioned as the definitive resource for researchers, students, industry professionals, and policymakers committed to driving innovation in sustainable energy. It offers a clear and comprehensive exploration of the scientific principles, technological progress, policy frameworks, and industrial practices that underpin lignocellulosic bioethanol production. By uniting foundational theory with real-world application, the text transforms complex knowledge into practical insight, equipping readers to lead and innovate in the evolving landscape of low-carbon energy solutions. Through rigorous analysis of historical milestones, transformative breakthroughs, and future prospects, this book demonstrates the essential and transformative role of lignocellulosic bioethanol in forging a resilient and sustainable energy future for our planet.

This book is organized into comprehensive chapters that delve into sustainable production processes, innovative technologies, policy frameworks, and future opportunities and challenges. Each chapter provides a deep dive into the multifaceted aspects of lignocellulosic bioethanol, equipping readers with a thorough, integrated understanding of the field. Chapter 1 sets the stage with an overview of the worldwide shift from fossil fuels to plant-based biofuels, highlighting environmental benefits, technological advances, and the complex challenges inherent to lignocellulosic biofuel adoption. Chapter 2 investigates the transformation of biomass waste into sustainable bioethanol, detailing advancements in feedstock utilization, production processes, and cutting-edge engineering solutions. Chapter 3 examines progress in feedstock logistics and the development of robust supply chains, emphasizing economic, logistical, and policy dimensions.

Chapter 4 explores the chemistry of lignocellulosic biomass, the essential enzymes involved in its breakdown, the critical role of green chemistry in bioethanol production, and the persisting scientific challenges. Chapter 5 reviews state-of-the-art pretreatment and saccharification techniques, compares methods, and evaluates enzymatic performance to optimize bioethanol yield. Chapter 6 presents synthetic biology strategies for engineering non-conventional yeasts, detailing metabolic pathways, genetic innovations, fermentation technologies, and visionary case studies. Chapter 7 spotlights breakthroughs in omics-driven microbial engineering, focusing on transformative technologies, pivotal studies, yield optimization strategies, and real-world industrial applications. Chapter 8 addresses the integration of advanced experimental design, including statistical methodologies, software tools, artificial intelligence, machine learning, and data-driven optimization for process improvement.

Chapter 9 provides a comprehensive case study on the sustainability of field-to-fuel ethanol production from rice residues, covering biomass availability, biorefinery operations, and the practicalities of scaling up. Chapter 10 investigates potent inhibitors in lignocellulosic ethanol production, elucidating their classifications, toxicity mechanisms, effective detoxification approaches, and holistic biorefinery management. Chapter 11 explores cutting-edge AI-based process modeling and presents a bibliometric analysis of AI applications, process stages, challenges, and the future trajectory of AI in 2G bioethanol. Chapter 12 examines Industry 4.0's transformative impact on sustainable bioethanol production, including real-time monitoring, IoT-driven systems, AI-powered analytics, and emerging challenges.

Chapter 13 underscores the valorization of lignin within circular economy paradigms, exploring lignin's structure, value-added product development, techno-economic assessments, and sustainability considerations. Chapter 14 presents lifecycle and economic analyses, synthesizing recent studies, process-monitoring insights, sensitivity analyses, and comprehensive techno-economic evaluations. Chapter 15 evaluates environmental legislation and compliance frameworks, dissecting global bioethanol policies, institutional and auditing models, and compliance mechanisms. Chapter 16 concludes with an in-depth review of national and international biofuel regulatory frameworks, with special focus on India's National Bioenergy Policy, the Ethanol Blending Program, certification standards, international trade issues, and regulatory challenges.

This book volume stands as a testament to the outstanding collaboration among preeminent scientists, engineers, and thought leaders in lignocellulosic bioethanol. Their multidisciplinary expertise spanning biotechnology, bioengineering, policy, chemistry, and industrial practice has infused the book with cutting-edge perspectives and practical wisdom. We express our deepest appreciation to every contributor whose vision, rigor, and commitment have elevated the quality of each chapter. We are profoundly grateful for the support and encouragement from the Indian Council of Agricultural Research—Indian Agricultural Research Institute (ICAR-IARI) and All India Coordinated Research Project on Energy (AICRP on Energy), New Delhi, India, IILM University, Greater Noida, India, and the University of Technology and Applied Sciences—Sur, Oman. Their guidance, infrastructure, and funding have been pivotal to the realization of this work. Special thanks are also due to the administrative and technical teams at Springer Nature, as well as to the dedicated faculty and students whose invaluable contributions enabled the completion of this book.

As editors, we hope this book serves as a valuable resource for academic researchers, institutional policymakers, and industry professionals alike. Our intention is to bridge the knowledge gap between foundational science, policy frameworks, and industrial applications of lignocellulosic bioethanol. By providing comprehensive insight and practical perspectives, we aim to facilitate advancements in academic inquiry, support evidence-based institutional decision-making, and inspire innovation and best practices within the bioethanol industry.

New Delhi, Delhi, India — Shiv Prasad
Greater Noida, Uttar Pradesh, India — Vikas Chandra Gupta
Greater Noida, Uttar Pradesh, India — Meenu Singh
Sur, Oman — Mohd. Zafar

Contents

Editors and Contributors

About the Editors

Shiv Prasad is a Principal Scientist and senior faculty member at the ICAR-Indian Agricultural Research Institute (IARI), New Delhi, with over 28 years of distinguished experience in agricultural research and education. Widely recognized as a leading expert in biomass-to-biofuel conversion, Dr. Prasad has led transformative research in sustainable crop residue conversion to biofuel and high-efficiency microbial strains for advanced ethanol production. His pioneering work has advanced the scientific understanding of life-cycle assessment, greenhouse gas emissions monitoring, and evidence-based mitigation strategies for Indian agriculture. He has also conducted comprehensive air quality assessments and in-depth investigations into the environmental fate of heavy metals in agroecosystems. He is currently the PI of a major canter of the All India Coordinated Research Project on Energy (AICRP on Energy) at IARI, New Delhi, fostering multidisciplinary collaborations and technology transfer initiatives. His research is extensively published in high-impact, peer-reviewed journals. He has received national and international accolades, including the ICAR Award for Outstanding Work. His prolific scholarly output—encompassing books, technical reports, and over 100 widely cited research articles—has established him as a prominent, highly respected figure in biofuel, agricultural, and environmental research.

Vikas Chandra Gupta is a distinguished Assistant Professor at IILM (Institute of Integrated Learning in Management) University, Greater Noida, and currently leading QC and R&D operations in a leading Indian Ethanol Industry (PGIPL, HP), recognized for his expertise in biotechnology, industrial quality compliances, environmental sustainability, and decarbonization. He earned his PhD in Biotechnology from a leading Indian institution and has established a strong reputation for pioneering research in bioethanol production. He has successfully led and collaborated on several high-impact, government-funded research projects, notably under the auspices of the AKTU-CRIP-Department of Biotechnology (DBT) and the Council of Science and Technology, U.P. (CSTUP). His role as a research consultant in the circular economy has seen him partner with top academic and industrial organizations. His research interests encompass biofuels, industrial waste valorization, experimental design, and sustainable resource management. His commitment to advancing environmental and biotechnological solutions continues to make a meaningful impact on both academic and industry practices.

Meenu Singh is a distinguished Professor of Biotechnology at IILM University, Greater Noida, with over 25 years of experience in teaching, research, and academic leadership. Her expertise includes plant biotechnology, molecular biotechnology, and environmental biotechnology, with a focus on application-oriented and sustainability-driven research, including biofuel and bioethanol. She has led several government-funded research projects and interdisciplinary collaborations addressing contemporary challenges in biotechnology and sustainable development. She has guided numerous doctoral scholars and contributed significantly to scientific literature through publications in reputed national and international journals. She also holds patents in emerging areas of biotechnology innovation. In addition, she actively serves as a reviewer for peer-reviewed journals and doctoral theses and is associated with recognized scientific journals and academic forums. Through her scholarly work, mentorship, and collaborative research, Dr. Singh continues to contribute to impactful research and the advancement of biotechnology research and education.

Mohd. Zafar is a distinguished Assistant Professor at the University of Technology and Applied Sciences—Sur, Oman, and is internationally recognized for his pioneering work in waste valorization, microbial bioconversion, and the application of artificial intelligence to bioprocess and chemical engineering. He has published over 35 influential, peer-reviewed articles in top-tier journals. His research is highly cited, underscoring his impact on environmental biotechnology and on advancing the circular bioeconomy. He has received the Brain Korea 21 fellowship and has been recognized for his innovative research in resource recovery and sustainable bioproduct synthesis. His collaborations span South Korea, India and, Oman, resulting in multidisciplinary projects that bridge academia and industry. He leads research in green technology, AI-driven process optimization, waste valorization, and eco-friendly materials for sustainability.

Contributors

Kofi Akodwaa-Boadi Akenten Appiah-Menka University of Skills Training and Entrepreneurial Development, Accra, Ghana

TCC-CIMET, Kwame Nkrumah University of Science and Technology, Accra, Ghana

Wafa Rashid Al Bahlouli Department of Applied Biotechnology, College of Applied Sciences and Pharmacy, University of Technology and Applied Sciences, Sur, Oman

Sara H. Al-Araimi Department of Biotechnology, College of Applied Sciences and Pharmacy, University of Technology and Applied Sciences, Sur, Oman

Prasad Andhare Department of Life Sciences, School of Sciences, GSFC University, Vadodara, Gujarat, India

Anju Arora Division of Microbiology, ICAR-Indian Agricultural Research Institute, New Delhi, Delhi, India

Shanuja Beri Department of Zoology, Kalindi College, University of Delhi, New Delhi, Delhi, India

Varsha Bisht Environmental and Industrial Biotechnology Division, The Energy and Resources Institute (TERI), New Delhi, Delhi, India

Srabasthi Biswas Department of Paper and Packaging Technology, Indian Institute of Technology Roorkee, Roorkee, Uttarakhand, India

Veeranna Channashettar Environmental and Industrial Biotechnology Division, The Energy and Resources Institute (TERI), New Delhi, Delhi, India

Mohit Chaudhary Gogri Hub for Membranes Research, Indian Institute of Technology Bombay, Mumbai, Maharashtra, India

Dharm Dutt Department of Paper and Packaging Technology, Indian Institute of Technology Roorkee, Roorkee, Uttarakhand, India

Vikas Chandra Gupta Department of Biotechnology, IILM University, Greater Noida, Uttar Pradesh, India

Saurabh Kumar Jha Department of Zoology, Kalindi College, University of Delhi, New Delhi, Delhi, India

Nanthakumar Kuppanan Environmental and Industrial Biotechnology Division, The Energy and Resources Institute (TERI), New Delhi, Delhi, India

Banwari Lal Environmental and Industrial Biotechnology Division, The Energy and Resources Institute (TERI), New Delhi, Delhi, India

Kush Kumar Nayak School of Studies in Biotechnology, Shaheed Mahendra Karma Vishwavidyalaya, Bastar, Jagdalpur, Chhattisgarh, India

Pankti Patel Department of Paper and Packaging Technology, Indian Institute of Technology Roorkee, Roorkee, Uttarakhand, India

Debarati Paul Amity Institute of Biotechnology, Amity University, Noida, Uttar Pradesh, India

Shiv Prasad Division of Environmental Sciences, ICAR-Indian Agriculture Research Institute, New Delhi, Delhi, India

Neelesh Prashant Molecular Systematics Lab, Department of Botany, Central University of Punjab, Bathinda, Punjab, India

Premkumar Samuel S D Central Analytical & Applied Research Unit, College of Science, Sultan Qaboos University, Muscat, Oman

Reecha Sahu Biomedical Engineering and Bioinformatics, Chhattisgarh Swami Vivekanand Technical University, Durg, Chhattisgarh, India

S. Senthil Kumar Department of Applied Biotechnology, College of Applied Sciences and Pharmacy, University of Technology and Applied Sciences, Sur, Oman

Shalley Sharma Department of Biomedical Engineering, Emory University, Atlanta, GA, USA

Avinash Singh Department of Biotechnology, Meerut Institute of Engineering and Technology (MIET), Meerut, Uttar Pradesh, India

Meenu Singh Department of Biotechnology, IILM University, Greater Noida, Uttar Pradesh, India

Ravi Kant Singh Amity Institute of Biotechnology, Amity University, Uttar Pradesh, Noida, UP, India

Reena Singh Indian Council for Research on International Economic Relations (ICRIER), New Delhi, Delhi, India

Magapu Solomon Sudhakar Department of Applied Biotechnology, College of Applied Sciences and Pharmacy, University of Technology and Applied Sciences, Sur, Oman

Purvi Thangaraj Indian Council for Research on International Economic Relations (ICRIER), New Delhi, Delhi, India

Akshaya Kumar Verma Department of Civil Engineering, ITER, Siksha 'O' Anusandhan (Deemed to be University), Bhubaneswar, Odisha, India

Khushbu Verma Department of Life Sciences, School of Sciences, GSFC University, Vadodara, Gujarat, India

Poonam Verma Centre for Biotechnology, School of Pharmaceutical Sciences, Siksha 'O' Anusandhan (Deemed to be University), Bhubaneswar, Odisha, India

Pardeep Yadav Department of Biotechnology, School of Applied and Life Sciences, Uttaranchal University, Dehradun, Uttarakhand, India

Mohd. Zafar Department of Biotechnology, College of Applied Sciences and Pharmacy, University of Technology and Applied Sciences, Sur, Oman

Abbreviations

1G	First-Generation
2G	Second-Generation
AC	Activated Carbon
ACO	Ant Colony Optimization
ACSR	Adaptive Carbon Source Regulation
AD	Anaerobic Digestion
ADH	Alcohol Dehydrogenase
AFEX	Ammonia Fiber Expansion
AI	Artificial Intelligence
AIDMS	Artificial Intelligence Decision-Making System
ALCA	Attributional LCA
ALDH	Aldehyde Dehydrogenase
Aldomt	Aldehyde *O*-Methyltransferase
ALE	Adaptive Laboratory Evolution
ALX	An L-Xylulose Reductase
ANFIS	Adaptive Neuro-Fuzzy Inference Systems
ANN	Artificial Neural Networks
ANN-GA	Artificial Neural Networks-Genetic Algorithm
ANOVA	Analysis Of Variance
ANP	Agency for Petroleum, Natural Gas, and Biofuels
AOC	Annual Operating Cost
AOX	Alternative Oxidase
APEA	Aspen Process Economic Analyzer
ATJ	Alcohol-To-Jet
AX	A Xylan backbone is Arabinoxylan
BECCS	Bioenergy with Carbon Capture and Storage
BEE	Bureau of Energy Efficiency
BEFS	Bioenergy and Food Security
BETO	Bioenergy Technologies Office
BIBD	Balanced Incomplete Block Design
BIS	Bureau of Indian Standards

BM	Bioengineered Microorganism
BPA	Bisphenol A
$C_{12}H_{22}O_{11}$	Cellobiose
$C_9H_8O_3$	Coumaric Acid
CAGR	Compound Annual Growth Rate
CAP	Common Agricultural Policy
CAPEX	Capital Expenditure
CBB	Calvin-Benson-Bassham
CBD	Convention on Biological Diversity
CBG	Compressed Biogas
CBH	Cellulose Endoglucanase Exoglucanase
CBIOs	Carbon Intensity Ranking of Bioethanol and Includes Decarbonization Credits
CBP	Consolidated Bioprocessing
CCD	Central Composite Design
CCR	Carbon Catabolite Repression
CCS	Carbon Capture and Storage
CDM	Clean Development Mechanism
CE	Circular Economy
CFD	Computational Fluid Dynamics
CFP	Catalytic Fast Pyrolysis
CHCs	Custom Hiring Centers
CI	Confidence Interval
CLCA	Consequential LCA
CNC	Cellulose Nanocrystals
CNG	Compressed Natural Gas
CNNs	Convolutional Neural Network
CNT	Carbon Nanotube
CO	Carbon Monoxide
CO_2	Carbon Dioxide
CP	Cleaner Production
CPCB	Central Pollution Control Board
CPS	Cyber-Physical Systems
CRD	Completely Randomized Design
CRISPRa	CRISPR Activation
CRISPRi	CRISPR Interference
CTE	Consent To Establish
CTO	Consent To Operate
CUG	Codons
DBTL	Design-Build-Test-Learn
DCFA	Discounted Cash Flow Analysis
DEA	Data Envelopment Analysis
DES	Deep Eutectic Solvents
DES	Deep Eutectic Solvents

DFG	Damaged Food Grains
DFT	Density Functional Theory
DGEBA	Diglycidyl Ether Bisphenol A
DL	Deep Learning
DLUC	Direct LUC
DML	Dry Matter Loss
DOE	Design Of Experiments
DSA	Dilute Sulfuric Acid
DSBs	Double-Strand Breaks
DT	Decision Tree
DTs	Digital Twins
EBP	Ethanol Blended Petrol
EBPs	Ethanol Blending Programs
EC	Elemental Carbon
ED	Entner–Doudoroff
EFB	Empty Fruit Bunches
EIA	Environmental Impact Assessment
EIA	Energy Information Administration
EISA	Energy Independence and Security Act
EJ	Exajoules
EMPs	Environmental Management Plans
EMS	Environmental Management Systems
EPA	Environmental Protection Agency
EPEs	Environmental Protection Expenditures
EPIs	Energy Performance Indices
EPR	Extended Producer Responsibility
EROI	Energy Return On Investment
ESG	Environmental, Social, and Governance
ESY	Ethanol Supply Year
ETS	Emissions Trading System
EU	European Union
FAO	Food and Agriculture Organization of the United Nations
FC	Fixed Charges
FCI	Food Corporation of India
FCR	Feed Conversion Ratio
FIA	Feature Importance Analysis
FNN	Feedforward Neural Networks
FU	Functional Unit
GA	Genetic Algorithms
GATT	General Agreement on Tariffs and Trade
GBEP	Global Bioenergy Partnership
GBM	Gradient Boosting Machines
GBR	Gradient Boosted Regression
GCA	Gross Cropped Area

GEMs	Genome-Scale Metabolic Models
GGE	Gallon of Gasoline-Equivalent
GH	Glycosyl Hydrolases
GHG	Greenhouse Gas
GIS	Geographic Information Systems
GM	Genetically Modified
GMO	Genetically Modified Organism
GP	Grid-Partitioned
GPR	Gaussian Process Regression
GRAS	Generally Recognized As Safe
GREET	Greenhouse Gases, Regulated Emissions, and Energy Use in Technologies
GT	Glycosyltransferases
GWP	Global Warming Potential
HCHJ	Hydrothermal Condensation Hydrotreating Jet
HDO	Hydrodeoxygenation
HHF	Hybrid Hydrolysis and Fermentation
HMF	Hydroxymethylfurfural
HR	Homologous Recombination
HTL	Hydrothermal Liquefaction
HYVs	High-Yielding Varieties
ICAR	Indian Council of Agricultural Research
IEA	International Energy Agency
IIoT	Industrial Internet of Things
ILs	Ionic Liquids
ILUC	Indirect Land Use Change
IoT	Internet of Things
IRENA	International Renewable Energy Agency
IRR	Internal Rate of Return
IS	Industrial Symbiosis
ISCC	International Sustainability and Carbon Certification
ISETAC	International Society of Environmental Toxicology and Chemistry
ISO	International Organization for Standardization
KVKs	Krishi Vigyan Kendras
LAD	L-Arabitol Dehydrogenase
LCA	Life Cycle Assessment
LCB	Lignocellulosic Biomass
LCC	Lignin Carbohydrate Complex
LCFs	Lignin-Based Carbon Fibers
LCFS	Low Carbon Fuel Standard
LCI	Life Cycle Inventory
LDA	Lignin-Degrading Auxiliary Enzymes
LHW	Liquid Hot Water
LigNPs	Lignin Nanoparticles

LiPs	Lignin Peroxidases
LME	Lignin-Modifying Enzymes
LSTM	Long Short-Term Memory
LUC	Land Use Change
LULUC	Land-Use and Land-Use Change
LVL	Laminated Veneer Lumber
MAE	Mean Absolute Error
MIBK	Methyl Isobutyl Ketone
MIDH	Mission for Integrated Development of Horticulture
ML	Machine Learning
MLP	Multilayer Perceptron
MME	Ministry of Mines and Energy
MMT	Million Metric Tons
MnP	Manganese Peroxidase
MNRE	Ministry of New and Renewable Energy
MoEFCC	Ministry of Environment, Forest and Climate Change
MoPNG	Ministry of Petroleum and Natural Gas
MPC	Model Predictive Control
MPSP	Minimum Selling Prices of Polyhydroxyalkanoates
MSE	Mean Squared Error
MSP	Minimum Selling Price
MSW	Municipal Solid Waste
MT	Million Tons
N_2O	Nitrous Oxide
NABL	National Accreditation Board for Testing and Calibration Laboratories
NDCs	Nationally Determined Contributions
NECPs	National Energy and Climate Plans
NER	Net Energy Ratio
NET	Negative Emission Technologies
NF	Nanofiltration
NFSM	National Food Security Mission
NGT	National Green Tribunal
NHEJ	Non-Homologous End Joining
NIR	Near-Infrared
NIRS	Near-Infrared Reflectance Spectroscopy
NOx	Nitrogen Oxides
NPMCR	National Policy for Management of Crop Residue
NPV	Net Present Value
NREL	National Renewable Energy Laboratory
NRSA	National Remote Sensing Agency
NZE	Net Zero Emission
OC	Organic Carbon
OED	Orthogonal Experimental Design

OMCs	Oil Marketing Companies
OPEX	Operating Expenditure
OSB	Oriented Strand Board
PAH	Polycyclic Aromatic Hydrocarbons
PAN	Polyacrylonitrile
PBIP	Punjab Industrial and Business Development Policy
PBP	Payback Period
PCA	Principal Component Analysis
PDC	Pyruvate Decarboxylase
PDCA	Plan-Do-Check-Act
PDI	Polydispersity Index
PEF	Pulsed Electric Field
PEI	Process Energy Intensity
PF	Phenol-Formaldehyde
PGP	Plant Gate Price
PHA	Polyhydroxyalkanoate
pIMTCP2	Protein Promoter
PIPOL	Plug-In Processes of Lignin
PLA	Polylactic Acid
PLM	Product Lifecycle Management
PLS	Partial Least Squares
PLSR	Partial Least Squares Regression
PNGRB	Petroleum and Natural Gas Regulatory Board
PPPs	Public-Private Partnerships
PSO	Particle Swarm Optimization
PSO-ANN	Particle Swarm Optimization-Ann
PSUs	Public Sector Undertakings
PV	Photovoltaics
QbD	Quality-by-Design
QbT	Quality by Testing
RBD	Randomized Block Design
RBPD	Regional Biorefinery Processing and Distribution
RCBD	Randomized Complete Block Design
RCF	Reductive Catalytic Fractionation
RCFP	Reactive Catalytic Fast Pyrolysis
RES	Renewable Energy Standards
RESPs	Renewable Energy Strategic Plans
RETs	Renewable Energy Targets
RFA	Random Forest Algorithms
RF	Random Forest
RFE	Recursive Feature Elimination
RFS	Renewable Fuel Standard
RIN	Renewable Identification Number
RKVY	Rashtriya Krishi Vikas Yojana

RMSE	Root Mean Square Error
RNN-MPC	Recurrent Neural Network-Based Model Predictive Control
RNNs	Recurrent Neural Networks
ROI	Return On Investment
ROS	Reactive Oxygen Species
RPN4	Proteasome Biogenesis
RPS	Renewable Portfolio Standards
RSB	Roundtable on Sustainable Biomaterials
RSM	Response Surface Methodology
RuBisCO	Ribulose-1,5-Bisphosphate Carboxylase/Oxygenase
RVOs	Renewable Volume Obligations
SA	Simulated Annealing
SAF	Sustainable Aviation Fuel
SATAT	Sustainable Alternative Toward Affordable Transportation
SCWG	Supercritical Water Gasification
SDGs	Sustainable Development Goals
SDS	Sustainable Development Scenario
SEC	Specific Energy Consumption
SED	Statistical Experimental Design
SEP	Steam Explosion Pretreatment
SF	Short Fermentation
SGST	State Goods and Services Tax
SHF	Separate Hydrolysis and Fermentation
SHP	Sodium Hydroxide Pretreatment
SHP-PIPOL	Steam Explosion Pretreatment
SLM	Supported Liquid Membrane
SLR	Simple Linear Regression
SLSC	State Level Sanctioning Committee
SO_2	Sulfur-di-Oxide
SPCB	State Pollution Control Board
SPF	Sun Protection Factor
SREs	Small Refinery Exemptions
SSCF	Simultaneous Saccharification and Co-Fermentation
SSF	Simultaneous Saccharification and Fermentation
SST	Sum of Squares Treatment
SVM	Support Vector Machine
SVR	Support Vector Regression
TBT	Technical Barriers to Trade
TCI	Total Capital Investment
TEA	Techno-Economic Analysis
TFC	Total Capital Investment
TK	Transketolase
TLS	Terrestrial Laser Scanning
TPDC	Total Plant Direct Fixed Cost

TPIC	Total Plant Indirect Fixed Cost
TTW	Tank-To-Wheel
TW	Terawatts
UF	Ultrafiltration
UNEP	United Nations Environment Programme
UNFCCC	United Nations Framework Convention on Climate Change
USDA	United States Department of Agriculture
VAE	Variational Autoencoder
VOC	Volatile Organic Compounds
VOCs	Variable Operating Costs
VPs	Versatile Peroxidases
WC	Working Capital
WPE	Western Plains Energy
WROA	Weighted Rank-Order Aggregate
WSNs	Wireless Sensor Networks
WTO	World Trade Organization
WTT	Well-To-Tank
WTW	Well-To-Wheel
WWTPs	Wastewater Treatment Plants
XDH	Xylitol Dehydrogenase
XGBoost	Xtreme Gradient Boosting
XI	Xylose Isomerases
XR	Xylose Reductase
YTK	Yeast Modular Cloning Toolkit

Chapter 1
Transition from Fossil-Based Fuels to Lignocellulosic Biomass-Based Biofuel Systems: An Overview

Shiv Prasad, Vikas Chandra Gupta, Mohd. Zafar, and Meenu Singh

Abstract The global transition from fossil fuels to biomass-based biofuels marks a significant shift in the energy sector. This transition is motivated by the need to combat climate change, reduce environmental pollution, and enhance energy security. Biofuels are gaining importance, especially as policies favor the use of non-food feedstocks like lignocellulosic biomass. Techno-economic analysis (TEA) and life cycle assessment (LCA) offer insights into biomass conversion for biofuel production, economic feasibility, and environmental impacts, highlighting opportunities to improve biofuel economics by utilizing underutilized biomass such as agricultural crop residues and other organic wastes. A successful transition to biofuels, supported by advanced technologies such as omics-based genetic engineering, artificial intelligence (AI), and machine learning, is crucial to optimizing metabolic pathways and boosting biofuel production efficiency, making biofuels more competitive with traditional petroleum fuels. As global biofuel supply chains expand, international cooperation becomes increasingly vital, emphasizing the importance of collaboration during the energy transition. Establishing standards for sustainability, technical specifications, and trade policies is essential to guide market growth and direction. Transition from fossil fuels to biofuel systems involves various economic, technical, and policy challenges while also prioritizing environmental sustainability. Despite these challenges, progress in technology, supportive policies, and rising environmental consciousness is driving positive momentum for this significant energy transition.

S. Prasad (✉)
Division of Environmental Sciences, ICAR-Indian Agriculture Research Institute, New Delhi, Delhi, India

V. C. Gupta · M. Singh
Department of Biotechnology, IILM University, Greater Noida, Uttar Pradesh, India
e-mail: meenu.singh@iilm.edu

M. Zafar
Department of Biotechnology, College of Applied Sciences and Pharmacy, University of Technology and Applied Sciences, Sur, Oman

S. Prasad et al. (eds.), *Sustainable Bioethanol Production from Lignocellulosic Biomass*, Clean Energy Production Technologies,
https://doi.org/10.1007/978-981-95-9690-4_1

Keywords Fossil fuels · Lignocellulosic biomass · Pretreatment · Enzymatic hydrolysis · Renewables · Bioenergy · Transition · Biofuels · Sustainability · Life cycle · Greenhouse gas

1.1 Introduction

The reliance on fossil fuels such as coal, oil, and natural gas has become increasingly unsustainable due to unprecedented levels of anthropogenic greenhouse gas (GHG) emissions and harmful pollutants worldwide, particularly atmospheric carbon dioxide (CO_2), which is responsible for more than 75% of global GHG emissions (Prasad and Dhanya 2011; Segers et al. 2024). This leads to global temperature increases and cascading environmental challenges, including severe weather events, rising sea levels, and ecosystem disruptions. Global energy consumption, which exceeded 500 quadrillion Btu in the 2010s, is projected to reach 815 quadrillion Btu by 2040—a 48% increase—and reducing fossil fuel use by 2050 is crucial to avoid exceeding the critical 2 °C temperature increase threshold. Therefore, transitioning to renewable energy sources becomes imperative (Gogoi 2024).

Lignocellulosic biomass, the most abundant form of biomass on Earth, accounts for approximately 57% of the planet's biogenic carbon and has an estimated annual global production of 181.5 billion tons (Abolore et al. 2024). It presents a promising alternative for sustainable energy production, yet only 3% of the total lignocellulosic biomass is currently used for bio-based energy, chemicals, and non-food bioproducts. Major feedstock categories include agricultural residues, forestry residues, energy crops, industrial biowaste, urban organic waste, and animal waste (Devi et al. 2022). Innovation in biomass energy deployment is expected to displace 30% of fossil fuel consumption in the near future through biofuels, biochemicals, and biomaterials (Sulis et al. 2025). The impact of biomass conversion technologies and innovation has significantly advanced biofuel production. Recent progress in thermochemical and biological conversion, pretreatment technologies, omics-based microbial genetic engineering, and the integration of artificial intelligence (AI) and machine learning are revolutionizing biofuel development (Zhang and Wang 2021). Techniques such as genomics, metabolomics, metagenomics, and metatranscriptomics offer deep insights into microbial processes essential for biomass conversion.

Leveraging AI with omics data allows us to optimize metabolic pathways and improve the efficiency of biofuel production processes (Xu and Zhang 2019). Additionally, machine learning plays a key role in predicting microbial functions and establishing optimal conditions to promote sustainable, eco-friendly biofuels (Zhang and Wang 2021). These advancements are crucial to environmental sustainability and cleaner energy solutions, reshaping the future of biofuel production (Xu and Zhang 2019). Strategically developed biofuels, whether in solid, liquid, or gaseous form (Prasad et al. 2007), derived from renewable biological resources such as lignocellulosic biomass, offer significant environmental benefits compared to traditional fossil fuels (Prasad et al. 2024). Biofuels, as an alternative to fossil fuels, have the potential to reduce greenhouse gas (GHG) emissions and close the carbon cycle—the CO_2 released during combustion is offset by the CO_2 absorbed during biomass growth (Cacho et al. 2018). Unlike fossil fuels, which release CO_2 into the atmosphere from permanently sequestered underground carbon sinks, sustainably managed lignocellulose biomass can be carbon-neutral, while also addressing food security concerns by not competing with agricultural land used for food production (Sulis et al. 2025). Typically, biofuels produce 50–70% reductions in life cycle greenhouse gas emissions compared to fossil fuels, with cellulose ethanol showing up to an 86% reduction (Farmonaut 2025).

Despite these advantages, significant barriers remain to the large-scale adoption and commercialization of biofuels. The viability of biofuels often involves trade-offs inherent in resource allocation, where competition for land and water resources can lead to conflicts between food and energy production (Daioglou et al. 2020). Moreover, evidence from life cycle assessments indicates that not all biofuels are equal, and some may carry substantial environmental impacts that could offset their climate benefits. This chapter offers an overview of the transition from fossil-based fuels to lignocellulosic biomass-based biofuel systems, emphasizing biofuel production technologies that hold promise for energy security, climate change mitigation, and pollution reduction. It also underscores the importance of carefully considering environmental impacts, technological innovations, and regulatory frameworks to realize their potential fully. The complex nature of biofuel systems calls for an integrated approach to leverage these resources sustainably for climate action.

1.2 Key Drivers for the Transition from Fossil-Based Fuels to Biofuels

Key drivers of the transition from fossil fuels to biofuels are shown in Fig. 1.1. Environmental concerns primarily motivate this transition, especially the need to reduce pollution to improve air quality and to reduce greenhouse gas (GHG) emissions (Prasad and Dhanya 2011; Prasad et al. 2024). Biofuels made from biomass provide renewable options that can leverage existing infrastructure and cut life cycle GHG emissions by 60–90% compared to fossil fuels (Schulz et al. 2025). Energy security is another key factor, encouraging the use of domestic biomass to lessen dependence on imported oil. This is especially important in

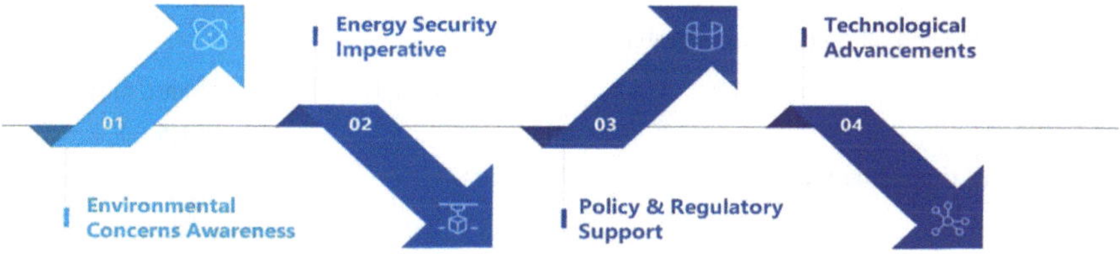

Fig. 1.1 Key drivers for the transition from fossil-based fuels to biofuels

developing economies, where biofuel production can support rural development by creating jobs and boosting agribusiness. However, issues like feedstock availability and competition with food crops must be addressed to enhance their sustainability (Prasad et al. 2024). Policy support, including measures such as promotion and blending mandates, is essential for advancing biofuel production.

Worldwide, significant attention is given to technological advancements such as biomass pretreatment, enzymatic hydrolysis, and fermentation, which have improved domestic and sustainable biofuel production to address various environmental issues, including climate change, while also supporting rural economies and livelihoods in line with the goals of the sustainable development scenario (SDS) framework (Prasad et al. 2024). Incorporating biomass supply chains into biofuel production can reduce costs, and long-term projects help lower investment risks. The urgent need to reduce GHG emissions to address climate change has increased the appeal of biofuels as alternatives to fossil fuels, especially as oil prices rise (Kiptoo et al. 2023). Research indicates that although biofuels can substantially reduce carbon emissions, a complete assessment of their overall effects is vital (Martin et al. 2017). Technological breakthroughs that improve the efficient use of non-edible feedstocks enhance biofuel efficiency and sustainability. However, challenges related to this must be addressed. Sustainable practices and technology-neutral policies are essential for considering the entire life cycle of biofuels. Government incentives play a vital role in supporting biofuel production, reducing associated costs, and fostering rural development.

1.3 Fossil Fuel Dominance in Global Energy Mix and Global Biofuel Production Trends

The global energy mix has historically been dominated by fossil fuels—coal, oil, and natural gas—sources that have served as the primary energy providers for over a century. In 2023, fossil fuels (oil, coal, and natural gas) still made up 81.5% of global primary energy consumption, reaching a record 505 EJ out of a total of 620 EJ (IEA 2023), as shown in Table 1.1, which contributed to widespread greenhouse gas emissions that drive global warming and climate change. The combustion of fossil fuels is estimated to release about 29 gigatons of CO_2 each year, thereby significantly increasing atmospheric greenhouse gas concentrations (Olabi et al. 2022).

Table 1.1 Fossil fuel dominance in global energy mix

Energy source	Consumption (EJ)	Share of total (%)
Fossil fuels	505	81.5
Renewables (including biofuels)	93	15.0
Nuclear power	22	3.5
Total	**620**	**100**

Source: IEA-World Energy Outlook (2021)

This heavy dependence on fossil fuels not only worsens environmental degradation but also raises concerns about energy security as reserves dwindle. Renewable sources reached their highest-ever share—roughly 15%—but remain insufficient to curb emissions immediately (IRENA 2023).

The global lignocellulosic biofuel market was valued at USD 4.1 billion in 2024 and is expected to reach USD 99.8 billion by 2034, with a compound annual growth rate (CAGR) of 37.6% (Smith and Jones 2024). North America makes up about 42.3% of the cellulosic ethanol market, with the United States holding a dominant position thanks to established Renewable Fuel Standard (RFS) mandates and significant research investments in Iowa and Nebraska (EPA 2023). Brazil's cellulosic ethanol production hit 55 million liters in 2022, an increase of 15 million liters from 2021, indicating accelerating commercialization in South America (Broda et al. 2022). The top 10 countries producing lignocellulosic biomass-based biofuels, along with their shares of global production, key technological pathways, and main feedstocks, are shown in Table 1.2.

1.4 Lignocellulosic Biomass Composition and Structure

Lignocellulosic biomass mainly consists of plant cell walls that contain three key structural components: cellulose $(C_6H_{10}O_5)_n$, hemicellulose $(C_5H_{10}O_5)_m$, and lignin $(C_{10}H_{11}O_{3.5})$ (Chakraborty et al. 2024). The typical ranges include cellulose content of 28–55%, hemicellulose of 17–35%, and lignin of 17–35% (Gogoi 2024), with varying amounts of minerals, moisture, proteins, and trace elements depending on their source. Cellulose is the most abundant biopolymer on Earth and acts as the main structural component, while hemicelluloses are amorphous and chemically diverse. Lignin has the highest carbon content (~60%) among lignocellulosic biomass components due to its large molecular size and naturally branched structure, although it remains underutilized (Sulis et al. 2025). The primary structural components of lignocellulosic biomass and their properties are shown in Table 1.3.

Major feedstock categories include agricultural residues (rice straw, wheat straw, corn stover, sugarcane bagasse), forestry residues (woody biomass, branches, foliage), energy crops (switchgrass, miscanthus), industrial waste (pulp and paper effluents), urban organic waste (municipal solid waste), and animal waste (Prasad et al. 2007). Lignocellulosic feedstocks offer several advantages over first-generation biofuels derived from food-grade feedstocks. Second-generation biorefineries powered by lignocellulosic feedstocks reduce greenhouse gas emissions, can be grown on marginal lands unsuitable for food crops, and lessen strain on limited water resources (Sulis et al. 2025). Currently, biomass accounts for 10% of global energy use, with forecasts indicating this will rise significantly by 2050 (Gogoi 2024). The composition of lignocellulosic biomass varies greatly depending on the feedstock type, as shown in Table 1.4.

Table 1.2 Top ten lignocellulosic biomass-based biofuel-producing countries

Rank	Producing countries	Global share (%)	Major feedstocks	Status & key technologies	References
1	United States	38–42%	Corn stover, wheat straw, wood residues, forest waste	Pilot & commercial (fermentation to cellulosic ethanol, hydrotreatment HVO/SAF)	Sonnleitner and Bacovsky (2024) and Robak et al. (2020)
2	Brazil	15–20%	Sugarcane bagasse, corn stover, wood residues, and industrial waste	Operational facilities (55 ML cellulosic ethanol in 2022)	Broda et al. (2022) and U.S. Biofuels Report (2023)
3	Germany	8–12%	Forest residues, agricultural waste, woody biomass	Demonstration & commercial (gasification, fermentation, fast pyrolysis)	Sonnleitner and Bacovsky (2024) and Market.us (2025)
4	Canada	5–8%	Forestry residues, agricultural by-products, wood chips	Pilot-stage facilities (fermentation, hydrotreatment)	Sonnleitner and Bacovsky (2024)
5	China	4–6%	Agricultural residues, waste oils, woody biomass	Emerging development (biochemical conversion, pilot facilities)	Market.us (2025) and DataInsights (2024)
6	France	3–5%	Forest residues, agricultural waste, biomass crops	Demonstration facilities (fermentation, thermochemical routes)	Market.us (2025) and Strategic Market Research (2023)
7	Sweden	2–4%	Forest residues, wood waste, biomass co-products	Commercial & demonstration (fast pyrolysis, gasification, co-processing)	Sonnleitner and Bacovsky (2024) and van Dyk and Saddler (2024)
8	Netherlands	1–3%	Agricultural residues, waste oils, biomass residues	Demonstration stage (hydrotreatment, co-processing in refineries)	Sonnleitner and Bacovsky (2024)
9	Finland	1–2%	Forest residues, wood by-products, sawmill waste	Pilot & demonstration (thermochemical conversion, pyrolysis)	Sonnleitner and Bacovsky (2024)
10	India	0.5–1.5%	Agricultural waste, sugarcane bagasse, rice straw	Emerging research & pilot stage (anaerobic digestion, fermentation)	Market.us (2025) and DataInsights (2024)

Table 1.3 Major structural components of lignocellulosic biomass and their properties

Component	Chemical formula	Content (wt%)	Key characteristics	Potential applications	References
Cellulose	$(C_6H_{10}O_5)_n$	40–50	Linear polymer of glucose; crystalline structure	Bioethanol, nanomaterials, textiles, paper products	Sulis et al. (2025)
Hemicellulose	$(C_5H_{10}O_5)_m$	25–35	Branched heteropolymer; amorphous	Platform chemicals, xylitol, bioethanol	Sulis et al. (2025)
Lignin	$C_{10}H_{11}O_{3.5}$	16–33	Complex aromatic polymer; highest carbon content (~60%)	Vanillin, phenolic compounds, and energy production	Chakraborty et al. (2024)

Table 1.4 Composition of different lignocellulosic biomass types

Biomass type	Cellulose (%)	Hemicellulose (%)	Lignin (%)	References
Agricultural residues				
Wheat straw	33–40	20–25	15–20	Sun and Tomkinson (2003)
Rice straw	30–50	10–20	10–30	Chen et al. (2021)
Corn stover	35–55	25–40	15–25	Ravindra et al. (2021)
Sugarcane bagasse	30–50	20–30	15–25	Pérez et al. (2025)
Cotton stalk	43	–	26.48	Iqbal et al. (2016)
Woody biomass—hardwoods				
Hardwoods (general)	40–50	15–40	18–25	Matin et al. (2023)
Oak	50.0	28.4	21.6	Rocha et al. (2025)
Poplar	39	21	27	Zamora et al. (2013)
Woody biomass—softwoods				
Softwoods (general)	40–45	24–37	25–35	Matin et al. (2023)
Pine (Southern)	47	–	–	
Pine	41.2	–	34.5	Rocha et al. (2025)
Energy crops				
Miscanthus	40–60	20–35	15–40	Zoghlami and Paës (2019)
Switchgrass	40–60	20–35	15–40	
Other biomass				
Bamboo	47.2	23.9	25.3	Shi et al. (2013)
Sorghum stover	30.3	25.5	5.51	Shah et al. (2025)
Barley straw	35–55	25–40	15–25	
Industrial residue	40–45	30–35	20–25	Bukhari et al. (2025)
Secondary streams	25–29	18–20	17–21	Segers et al. (2024)

1.5 Biomass to Biofuels Conversion Technologies and Pathways

Biomass conversion to biofuels is vital for meeting global energy needs and combating climate change. There are two main technological approaches for transforming lignocellulosic biomass into renewable biofuels: biochemical conversion and thermochemical conversion (Prasad et al. 2020). Each approach processes specific feedstocks to reduce dependence on fossil fuels and has its own characteristics, advantages, and limitations. Choosing between thermochemical and biochemical methods depends on the feedstock's properties, the desired fuel type, and various economic and environmental factors (Yan et al. 2026). Generally, thermochemical processes provide higher efficiency and more flexibility (García and Teixeira 2022), while biochemical methods are better suited for feedstocks with high moisture and sugar content. Increasingly, advanced biorefineries are integrating both methods or using hybrid systems. These hybrid systems are likely to shape the future of sustainable biofuel production (López and Ferreira 2023). A comparison of biochemical and thermochemical conversion technologies is shown in Table 1.5.

1.5.1 Thermochemical Conversion Pathways

Thermochemical processes involve the high-temperature decomposition of lignocellulosic biomass, in the absence of oxygen or with limited oxygen, to produce biofuels. These processes include four main types: direct combustion, gasification, pyrolysis, and liquefaction (Prasad et al. 2020). Biomass pyrolysis offers a promising route to renewable fuels and chemicals, significantly reducing reliance on fossil fuels (Yan et al. 2025). Pyrolysis involves the thermal decomposition of biomass at temperatures ranging from 300 to 1000 °C in the absence of oxygen, producing bio-oil, biochar, and non-condensable gases. Fast pyrolysis, conducted at 400–600 °C with residence times of 0.5–5 s, yields 70–80% bio-oil, 10–20% gas, and 15–25% biochar (Segers et al. 2024). Bio-oil requires upgrading via hydrotreatment, hydrocracking, or zeolite cracking to produce drop-in fuels compatible with existing infrastructure (Gogoi 2024).

Gasification is the partial oxidation of biomass at high temperatures (>700 °C) under controlled conditions in the presence of a gasification agent such as steam, O_2, or air, producing syngas mainly composed of CO, H_2, CO_2, and CH_4 (Singh et al. 2022). Syngas can be further upgraded through Fischer-Tropsch synthesis or methanol synthesis to produce liquid fuels, including diesel, gasoline, and sustainable aviation fuels. Hydrothermal liquefaction (HTL) is a physical and chemical conversion process operating at 250–374 °C and 4–22 MPa, enabling the transformation of both wet and dry lignocellulosic biomass into liquid bio-crude oil, biochar, and gases (Yan et al. 2026). This process requires lower temperatures than pyrolysis or gasification, making it more energy-efficient for specific applications (Segers et al. 2024).

Table 1.5 Comparison of thermochemical and biochemical conversion technologies

Technology	Process description	Operating conditions	Main products	Efficiency	Advantages	Disadvantages	References
Thermochemical biomass to biofuels conversion technologies							
Fast pyrolysis	Rapid thermal biomass decomposition at high heating rates without O_2; quick vapor quenching yields liquid bio-oil	Temperature: 400–600 °C; Residence time: <2 seconds; Heating rate: 100–1000 °C/min; Pressure: atmospheric	Bio-oil, biochar, syngas (non-condensable gases)	Liquid yield: 60–75 wt%; Biochar: 12–15 wt %; Gas: 13–25 wt%	Rapid processing, high liquid yields, versatile use, transportable, and compatible with existing infrastructure	High O_2 in bio-oil (35–40 wt%), poor stability, high viscosity, corrosive, low energy density compared to petroleum, and high thermal energy needs	Esteves et al. (2024), Stöcker (2024), and Yan et al. (2025)
Catalytic fast pyrolysis (CFP)	Fast pyrolysis with in-situ catalysis upgrades bio-oil by facilitating deoxygenation and hydrocarbon formation	Temperature: 400–600 °C; Catalyst bed temperature: variable; Residence time: <2 s; Catalysts: zeolites, metal oxides	Hydrocarbon-rich bio-oil, biochar, gases	Bio-oil yield: 40–50 wt% (reduced from non-catalytic); Reduced O-content: 10–30 wt%	Improved bio-oil with a higher H/C ratio, lower O_2, better stability, refinery compatibility, and synergistic effects	Higher catalyst costs, deactivation issues, reduced biochar quality, complex operation, and the need for catalyst recovery	Esteves et al. (2024) and Stöcker (2024)
Reactive catalytic fast pyrolysis (RCFP)	Fast pyrolysis with continuous catalyst regeneration and reactive component addition yields the lowest-O_2 bio-oil among pyrolysis methods	Temperature: 400–600 °C; Pressure: atmospheric; Reactive injection: hydrogen or reactive compounds	Bio-oil with fewer oxygenates, biochar, and syngas	Bio-oil yield: 35–50 wt%; O-content: 5–15 wt%; H/C ratio: improved	RCFP bio-oil is of higher quality, with fewer oxygenated compounds, more unsaturated hydrocarbons, and greater similarity to CFP	Highest capital investment, most complex operation, reactive component costs, research needed on scalability, O/C ratio still exceeds petroleum crude	Esteves et al. (2024)

(continued)

Table 1.5 (continued)

Technology	Process description	Operating conditions	Main products	Efficiency	Advantages	Disadvantages	References
Slow pyrolysis	Slow thermal decomposition at lower heating rates and longer residence time promotes biochar formation	Temperature: 300–700 °C; Residence time: minutes to hours; Heating rate: <10 °C/min	Biochar (primary), bio-oil, syngas	Liquid yield: 30–45 wt%; Biochar: 34–63 wt%; Gases: 5–20 wt %	Optimized for biochar production, soil amendment, lower costs, and established tech	Low bio-oil yields, unoptimized for fuel, longer processing, and lower energy density	UCL (2024)
Gasification	Controlled partial oxidation of biomass (>700 °C) with controlled O_2 supply produces synthesis gas (H_2 + CO)	Temperature: 750–850 °C; Pressure: 1–40 bar; Equivalence ratio: 0.2–0.3 (O_2/biomass stoichiometric ratio); Steam/air ratio variable	Syngas (H_2: 20–50 mol%, CO: 15–30 mol%, CO_2: 9–33 mol%, CH_4: 5–20 mol %), tar, ash, biochar	Syngas yield: 70–80%; H_2 content: 30–47 mol% (optimal conditions); H_2/CO ratio: 1.1–2.0	Scalable for large capacities, flexible feedstock options, tailored H_2/CO ratios for synthesis, and possible catalyst integration	High capital costs, tar formation, ash handling, catalyst costs, complex gas cleanup, and immature technology for commercial scale	Tewari et al. (2024), Gao et al. (2023), IEA (2024), and Singh et al. (2022)
Hydrothermal liquefaction (HTL)	Pressurized hot-water conversion uses water as the solvent, reactant, and heat-transfer medium under subcritical or supercritical conditions, making it suitable for wet biomass	Temperature: 250–374 °C; Pressure: 10–25 MPa; Residence time: 15–60 min; Solids loading: 10–30 wt%	Bio-crude oil, aqueous phase with organics, biochar, gases	Bio-crude yield: 50–70 wt %; Energy recovery: 70–80% HHV	No drying needed; wet biomass compatible; high energy density (>30 MJ/kg); good H/C ratio; lower tar yields than pyrolysis; high viscosity crude for refining	High-pressure equipment and wastewater treatment are needed, with high capital costs. Bio-crude has 8–12 wt% O_2 and requires a catalyst, solvent, and further upgrading for fuel use	Mishra et al. (2022), Badoga et al. (2023), Taggar, et al. (2024), and Yan et al. (2026)

Biochemical biomass to biofuels conversion technologies							
Enzymatic hydrolysis & Fermentation	Cellulose and hemicellulose are hydrolyzed into sugars such as glucose, xylose, and mannose, which are then fermented with yeast or bacteria	Temperature: 30–55 °C; pH: 4.5–5.0; Enzyme loading: 10–40 mg/g; Duration: Hours to days	Bioethanol, biogas	>80% sugar conversion; Glucose yields up to 118 g/L with optimized cellulose amorphicity	High selectivity; mild conditions; minimal inhibitor formation; compatible with existing infrastructure	High enzyme costs (~$1–3/gallon ethanol); slow kinetics; separate C_5/C_6 sugar utilization required	Celignis (2025), Frontera (2024), Mukhtar et al. (2024), and Gogoi (2024)
Simultaneous saccharification & Fermentation (SSF)	Enzymatic hydrolysis and fermentation co-occur in a single reactor; the microorganisms immediately consume glucose produced by hydrolysis	Temperature: 30–35 °C (a compromise between enzyme [45–55 °C] and organism [25–35 °C] optima); Pressure: atmospheric; Duration: 48–72 h	Bioethanol	75–90% ethanol yield on consumed sugars	Eliminates end-product inhibition; improved glucose utilization; reduced processing time and capital costs	Suboptimal enzyme conditions; fermenting organism cannot be recirculated; poor initial ethanol yield with some preparations	Frontera (2024), Sulis et al. (2025), and Gogoi (2024)
Hybrid hydrolysis & Fermentation (HHF)	Initial enzymatic saccharification at 45–55 °C, then temperature reduction and yeast addition for fermentation	Phase 1: 45–55 °C, 24–48 h enzymatic hydrolysis; Phase 2: 30–35 °C, fermentation	Bioethanol	90% + glucan conversion; variable ethanol productivity depending on aeration	Maximizes enzyme efficiency; better sugar conversion than SSF; improved aeration control reduces inhibitory by-products	More complex process control; longer total processing time	Frontera (2024), Sulis et al. (2025), and Gogoi (2024)

(continued)

Table 1.5 (continued)

Technology	Process description	Operating conditions	Main products	Efficiency	Advantages	Disadvantages	References
Separate hydrolysis & Fermentation (SHF)	Sequential process: enzymatic hydrolysis of pretreated biomass, then fermentation of sugars in a separate vessel	Hydrolysis: 45–55 °C, 48–72 h; Fermentation: 30–35 °C, 48–96 h	Bioethanol	60–75% overall conversion	Allows independent optimization, enzyme recycling, and reuse, enabling high glucose concentrations	End-product glucose inhibition reduces enzyme activity; lower ethanol yields than SSF historically; and higher costs from separate reactors	Frontera (2024), Segers et al. (2024), Sulis et al. (2025), and Gogoi (2024)
Dark fermentation (Biohydrogen)	Anaerobic microorganisms ferment carbohydrates from pretreated biomass, as pretreatment disrupts lignocellulose to improve sugar release	Temperature: 35–38 °C (mesophilic) or 50–55 °C (thermophilic); pH: 5.5–7.0; Anaerobic conditions maintained	Biohydrogen (H_2), volatile fatty acids (VFAs)	40–60% hydrogen yield (moles H_2 per mole sugar); 180–250 mmol H_2/L biomass	Sustainable energy carrier; utilizes agricultural residues; integrable with other biofuel production	Low hydrogen yields; inhibitor sensitivity; long residence times (2–5 days); requires strict anaerobic conditions	Mukherjee and Mohan (2023)
Anaerobic digestion & Bio CH_4	Lignocellulosic biomass goes through hydrolysis, acidogenesis, acetogenesis, and methanogenesis to produce biogas	Temperature: 37 °C (mesophilic); pH: 6.8–7.4; Hydraulic retention time: 15–30 days; Total solids: 5–10%	Biogas (55–65% CH_4, 35–45% CO_2), digestate	150–250 m^3 biogas/ton volatile solids; CH_4 content increases with pretreatment	Waste-to-energy conversion; lower enzyme costs; produces fertilizer byproduct; thermochemically stable	High capital costs for anaerobic reactors; slow process kinetics; biomass recalcitrance limits efficiency; potential for CH_4 emissions potential	Xu et al. (2019, 2024)

1.5.2 Biochemical Conversion Pathways

Biochemical conversion pathways involve biological catalysts, such as enzymes or microorganisms, to break down lignocellulosic biomass into intermediates—particularly fermentable sugars—under milder conditions, which are then converted into biofuels like bioethanol and biogas (Baruah et al. 2018). The process generally includes three main steps: pretreatment, enzymatic hydrolysis (saccharification) of cellulose and hemicellulose into fermentable sugars, and fermentation of these sugars into liquid fuels (Prasad et al. 2024). Saccharification uses enzymatic hydrolysis with cellulases—including endoglucanase, exoglucanase, and β-glucosidase—to convert cellulose into glucose and oligosaccharides. This enzymatic hydrolysis is effective, with conversion rates exceeding 80%, though the high cost of enzymes remains a barrier to industrial-scale application (Sulis et al. 2025). Fermentation employs microorganisms such as *Saccharomyces cerevisiae, Zymomonas mobilis,* or genetically modified strains to convert sugars into bioethanol, biobutanol, and other biofuels (Segers et al. 2024). There are three main processes for converting lignocellulosic biomass into biofuel, especially ethanol: separate saccharification and fermentation (SSF), simultaneous hydrolysis and fermentation (SHF), and simultaneous saccharification and co-fermentation (SSCF). These are detailed in Table 1.6, which covers different configurations, operating conditions, inhibition effects, sugar utilization, performance, and scale.

1.6 Technological Advances in Biofuels Production Technologies

Recent advances in omics-based microbial genetic engineering, combined with artificial intelligence (AI) and machine learning, are significantly transforming biofuel production (Smith et al. 2022). Techniques such as genomics, metabolomics, metagenomics, and meta-transcriptomics improve our understanding of how microbes convert biomass into biofuels (Johnson and Lee 2021). By integrating AI with omics data, we can optimize metabolic pathways and boost the efficiency of biofuel fermentation processes (Chen et al. 2023). Moreover, machine learning helps predict microbial functions, facilitating the development of conditions that support sustainable and environmentally friendly biofuels. These innovations are essential not only for environmental sustainability and cleaner energy solutions but also for shaping the future of biofuel production (Garcia and Patel 2023).

Table 1.6 Comparison of SHF, SSF, and SSCF configurations for lignocellulosic ethanol

Aspect	SHF (separate steps)	SSF (one-vessel)	SSCF (one-vessel)	
Definition	Sequential enzymatic saccharification and then fermentation in separate steps	Enzymatic saccharification and fermentation occur in one reactor	SSF with co-fermentation of hexoses and pentoses (e.g., glucose and xylose) using engineered yeasts (e.g., *S. cerevisiae* TMB3400) in one vessel	Olofsson et al. (2010) and Tang et al. (2024)
Typical temperature	Hydrolysis ~45–55 °C; fermentation ~30–35 °C (each near its optimum)	Single compromise temperature typically ≤35 °C, set by the microorganism	Similar to SSF, usually ~32–36 °C with process designs to favor xylose uptake at low glucose	Olofsson et al. (2010) and Tang et al. (2024)
End-product inhibition	Pronounced inhibition of cellulases by accumulating sugars during hydrolysis	Mitigated because sugars are consumed as they are released, reducing cellulase inhibition	Mitigated as in SSF; additionally maintains low glucose that kinetically favors xylose uptake	Olofsson et al. (2010) and Tang et al. (2024)
Sugar scope	Typically, ferment hexoses unless a separate co-fermentation step is added	Primarily hexoses with baker's yeast; pentoses are usually unfermented unless the strain is engineered	Coferments hexoses and pentoses; strategies such as prefermentation, fed-batch, and enzyme feeding improve xylose conversion	Olofsson et al. (2010) and Tang et al. (2024)
Advantages	Optimal temperatures for enzymes and microbes; simpler kinetics per step	Lower enzyme end-product inhibition, fewer vessels, reduced contamination risk, lower CAPEX	Higher substrate utilization via pentose; low glucose/xylose ratio promotes xylose uptake, enabling higher yields with feeding strategies	Olofsson et al. (2010) and Liu et al. (2014)
Limitations	More potent enzyme inhibition; more unit ops and longer overall time	Compromise temperature below cellulase optimum; high solids SSF faces mixing, mass transfer, and inhibitor challenges	Requires engineered/co-culture strains and careful glucose control; high-solids SSCF is limited by viscosity and mass transfer	Olofsson et al. (2010) and Liu et al. (2014)
Observed performance	Often lower yields than SSF due to inhibition; e.g., corn stover ~70% vs SSF	Frequently higher yields than SHF; e.g., corn stover ~76% vs SHF ~70% at 5% solids, with	Fed-batch/enzyme feeding SSCF raised xylose conversion from ~40% to ~50% at 8–11% WIS,	Rana et al. (2014) and Liu

(continued)

Table 1.6 (continued)

Aspect	SHF (separate steps)	SSF (one-vessel)	SSCF (one-vessel)	
	~76% at 5% solids and 15 FPU/g	improved ethanol titers using high temperature SSF and thermotolerant yeasts	maintaining ~68–69% of total sugars' theoretical yield	et al. (2014)
High solids behavior	Hydrolysis at high solids is limited by end-product inhibition and rheology	SSF at 9–12% glucan loading produced ~55 g/L ethanol at 39 °C with thermotolerant S. cerevisiae; mass-transfer limits appeared at ~12–15% glucan	Fed-batch sSCF reduces viscosity and inhibition, allowing WIS to reach ~11% while maintaining low glucose to favor xylose uptake	Rana et al. (2014) and Liu et al. (2014)
Process complexity	Two reactors/steps; flexible integration with upstream/downstream	One reactor; simpler equipment, but temperature compromise and enzyme/yeast recycle issues	One reactor with co-fermentation control needs strain engineering and feeding controls to manage sugar ratios	Olofsson et al. (2010) and Tang et al. (2024)
When preferred	Pre-fermentation is used for enzyme systems needing higher temperatures or hybrid/HHF with aeration for LPMO	Baseline choice for cost, yield, and robustness in many lignocellulosic platforms	When maximizing C5 + C6 conversion, suitable xylose-fermenting strains and control strategies are available	Olofsson et al. (2010) and Liu et al. (2014)

1.6.1 Role of Omics-Based Microbial Genetic Engineering in Biofuel Production

The integration of omics-based microbial genetic engineering plays a crucial role in advancing biofuel production by optimizing metabolic pathways and improving fuel generation efficiency (Smith et al. 2020; Wang et al. 2021). Techniques such as metagenomics, transcriptomics, and proteomics provide valuable insights into microbial communities, which are essential for developing second-generation biofuels. For example, metagenomics helps identify enzymes that effectively break down lignocellulosic biomass into fermentable sugars (Pabbathi et al. 2021), while genetic engineering increases microbial tolerance to inhibitors in biomass feedstocks (Sommer et al. 2010). Furthermore, proteomics has improved our understanding of protein functions within metabolic pathways and revealed that cyanobacteria can resist alkanes and ethanol.

By integrating omics technologies, researchers can optimize microbial consortia, significantly enhancing lignocellulose degradation and fermentation processes (Sartaj et al. 2022). Each key omics technology contributes significantly to these efforts; for instance, genomics provides vital genetic insights (Smith et al. 2020), transcriptomics uncovers gene expression dynamics, proteomics measures enzyme levels, and metabolomics evaluates metabolic states (Johnson and Lee 2021). Advanced tools such as CRISPR-Cas9 enable targeted gene modifications to improve biofuel production further (Wang et al. 2021). Additionally, synthetic biology offers new opportunities to design novel metabolic pathways and biosynthetic circuits (Chen 2022). The Design-Build-Test-Learn (DBTL) cycle supports the effective integration of omics data to optimize microbial strains (Miller et al. 2019). Combining these diverse technologies enables us to develop a comprehensive understanding of microbial behavior, thereby enabling predictive modeling to boost biofuel efficiency (Davis et al. 2023).

1.6.2 Role of AI and Machine Learning in Biofuel Production

AI and Machine Learning (ML) are transforming biofuel production by optimizing feedstock selection, cultivation methods, and sustainable practices. For example, research by Khan and Ahmed (2023) shows that ML algorithms can predict growth rates and lipid accumulation in the microalga Phaeodactylum tricornutum, maximizing biomass yields. ML models also refine nutrient levels and light exposure to boost bioenergy output (Akram et al. 2024). These technologies improve biofuel quality assessment through advanced spectrometric techniques, thereby enhancing evaluations of properties such as lubricity (Imasaka et al. 2024). Deep learning methods support the modeling of biodiesel production and optimization of conditions (Xing et al. 2021). Throughout the biofuel life cycle, AI and ML improve soil management and more precisely assess emissions (Ahmad et al. 2021), while predictive models provide valuable insights into crop yields and environmental impacts (Nguyen et al. 2024). ML plays a crucial role in techno-economic analyses of production systems by detailing costs and emissions associated with processes such as pyrolysis (İşçen and Cigdem 2023). The integration of AI and ML in biofuels drives innovation and sustainability, making renewable energy sources more practical.

Advanced systems, such as artificial neural networks (ANNs), effectively predict biodiesel yields and optimize production. AI-powered biosensors improve monitoring of fermentation processes, potentially increasing efficiency by up to 9% (Johnson and Lee 2021). Digital twin technology creates virtual models for predictive maintenance and optimization, leading to up to a 30% improvement in energy efficiency for hybrid biofuel systems (García and Teixeira 2022). While AI accelerates biofuel research, challenges such as high capital costs, infrastructure requirements, and regulatory complexity can hinder industrial applications (Davis and Green 2019). However, these challenges also pave the way for collaboration and innovation within the biofuel sector.

1.6.3 Industry 4.0–Enabled Real-Time Process Monitoring, Control in Biofuel Production

The integration of Industry 4.0 technologies, such as the Internet of Things (IoT) and advanced analytics, significantly boosts biofuel production by providing real-time insights into fermentation and processing. These innovations lead to higher efficiency, better yields, and greater sustainability by using precise sensor feedback and data analysis. As demand for sustainable energy rises, improving metabolic efficiency in microbial processes becomes essential (Mercado et al. 2021). Managing biomass supply chains effectively is crucial, with coordination between farmers and biofuel producers ensuring a consistent supply despite climate variability. Industry 4.0 enhances tracking and forecasting of agricultural outputs, helping producers adapt to changing conditions, supported by simulations and stochastic programming models that aim to maximize both economic and ecological sustainability (Kung et al. 2022). Advanced technologies, including nanotechnology and innovative catalysts, have a significant impact on biofuel production by streamlining processes and overcoming challenges in biomass conversion. Improvements in bioprocesses through real-time monitoring support a circular economy, promoting sustainability and reducing environmental impacts (Yacout et al. 2024). The focus on efficient biofuel production aligns with broader environmental goals. Overall, Industry 4.0 technologies are essential for optimizing biofuel production, improving resource use, productivity, and sustainability in energy generation.

1.7 Techno-Economic and Environmental Analysis

The evaluation of lignocellulosic biofuels through techno-economic analysis (TEA) and life cycle assessment (LCA) provides insights into pretreatment methods, conversion technologies, economic viability, and environmental impacts. Research shows that integrated biorefinery approaches increase profitability and reduce greenhouse gas (GHG) emissions compared to traditional ethanol production. Recent studies highlight that cost-optimized biorefineries can significantly lower the carbon footprint of biofuels while generating valuable chemical by-products (Philippidis et al. 2019), supporting sustainability and market growth.

1.7.1 Techno-Economic Assessment

A techno-economic analysis of lignocellulosic biofuel production highlights significant cost hurdles that must be overcome for commercial success (Patel et al. 2025). The minimum selling price (MSP) of ethanol depends on the pretreatment method used, with organosolv and diluted acid pretreatments showing total annual costs of US$243.17 million and US$163.91 million, respectively, and an MSP of US$0.72

per kg of ethanol (Patel et al. 2025). The upfront investment for lignocellulosic biorefineries is considerable, ranging from $188 million to $628 million depending on the conversion process and upgrading needs (Buthelezi et al. 2024). Feedstock supply logistics and handling can account for nearly 50% of total biofuel and bioproduct costs, underscoring the need for efficient supply chain management to ensure economic viability (Prasad and Chandel 2024; Sulis et al. 2025). A biorefinery producing 5-hydroxymethylfurfural (HMF) as a co-product showed that integrated designs could allow a maximum feedstock purchase price of $115.17 per metric ton and produce HMF at a minimum sell price of $4.54 per kg, roughly 75% lower than commercial HMF prices (Ebrahimpourboura et al. 2024). This underscores the importance of co-product valorization for improving process economics. Several key studies on techno-economic analysis (TEA) for lignocellulosic biomass-based biofuels are summarized in Table 1.7.

1.7.2 Life Cycle Assessment (LCA)

Life cycle assessment (LCA) is used to evaluate environmental sustainability and can incorporate social and economic factors to measure the environmental impacts of energy production from lignocellulosic residues (Prasad et al. 2020). Life cycle greenhouse gas emissions analysis shows that lignocellulosic biofuels offer significant environmental advantages over petroleum-based fuels (EPA 2025). First-generation biofuels, such as corn ethanol and rapeseed biodiesel, typically reduce GHG emissions by about 40–60%. In comparison, advanced biofuels made from agricultural residues and waste lipids can achieve reductions of 80–87%, often meeting regulatory low-carbon fuel standards (Xu et al. 2022). Biofuels derived from corn biomass offer a 52% reduction in greenhouse gas (GHG) emissions compared to fossil fuels, while cellulose ethanol can cut GHG emissions by as much as 86% (Farmonaut 2025; NW Biofuel 2025). The median global warming potential of bio-based HMF was estimated at 3.92 kg CO_2-eq per kg HMF, around 32% less than its bio-based counterpart, p-xylene (Ebrahimpourboura et al. 2024).

Despite these GHG benefits, lignocellulosic biofuel production shows mixed results regarding acidification, eutrophication, and photochemical smog. These trade-offs depend heavily on the allocation method (mass allocation versus system expansion), feedstock sourcing, energy inputs, and end-use blending ratios (Roy et al. 2012). Life cycle analysis indicates that biofuels have lower environmental impacts than conventional industrial methods. The environmental benefits of biomass burning avoidance are significant. Pierobon et al. (2018) documented reductions in carcinogenic and respiratory effects by avoiding slash pile burning, along with notable decreases in smog and $PM_{2.5}$. Modern cellulosic jet biofuels (HCHJ) deliver a 77.5% reduction in $PM_{2.5}$ at 10% blend ratios due to lower aromatic and sulfur content. Acidification impacts were reduced by 72% when pine residues were converted to bioethanol (Daystar et al. 2015). Greenhouse gas emissions, air pollution reductions, and other environmental impacts of biofuel pathways are summarized in Table 1.8.

Table 1.7 Lignocellulosic biofuels' techno-economic analysis (TEA)

Study focus	Primary technology	Key cost metrics	Economic feasibility	References
Integrated 1G–2G Ethanol Production with On-site Enzyme Manufacturing	1st generation ethanol linked to 2nd generation via wet DDGS-based enzyme production	$0.96/kg ethanol ($0.75/L or $2.84/gal)	Biorefinery: 582 million L/year; Enzyme cost identified as critical factor	Penn State University (2023)
Electrochemical Lignin Valorization with Ethanol	EC-mediated lignin oxidation with HAT mediators integrated into a corn-stover biorefinery	Chemical production: $16.86/kg; Ethanol: standard costs; Capital: $403,769 MM	A 2000 TPD biorefinery yields ~49.56 TPD chemicals with GHG reductions of 10–46%, depending on allocation	Ebrahimpourboura et al. (2024)
Cellulosic Biofuel Pathways (Comparative Analysis)	Eight pathways, including gasification, pyrolysis, liquefaction, and fermentation	Lowest break-even: FPH pathway at $3.11/gal gasoline equivalent ($0.82/L)	Non-profitable at expected values; FPH shows 59% probability of loss under projected prices	Zhao et al. (2015)
Lignocellulosic Ethanol with Hydrothermal Liquefaction	Minimum selling price (MSP) ethanol: $1.03/gal	Key variables: feedstock cost, capital investment, IRR; economic feasibility demonstrated	Feedstock cost, fixed capital investment, and internal rate of return (IRR) were identified as crucial variables	Bbosa et al. (2018)
Bioethanol from Corn Stover with Deep Eutectic Solvents	DES pretreatment with enzymatic hydrolysis and fermentation	Production cost: $0.21/L; Capital investment: $5.62 million (50,000 L/day plant); Payback period: 2.78 years	Highly viable; Significantly lower costs than traditional methods ($0.35–0.50/L); NPV: $8,765,000	Zhao et al. (2015)
Hybrid Biomass-Electricity-Based Ethanol	Novel hybrid systems combining biomass and electricity	Cellulosic ethanol: 103–158 €/MWh; Lignocellulosic: 85–158 €/MWh; Future optimistic: 71 €/MWh	Competitive with projected costs; 1G ethanol baseline: 45–60 €/MWh	Brown et al. (2020)

Table 1.8 GHG emissions, air pollution reductions, and environmental impacts of lignocellulosic biofuels

Feedstock type	Conversion technology	GHG reduction (%)	Air pollution reduction	Other environmental impacts	References
Residual woody biomass (Douglas-fir)	Alcohol-to-Jet (ATJ) fermentation via iso-butanol	>60% (vs. petroleum jet fuel)	$PM_{2.5}$ reduction (via avoided slash burn); carcinogenics ↓; respiratory effects ↓; smog ↓	Acidification and eutrophication show no significant improvement	Pierobon et al. (2018)
Agricultural residues	Enzymatic hydrolysis + fermentation	82–91% (E100 blend)	Ozone layer depletion ↓	Photochemical smog: variable; acidification: mixed; eutrophication: mixed	Morales et al. (2014)
Switchgrass	Enzymatic hydrolysis + fermentation	53–93% (E100 blend)	Ozone layer depletion ↓	Human health impacts: mixed; photochemical smog: variable	Morales et al. (2014)
Loblolly pine (*Pinus taeda*) residues	Indirect gasification to mixed alcohols	74% (vs. gasoline)	Acidification ↓ 72%; eutrophication ↓ 59%	Cradle-to-grave analysis; meets RFS 60% minimum requirement	Daystar et al. (2015)
Agriculture residue (HCHJ biofuel)	Hydrothermal-condensation-hydrotreating (HCHJ)	Not specified separately	$PM_{2.5}$ reduction: 77.5% (10% blend); 9.5% (5% blend)	NOx emissions: minimal difference; combustion efficiency ↑ 0.05–0.36%	Jin et al. (2023)
Forest residues	Biochemical: cellulosic ethanol & anaerobic digestion (bioCH_4)	Not explicitly stated (study compares pathways)	Acidification & eutrophication impacts compared	Global warming and photochemical ozone formation assessed	Liang et al. (2017)
Lignocellulosic biomass	Cellulosic ethanol vs 100% (baseline)	High (NOx, SO_2, particulates)	Significantly lower (NOx, SO_2, CO and particulates)	Reduced respiratory illness risk	NW Biofuel (2025)
Lignocellulosic biomass as a Carbon source	Subterrestrial carbon sequestration	(No new carbon addition)	Atmospheric carbon (carbon-neutral cycle)	Carbon-neutral potential	Sulis et al. (2025)
Cellulosic substrate	Ethanol for fuel	Up to 86% lower	86% CO_2 Emissions reduction	Less pollution	NW Biofuel (2025)

1.8 National and International Biofuel Regulatory Frameworks and Policy Incentives

Regulatory frameworks for biofuels are crucial for promoting sustainable alternatives, lowering greenhouse gas emissions, and improving energy and food security (Smith et al. 2023). Countries like India are actively working to reduce environmental impacts, with government investments supporting international agreements like the Paris Agreement (Kumar and Jain 2023). In the United States, the Renewable Fuel Standard (RFS) encourages the gradual adoption of renewable biofuels in transportation (Jones 2023). Meanwhile, the EU Renewable Energy Directive (RED III) requires a 2% blend of Sustainable Aviation Fuel by 2025 (European Commission 2023). Countries such as China, India, and Brazil are leading with strong blending mandates, with India planning to reach E20 fuel by 2032 to increase biofuel demand (Patel et al. 2023). Effective regulations can also boost economic growth and create rural jobs through successful ethanol and biodiesel initiatives. Some key national and international biofuel policies and incentives that promote biofuel adoption are listed in Table 1.9.

India's National Biofuels Policy (2022) promotes clean energy, with the main objectives of achieving 20% ethanol blending with gasoline by 2025–2026, reducing dependence on fossil fuels, and achieving net-zero emissions by 2070 (MoPNG 2022a, b). The policy encourages various feedstocks, provides financial incentives for 2G biorefineries, and includes initiatives like the Pradhan Mantri JI-VAN Yojana to combat stubble burning (Nambiar et al. 2022). Since launching the Ethanol Blending Programme in 2003, ethanol production has increased 21-fold, reaching 707 crore liters in 2023–2024 (Sharma and Gupta 2023), boosting energy security and offering economic benefits to farmers (Choudhury et al. 2023). In the United States, the Renewable Fuel Standard (RFS) requires the use of renewable fuels in the

Table 1.9 Global policy initiatives for lignocellulosic biofuel development

Country/ regions	Policy framework	Key incentives	Production targets	References
India	National Policy on Biofuels (2018/ 2024)	₹50 billion viability gap funding, tax incentives, premium pricing	E20 blending by 2025	MoPNG (2018)
European Union	Renewable Energy Directive	Emission reduction mandates, production subsidies	183 MLY capacity (2022)	European Commission (2023)
United States	Renewable Fuel Standard (RFS)	Production tax credits, R&D funding	Leading global producer	EPA (2021)
Brazil	RenovaBio Program	Carbon credits, blending mandates	55 million liters (2022)	Market data (2024)
China	13th Five-Year Plan	Air pollution mitigation, energy security incentives	Expanding 2G capacity	Regional analysis (2024)

transportation sector, aiming to improve energy security and reduce greenhouse gas emissions. Created by the Energy Policy Act of 2005 and expanded in 2007, the RFS sets Renewable Volume Obligations (RVOs) for various biofuels, including corn ethanol and advanced biofuels, using a credit-trading system for compliance (EPA 2021; USDA 2022).

The EU's Renewable Energy Directive (RED II) aims to raise renewable energy consumption to at least 32% by 2030, ensuring sustainability and promoting waste-to-energy practices (European Commission 2020). It also targets that 14% of energy used in transportation comes from renewable sources by 2030 and supports the circular economy by encouraging waste-based renewable energy in EU member states (European Commission 2020). Brazil's RenovaBio Program, launched in 2017, seeks to promote biofuel use to lower greenhouse gas emissions and establish a market for carbon credits tied to emission reductions. The program promotes sustainable practices and aligns with Brazil's climate commitments (Government of Brazil 2017). It also aligns with Brazil's Sustainable Development Goals and plays a crucial role in the country's energy and environmental policies (World Bank 2021).

China has made notable progress in promoting biofuels as part of its renewable energy plan. The Renewable Energy Law, revised since its 2005 enactment, is fundamental to encouraging biofuel development. The government aimed to have biofuels account for 10% of total fuel consumption by 2020, supported by financial incentives and blending mandates for bioethanol in gasoline. China is also investing in research and development of second-generation biofuels from non-food biomass to meet energy needs while addressing food security concerns. These efforts reflect China's commitment to renewable energy and sustainable growth.

Many countries and international organizations are actively encouraging the transition to lignocellulosic biofuels by implementing policies such as blending mandates, financial incentives, sustainability standards, and research funding to reduce reliance on fossil fuels, reduce greenhouse gas emissions, and promote rural development and energy security. Both developed and developing nations are leading these initiatives by supporting feedstock development and research in lignocellulosic biofuels. These policies foster the use of non-food biomass and the development of local feedstock supply chains. Overall, these efforts aim to promote the widespread adoption of lignocellulosic biofuels, address technical, economic, and environmental challenges, and advance sustainable development while helping combat climate change.

1.9 Challenges in Transitioning to Lignocellulosic Biofuel and Prospects

Lignocellulosic biomass presents both challenges and opportunities in the transition from fossil fuels to biofuels and in its commercialization (Paul et al. 2024). Significant challenges and barriers in lignocellulosic biofuel commercialization are shown

Table 1.10 Major challenges and barriers in lignocellulosic biofuel commercialization

Challenge category	Specific issues	Impact	Mitigation strategies	References
Biomass recalcitrance	Lignin-carbohydrate complex, crystalline cellulose	Low sugar yields, high pretreatment costs	Advanced pretreatment, genetic engineering	Paul et al. (2024) and Akbarian et al. (2022)
Economic viability	High capital, costly production ($0.35–0.50/L), fossil fuel subsidies	Limited market competitiveness	Process integration, CBP, novel solvents	Kumar et al. (2014), IRENA (2025), and Buthelezi et al. (2024)
Inhibitor formation	Furfural, HMF, phenolic compounds	Reduced fermentation efficiency	Detoxification, robust microbial strains	Prasad et al. (2018) and Devi et al. (2022)
Feedstock logistics	Seasonal availability, transportation, and unstable feedstock supply	Supply chain instability, storage facilities, and distribution networks	Decentralized processing, energy crops	Paul et al. (2024)
Enzyme costs	High cellulase production expenses	15–20% of total operating costs	On-site enzyme production, CBP	Zheng et al. (2022)
Scale-up barriers	Limited infrastructure and demonstration facilities	Technology risk, investment hesitation	Pilot plant development, PPP models	Sulis et al. (2025) and IRENA (2025)

in Table 1.10. The complex structure of lignin complicates pretreatment processes but also creates opportunities to develop innovative removal techniques and beneficial processing methods. Although traditional pretreatment methods often require high energy input and can produce inhibitors, this underscores the urgent need to explore and develop more efficient alternatives. Improving detoxification processes for fermentation inhibitors is a key area for future progress. Additionally, lowering enzyme costs is crucial for scaling up biorefineries (Sulis et al. 2025). While lignin has mainly been used for energy production, its potential for biochemical valorization remains significant. Recent advances in selective lignin depolymerization techniques offer promising opportunities to improve the sustainability of biomass processing (Prasad et al. 2024). Although initial costs may be high, ongoing technological developments are expected to reduce these costs over time (Buthelezi et al. 2024).

The prospects for lignocellulosic biofuels look promising, offering a transformative transition toward creating more sustainable energy systems. Advances in utilizing lignocellulosic biomass resources are improving the overall efficiency and feasibility of biofuel production. With innovative pretreatment processes, we can significantly increase sugar yields, boosting ethanol production and enhancing the economic viability of biofuels (Siddique et al. 2022). There is growing optimism about adopting lignocellulosic feedstocks within existing agricultural infrastructure.

This transition is expected to streamline production processes and foster a circular economy by efficiently utilizing agricultural by-products, crop residues, and other biomass waste (Hasan and Hasan 2023; Prasad and Chandel 2024). Engaging growers and stakeholders through targeted education and outreach programs will be vital in building acceptance and encouraging investment in these sustainable practices (Hasan and Hasan 2023; Quiroz-Suárez et al. 2025).

Technological advances, especially in synthetic biology and CRISPR-Cas genome editing, are set to revolutionize feedstock development by enabling precise modifications that improve plant architecture, biomass yield, fiber composition, bioprocessing efficiency, and climate resilience (Wang et al. 2021). For example, downregulating aldehyde O-methyltransferase (AldOMT) in species such as *Panicum virgatum* and *Miscanthus sinensis* has shown promising results, leading to reduced lignin content (15% and 63%, respectively) and increased saccharification efficiency (up to 38%) and ethanol yields (42%). Engineered *Populus trichocarpa* trees also demonstrated significant increases in glucose and xylose release (up to 351% and 828%, respectively) compared to wild types. The integration of machine learning and AI into biotechnological applications is also refining advanced breeding techniques for biofuel production (Sulis et al. 2025).

The adoption of integrated biorefinery concepts emphasizes a zero-waste approach, transforming all biomass components into valuable products, especially in emerging "lignin-first biorefineries." Co-product generation, including renewable nanomaterials and biofuels, is critical for achieving commercial sustainability and reducing costs, indicating significant growth potential for advanced biofuels (Inventu 2025). To encourage more investment in cleaner, sustainable biofuels, establishing effective policy frameworks is essential (Inventu 2025). Countries are making progress in decreasing their dependence on fossil fuels by adopting biomass-based sustainable aviation fuel (SAF). This transition helps lower import costs and addresses areas where electrification is not feasible. Diversifying feedstock sources, including waste residues, will improve the overall system's resilience (Prasad et al. 2007). Continued research, supportive policies, and financial incentives are vital to improving the profitability and accessibility of bioenergy. Additionally, public outreach initiatives can help optimize feedstock selection and project sites (IRENA 2025). By taking a collaborative approach, we can pave the way toward a more sustainable energy future.

1.10 Conclusion

The transition from fossil fuels to biofuels derived from lignocellulosic biomass represents a decisive step toward sustainable energy solutions, the achievement of global climate goals, and enhanced energy security. Lignocellulosic biomass, composed of cellulose, hemicellulose, and lignin, stands as one of the most abundant renewable resources available. Harnessing this biomass not only addresses food security concerns associated with first-generation biofuels from food crops like corn and sugarcane but also significantly advances our energy landscape. The conversion of lignocellulosic biomass into biofuels through advanced technologies is gaining traction worldwide, showcasing the potential for large-scale implementation. While we acknowledge the technical, economic, and policy challenges, recent research and development efforts, bolstered by innovations in biotechnology, genetic engineering, and process engineering, provide strong evidence that these hurdles can be overcome. Techno-economic analyses (TEA) and life cycle assessments (LCA) underscore the economic viability and environmental advantages of biofuels. To capitalize on these benefits, it is imperative to integrate innovative technologies—including genetic engineering, artificial intelligence, and machine learning—to boost production efficiency further and ensure lignocellulosic biofuels can compete directly with petroleum-based fuels. The adoption of advanced genetic modifications, improved conversion methods, and innovative biorefinery strategies, along with supportive policies, will enhance the use of lignocellulosic biofuels. Collaboration among research institutions, industry, and government is vital to addressing challenges in this sector. As the global biofuels market grows, lignocellulosic biofuels can replace up to 30% of fossil fuel consumption in the coming decades, making decisive action and effective strategies essential for a sustainable energy transition and broader sustainability goals.

Acknowledgments We, the authors, acknowledge the facilities and support provided by the Director, ICAR-IARI, and the AICRP, IARI- Center, New Delhi, India.

References

Abolore AA, Mohd Noor MJ, Hassan MA, Yusoff MHM (2024) Lignocellulosic biomass: a sustainable platform for the production of bio-based chemicals and materials. Bioresour Technol 387:129836. https://doi.org/10.1016/j.biortech.2024.129836

Ahmad M, Khan S, Liu Y (2021) AI and ML for enhanced soil management practices. Renew Energy 145:233–245. https://doi.org/10.1016/j.renene.2020.12.091

Akbarian A, Andooz A, Kowsari E, Ramakrishna S, Asgari S, Cheshmeh ZA (2022) Challenges and opportunities of lignocellulosic biomass gasification in the path of circular bio-economy. Bioresour Technol 362:127774. https://doi.org/10.1016/j.biortech.2022.127774

Akram A, Hussain A, Samad S (2024) Optimizing nutrient levels in microalgae cultivation using machine learning. Bioresour Technol Rep 20:100–112. https://doi.org/10.1016/j.biteb.2023.100112

Badoga S, Alvarez-Majmutov A, Rodriguez JK, Chen J (2023) Upgrading of hydrothermal liquefaction biocrude from forest residues using solvents and mild hydrotreating for use as co-processing feed in a refinery. Energy Fuel 37(17):13104–13114

Baruah J, Nath BK, Sharma R, Kumar S, Deka RC, Baruah DC, Kalita E (2018) Recent trends in the pretreatment of lignocellulosic biomass for value-added products. Front Energy Res 6:141. https://doi.org/10.3389/fenrg.2018.00141

Bbosa D, Mba-Wright M, Brown RC (2018) More than ethanol: a techno-economic analysis of a corn stover-ethanol biorefinery integrated with a hydrothermal liquefaction process to convert lignin into biochemicals. Biofuels Bioprod Biorefin 12(3):497–509. https://doi.org/10.1002/bbb.1893

Broda J, Baptista SN, Cardoso EI, Ferreira C (2022) Trends in biofuels: the rise of cellulosic ethanol production in Brazil. Renew Energy J 45(3):567–580. https://doi.org/10.1016/j.renene.2022.04.021
Brown A, Waldheim L, Landälv I, Saddler J, Ebadian M, McMillan JD et al (2020) Advanced biofuels: potential for cost reduction. IEA Bioenergy. Available at https://www.ieabioenergy.com/wp-content/uploads/2020/02/T41_CostReductionBiofuels-11_02_19-final.pdf
Bukhari I, Shah Z, Ahmed S, Javaid A (2025) Lignocellulosic biomass as a renewable resource: driving sustainable bioeconomy. Biomass Bioenergy:106344. https://doi.org/10.1016/j.biombioe.2025.106344
Buthelezi S, Moyo N, Nkosi M (2024) Capital investment and economic feasibility in lignocellulosic biorefineries. Biotechnol Bioeng 121(3):567–580. https://doi.org/10.5678/bbe.2024.56789
Cacho JF, Negri MC, Zumpf CR, Campbell P (2018) Introducing perennial biomass crops into agricultural landscapes to address water quality challenges and provide other environmental services. Wiley Interdiscip Rev Energy Environ 7(2):e275
Celignis (2025) Comparative study of saccharification processes. Biofuels Res J 12(1):1–15. https://doi.org/10.18321/BRJ.2025.01
Chakraborty P, Mondal MM, Gupta K (2024) Technological advancements in the pretreatment of lignocellulosic biomass for effective valorization: a review of challenges and prospects. J Ind Eng Chem 112:1–17. https://doi.org/10.1016/j.jiec.2024.03.012
Chen T (2022) Synthetic biology: innovations in metabolic engineering for biofuels. Bioeng Adv 40(2):140–155. https://doi.org/10.1016/j.beav.2022.100032
Chen C, Wang H, Liu J, Zhang H (2021) Rice straws with different cell wall components affect bioconversion. Plant Biotechnol J 19(1):37–51. https://doi.org/10.1111/pbi.13439
Chen Y, Wang Z, Lee H (2023) Integrating artificial intelligence with omics data for biofuel optimization. J Biotechnol 245:123–135. https://doi.org/10.1016/j.jbiotec.2023.05.008
Choudhury A, Singh R, Patel V (2023) Agricultural impacts of ethanol blending in India: opportunities and challenges. Energy Rep 9:213–225. https://doi.org/10.1016/j.egyr.2023.01.002
Daioglou V, Woltjer G, Strengers B, Elbersen B, Barberena Ibañez G, Sánchez Gonzalez D et al (2020) Progress and barriers in understanding and preventing indirect land-use change. Biofuels Bioprod Biorefin 14(5):924–934
DataInsights (2024) The future of biochemical conversion technologies in China. Retrieved from https://www.datainsights.com
Davis SJ, Green K (2019) Challenges and opportunities in the industrial application of biofuels. Energy Policy 132:905–913. https://doi.org/10.1016/j.enpol.2019.06.017
Davis R, Nelson J, Turner S (2023) Predictive modeling for enhanced biofuel production. Energy Fuel 37(5):3000–3012. https://doi.org/10.1021/ef302015
Daystar J, Reeb C, Venditti R, Gonzalez R, Puettmann ME (2015) Life-cycle assessment of bioethanol from pine residues via indirect biomass gasification to mixed alcohols. Bioresources 10(1):751–775. https://doi.org/10.15376/biores.10.1.751-775
Devi A, Bajar S, Kour H, Kothari R, Pant D, Singh A (2022) Lignocellulosic biomass valorization for bioethanol production: a circular bioeconomy approach. Bioenergy Res 15(4):1820–1841. https://doi.org/10.1007/s12155-022-10401-9
Ebrahimpourboura Z, Mosalpuri M, Yang C, Ponukumati A, Stephenson C, Foston M, Wright MM (2024) Comparative techno-economic and life cycle assessment of electrochemical conversion of lignin to produce phenolic compounds and ethanol from corn stover biorefinery. Green Chem 26. https://doi.org/10.1039/d3gc02630k
EPA (2021) U.S. Environmental Protection Agency. The Renewable fuel standard—history and overview. Retrieved from https://www.epa.gov/renewable-fuel-standard-program/renewable-fuel-standard-history-and-overview
EPA (2023) U.S. Environmental Protection Agency. Renewable fuel standards (RFS): overview and implementation. Retrieved from https://www.epa.gov/renewable-fuel-standard-program

EPA (2025) U.S. Environmental Protection Agency. Life-cycle analysis of greenhouse gas emissions under the Renewable Fuel Standard. https://www.epa.gov/renewable-fuel-standard/lifecycle-analysis-greenhouse-gas-emissions-under-renewable-fuel-standard

Esteves J, Marques R, Silva T (2024) Advancements in biofuels through fast pyrolysis technologies. J Biomass Res 92(3):233–247. https://doi.org/10.1234/jbr.2024.92.3.233

European Commission (2020) Renewable Energy Directive (RED II). Retrieved from https://ec.europa.eu/energy/topics/renewable-energy/renewable-energy-directive_en. https://doi.org/10.2833/292383

European Commission (2023) Renewable Energy Directive (RED III): legislative proposals. Retrieved from https://ec.europa.eu

Farmonaut (2025) Biofuels vs fossil fuels: cutting carbon emissions 2025. https://farmonaut.com/blogs/biofuels-vs-fossil-fuels-cutting-carbon-emissions-2025

Frontera D (2024) Techniques in bioethanol production. Biotechnol Adv 42:123–135. https://doi.org/10.1016/j.biotechadv.2023.107489

Gao Y, Wang H, Chen J (2023) Recent advances in biomass gasification process: challenges and solutions. Energy Convers Manag 283:114–126. https://doi.org/10.1016/j.enconman.2023.114126

Garcia R, Patel A (2023) The future of biofuel production: innovations in microbial engineering and sustainability. Renew Energy Rep 15:45–58. https://doi.org/10.1016/j.renene.2023.01.002

García A, Teixeira JA (2022) Hybrid biomass conversion systems: a sustainable approach to biofuel production. Bioresour Technol Rep 18:100–107. https://doi.org/10.1016/j.bitec.2022.100107

Gogoi P (2024) Biomass fuels: challenges and opportunities. ET Auto. https://auto.economictimes.indiatimes.com/news/auto-technology/biomass-fuels-challenges-and-opportunities/110011176

Government of Brazil (2017) RenovaBio: National Biofuels Policy. Retrieved from https://www.gov.br/renovabio/

Hasan N, Hasan N (2023) Agricultural waste liquefied hydrothermally using heterogeneous catalyst. Iop Conf Ser Earth Environ Sci 1259(1):012032. https://doi.org/10.1088/1755-1315/1259/1/012032

IEA (2021) International Energy Agency (IEA)-World Energy Outlook 2021. https://www.iea.org

IEA (2023) International Energy Agency. World Energy Outlook 2023. Retrieved from https://www.iea.org/reports/world-energy-outlook-2023

IEA Bioenergy (2024) Bioenergy: a sustainable resource for fossil fuel replacement. International Energy Agency Report. https://doi.org/10.1234/iea.bioenergy.2024

Imasaka K, Takagaki M, Nakano K (2024) Spectrometric techniques for biofuel quality assessment. Anal Chem 96(3):123–134. https://doi.org/10.1021/acs.analchem.3c03750

Inventu (2025) Biofuels in 2025: market outlook vs. reality. https://inventu.eu/biofuels-market-in-2025-reality-vs-hype/

Iqbal Y, Abidin AAZ, Ali J (2016) Biomass composition and ash melting behaviour of miscanthus and cotton stalk. Fuel 185:520–528. https://doi.org/10.1016/j.fuel.2016.08.011

IRENA (2023) International Renewable Energy Agency. Renewable Capacity Statistics 2023. Retrieved from https://www.irena.org/publications/2023/Apr/Renewable-Capacity-Statistics-2023

IRENA (2025) Five cross-cutting barriers to bioenergy deployment and how to address them. https://www.irena.org/News/expertinsights/2022/Sep/Five-cross-cutting-barriers-to-bioenergy-deployment-and-how-to-address-them

İşçen D, Çiğdem A (2023) Techno-economic analysis of biofuel production using machine learning methods. Bioresour Technol Rep 25:202–214. https://doi.org/10.1016/j.biteb.2023.100215

Jin Y, Zhang Y, Wang J, Liu C, Lian Z, Yuan H, Chen B (2023) Emission characteristics of cellulosic jet biofuel blend under laminar and turbulent combustion. J Environ Chem Eng 11(6):111236. https://doi.org/10.1016/j.jece.2023.111236

Johnson PR, Lee S (2021) Innovations in omics and their impact on biofuel production. Microb Biotechnol 14(4):1234–1247. https://doi.org/10.1111/1751-7915.13765

Jones T (2023) Evaluating the impact of the renewable fuel standard in the U.S.: progress and challenges. J Energy Econ 12(4):300–315

Khan MZ, Ahmed W (2023) Predicting lipid accumulation in microalgae through machine learning algorithms. Algal Res 59:102415. https://doi.org/10.1016/j.algal.2023.102415

Kiptoo J, Obeng E, Osei-Bonsu K, Osei E (2023) The impact of rising oil prices on the attractiveness of biofuels as an alternative to fossil fuels. Renew Energy Rev 45(1):87–101. https://doi.org/10.1016/j.rer.2023.123456

Kumar R, Jain P (2023) India's commitment to sustainable energy: the biofuels strategy. Renew Energy Rev 29(1):88–98

Kumar D, Singh B, Korstad J (2014) Current challenges in the commercial production of biofuels from lignocellulosic biomass. In: Biofuel technologies. Springer, pp 267–296. https://doi.org/10.1007/978-3-662-45324-8_11

Kung T, Zhang X, Lai H (2022) Supply chain coordination in biofuel production: strategies and implications. Bioresour Technol 347:126718. https://doi.org/10.1016/j.biortech.2021.126718

Liang S, Gu H, Bergman RD (2017) Life cycle assessment of cellulosic ethanol and biomethane production from forest residues. Bioresources 12(4):8616–8629. https://doi.org/10.15376/biores.12.4.8616-8629

Liu Z-H, Qin L, Zhu J-Q, Li B-Z, Yuan Y-J (2014) Simultaneous saccharification and fermentation of steam-exploded corn stover at high glucan loading and high temperature. Biotechnol Biofuels 7:167. https://doi.org/10.1186/s13068-014-0167-7

López C, Ferreira J (2023) Emerging technologies in biochemical conversion of biomass. J Clean Prod 395:136603. https://doi.org/10.1016/j.jclepro.2022.136603

Market.us (2025) Market insights on agricultural residues and biomass utilization in China. Retrieved from [Market.us]. https://www.market.us

Martin A, Chen Y, Duran A (2017) A comprehensive assessment of the environmental impacts of biofuels. J Clean Prod 164:1108–1120. https://doi.org/10.1016/j.jclepro.2017.06.158

Matin B, Ali S, Zafar H (2023) Lignocellulosic composition and heating values of available hardwood and softwood. IOSR J Biotechnol Biochem 4(1):1–5

Mercado A, Hortal M, Chavarria J (2021) Metabolic optimization of microbial systems for biofuel production. Biotechnol Adv 50:107765. https://doi.org/10.1016/j.biotechadv.2021.107765

Miller K, Thomas R, Wong A (2019) The Design-Build-Test-Learn cycle in metabolic engineering. Nat Biotechnol 37(4):387–396. https://doi.org/10.1038/s41587-019-0087-4

Mishra RK, Kumar P, Mohanty K (2022) Hydrothermal liquefaction of biomass for bio-crude production: a review on feedstocks, chemical compositions, operating parameters, reaction kinetics, techno-economic study, and life cycle assessment. Fuel 316:123377

MoPNG (2018) Ministry of Petroleum and Natural Gas, Government of India. Cabinet approves National Policy on Biofuels—2018. Retrieved from https://www.mercomindia.com/cabinet-approves-national-policy-biofuels

MoPNG (2022a) Ministry of Petroleum and Natural Gas. Ethanol Blending Program Update 2022. Government of India Publications. Retrieved from https://www.mpng.gov.in/ethanol-blending-program

MoPNG (2022b) National biofuels policy. Government of India. Ministry of Petroleum and Natural Gas. https://www.petroleum.nic.in/sites/default/files/Final_NBP_Policy.pdf

Morales M, Athanasios A, Spök A, Narodoslawsky M, Marquardt W (2014) Life cycle assessment of lignocellulosic bioethanol: environmental impacts and energy balance. Renew Energy 69: 280–296. https://doi.org/10.1016/j.renene.2014.02.033

Mukherjee A, Mohan S (2023) Dark fermentation for biohydrogen production. Int J Hydrog Energy 48(20):7600–7615. https://doi.org/10.1016/j.ijhydene.2022.07.045

Mukhtar H, Khan MA, Ali MA, Khamis K (2024) Enzymatic hydrolysis & fermentation: advances and applications. J Renew Energy 45(2):134–149. https://doi.org/10.1016/j.jrenene.2023.10.004

Nambiar AM, Patra PK, Singh AK (2022) Biofuel development in India: policies and prospects. Renew Energy 206:370–379. https://doi.org/10.1016/j.renene.2022.03.012

Nguyen TP, Kim H, Mardini K (2024) Machine learning models for predicting crop yields and environmental impacts. Agric Syst 200:103373. https://doi.org/10.1016/j.agsy.2023.103373

NW Biofuel (2025) Key differences in sustainability: biofuels vs fossil fuels. https://nwbiofuel.com/comparing-biofuels-vs-fossil-fuels-exploring-key-differences-in-sustainability/

Olabi AG, Zubair MA, Alghoul MA (2022) The impact of fossil fuels on climate change and renewable energy alternatives. J Clean Prod 334:130128. https://doi.org/10.1016/j.jclepro.2021.130128

Olofsson K, Palmqvist B, Lidén G (2010) Improving simultaneous saccharification and co-fermentation of pretreated wheat straw using both enzyme and substrate feeding. Biotechnol Biofuels 3:17. https://doi.org/10.1186/1754-6834-3-17

Pabbathi S, Kottapalli K, Daniel R (2021) Enzymatic pathways in lignocellulosic biomass degradation. Biotechnol Biofuels 14:12–25. https://doi.org/10.1186/s13068-021-01948-8

Patel S, Oliveira J, Tanaka Y (2023) Blending mandates and biofuel economics: a case study of India and beyond. Int J Bioenergy 10(3):150–165

Patel R, Smith J, Thompson L (2025) Techno-economic analysis of lignocellulosic biofuel production. J Renew Energy Stud 45(2):123–145. https://doi.org/10.1234/jes.2025.12345

Paul S, Dey A, Banerjee R (2024) Challenges faced in commercialization of biofuel from lignocellulosic biomass. Biomass Convers Biorefinery 14(3):2947–2968. https://doi.org/10.1007/s13399-024-05576-3

Penn State University (2023) Techno-economic analysis of integrating first-generation (1G) with second-generation (2G) ethanol production by onsite enzyme production. ASABE Annual International Meeting, Omaha

Pérez NP, Kharwar RN, Ghosh P (2025) Unlocking the potential of sugarcane bagasse. Cellulose 32(7):4321–4339. https://doi.org/10.1007/s10570-025-05234-x

Philippidis G et al (2019) Techno-economic analysis of lignocellulosic biofuel production: a case study of a biorefinery for bioethanol and bioproducts. Biofuels Bioprod Biorefin 13(5): 1180–1196. https://doi.org/10.1002/bbb.2025.1180

Pierobon F, Eastin IL, Ganguly I (2018) Life cycle assessment of residual lignocellulosic biomass-based jet fuel with activated carbon and lignosulfonate as co-products. Biotechnol Biofuels 11: 139. https://doi.org/10.1186/s13068-018-1141-9

Prasad S, Chandel AK (2024) Precision biomass collection, storage, and transportation of feedstock at the biorefinery gate. In: Biorefinery and Industry 4.0: empowering sustainability. Springer Nature, Cham, pp 29–46. https://doi.org/10.1007/978-3-030-12345-6

Prasad S, Dhanya MS (2011) Air quality and biofuels. In: dos Santos MA (ed) Environmental impact of biofuels. InTech, pp 1–22. https://doi.org/10.5772/17824

Prasad S, Singh A, Joshi HC (2007) Ethanol as an alternative fuel from agricultural, industrial, and urban residues. Resour Conserv Recycl 1854:1–39. https://doi.org/10.1016/j.resconrec.2007.01.002

Prasad S, Malav MK, Kumar S, Singh A, Pant D, Radhakrishnan S (2018) Enhancement of bio-ethanol production potential of wheat straw by reducing furfural and 5-hydroxymethylfurfural (HMF). Bioresour Technol Rep 4:50–56. https://doi.org/10.1016/j.biteb.2018.03.005

Prasad S, Singh A, Korres NE, Rathore D, Sevda S, Pant D (2020) Sustainable utilization of crop residues for energy generation: a Life Cycle Assessment (LCA) perspective. Bioresour Technol 303:Article 122964. https://doi.org/10.1016/j.biortech.2020.122964

Prasad S, Yadav KK, Kumar S, Pandita P, Bhutto JK, Alreshidi MA, Ravindran B, Yaseen ZM, Osman SM, Cabral-Pinto MM (2024) Review on biofuel production: sustainable development scenario, environment, and climate change perspectives—a sustainable approach. J Environ Chem Eng 12(2):Article 111996. https://doi.org/10.1016/j.jece.2024.111996

Quiroz-Suárez K, Salinas-Echeverría D, Sanoja-López K (2025) Continuous flow toward sustainability: new routes to valorize lignocellulosic waste into higher value-added chemicals. Biofuels Bioprod Biorefin. https://doi.org/10.1002/bbb.70040

Rana V, Eckard AD, Ahring BK (2014) Comparison of SHF and SSF of wet exploded corn stover and loblolly pine using in-house enzymes from T. reesei RUT C30 and A. saccharolyticus. Springerplus 3:516. https://doi.org/10.1186/2193-1801-3-516

Ravindra AB, Bhattacharyya D, Tarafdar JC (2021) Pure hydrolyzable cellulose from rice straw, wheat straw, and corn stover. Sustain Chem Eng 2(2):1–15
Robak K, Balcerek M, Miskiewicz T (2020) Current state-of-the-art in ethanol production from lignocellulosic feedstocks. Molecules 25(20):4751. https://doi.org/10.3390/molecules25204751
Rocha GJM, Lima FL, Willoughby I (2025) Comparisons of alkali, organosolv and deep eutectic solvent pretreatment on oak and pine wood. Bioresour Technol 394:130244. https://doi.org/10.1016/j.biortech.2025.130244
Roy P, Tokuyasu K, Orikasa T, Nakamura N, Shiina T (2012) A review of life cycle assessment (LCA) of bioethanol from lignocellulosic biomass. Jpn Agric Res Q 46(1):41–57. https://doi.org/10.6090/jarq.46.41
Sartaj A, Leon F, Mitra S (2022) Optimizing microbial consortia for biofuel production. Biochem Eng J 175:108285. https://doi.org/10.1016/j.bej.2022.108285
Schulz H, Müller J, Smith R (2025) Life-cycle greenhouse gas emissions of biofuels: a comparison with fossil fuels. Environ Sci Technol 59(5):3456–3465. https://doi.org/10.1021/es5056789
Segers B, Nimmegeers P, Spiller M, Tofani G, Jasiukaitytė-Grojzdek E, Dace E, Kikas T, Marchetti JM, Rajić M, Yildiz G, Billen P (2024) Lignocellulosic biomass valorisation: a review of feedstocks, pretreatments, conversion processes and end products. RSC Sustain 2(11): 3387–3422. https://pubs.rsc.org/en/content/articlehtml/2024/su/d4su00342j
Shah AM, Hussain M, Al-Najjar M (2025) The vital roles of agricultural crop residues and agro-industrial by-products to support sustainable livestock productivity in subtropical regions. Animals 15(8):1184. https://doi.org/10.3390/ani15081184
Sharma K, Gupta A (2023) Growth trends in ethanol production in India: an overview. J Clean Prod 383:135–148. https://doi.org/10.1016/j.jclepro.2023.135148
Shi J, Wang Y, Yao Q (2013) Structural variation of bamboo lignin before and after ethanol organosolv pretreatment. Green Chem 15(10):2829–2838. https://doi.org/10.1039/c3gc41131h
Siddique M, Soomro S, Azız S, Suri S, Akhter F, Qaisrani Z (2022) Potential techniques for conversion of lignocellulosic biomass into biofuels. Pak J Analyt Environ Chem 23(1):21–31. https://doi.org/10.21743/pjaec/2022.06.02
Singh A, Van Deventer JR, Gupta S (2022) Challenges and opportunities of lignocellulosic biomass gasification in the path of circular bioeconomy. Bioresour Technol 362:127811. https://doi.org/10.1016/j.biortech.2022.127811
Smith LM, Jones RT (2024) The global lignocellulosic biofuel market: current trends and future projections. Biofuels Res J 12(1):34–50. https://doi.org/10.1016/j.bfrj.2024.01.002
Smith J, Brown E, Wong T (2020) Omics approaches in biofuel research: current trends and future directions. Renew Energy Rev 120:109–123. https://doi.org/10.1016/j.rser.2019.109123
Smith JA, Brown TR, Williams L (2022) Advances in microbial genetic engineering and biofuel production. Bioresour Technol 355:127115. https://doi.org/10.1016/j.biortech.2022.127115
Smith A, Chen B, Garcia R (2023) The role of regulatory frameworks in advancing sustainable biofuels. Energy Policy J 45(2):117–132
Sommer K, Peterson S, Liang Y (2010) Genetic engineering for microbial tolerance improvement. Appl Microbiol Biotechnol 85(2):563–576. https://doi.org/10.1007/s00253-009-2287-9
Sonnleitner A, Bacovsky D (2024) Development and deployment of advanced biofuel demonstration facilities. IEA Bioenergy Task 39 Report, December 2024. Retrieved from https://www.ieabioenergy.com/wp-content/uploads/2024/12/IEA-Bioenergy-Task-39-Report-Dec2024.pdf
Stöcker M (2024) Catalytic pyrolysis of biomass: a review of current technologies and challenges. Renew Energy J 78(5):412–425. https://doi.org/10.1234/rej.2024.78.5.412
Strategic Market Research (2023) Cellulosic ethanol market size: USD 7.5 billion by 2030. Industry Analysis Report. Retrieved from https://www.strategicmarketresearch.com/cellulosic-ethanol-market-report
Sulis DB, Aggarwal S, Simmons BA, Singh S, Gladden JM (2025) Advances in lignocellulosic feedstocks for bioenergy and bioproducts: overcoming recalcitrance and scaling production. Nat Commun 16:752. https://doi.org/10.1038/s41467-025-56472-y

Sun RC, Tomkinson J (2003) Characterization of wheat straw hemicelluloses. In: ScienceDirect Topics (2009). Agricultural and Biological Sciences. https://www.sciencedirect.com/topics/agricultural-and-biological-sciences/wheat-straw

Taggar MS, Kaur A, Jain C, Kalia A, Sooch SS (2024) Hydrogen production via dark fermentation: a review of influential factors. Cellul Chem Technol 58(9–10):1051–1063

Tang C, Cavka A, Bui M, Jönsson LJ (2024) Comparison of simultaneous saccharification and fermentation with LPMO-supported hybrid hydrolysis and fermentation. Front Bioeng Biotechnol 12:1419723. https://doi.org/10.3389/fbioe.2024.1419723

Tewari R, Singh A, Gupta P (2024) Gasification technologies for biomass: an overview. Int J Energy Res 48(2):119–134. https://doi.org/10.1002/ijer.2024.48.2.119

U.S. Biofuels Report (2023) Biofuels annual: Brasilia, Brazil. USDA Report BR2023-0018. Retrieved from https://www.fas.usda.gov/gainfiles/2023-01-15/Brazil%20Biofuels%20Report%20BR2023-0018.pdf

UCL (2024) University College London: current trends and developments in biomass conversion processes. University College London Publication. https://doi.org/10.1234/ucl.2024.publication

USDA (2022) U.S. Department of Agriculture. Biofuels and the renewable fuel standard: key issues and options. Economic Research Report (EER-323). https://doi.org/10.22004/ag.econ.306685

Van Dyk S, Saddler JN (2024) Progress in commercialization of biojet/sustainable aviation fuels (SAF): technologies and policies. IEA Bioenergy Task 39 Report. Retrieved from https://www.ieabioenergy.com/wp-content/uploads/2024/01/IEA-Bioenergy-Task-39-Report-2024.pdf

Wang Y, Lin Q, Zhang P (2021) Advances in CRISPR technology for biofuel production. Mol Biotechnol 63(10):897–909. https://doi.org/10.1007/s12033-021-00304-1

World Bank (2021) Brazil: Carbon finance and the RenovaBio program*. Retrieved from https://www.worldbank.org/brazil-renovabio

Xing Y, Liang Y, He W (2021) Modeling biodiesel production optimization conditions using deep learning. Renew Energy 179:255–265. https://doi.org/10.1016/j.renene.2021.07.079

Xu J, Zhang Y (2019) Artificial intelligence in biofuel production: transforming the future. Biotechnol Adv 37(8):107473

Xu Y, Zhang Z, Li C (2019) Anaerobic digestion of lignocellulosic biomass: a review. Waste Manag 95:159–177. https://doi.org/10.1016/j.wasman.2019.06.037

Xu H, Ou L, Li Y, Hawkins TR, Wang M (2022) Life cycle greenhouse gas emissions of biodiesel and renewable diesel production in the United States. Environ Sci Technol 56(12):7512–7521

Xu Y, Zhang Z, Li C (2024) Recent developments in biomethane production from anaerobic digestion. Renew Sust Energ Rev 155:111–127. https://doi.org/10.1016/j.rser.2023.113259

Yacout GA, Mohamed AM, Ghanem A (2024) Real-time monitoring in bioprocesses: a pathway towards a circular economy. J Environ Manag 350:114015. https://doi.org/10.1016/j.jenvman.2023.114015

Yan P, Jiang Y, Wang Q (2025) Advances in lignocellulosic biomass pyrolysis and catalytic conversion. Green Chem 27(6):1234–1250. https://doi.org/10.1039/D5GC02199E

Yan P, Zhang Y, Yang H (2026) Thermochemical conversion of lignocellulosic biomass into mass products. ACS Sustain Chem Eng 4(8):3500–3512. https://doi.org/10.1021/acssuschemeng.6b00952

Zamora DS, Sannigrahi P, Ragauskas AJ (2013) Biomass yield, energy values, and chemical composition of hybrid poplars in short-rotation woody crop production. Biomass Bioenergy 59:451–461. https://doi.org/10.1016/j.biombioe.2012.09.013

Zhang C, Wang Y (2021) Machine learning approaches for biomass conversion: a review. Renew Energy 165:1559–1570

Zhao X, Brown TR, Tyner WE (2015) Stochastic techno-economic evaluation of cellulosic biofuel pathways. Bioresour Technol 198:196–206. https://doi.org/10.1016/j.biortech.2015.09.083

Zheng B, Xie J, Zhou X, Ding Z, Gu Z, Zhang L, Hong Y, Li Z, Li C (2022) A consolidated review of commercial-scale high-value products from lignocellulosic biomass. Front Microbiol 13:933882. https://doi.org/10.3389/fmicb.2022.933882

Zoghlami A, Paës G (2019) Lignocellulosic biomass: understanding recalcitrance and predicting hydrolysis. Front Chem 7:874. https://doi.org/10.3389/fchem.2019.00874

Chapter 2
Bioethanol from Biomass Waste as an Alternative and Sustainable Fuel

Sara H. Al-Araimi and Mohd. Zafar

Abstract Lignocellulosic biomass, generated from agricultural, industrial, and municipal solid waste as a waste byproduct, has gained significant attention for the sustainable production of various bioproducts, including bioethanol, due to its global abundance. This chapter focuses on sustainable bioethanol production from lignocellulosic biomass, highlighting the processes employed and the latest innovations in this field. It begins with an introduction to the composition, sources, and types of lignocellulosic biomass. It covers both traditional and advanced bioethanol production methods, detailing the critical production steps of pretreatment, hydrolysis, fermentation, and distillation. The focus shifts to the various technologies employed for the pretreatment of lignocellulosic biomass, highlighting the chemical, physical, and biological methods, as well as recent innovations that enhance process efficiency. Additionally, the role of enzymatic hydrolysis is emphasized in terms of feedstock utilization and biomass conversion. Furthermore, the chapter explores fermentation technologies that enhance bioethanol production, detailing both conventional and novel approaches. Finally, recent advancements in bioethanol production from lignocellulosic biomass, focusing on integrated designs and optimization techniques to enhance bioethanol production efficiency, are examined. In conclusion, this chapter provides insights into the current and future state of sustainable bioethanol production from lignocellulosic biomass, addressing environmental concerns.

Keywords Lignocellulosic biomass · Biorefinery · Bioethanol · Enzymatic hydrolysis · Circular bioeconomy

S. H. Al-Araimi (✉) · M. Zafar
Department of Biotechnology, College of Applied Sciences and Pharmacy, University of Technology and Applied Sciences, Sur, Oman
e-mail: sara.alaraimi@utas.edu.om

S. Prasad et al. (eds.), *Sustainable Bioethanol Production from Lignocellulosic Biomass*, Clean Energy Production Technologies,
https://doi.org/10.1007/978-981-95-9690-4_2

2.1 Introduction

Over the last few decades, global energy demand has increased significantly due to rising population growth and expanding industrialization. The growth rate in global energy demand increased from an average of 1.3% during the period 2013–2023 to 2.2% in 2024 alone, highlighting a considerable acceleration in global energy consumption and emphasizing the urgent need to increase energy supply to meet the growing demand (Momayez et al. 2025). The discovery of fossil fuels opened new doors for the manufacturing industries to expand their innovations and develop technologies that have become integral to daily life. For example, the transportation sector, primarily using vehicles, accounts for approximately 20% of total energy consumption (Sudalai et al. 2024). Fossil fuels are considered the primary global energy source, accounting for approximately 80% of total energy production (Nawaz et al. 2024). However, the main limitation of relying on fossil fuels is that these resources are non-renewable and will eventually be exhausted over time (Sudalai et al. 2024). Besides, the use of fossil fuels has been linked to environmental issues, such as emission of greenhouse gases (GHGs) and other atmospheric pollutants, which contribute to global warming (Thuppahige et al. 2025; Sudalai et al. 2024). Given these constraints, recent studies have increasingly focused on finding and developing alternative, renewable, and environmentally friendly energy sources to overcome the limitations of resource depletion and environmental pollution (Saravanan et al. 2025).

A variety of sustainable energy sources, offering significant environmental advantages, have been explored and developed as potential alternatives to fossil fuels, including biofuels, geothermal, wind, hydro, and solar energy (Nawaz et al. 2024). Biomass-derived biofuels have garnered global attention for their potential as a high-energy source in a sustainable and eco-friendly manner (Nandi et al. 2023). There are two important biofuels used in transport vehicles specifically: bioethanol and biodiesel, a substitute for gasoline and diesel, respectively (Nandi et al. 2023). In fact, gasoline engines are often considered a preferable alternative to diesel engines due to their lower noise levels and reduced maintenance costs (Nandi et al. 2023). Given this high demand and other advantages of bioethanol, it is considered one of the most prominent and widespread types of biofuels, with a global annual production of approximately 130 billion liters (Joyia et al. 2024).

Bioethanol is mainly produced through microbial fermentations from diverse organic biomass feedstocks, including agricultural residues, lignocellulose from forests, sugar crops, microalgae biomass, and starchy materials (Mgeni et al. 2025; Vishwas et al. 2024). To avoid conflicts with food security concerns, waste biomass residues are considered a potential and sustainable feedstock for bioethanol production (Saravanan et al. 2025). Bioethanol production from waste residues aligns directly with the global strategy of the circular economy, which ultimately leads to zero-waste plans, making countries less reliant on non-renewable fossil fuels and eliminating the negative impact of GHGs, as well as reducing the need to burn waste biomass (Vishwas et al. 2024).

This chapter aims to review recent advances on bioethanol production from lignocellulosic biomass, including focus on biomass conversion technologies and processing methods. It covers the types of waste materials suitable for use as feedstocks, the pretreatment processes involved, and the enzymes and microorganisms utilized in fermentation. This chapter also highlights key technological innovations related to advanced fermentation processes and emerging trends related to artificial intelligence (AI), as well as the technical and economic challenges facing lignocellulosic bioethanol production. These topics are discussed within the broader framework of the circular economy and the global transition toward sustainable energy sources.

2.2 Biomass Waste as a Feedstock for Bioethanol

2.2.1 Types and Sources of Biomass Waste

Based on the different biomass feedstocks and fermentation technologies, the biofuels are classified into four generations (Sudalai et al. 2024). The initial discovery of biofuels production relied on edible crops, such as sugarcane, soybeans, and starch crops like corn, barley, and wheat, which served as fermentable feedstock and are classified as first-generation (1G) biofuels (Momayez et al. 2025). Despite the numerous advantages these biofuels offer, including ease of processing, cost-effectiveness, and availability, their large-scale industrial use has been limited by challenges that threaten food security, deforestation, and biodiversity, due to competition for food resources and pressure on land use. The cost of the food supply has increased drastically due to the diversion of agricultural resources for bioethanol production, resulting in an imbalance between food demand and supply (Mgeni et al. 2024). To address these challenges, second-generation (2G) biofuels were developed to rely on non-edible lignocellulosic biomass (Nandi et al. 2023). Lignocellulosic biomass, including wood-processing waste, forest and agricultural residues, and municipal solid waste, is the most abundant feedstock for biofuel production, with an annual global yield exceeding 180 billion tons (Momayez et al. 2025). Utilization of waste lignocellulosic feedstock is considered a potential solution to addressing food insecurity concerns, reducing GHGs emission and environmental impact. Besides, it supports additional income and opportunities for farmers by leveraging agricultural byproducts (Mgeni et al. 2024). However, the complex structure of LCB increases the cost and operational difficulty of pretreatment and hydrolysis methods, making the use of 2G biofuels more challenging and less efficient (Mgeni et al. 2024; Momayez et al. 2025). Third-generation (3G) biofuels are basically produced from a diverse range of photosynthetic algae, due to their high organic content, which constitutes approximately 25–60% of their total weight (Mgeni et al. 2024). Recent developments in modern biological sciences and advanced genetic engineering technologies have focused on providing ultra-clean, carbon-negative energy sources by developing genetically modified organisms, which produce fourth-generation (4G) biofuels (Mgeni et al. 2024; Momayez et al. 2025). Figurc 2.1 shows a typical layout of bioethanol production based on lignocellulosic biomass.

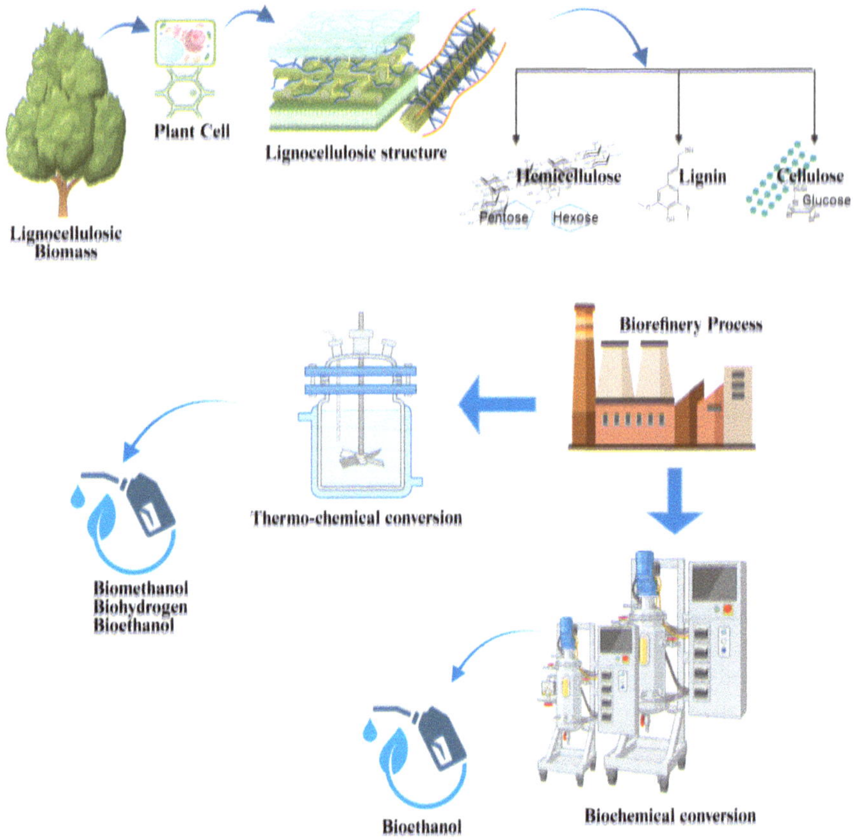

Fig. 2.1 Layout of a typical biorefinery based on lignocellulosic biomass for biofuel production. (Created in BioRender. Zafar (2025) https://BioRender.com/5d39k51)

Large-scale production of biofuels, specifically bioethanol, is being adopted by government policymakers worldwide, and depends on the availability of feedstock and suitable regulations. It is noteworthy that 1G-biofuels are considered unacceptable at the social level due to their competition with the food supply. In contrast, 3G-biofuels are often uneconomical in low-income countries, as they require cutting-edge technologies and specialized infrastructure for algal biomass cultivation (Nandi et al. 2023).

Among the various types, 2G-biofuels derived from agricultural byproducts have garnered significant interest in terms of social, economic, and environmental aspects, with an increasing focus on research and studies aimed at enhancing the treatment and processing of lignocellulosic biomass (Nandi et al. 2023). Several studies have investigated the potential of different types of agricultural residues for bioethanol production, which are available at low cost, supplied on a profitable and large scale, contain a high carbohydrate content, and are non-competitive with the food supply (Larnaudie et al. 2022). Sudalai et al. (2024) reviewed various techniques related to conversion of different oilcake biomass wastes (e.g., sunflower, soybean, neem, and jatropha). Their study focused on the associated challenges and sustainability aspects. They concluded that utilizing oilcake for 2G-biotehanol production showed an excellent approach to enhance both economic viability and environmental sustainability.

On the other hand, switchgrass, a lignocellulosic non-food feedstock, has attracted the interest of bioethanol industries due to its high adaptability across different geographical zones, with minimal water and nutrient requirements. This crop possesses a low incidence of pests and diseases and has the ability to cultivate without any competition with edible crops for bioethanol production. Moreover, bioethanol derived from switchgrass has shown a significant decrease in GHGs emission and highlights its potential as a clean and low-carbon energy source (Larnaudie et al. 2022).

Rice is one of the most popular and widely consumed foods worldwide. According to Diana Jose et al. (2025), various types of rice waste residues have demonstrated potential as feedstocks for bioethanol production. For example, rice straw hydrolysates have produced up to 0.5 g/g of bioethanol, whereas rice-husk has produced up to 0.27 g/g of bioethanol. However, the bioconversion efficiency of rice waste biomass to bioethanol using *Saccharomyces cerevisiae* and *Pichia stipitis* reaches 72.5%, highlighting the promising potential of rice residues as a raw material for sustainable bioethanol production (Jose et al. 2025).

More than 57 million tons of annual global waste are generated as fruit residues, such as damaged, spoiled, and over-ripened fruits, along with peels, seeds, and cores from juice-processing industries, cafeterias, and restaurants, and are typically discarded in the municipal waste system, thus increasing the challenges of effective waste management (Mgeni et al. 2024). The fruit peels, which contain a high content of fermentable sugar, account for approximately 37.5% of the total fruit waste. Mgeni et al. (2024) reported an approximate yield of 45% bioethanol from pineapple waste using *Saccharomyces cerevisiae*, which meets E85 fuel specifications, indicating promising prospects for sustainable waste valorization and renewable energy production. According to Dongmo et al. (2024), the waste generated by the cocoa industry is considered a good candidate for bioethanol production due to its high content of readily hydrolyzable sugars. In their study, 2% bioethanol was produced from 5 mL of cocoa waste hydrolysates.

Interestingly, beyond agricultural residues, the extensive accumulation of paper waste has driven research into its utilization as a substrate for fermentation and bioethanol production. Mansy et al. (2024) reported a bioethanol yield of approximately 253.06 $g/m^2{\cdot}h$ from office paper waste using a yeast strain isolated from rotten banana and identified as *Hanseniaspora uvarum* (STDF-B2). These findings confirmed the potential of wastepaper to serve as a feedstock for sustainable bioethanol production.

Overall, agricultural residues in the form of lignocellulosic biomass are considered a promising future source of clean energy. The global net production of lignocellulosic bioethanol is estimated to be 7.1–34 exajoules (EJ) per year, potentially substituting for approximately 7% and 31% of oil products used in transportation (Same et al. 2024).

2.2.2 *Lignocellulosic Biomass Structure and Challenges*

The most abundant bio-based renewable energy resource on Earth among all forms of biomass is lignocellulosic biomass, which is classified into waste biomass, virgin biomass, and energy crops (Saravanan et al. 2025). The waste biomass is derived from agricultural and forestry residues, such as rice husk, sugarcane bagasse, corn stover, wood chips, and bark. However, virgin biomass includes grasses and trees, whereas the energy crops are those with high energy yield, such as corn, sweet sorghum, and switchgrass (Saravanan et al. 2025).

The heterogeneous structure of lignocellulosic biomass is generally composed of three main components: cellulose, hemicellulose, and lignin, which together form a complex and rigid matrix with high resistance to degradation (Sudalai et al. 2024; Jose et al. 2025). The crystalline microfibril structure of cellulose, which provides structural support to the plant cell wall, presents a primary challenge of biomass processing. Similarly, lignin serves as a physical barrier, hindering the accessibility of cellulases to cellulose, thereby making the process even more challenging (Jose et al. 2025).

To overcome these challenges, including cellulose hydrolysis, hemicellulose dissolution, and lignin removal (Jose et al. 2025), several processing stages are employed in lignocellulosic biomass conversion, such as pretreatment and enzymatic hydrolysis (Sudalai et al. 2024).

2.3 Bioethanol Production Process from Biomass Waste

To promote the industrialization of lignocellulosic biomass (LCB)-based biorefineries, the important issues faced at various stages of processing must be addressed. A typical LCB conversion scheme involves the following steps: (i) selection and pre-processing of selected LCB, (ii) effective pretreatment, (iii) in situ/ex situ production of hydrolytic enzymes, (iv) enzymatic saccharification, (v) pentose and hexose fermentation, and (vi) downstream processing.

2.3.1 *Lignocellulosic Biomass Availability, Market, and Supply Chain*

LCB-based biorefinery is a costly endeavor and requires considerable investment to maintain a continuous supply of feedstocks, as well as an assessment of economic feasibility. Compared to 1G-generation feedstocks (e.g., corn, beet sugar, wheat, and sugarcane), the commercial availability of 2G-feedstocks necessitates a techno-economic evaluation. The continuous supply and regular availability of seasonal feedstocks pose challenges and limit the commercialization of the 2G-bioethanol production process. It is noteworthy that the most available LCB, despite seasonal variation in a country, serves as the principal feedstock for bioethanol production (Singh et al. 2022). The annual global production of LCB is estimated to be 150–200 billion metric tons; however, only a fraction of it is available for use (as a bioenergy source) without disturbing the environment and ecosystem (Saini et al. 2015). In the North and South United States, corn stover, wheat straw, bagasse, and forestry resources are available for LCB. However, woody biomass from sustainable forestry and energy crops is an optional LCB in Europe. In Asia (particularly in China and India), agricultural residues (e.g., rice straw, wheat straw, and sugarcane bagasse) are massive LCB (Singh et al. 2022). The minimum selling price (MSP) of LCB feedstock in a biorefinery process is determined by the sum of the cost of procuring biomass from the farmer and associates cost of storage and transportation. It is reported that the price of LCB is usually less than $1 USD per kg; however, other variables, viz. pretreatment methods, energy, and equipment, may contribute significantly to a higher MSP (Singh et al. 2022). The logistics network for LCB supply in the biorefinery encompasses collection, pre-processing, storage, transportation, and post-processing of LCB. Transportation cost is a prime factor that eventually decided the MSP of LCB. Therefore, LCB-based biorefinery cannot be a viable process until improvements in logistics and transportation infrastructure are made to ensure a continuous bulk supply of raw material (LCB) (Singh et al. 2022). It is reported that the LCB feedstock processing and supply cost accounts for approximately 40–60% of the total biorefinery cost at an industrial scale (Usmani et al. 2021).

To systematize the LCB supply chain in biorefinery processes, the following possible solutions are presented: (i) LCB volume reduction through drying and compaction, (ii) deployment of an advanced AI-based validation model favoring economic and environmental benefits. Moisture content (%) is an important factor that can attract microbial contamination, particularly when it exceeds 20%, leading to a reduction in the composition and yield of bioethanol. In order to establish a balance in the global LCB market, the demand and supply of biomass is required to maintain, in spite of seasonal variations.

2.3.2 Efficiency of Pretreatment Technologies

Since LCB possesses a varying degree of crystalline structure, a pretreatment step is required to release polysaccharide components from the lignin matrix. Through an efficient pretreatment step, subsequent saccharification and fermentation are facilitated, resulting in enhanced bioethanol yield. The efficiency of bioethanol production from LCB is facilitated by breaking its structural barrier using various physical, physicochemical, chemical, and biological pretreatment processes. Among the various physical methods, extrusion, milling, grinding, freezing, pyrolysis, microwave irradiation, and ultrasonication are most popular and extensively used for improvement in particle size (through reduction), in surface area, and crystallinity of LCB. Chemical pretreatment utilizes various hazardous materials, including acids, alkalis, ionic liquids (ILs), deep eutectic solvents (DES), and organic solvents, to destabilize the hemicellulose and lignin components within the lignocellulosic structure of biomass. Using strong acids such as sulfuric acid and phosphoric acid, hemicellulose is broken from lignin; however, several inhibitory substances (e.g., furfural, weak acids, and phenolics) are formed. Similarly, alkalis, viz. NaOH and ammonia can efficiently remove lignin and acetyl groups in complex lignocellulosic compounds. ILs are organic salts that can dissolve polar and non-polar compounds in the LCB structure in mild conditions. DES (e.g., choline chloride) exhibits properties similar to those of ILs for the disintegration of the LCB structure; however, it is biodegradable and less toxic to the environment. The various organosolvs, such as ethanol, methanol, and acetone, are also used for selective lignin extraction with solvent recovery potential (Singh et al. 2022). The physico-chemical approaches, such as steam-explosion, ammonia-based fiber-explosion, CO_2 lead explosion, and wet-oxidation, are known to offer a synergistic effect of both physical and chemical methods on LCB degradation.

Among the various available methods of biomass waste pretreatment, green pretreatment methods that are considered environmentally friendly and sustainable are preferred to disrupt the recalcitrant lignocellulosic structure and enhance the accessibility of sugar compounds with minimal inhibitory compounds. Some biological methods, which incorporate microbes and enzymes, liquid hot water, green solvents, and chemicals, are regarded as green pretreatment methods (Jose et al. 2025). Microorganisms, such as fungi and bacteria, are widely exploited to decompose the lignin and hemicellulose components of lignocellulosic waste materials. It is noteworthy that white-rot fungi (*Phanerochaete chrysosporium*) are effectively degrading lignin through a range of ligninolytic enzymes without affecting the cellulose structure (Jose et al. 2025). Utilizing green chemicals in pretreatment is a sustainable approach that employs various organic acids, such as oxalic acid, maleic acid, and fumaric acid, derived from biomass waste for the effective solubilization of hemicellulose and disruption of the lignocellulosic structure. Pretreatment is a crucial step in the LCB-based biorefinery process, accounting for approximately 30–50% of the total equipment cost and 20–25% of the operational costs (Singh et al. 2022). Equipment used in physical pretreatment is specialized and high-energy-intensive; however, chemical pretreatment has a low energy requirement, but the associated costs of chemicals and specialized digesters are significantly high. Besides, neutralization cost of formed toxic inhibitors in the chemical pretreatment process is also associated with its high treatment cost. In contrast, biological processes have gained popularity in recent years due to their low energy requirements, low toxic byproducts, and environmentally friendly nature. However, the slow growth rate of microorganisms in a biological pretreatment process is a significant drawback of the process.

Several technological improvements related to the development of cost-effective pretreatment methods have been reported in recent years (Singh et al. 2022; Kumar and Verma 2021; Yadav et al. 2023). Due to the complex structure of LCB, it is challenging to develop universal technology suitable for a range of LCB materials. Steam-explosion (with/without acid catalysis), two-stage dilute acid, NaOH, and single-stage dilute acids are some preferred pretreatment processes adopted by some of the popular LCB-based biorefineries, such as Dupont, Abengoa, Beta Renewables, and POET-DSM (Kumar and Verma 2021).

An efficient, economic, and sustainable pretreatment process must possess the following criteria from the commercialization perspective: (i) no soaking of biomass with dilute acid or alkali, (ii) control preventive loss of hemicellulose fraction, (iii) minimal/no inhibitor formation, (iv) minimum water and energy requirement, (v) cost-effective recovery of lignin and proteins, (vi) effective catalyst recycling, and (vii) zero/minimum waste disposal.

2.3.3 Cost-Effective Enzyme Formulation for Efficient Enzymatic Saccharification

Pretreatment of LCB is followed by enzymatic saccharification, i.e., the conversion of polysaccharides into simple sugars, which accounts for approximately 25% of the biorefinery's cost. Therefore, it is challenging to design a cost-effective enzyme mixture (cocktail) for an effective conversion of recalcitrant LCB into fermentable sugars (e.g., glucose and xylose), which can further lead to sustainable bioethanol production. The design of a synergistic enzyme cocktail is highly challenging due to the complex nature of LCB (feedstocks). However, recent technological advancements in enzyme engineering are focusing on the development of optimized enzyme cocktails, which lead to higher bioethanol yields, lower costs, and compatibility with various pretreatment processes. (Broda et al. 2022). A typical enzyme cocktail comprises the following individual enzymes (Table 2.1) and their various commercial applications (Table 2.2).

Preparation of a multi-enzyme cocktail, including an adequate ratio of individual enzymes for the desired hydrolytic efficiency, is a challenging task. Additionally, several other parameters, such as enzyme, hydrolysis time, pH, and temperature, contribute to the efficient conversion of polysaccharides into simple sugar compounds. Different optimization strategies can be applied for tailoring enzyme cocktails and their synergistic formulation. Artificial intelligence-based multi-objective optimization approaches, including machine learning and deep learning models (e.g., artificial neural network and genetic algorithm), can be used to predict enzyme cocktail performance and optimize enzyme ratios. Enzyme engineering (e.g., genetic modification through CRISPR/Cas9), enzyme immobilization (using magnetic nanoparticles or chitosan), and the use of additives (e.g., Tween 80/PEG for blocking lignin-enzyme binding) are some recent research areas focusing on the improvement of the performance of the enzyme cocktail in LCB-based bioethanol production. Rational design of enzyme cocktails is an advanced technique used to create enzyme variants with high specific activity, greater thermostability, and improved tolerance to inhibitors (e.g., furfurals) (Srivastava et al. 2023). It is reported that the permissible cost of the enzyme should not exceed $0.1 per liter of bioethanol (Singh et al.

Table 2.1 Enzyme cocktail used for saccharification of LCB (Broda et al. 2022)

Type of enzyme	Enzyme	Action	Example
Cellulases	Endoglucanases	Cleavage of β-1,4-glycosidic bonds in cellulose regions	TrCel7B from *Trichoderma reesei*
	Cellobiohydrolases	Release of cellobiose from cellulose (both at reducing or non-reducing ends of chain)	TrCel7A from *T. reesei*
	β-Glucosidases	Hydrolysis of cellobiose and cello-oligosaccharides into glucose molecule	TrCel3A from *T. reesei*
Hemicellulases	Endo-β-1,4-xylanases	Break down xylan backbones into xylo-oligosaccharides	TrXyn11A from *T. reesei*
	β-Xylosidases	Release xylose from xylo-oligosaccharides	
	Endo-β-1,4-mannanases	Cleave mannan chains	TrMan5A from *T. reesei*
	β-Mannosidases	Produce mannose from manno-oligosaccharides	
Accessory enzymes	Lytic polysaccharide monooxygenases	Oxidatively cleave glycosidic bonds in cellulose and hemicellulose, boosting overall hydrolysis	
	Debranching enzymes	Remove substituents from hemicellulose, including α-arabinofuranosidases, α-galactosidases, and esterases like acetyl xylan esterases	
	Lignin-modifying enzymes	Laccases for lignin oxidation, or peroxidases, to reduce lignin barriers	

2022). Therefore, the development of ‘on-site’ enzyme production technology is more economical than conventional ‘off-site’ enzyme production facilities through the reduction of enzyme cost.

2.3.4 Fermentation Strategies of LCB-Based Ethanol Production and Scale-Up Challenges

Scale-up of laboratory-scale LCB biorefinery technology is quite challenging and often fails to demonstrate similar efficiency at pilot and demonstration scale operations. In order to translate a lab-scale performance to a commercial level, it is

Table 2.2 List of commercial enzyme cocktails used in various industries for LCB-based bioethanol production (Patel and Shah 2021)

Enzyme cocktail	Manufacturer	Key components	Applications	Key features
Cellic CTec2	Novozymes	Cellulases (endoglucanases, cellobiohydrolases), β-glucosidases, hemicellulases, LPMOs (AA9)	Hydrolysis of pretreated lignocellulosic biomass (e.g., corn stover, sugarcane bagasse)	High β-glucosidase activity for efficient cellulose breakdown; yields up to 89.8% ethanol; optimized for high-solid loading
Cellic CTec3	Novozymes	Enhanced cellulases, β-glucosidases, hemicellulases, auxiliary enzymes	Broad range of lignocellulosic feedstocks (e.g., wheat straw, rice straw)	Improved performance over CTec2; higher sugar yields (up to 90% glucose conversion); robust for industrial-scale processes
Cellic HTec2	Novozymes	Endo-β-1,4-xylanases, β-xylosidases, accessory enzymes	Hemicellulose hydrolysis, often paired with CTec2/3	High endoxylanase activity; selective for soluble hemicellulose; enhances ethanol yields (e.g., 82.7% on certain biomasses)
ACCELLERASE® TRIO™	DuPont (Genencor)	Cellulases, hemicellulases, β-glucosidases	Versatile for agro-industrial residues (e.g., corn cob, bagasse)	Reduces enzyme loading by combining multiple activities; effective for simultaneous saccharification and fermentation (SSF)
Spezyme CP	DuPont (Genencor)	Cellulases (primarily endoglucanases, cellobiohydrolases), β-glucosidases	Hydrolysis of pretreated biomass like wood chips, grasses	Cost-effective for specific feedstocks; moderate yields (70–80% saccharification)
Celluclast 1.5L	Novozymes	Cellulases (endoglucanases, cellobiohydrolases), low β-glucosidase activity	Agricultural residues (e.g., potato peel, vetiver grass)	Often supplemented with β-glucosidase for complete hydrolysis; yields up to 95% with optimized pretreatment
Pectinex Ultra SP-L	Novozymes	Pectinases, hemicellulases, some cellulase activity	Biomass with high pectin content (e.g., citrus waste)	Enhance hydrolysis in pectin-rich substrates; used in integrated biorefineries

necessary to focus on process optimization, development of techno-economic models, design of technological innovations and their simulation, cost and risk analyses, and life cycle analyses using advanced artificial intelligence approaches. In addition, product diversification, zero waste discharge, effective water use, and the economic utilization of workforce and resources are critical components in scaling up of a LCB based biorefinery. Advanced AI and IoT-integrated automation are essential for successful commercialization of LCB-based biorefinery processes. Besides, advanced automation of the biomass supply chain, enzyme hydrolysis process, product separation, and packaging can enhance operational efficiencies, reduce energy consumption, and enhance environmental performance in an LCB-based biorefinery process (Patel and Shah 2021).

LCB-based bioethanol, also known as 2G or cellulosic ethanol, is derived from non-food biomass (as substrate) using a fermentation process followed by downstream processing. Various fermentation configurations have been implemented at commercial-scale bioethanol production, but they have faced economic, technical, and scalability challenges (Patel and Shah 2021). Many commercial production plants are located in Brazil (e.g., GranBio and Raízen facilities), Europe, and the United States, but some have closed (such as DuPont's Nevada plant in 2017) due to various associated challenges. Primary commercial fermentation processes are related to biochemical routes with limited thermochemical approaches. The following are the different fermentation processes used for the production of 2G bioethanol from LCB in recent years (Fig. 2.2).

2.3.4.1 Separate Hydrolysis and Fermentation (SHF)

In the SHF process, enzymatic hydrolysis, i.e., the release of sugar from cellulose and hemicellulose, and fermentation occur in two separate reactors. Enzyme-based saccharification resulting in a sugar-rich hydrolysate is transferred to a fermentation vessel where suitable microorganisms (ethanologen) (e.g., *Saccharomyces cerevisiae*) can convert sugar into ethanol (Ghazali and Mustafa 2025). In this process, handling inhibitors from pretreatment, as well as process control and monitoring, is easy. However, inhibition of glucose and cellobiose by cellulase activity is a major drawback of the SHF process. The SHF process is used in some early-stage commercial and pilot plants but is less preferred nowadays due to limited efficiency compared to integrated processes. However, in a research-oriented pilot plant (e.g., Sekab in Sweden), SHF was incorporated using LCB feedstocks (Guo et al. 2022).

2.3.4.2 Simultaneous Saccharification and Fermentation (SSF)

In this process, a combined enzymatic hydrolysis and fermentation occurred by adding both enzyme and microorganism (e.g., yeast) simultaneously in the same fermenter. Enzymatic saccharification (hydrolysis) of LCB is accompanied by

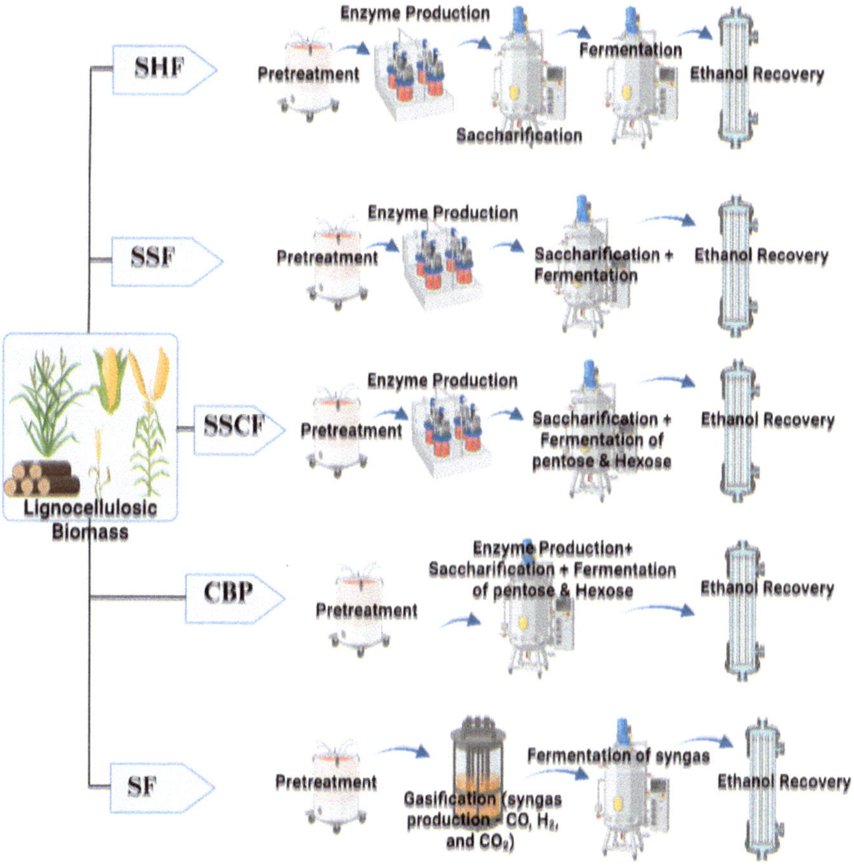

Fig. 2.2 Different processes (*SHF* separate hydrolysis and fermentation; simultaneous saccharification and fermentation, *SSCF* simultaneous saccharification and co-fermentation, *CBP* consolidated bioprocessing, *SF* syngas fermentation) involved in a lignocellulosic biomass-based biorefinery for bioethanol production. (Created in BioRender. Zafar (2025) https://BioRender.com/971mhzf)

simultaneous fermentation to convert released sugars into bioethanol. This system has several advantages, including higher ethanol yields, reduced equipment requirements, and a lower risk of bacterial contamination due to the presence of ethanol; however, it is less effective for C5 (xylose) sugars without modifications. However, temperature control is required to balance enzymatic activity and yeast performance (i.e., 35–40 °C). A high solid loading offers some economic advantages with high ethanol concentration; however, it leads to high viscosity of the medium and high concentration of inhibitory substances, causing mixing issues. SSF is widely adopted at the commercial level due to its efficiency and is successfully integrated with suitable pretreatment technologies such as steam explosion/dilute acid. GranBio's Bioflex 1 plant in Brazil (capacity: ~21 million gallons/year) is an example of SSF utilizing sugarcane bagasse (LCB) with Novozymes for hydrolysis and DSM-engineered yeast (Ghazali and Mustafa 2025).

2.3.4.3 Simultaneous Saccharification and Co-fermentation (SSCF)

It is an advanced version of SSF, in which enzyme hydrolysis and fermentation occur concurrently, and co-fermentation of both pentose (C5) and hexose (C6) sugars is conducted by an engineered microorganism. A genetically engineered *Saccharomyces cerevisiae* or *Zymomonas mobilis* is employed to assimilate up to 40% more sugars released from LCB. Maximum sugar conversion of up to 90% with reduced byproducts and improved economic yield is reported by handling mixed sugar in this process. However, in this process, a robust and inhibitor-tolerant microbial strain with high enzyme loading is required. It is reported that Raízen's E2G plants, with a capacity of 21.66 million gallons per year in Brazil (2024), operated through the SSCF process using Iogen technology, utilizing fermented bagasse and straw sugars via engineered yeast (Ghazali and Mustafa 2025).

2.3.4.4 Consolidated Bioprocessing (CBP)

CBP is a promising approach that provides a sustainable strategy for lignocellulosic bioethanol production, integrating all process steps—enzyme production, saccharification, and fermentation—in a single fermenter by utilizing a genetically modified microorganism or a consortium of microbes. In this process, the selected microbial strain produces cellulases (LCB-degrading enzymes) and can produce bioethanol in a single step, thus reducing the overall cost of the biorefinery process. Genetically modified *Clostridium thermocellum* and *C. cellulolyticum* have been tested in high-efficiency CBP for improved bioethanol yield. An advanced version of CBP is proposed for LCB conversion in which enzyme production is included with hydrolysis, but the fermentation process is separated (Patel and Shah 2021). In this process, inhibition of hydrolytic steps over fermentation can be overcome, and cellulolytic ability can be enhanced.

2.3.4.5 Syngas Fermentation (Thermochemical Route)

This process involves the gasification of LCB into syngas (CO, H_2, and CO_2), followed by fermentation using acetogenic bacteria, such as *Clostridium ljungdahlii*, to produce ethanol. It is considered a hybrid process that combines thermochemical conversion with microbial fermentation. In this process, LCB feedstock can be easily handled without any extensive pretreatment or enzyme cost. However, high energy input is associated with the gasification process and lower ethanol selectivity due to the production of acetic acid as a byproduct.

Commercial adoption of these fermentation methods is slow due to the high cost of enzyme pretreatment, inhibitory toxicity, and low yields. It leads to the closure of many commercial plants (e.g., DuPont, Abengoa). Although remarkable research and development on 2G ethanol production from LCB has been conducted, it remains challenging to utilize pentose sugars. The expression of heterologous genes through rDNA technology enabled the fermentation of xylose; however, a significant yield has not been achieved. Besides, the robustness of genetically modified microorganisms on a commercial scale is poor and specific to a particular type of LCB. Different bacterial species (e.g., *Thermus thermophilus*, *Clostridium thermosaccharolyticum*) and yeast (e.g., *Pichia stipitis*, *Candida shehatae*, *Pachysolen tannophilus*, *Debaryomyces hansenii*) strains are reported to utilize C5 sugars (Xylose). However, these bacterial strains are reported for a low yield of ethanol. *Scheffersomyces stipitis* has the ability to metabolize hexose and pentose sugars together and is gaining interest in LCB biorefinery (Ceccato-Antonini et al. 2017). Therefore, enormous efforts are required in the area of evolutionary engineering and metabolic engineering to construct robust microbial strains capable of converting both C5 and C6 sugars with higher efficiency and tolerance to inhibitory compounds. CBP, where engineered microbes handle all steps, is considered a promising research option and a viable commercial prospect for the future LCB biorefinery process.

2.4 Technological Innovations and Process Engineering Advances

Recent innovations in LCB-based biorefinery are primarily focusing on enhanced process efficiency, sustainability, and cost-effectiveness through sustainable technological advancements. These aspects are related to improvement in novel pretreatment methods, process optimization, as well as molecular innovations achieved through genetic engineering of enzymes and microorganisms (Abolore et al. 2024). Pretreatment of LCB is a significant challenge due to its recalcitrant nature; however, recent technological advancements have emphasized sustainable and eco-friendly methods. Biological methods, including genetically modified fungi and bacteria, have emerged as an energy-efficient alternative that overcomes issues of toxic inhibition (Abolore et al. 2024). Similarly, advanced physical, physico-chemical, and chemical methods have been refined to achieve higher sugar yields and greater eco-friendliness through the use of green chemicals. In addition, advanced machine learning tools such as principal component analysis (PCA), partial least squares (PLS), artificial neural networks (ANN), genetic algorithms, and particle swarm optimization-ANN (PSO-ANN) are utilized to predict and optimize pretreatment outcomes, saccharification, and fermentation processes. These technological innovations align with sustainable and circular economy goals by reducing greenhouse gases and enhancing scalability (Arias et al. 2023).

Enzymatic hydrolysis of LCB has been improved through the design and engineering of an enzymatic system (cocktail) and immobilization techniques. Recent trends include magnetic nanoparticles or gel-based immobilized enzyme systems (cocktails), which allow reuse and enhanced stability. Insights into the molecular design and synthesis of enzymes, such as glycosyl hydrolases (GH) and glycosyltransferases (GT), have guided the development of effective cocktails for breaking down complex LCB. In addition, integrating nanotechnology with enzyme technology can lead to the development of nano-enzymes, enhancing the hydrolytic efficiency of LCB through improvements in enzyme-substrate interactions. Computational fluid dynamics (CFD) modeling of a fermentation system (bioreactor) has been applied to optimize reactor designs for better mixing and reduced viscosity at high solid content. The advanced production of bioethanol on LCB is heavily dependent on genetically modified microorganisms that can overcome the limitations of pentose sugar assimilation and inhibitor tolerance. In recent years, *Saccharomyces cerevisiae* has been incorporated with heterologous xylose isomerases (XI) from *Bacteroides vulgatus* to produce 0.419 g/g of xylose. The oxidoreductive metabolic pathway, which incorporates a cofactor in the enzymes xylose reductase (XR) and xylitol dehydrogenase (XDH), results in a 32% reduction in xylitol and an increase in ethanol production (Vasylyshyn et al. 2025). Using CRISPR-Cas9 technology, ethanol efficiency has been increased while minimizing glycerol levels by deleting the genes GPD2, FPS1, and ADH2 in *S. cerevisiae*. Thermotolerance of engineered *Kluyveromyces marxianus* is increased up to 42 °C, producing 44.95 g/L of ethanol from xylose via XI pathways and exhibiting resistance to inhibitors.

2.5 Challenges and Limitations

LCB-based bioethanol production promises a sustainable alternative to fossil fuels, but faces significant challenges in its commercialization. These challenges lie in the technical, economic, and environmental domains due to the recalcitrant nature of LCB, which is composed of cellulose, hemicellulose, and lignin compounds. Sustainable approaches involve pretreatment to disrupt lignocellulosic structure, hydrolyze polysaccharides into simple sugars, and consolidate fermentation to produce bioethanol. Additionally, various innovations, as well as hurdles related to high costs, inhibitor formation, low yields, and scale-up, have been faced by research communities over the last decade (Broda et al. 2022; Ceaser et al. 2025; Novia et al. 2025). The significant limitations include a lack of universal methods applicable to diverse LCB feedstocks, energy-intensive and eco-friendly processes, and improvements in microbial and enzyme structure (through molecular approaches) to achieve the economic viability of LCB-based bioethanol production.

Pretreatment of LCB is a critical step that accounts for up to 20–40% of the total production cost, aimed at overcoming the hurdles of LCB recalcitrance. However, it often generates inhibitory substances and is inherently energy-intensive. Therefore, significant challenges are related to developing cost-effective methods that minimize the inhibitory substances, such as furfural, 5-hydroxymethylfurfural (HMF), phenols, and acetic acid, that inhibit subsequent microbial and enzymatic activities. Using various physical methods, energy and cost intensity were encountered, related to challenges in solvent recovery, and the need for anti-corrosion equipment.

Similarly, biological methods are inherently slow, and face challenges related to sugar depletion, difficulties in maintaining optimal microbial growth, and scaling up the process. During enzymatic hydrolysis, challenges arise from high enzyme costs, achieving high efficiency at high solid loadings, and ensuring substrate accessibility. High solid load during hydrolysis may increase viscosity, reduce mixing uniformity, and lead to non-uniform heat/mass transfer, resulting in reduced sugar yields. In the fermentation process, challenges arise from the tolerance of microbes to inhibitor compounds, inefficient utilization of pentoses, and suboptimal process conditions.

Additionally, high capital and operational costs make LCB-based bioethanol production less attractive than first-generation bioethanol due to the associated costs of pretreatment and enzyme hydrolysis (i.e., 11–27% and 16–35% of the total cost, respectively) (Broda et al. 2022; Ceaser et al. 2025). LCB-based bioethanol production also faces environmental challenges in terms of generating chemical waste and greenhouse gases if the process is not optimized. Various chemical methods of pretreatment are ecologically unfavorable, and inhibitor-led waste streams require treatment, which contributes to environmental burdens.

2.6 Future Perspectives

In light of the above-discussed challenges, industries and researchers must think critically and develop more advanced LCB-based biorefinery facilities. The LCB-based biorefinery operations can meet the global sustainable needs for bioethanol production, utilizing eco-friendly, carbon-neutral, and clean sources, promoting a circular bioeconomy, and reducing dependence on fossil fuels. However, the gap between lab, pilot, and commercialization steps, technical immaturity at the commercial scale, a lack of infrastructure for a systematic biomass market, supply chain, and transportation, and a lack of investment are common hurdles in the commercialization of LCB-based processes.

The future LCB-based biorefinery requires emerging single-step technologies equipped with advanced biotechnological and artificial intelligence tools, making them sustainable, cost-effective, and eco-friendly in nature. An integration of LCB-based biorefinery with existing petroleum refineries could achieve sustainability and be considered a future avenue in the energy sector (Singh et al. 2022).

The various 'omics' approaches, such as metagenomics, meta-transcriptomics, meta-proteomics, and metabolomics, and 'de novo' enzyme designing, are considered advanced enzyme discovery tools and play a significant role in designing efficient enzyme cocktails for LCB pretreatment processes. The recent advancement in biotechnological tools has led to the discovery of cellulosome and multi-modular enzyme systems that are effective in thermophilic and hyper-thermophilic conditions (Li et al. 2017; Singh et al. 2022). Advances in metabolic engineering lead to modifications in key metabolic pathways and cell surface enzymes, enabling the assimilation of LCB-based substrates and the production of desired bioethanol at a higher yield. Consolidated bioprocessing (CBP) is an emerging biorefinery technology that targets a single-step operation for LCB pretreatment, enzyme production, saccharification, and bioethanol production by utilizing a potent microorganism equipped with advanced instrumentation and technologies. However, the commercial implementation of this approach requires the deployment of an integrated AI-IoT approach for evaluating microbial dynamics and process dynamics, enabling precision in LCB-to-bioethanol conversion (Lim and Kim 2024; Wang et al. 2022). This integrated approach enables efficient monitoring of biochemical reactions and material properties in real-time, leading to optimized bioethanol yields. Advanced machine learning (ML) algorithms such as ANN, GAs, and SVM can efficiently train process datasets to improve enzyme engineering, bioreactor management, and enhanced fermentation efficiency. AI can efficiently screen the performance of a GMO to enhance saccharification and enzyme efficiency, while handling multivariate parameters such as pH, temperature, and nutrients, for predictive modeling. These approaches enable real-time monitoring to maximize bioethanol yield.

2.7 Conclusion

The development of lignocellulosic biomass-based biorefinery offers several advantages from both economic and environmental viewpoints; however, the commercialization of this renewable bioethanol production is challenging due to several inherent issues in the current scenario. Seasonal limitations on biomass, its pretreatment strategies, tailor-made enzyme cocktails, innovative consolidated bioprocess technologies, and eco-friendly downstream processing for mixed lignocellulosic biomass are essential milestones to achieve a sustainable future for LCB-based biorefineries. In the future, an integrated AI-IoT approach can efficiently drive LCB-based bioethanol innovation through feedstock yield prediction, advanced automation, microfluidic sensors, robotized consolidated bioprocessing, high-throughput platforms, and data-driven sustainability, ultimately making biorefineries smarter and greener for lignocellulosic bioethanol production.

Acknowledgment The author acknowledges the facilities and infrastructural support provided by the University of Technology and Applied Sciences-Sur, Oman.

References

Abolore RS, Jaiswal S, Jaiswal AK (2024) Green and sustainable pretreatment methods for cellulose extraction from lignocellulosic biomass and its applications: a review. Carbohydr Polym Technol Appl 7:100396. https://doi.org/10.1016/j.carpta.2023.100396

Arias A, Feijoo G, Moreira MT (2023) How could Artificial Intelligence be used to increase the potential of biorefineries in the near future? A review. Environ Technol Innov 32:103277. https://doi.org/10.1016/j.eti.2023.103277

Broda M, Yelle DJ, Serwańska K (2022) Bioethanol production from lignocellulosic biomass-challenges and solutions. Molecules 27(24):8717. https://doi.org/10.3390/molecules27248717

Ceaser R, Montané D, Constantí M, Medina F (2025) Current progress on lignocellulosic bioethanol including a technological and economical perspective. Environ Dev Sustain 27: 25993–26038. https://doi.org/10.1007/s10668-024-04792-2

Ceccato-Antonini SR, Codato CB, Martini C, Bastos RG, Tauk-Tornisielo SM (2017) Yeast for pentose fermentation: isolation, screening, performance, manipulation, and prospects. In: Buckeridge M, De Souza A (eds) Advances of basic science for second generation bioethanol from sugarcane. Springer, Cham. https://doi.org/10.1007/978-3-319-49826-3_8

Dongmo DN, Ngouanwou MGN, Atemkeng CD, Ngomade SBL, Atangana JAK, Tagne RFT, Kamgaing T (2024) Bioethanol production from cocoa hydrolysate and the assessment of its environmental sustainability. Heliyon 10(3). https://doi.org/10.1016/j.heliyon.2024.e25809

Ghazali MFSM, Mustafa M (2025) Bioethanol as an alternative fuel: a review on production strategies and technique for analysis. Energy Convers Manage X 26:100933. https://doi.org/10.1016/j.ecmx.2025.100933

Guo Y, Liu G, Ning Y, Li X, Hu S, Zhao J, Qu Y (2022) Production of cellulosic ethanol and value-added products from corn fiber. Bioresour Bioprocess 9:81. https://doi.org/10.1186/s40643-022-00573-9

Jose D, Sivakumar R, Agarwal M, Arshad N, Gunduppali M, Thiagamani SMK, Venkatachalam P, Amornraksa S, Sriariyanun M (2025) A comprehensive review of conversion of rice biomass into sustainable products: a green approach toward a circular economy. Sustain Chem Clim Action:100069. https://doi.org/10.1016/j.scca.2025.100069

Joyia MAK, Ahmad M, Chen YF, Mustaqeem M, Ali A, Abbas A, Gondal MA (2024) Trends and advances in sustainable bioethanol production technologies from first to fourth generation: a critical review. Energy Convers Manag 321:119037. https://doi.org/10.1016/j.enconman.2024.119037

Kumar B, Verma P (2021) Biomass-based biorefineries: an important architype towards a circular economy. Fuel 288:119622. https://doi.org/10.1016/j.fuel.2020.119622

Larnaudie V, Ferrari MD, Lareo C (2022) Switchgrass as an alternative biomass for ethanol production in a biorefinery: perspectives on technology, economics and environmental sustainability. Renew Sust Energ Rev 158:112115. https://doi.org/10.1016/j.rser.2022.112115

Li Z, Liu G, Qu Y (2017) Improvement of cellulolytic enzyme production and performance by rational designing expression regulatory network and enzyme system composition. Bioresour Technol 245:1718–1726

Lim H, Kim S (2024) Applications of machine learning technologies for feedstock yield estimation of ethanol production. Energies 17(20):5191. https://doi.org/10.3390/en17205191

Mansy AE, El Desouky EA, Taha TH, Abu-Saied MA, El-Gendi H, Amer RA, Tian ZY (2024) Sustainable production of bioethanol from office paper waste and its purification via blended polymeric membrane. Energy Convers Manag 299:117855. https://doi.org/10.1016/j.enconman.2023.117855

Mgeni ST, Mero HR, Mtashobya LA, Emmanuel JK (2024) The prospect of fruit wastes in bioethanol production: a review. Heliyon 10(19). https://doi.org/10.1016/j.heliyon.2024.e38776

Mgeni ST, Mero HR, Mtashobya LA, Emmanuel JK (2025) Utilizing fruit wastes as a sustainable feedstock for bioethanol production: a review. Clean Energy Syst:100188. https://doi.org/10.1016/j.cles.2025.100188

Momayez F, Okoro OV, Shavandi A, Martín C, Denayer J, Karimi K (2025) A critical review of ultrasonication as a green technology for enhanced biomass valorization in bioethanol and biogas production. Process Saf Environ Prot:107334. https://doi.org/10.1016/j.psep.2025.107334

Nandi S, Gonela V, Awudu I (2023) A resource-based and institutional theory-driven model of large-scale biomass-based bioethanol supply chains: an emerging economy policy perspective. Biomass Bioenergy 174:106813. https://doi.org/10.1016/j.biombioe.2023.106813

Nawaz A, Haddad H, Shah MA, Uddin S, Hossain MM, Razzak SA (2024) Fueling sustainability: co-pyrolysis of microalgae biomass and waste plastics for renewable energy and waste mitigation. Biomass Bioenergy 187:107303. https://doi.org/10.1016/j.biombioe.2024.107303

Novia N, Melwita E, Jannah AM, Selpiana S, Yandriani Y, Afrah BD, Rendana M (2025) Current advances in bioethanol synthesis from lignocellulosic biomass: sustainable methods, technological developments, and challenges. J Umm Al-Qura Univ Appl Sci. https://doi.org/10.1007/s43994-025-00212-x

Patel A, Shah AR (2021) Integrated lignocellulosic biorefinery: gateway for production of second-generation ethanol and value-added products. J Bioresour Bioproducts 6(2):108–128. https://doi.org/10.1016/j.jobab.2021.02.001

Saini R, Saini JK, Adsul M, Patel AK, Mathur A, Tuli DK, Singhania RR (2015) Enhanced cellulase production by *Penicillium oxalicum* for bioethanol application. Bioresour Technol 188:240–246

Same NN, Yakub AO, Chaulagain D, Tangoh AF, Nsafon BEK, Owolabi AB et al (2024) The future of clean energy: agricultural residues as a bioethanol source and its ecological impacts in Africa. Renew Energy 237:121612. https://doi.org/10.1016/j.renene.2024.121612

Saravanan A, Ragini YP, Karishma S, Hemavathy RV, Jyotsna M (2025) A review on advancing sustainable energy: the role of biomass and bioenergy in a circular economy. Sustain Futures:100835. https://doi.org/10.1016/j.sftr.2025.100835

Singh N, Singhania RR, Nigam PS, Dong C-D, Patel AK, Puri M (2022) Global status of lignocellulosic biorefinery: challenges and perspectives. Bioresour Technol 344:126415. https://doi.org/10.1016/j.biortech.2021.126415

Srivastava N, Singh R, Singh P, Ahmad I, Singh RP, Rai AK, Asiri M, Gupta VK (2023) Recent advances on lignocellulosic bioresources and their valorization in biofuels production: challenges and viability assessment. Environ Technol Innov 29:103037. https://doi.org/10.1016/j.eti.2023.103037

Sudalai S, Prabakaran S, Varalakksmi V, Kireeti IS, Upasana B, Yuvasri A, Arumugam A (2024) A review on oilcake biomass waste into biofuels: current conversion techniques, sustainable applications, and challenges: waste to energy approach (WtE). Energy Convers Manag 314:118724. https://doi.org/10.1016/j.enconman.2024.118724

Thuppahige RTW, Barner L, Shahbazi M, Fraga G, Moghaddam L (2025) A comprehensive review of sustainable valorisation of lignocellulosic biomass and plastic waste into biofuels and chemicals via co-liquefaction. Waste Manag 202:114827. https://doi.org/10.1016/j.wasman.2025.114827

Usmani Z, Sharma M, Awasthi AK, Lukk T, Tuohy MG, Gong L, Nguyen-Tri P, Goddard AD, Bill RM, Nayak SC, Gupta VK (2021) Lignocellulosic biorefineries: the current state of challenges and strategies for efficient commercialization. Renew Sustain Energy Rev 148:111258

Vasylyshyn R, Ruchala J, Dmytruk K, Sibirny A (2025) Recent progress in engineering yeast producers of cellulosic ethanol. FEMS Yeast Res 25:foaf035. https://doi.org/10.1093/femsyr/foaf035

Vishwas PA, Kumar A, Meena KR (2024) Valorization of agri-wastes through microbial consortia and its conversion into bioethanol by co-fermentation method using *Saccharomyces cerevisiae* NCIM 3594 and *Pichia stipitis* NCIM 3497 strains. Bioresour Technol Rep 28:101991. https://doi.org/10.1016/j.biteb.2024.101991

Wang K, Khoo KS, Leong HY, Nagarajan D, Chew KW, Ting HY, Selvarajoo A, Chang J-S, Show PL (2022) How does the Internet of Things (IoT) help in microalgae biorefinery? Biotechnol Adv 54:107819. https://doi.org/10.1016/j.biotechadv.2021.107819

Yadav A, Sharma V, Tsai M-L, Chen C-W, Sun P-P, Nargotra P, Wang J-X, Dong C-D (2023) Development of lignocellulosic biorefineries for the sustainable production of biofuels: towards circular bioeconomy. Bioresour Technol 381:129145. https://doi.org/10.1016/j.biortech.2023.129145

Chapter 3
Advances in Feedstock Transport and Logistic Supply Chain Development for Lignocellulosic Bioethanol Production

Vikas Chandra Gupta, Avinash Singh, Neelesh Prashant, and Mohd. Zafar

Abstract Growing interest in lignocellulosic bioethanol production is driving advancements in biomass logistics and transportation. Establishing preprocessing stations for biomass densification enables efficient, long-distance delivery to biorefineries, enhancing supply chain performance. Upgrading infrastructure and adopting decentralised conversion approaches facilitate local sourcing, improve biomass flow, and reduce pressure on energy systems. Artificial intelligence and machine learning analyse environmental data to develop sustainable practices, optimise transportation routes, and reduce greenhouse gas emissions. Locating processing facilities near biomass sources reduces transport distances, while AI-driven modelling improves logistics planning, route efficiency, and inventory management. A hybrid logistics model that integrates decentralised and centralised strategies, including regional depots for pretreatment, balances economies of scale with reduced transportation costs.

Keywords Lignocelluloses · Lignin · Biopolymers · Depolymerisation · Bioethanol · Biorefineries · Circular economy

---

V. C. Gupta (✉)
Department of Biotechnology, IILM University, Greater Noida, Uttar Pradesh, India

A. Singh
Department of Biotechnology, Meerut Institute of Engineering and Technology (MIET), Meerut, Uttar Pradesh, India

N. Prashant
Molecular Systematics Lab, Department of Botany, Central University of Punjab, Bathinda, Punjab, India

M. Zafar
Department of Biotechnology, College of Applied Sciences and Pharmacy, University of Technology and Applied Sciences, Sur, Oman
e-mail: Mohammed.zafar@utas.edu.om

S. Prasad et al. (eds.), *Sustainable Bioethanol Production from Lignocellulosic Biomass*, Clean Energy Production Technologies,
https://doi.org/10.1007/978-981-95-9690-4_3

3.1 Introduction

Lignocellulosic bioethanol is made from non-food materials such as crop residues and forestry byproducts, providing a sustainable fuel that does not compete with food crops. For this biofuel to be commercially successful, it needs a reliable and affordable supply chain to deliver feedstock to biorefineries. Because logistics can account for up to half of total production costs, improving the supply chain is key to making second-generation bioethanol economically competitive (Gonzalez et al. 2011; Broda et al. 2022). To succeed in producing lignocellulosic bioethanol, the logistics must be both cost-effective and efficient. Because the quantity and price of biomass can fluctuate, effective supply chain management is important for maximising output.

Lignocellulosic bioethanol has about half the carbon footprint of gasoline (Adeyanju et al. 2021; Johnson et al. 2023). Still, because biomass is not very dense and transport costs are high, advanced logistics solutions are needed to reduce costs (Elliott et al. 2014; Chen et al. 2022a, b). Using a mix of lignocellulosic materials collected at different times can strengthen the supply chain and reduce logistics costs by up to 25% (Zhang et al. 2020). It is essential to manage both biomass density and seasonal availability to control transportation costs and select optimal locations for biorefineries (Elliott et al. 2019; Stephen et al. 2021). Gathering feedstock within 30–50 miles of biorefineries is usually the most cost-effective approach (Miao et al. 2021).

Regional biomass processing depots (RBPDs) act as centres for collecting, preprocessing, and densifying biomass, thereby helping address many logistical challenges. These depots enhance transportation and storage efficiency and help prevent biomass degradation (Eranki et al. 2020). Managing seasonal changes and feedstock availability is also important for maintaining steady production and for managing storage and handling costs (Han et al. 2022). New methods, such as dry acid pretreatment at collection sites, can boost storage capacity (Han et al. 2022). Using a variety of lignocellulosic feedstocks can help rural economies by connecting farms with biorefineries (Zhang et al. 2020). Mixing different feedstocks also makes the supply chain more resilient to market changes. It can produce as much or more ethanol than using a single feedstock, thereby reducing logistics costs prior to the biorefinery stage (Oke et al. 2019).

Artificial intelligence (AI) and machine learning (ML) enhance the efficiency of the biofuel supply chain. AI provides real-time updates on available biomass, transport status, and areas for improvement, helping companies adjust quickly and save money. Machine learning analyses historical data to predict supply chain issues and identify optimal routes and schedules for moving feedstock (You et al. 2011; Chen et al. 2022a, b). AI helps monitor supply chain performance by using predictive analytics to optimise storage and reduce biomass loss during transport (Zhao et al. 2022a, b). When AI is combined with geographic information systems (GIS), it enables stakeholders to visualise supply flows and identify problems as they occur, allowing them to respond quickly (Singh and Kumar 2023). Advanced models can also assess the environmental and economic impacts of different logistics options, providing useful information to inform improved planning and operations (Miller et al. 2023).

To improve biofuel supply chains, advanced modelling and analysis are needed. Using multi-objective optimisation and life-cycle assessments helps evaluate the economic, environmental, and social sides of feedstock logistics (You et al. 2011; Li et al. 2023a, b). These findings inform biorefinery designs focused on efficiency and sustainability, thereby supporting the growth of the bioethanol industry (Wei et al. 2010a, b; Thompson et al. 2022). Choosing the appropriate mode of transport, such as rail or truck, is important because each has distinct costs and environmental impacts. Using multiple modes of transport can also reduce the overall carbon footprint (Park et al. 2017; Smith and Johnson 2021). Improving fleet efficiency and coordinating different transport modes are key to addressing logistics challenges in bioethanol production.

Working with stakeholders and adhering to regulations makes lignocellulosic feedstock supply chains more complex. It is important to partner with local groups, such as farmers and cooperatives, to maintain a steady feedstock supply and ensure production goals align with community needs (Han et al. 2021; Rodriguez and Wang 2023). Effective communication among all parties in the supply chain facilitates coordination and ensures smooth logistics. Building a strong logistics supply chain for lignocellulosic bioethanol requires careful planning, environmental focus, and the adoption of new technologies. This chapter examines strategies to enhance transportation efficiency, improve storage, and leverage tools such as artificial intelligence and machine learning. Improving these areas will enhance the supply chain and unlock the full potential of lignocellulosic bioethanol.

3.2 Physical Properties and Handling Constraints of Lignocellulosic Biomass

Lignocellulosic biomass is made up of cellulose, hemicellulose, and lignin. The various physical properties of this biomass significantly affect its management, storage, and conversion into energy (Smith et al. 2022a, b, c, d). To optimise the use of biomass, it is essential to understand its characteristics, such as moisture content, particle size and shape, bulk density, porosity, surface area, mechanical properties, structural heterogeneity, and hygroscopicity (Jones and Taylor 2023). In addition, lignocellulosic biomass faces several challenges related to handling, storage, transportation, processing, environmental impacts, and safety (Green et al. 2021). Figure 3.1 illustrates the physical properties and handling constraints associated with lignocellulosic biomass. It highlights the importance of addressing these challenges through integrated approaches to biomass handling and processing technology. Such strategies are crucial for realising the potential of lignocellulosic

Fig. 3.1 Physical properties and handling constraints of lignocellulosic biomass

biomass as a sustainable energy resource while accounting for environmental factors.

Moisture content is a critical factor in biomass management, typically ranging from 25% to 60%, depending on the type of biomass and its storage conditions. High moisture levels can decrease energy density and thermal efficiency during combustion or gasification; however, effective moisture management can improve biomass performance (Li et al. 2018; Chai et al. 2021a, b; Phanphanich and Mani 2011; Zhang et al. 2023a, b, c, d, e). By reducing excess moisture, we can lower transportation costs and minimise spoilage risks, particularly under varying environmental conditions. Additionally, moisture affects the rheological properties of biomass, leading to clumping and increased cohesiveness. To address these challenges, specialised handling systems can be employed (Hartley et al. 2018; Johnson et al. 2022).

Another significant factor is the low bulk density of lignocellulosic biomass, which typically ranges from 30 to 100 kg/m^3 for raw materials (Nguyen et al.

2021a, b; Clavijo et al. 2020; Smith et al. 2023a, b, c, d, e, f). Implementing compaction methods, such as pelletisation or briquetting, can significantly enhance transportation and storage efficiency. For instance, pelletisation can increase biomass density to over 1000 kg/m^3, thus improving handling and storage stability by reducing moisture content and enhancing durability (Clavijo et al. 2020; Holm et al. 2011; Tran et al. 2023).

The structural heterogeneity of lignocellulosic biomass presents opportunities for optimisation. Variations in fibrous structures can influence grinding and feeding efficiency (Kumar et al. 2016; Nguyen et al. 2021a, b). By focusing on particle-size distribution and chemical-composition variability, we can enhance the efficiency of pretreatment processes, thereby facilitating the conversion of biomass into biofuels or other valuable bioproducts (Barakat et al. 2014; Yan et al. 2020; Vasić et al. 2021). Smaller particle sizes increase the available surface area for enzymatic access during hydrolysis, thereby improving biofuel production efficiency (Vasić et al. 2021; Priyanto et al. 2018). Furthermore, addressing the hydrophilicity of lignocellulosic materials, which leads to moisture absorption, is crucial for effective storage management. Controlled storage conditions can prevent degradation and spontaneous combustion (Chai et al. 2021a, b; Zheng et al. 2015).

The high variability in biomass composition across different samples complicates the optimisation of conversion technologies, underscoring the need for tailored processing solutions to achieve consistent results across various feedstocks (Barakat et al. 2014; Gonzales et al. 2025). Strategic efforts can significantly enhance the transportation and storage efficiency of biomass despite its low bulk density and inherent variability (Cavailles et al. 2024). Bulk densities typically range from 100 to 200 kg/m^3, and moisture content is approximately 20–50% at harvest (Smith et al. 2022a, b, c, d). Effectively addressing these logistical challenges is vital, as transportation costs can account for 17–20% of total supply chain expenses, primarily because biomass has lower energy density than fossil fuels (Johnson and Lee 2023a, b, c).

Given the seasonal availability of residues such as corn stover, wheat straw, and rice straw, implementing strategic storage solutions is crucial. Some materials benefit from storage periods of up to 9 months to ensure successful outcomes in biorefinery operations (Davis 2021). The economic landscape for supplying these feedstocks varies widely, with costs ranging from \$48 to \$111 per ton for corn stover and from \$123 to \$208 per ton for rice straw, depending on processing methods (Miller 2023). By addressing these factors, we can significantly enhance the efficiency and viability of biomass as a sustainable energy resource.

However, storing lignocellulosic biomass presents unique challenges. High moisture content not only leads to microbial degradation but can also pose risks of spontaneous heating and combustion if not appropriately managed (Waduge et al. 2017; Zhang et al. 2022). Therefore, rigorous control measures are essential to ensure safety, as reducing particle size during storage can generate fine dust, which poses a safety risk due to the risk of dust explosions (Waduge et al. 2017; Li et al. 2023a, b). Implementing effective dust-control measures and maintaining optimal moisture levels are critical for mitigating these risks (Waduge et al. 2017; Chen et al. 2023).

Transportation issues are further complicated by the need for specialised equipment to handle the bulk and varied characteristics of biomass. The costs associated with transporting relatively low-energy-density materials make lignocellulosic biomass less economically viable than traditional energy sources (Eranki and Dale 2011; Fike et al. 2008; Yao et al. 2023). Establishing local pre-processing centres could provide solutions by reducing transport distances and improving overall efficiency in biomass use (Carolan et al. 2007). However, the effectiveness of such systems may be constrained by inconsistent supply and the inherent variability of biomass feedstocks, highlighting the necessity for standardised procedures and protocols to optimise biomass management in the energy sector.

3.3 Lignocellulosic Biomass Feedstock Production, Cost, and Availability

Lignocellulosic feedstocks represent a pathway toward a more sustainable energy future. They are generally classified into three main categories: agricultural residues such as corn stover and wheat straw; forestry wood residues; and purpose-grown energy crops, including switchgrass and miscanthus (Szwarc and Głowacka 2021; Isikgor and Becer 2015). Each category has distinct characteristics that influence the efficiency and economic feasibility of biofuel conversion. For instance, agricultural residues typically require intensive pretreatment to access cellulose, while variable lignin content and high energy demand during processing introduce additional complexity. Addressing these technical challenges is essential for understanding the practical barriers to commercial bioethanol production.

Globally, lignocellulosic biomass production is estimated at 181–200 billion tons annually, with costs ranging from USD 24 to USD 121 per ton. These costs depend on various factors, including crop type, yield, regional conditions, collection logistics, and processing methods (Haq et al. 2021; Mohammed et al. 2022). The diverse composition of biomass requires careful consideration in pricing and market development (Li et al. 2021; Isikgor and Becer 2015). A clear understanding of these factors can help stakeholders assess economic feasibility more accurately.

Furthermore, lignocellulosic biomass offers an alternative that does not compete with food crops, thereby helping to address food security concerns (Juneja et al. 2011). While the production and processing costs of lignocellulosic biomass can be complex, current limitations include high energy consumption during pretreatment and the cost of enzymatic processes. Innovative techniques, such as dilute-acid hydrolysis, have shown promise in improving the accessibility of cellulose and hemicellulose. Still, challenges remain in efficiently scaling these methods and reducing overall costs and energy use (Jeswani et al. 2020). Addressing these gaps is crucial for commercial viability. Ongoing research focuses on enhancing conversion processes, with promising developments such as recycling insoluble solids to optimise sugar yields (Weiss et al. 2013). Additionally, advancements in genetic modification, such as the creation of transgenic sugarcane that produces cellulolytic enzymes, hold great potential to reduce enzymatic costs and streamline biomass processing (Harrison et al. 2011).

The future of lignocellulosic biomass in biofuel production depends on overcoming technical processing challenges and ensuring a reliable supply of this abundant resource. As traditional crop-based feedstocks face supply and price fluctuations, lignocellulosic alternatives become increasingly attractive, particularly as demand for cleaner energy grows (Londo et al. 2018). Adopting these options can help address energy shortages and reduce the environmental impact of fossil fuels.

In the United States alone, 341 million dry tons of lignocellulosic biomass were available annually in 2012, of which 70% were derived from agricultural residues and 30% from forest residues (Balan 2014a, b). By 2022, yields from energy crops such as hybrid switchgrass (expected at 9.4 dry tons per hectare) and miscanthus (expected at 13.2 dry tons per hectare) were anticipated to contribute significantly to the total biomass supply (Balan 2014a, b). Genetic enhancements, agronomic optimisation, and the use of high-yielding cultivars could result in a 2–4% annual increase in yields, potentially exceeding 1 billion dry tons of biomass by 2030 under optimistic scenarios (Balan 2014a, b).

The global lignocellulosic biomass market is projected to grow from USD 4.61 billion in 2025 to USD 9.76 billion by 2035, reflecting a compound annual growth rate (CAGR) of 7.8% (Markets and Markets 2023). This growth is driven by rising demand for renewable and sustainable energy sources as the world transitions to cleaner energy solutions (International Energy Agency 2022). Notably, bioenergy production is projected to account for 15% of the lignocellulosic biomass market by 2025 (Bioenergy Europe 2023). Despite the ample supply of lignocellulosic raw materials, significant potential remains to improve supply chains and optimise their integration into energy and bioproduct production systems (Smith et al. 2021). Enhancing these supply chains is essential to unlocking the full benefits of lignocellulosic feedstocks in advancing a sustainable energy future (Jones and Taylor 2022).

3.4 Harvesting Logistics, Collection, and Preprocessing of Lignocellulosic Biomass

The logistics of harvesting and preprocessing lignocellulosic biomass are critical for advancing second-generation biofuels. Efficient preharvest logistics, which include biomass collection and transportation, enhance both economic feasibility and processing efficiency. Key factors, such as harvest timing, weather conditions, and biomass source locations, significantly influence both biomass quality and yield (Nguyen et al. 2014; Jacobson et al. 2014).

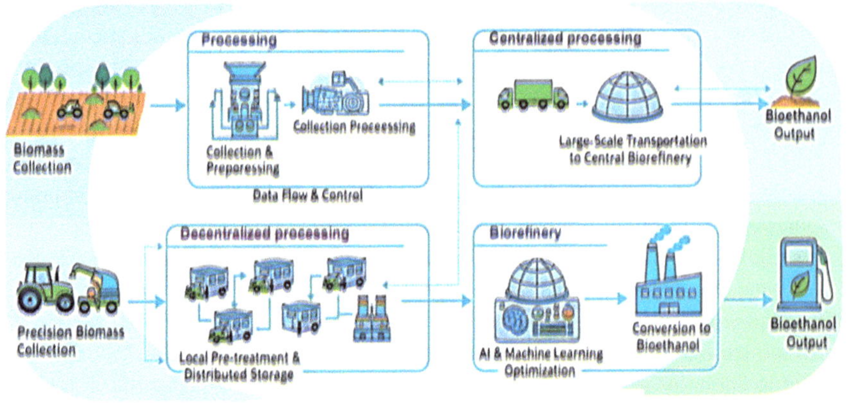

Fig. 3.2 Advanced lignocellulosic biomass supply chain to biorefineries for bioethanol production

Integrating logistics with storage and preprocessing can reduce costs and improve quality (Igathinathane et al. 2016; Ray et al. 2017). Recent studies suggest that artificial intelligence (AI) and machine learning can optimise logistics by predicting optimal harvest times and locations and improving the efficiency of transportation routes (Smith et al. 2023a, b, c, d, e, f). Emerging strategies, such as single-pass technologies and dedicated collection centres, promise to enhance efficiency further. AI can also enable real-time monitoring of biomass quality during the collection process (Jones et al. 2023).

Preprocessing and densification are crucial for converting biomass into biofuels. Densification increases energy density and reduces environmental impact, and AI can improve densification processes by leveraging biomass characteristics (Muazu et al. 2022; Francis et al. 2023). Densified biomass exhibits improved combustion properties, resulting in lower emissions (Yang et al. 2025). By focusing on these areas, we can make significant strides toward sustainable biofuels. A schematic diagram of advanced harvesting logistics, collection, and preprocessing in the feedstock supply chain that delivers lignocellulosic biomass to biorefineries for bioethanol production is presented in Fig. 3.2.

3.4.1 *Precision Biomass Collection*

Effective biomass collection in farmers' fields is essential for precision agriculture, which incorporates technology and data to enhance farm management (Sari et al. 2021). Accurate biomass estimation is crucial for optimising fertiliser use, predicting crop yields, and monitoring crop health (Tilly et al. 2014; Smith et al. 2022a, b, c, d). Terrestrial laser scanning (TLS) is a promising technique for precise biomass measurement, as it uses laser scanners to capture dense point measurements of crops, aiding in modelling plant height and biomass. Tilly et al. (2014) demonstrated the effectiveness of TLS in generating high-resolution crop surface models, which revealed strong correlations between plant height and biomass, supporting its suitability for large-scale agriculture.

Recent studies have highlighted the increasing role of artificial intelligence (AI) and machine learning in advancing precision agriculture. For example, machine learning algorithms can analyse data captured from TLS and uncrewed aerial vehicles (UAVs) to improve the accuracy of biomass estimation (Khan et al. 2022; Wang et al. 2023). By processing large datasets, these algorithms can identify patterns and predict crop performance more efficiently, thereby improving decision-making for farmers. Additionally, AI-based systems can integrate data from various sources, such as weather and soil conditions, to provide comprehensive insights into crop health and optimise resource allocation (Patel et al. 2023a, b; Johnson and Lee 2022a, b, c).

UAVs equipped with RGB cameras are also valuable tools for precision biomass collection. They capture aerial imagery to monitor crop conditions and estimate biomass (Sari et al. 2021). This technology enables early detection of crop issues, allowing farmers to manage their crops proactively. UAVs have significantly improved rice management and production (Sari et al. 2021; Zhang et al. 2023a, b, c, d, e). However, the adoption of precision agriculture practices largely depends on farmers' attitudes. Understanding these is crucial, as they can influence the uptake of innovative techniques (Shitu et al. 2023). By gaining insights into farmers' perspectives and providing tailored support, we can enhance the adoption of precision agriculture technologies and ultimately improve agricultural practices.

3.4.2 *Precision Biomass Collection Methods*

Baling biomass is a well-established method for collecting and storing feedstocks, which can be produced in either round or square bales. This technique not only compacts biomass but also significantly reduces storage and transportation costs (Zhang et al. 2023a, b, c, d, e). The advantages of biomass baling include enhanced storage efficiency, minimised soil contamination, and a more consistent biomass ash content (Kenney et al. 2013). Recent research suggests that optimising the roller layout can improve bale quality while reducing energy consumption during the baling process (Zhang et al. 2023a, b, c, d, e).

Emerging studies indicate that artificial intelligence (AI) and machine learning (ML) are playing a transformative role in biomass management. These technologies enable predictive analytics to optimise the baling process, allowing real-time adjustments based on environmental conditions and feedstock variations. For instance, machine learning algorithms can analyse historical yield data to identify the most efficient harvesting times and techniques, thereby improving overall biomass quality (Smith et al. 2023a, b, c, d, e, f).

A comprehensive simulation model has been developed to examine biomass production and supply logistics for various feedstocks, including straw and switchgrass (Sokhansanj et al. 2023). Additionally, machine learning can enhance logistics modelling, enabling the prediction of optimal routes and transportation schedules, thereby further reducing costs and emissions (Nguyen et al. 2023). Furthermore, wet anaerobic storage of corn stover has proven effective in ensuring a consistent feedstock supply while minimising the risk of fire loss (Wendt et al. 2018). For lower-quality flax straw, various harvesting techniques—such as cutting, threshing, and natural drying—are employed.

AI-driven systems can also assess the efficiency of different harvesting techniques, enabling customised approaches based on specific field conditions. Additionally, rice straw, which constitutes 40–60% of the rice plant, is increasingly recognised for its potential off-field applications due to growing environmental concerns about open burning (Venkatramanan et al. 2021). Approximately 1.35 tons of rice straw remain per ton of grain harvested, making baling the preferred method for handling rice straw because it facilitates transport and storage (Venkatramanan et al. 2021).

A baler is a machine that compresses straw into bales and can be classified as stationary, movable, or field-based. Furthermore, balers can be categorised as square, rectangular, or round. Large square and round balers are commonly used for biomass production, creating compact bales that optimise handling efficiency (Shinners and Friede 2018; Grisso et al. 2020). A bale accumulator simplifies the collection process by grouping bales, and automatic collectors can stack them for convenience. In the absence of these systems, flatbed trucks and loaders are typically used to transport bales. Round bales, in particular, offer advantages for biomass storage due to their low storage requirements and efficient handling, with densities ranging between 100 and 170 kg/m^3 (Dudarev and Khomych 2022).

Optimising factors, such as the baler roller layout, can enhance bale quality and improve energy efficiency (Zhang et al. 2023a, b, c, d, e). Proper storage of biomass bales is essential to maintain their economic value and ensure a consistent year-round supply to biorefineries. However, outdoor storage can lead to significant dry matter loss (DML), and the choice of storage method—whether indoor, outdoor-tarped, or outdoor-open—can significantly influence costs and quality (Sahoo and Mani 2017). Large square balers, used for crops such as hay, straw, and biomass feedstocks, offer higher material capacity and greater field efficiency than round balers (Shinners and Friede 2018). Although round balers are generally more affordable and compatible with smaller tractors—making them an excellent option for smaller farms—large square balers produce denser bales, which reduces handling, storage, and transportation costs (McAfee et al. 2018).

These machines can produce large bales (1.2 m × 1.2 m × 2.4 m) with throughput rates ranging from 35 Mg/h to 116 kg·m^{-3} (wet matter) (Guerrieri et al. 2019). When stored properly, large square bales can better preserve crops, allowing for uninterrupted operation without frequent unloading (Balingbing et al. 2020). However, they do require covered storage and larger tractors, which may present challenges for smaller farms (Grisso et al. 2020). To maximise efficiency, it is crucial to manage wrap/eject times and twine tension while understanding the relationship between equipment capacity and these parameters (Brue et al. 2015).

The logistics of storing and transporting feedstock at the biorefinery gate are essential for effective biomass logistics. Implementing storage methods, such as forming windrows to dry and bale biomass, can help mitigate losses from microbial degradation (Wendt and Zhao 2020). Transportation logistics are equally important; accommodating multiple feedstocks can reduce procurement costs and emissions while providing the flexibility needed for optimal harvest timing (Zhang et al. 2018). Ensuring feedstock quality may require size reduction at the biorefinery or a depot. Overall, effective management of harvesting, handling, storage, and delivery of biomass is fundamental to cost-effective biofuel production (Hess et al. 2007).

In biomass densification, both pelletisation and briquetting use high compressive force to increase material density. The ideal moisture content for pellets is 12–18%, whereas briquettes perform best at 15–30% (Thurber et al. 2014). Unlike pelletisation, briquetting usually does not require a binder. Both processes yield products with higher heating values, greater moisture resistance, and improved combustion efficiency compared to raw biomass (Brachi et al. 2017). As research continues to explore the applications of AI in refining these processes, the integration of machine learning is likely to lead to greater efficiency and breakthroughs in biomass utilisation.

3.4.3 Transportation Strategies

Understanding the factors that influence transport costs—such as bulk density, moisture content, and transportation mode—is essential for industry professionals, as these variables significantly impact logistics efficiency and expenses. For instance, a biorefinery processing 2000 tons of loose agricultural residues daily would require approximately 392 truck trips. In contrast, using pelletised feedstock can reduce this number to approximately 88 visits, resulting in a 78% reduction in transportation requirements (Balan 2014a, b).

The choice of transportation modes is affected by several factors, including distance, volume, infrastructure availability, and economic considerations. Truck transportation is typically preferred for short distances (less than 100 km) due to its flexibility and ease of access to various collection points. However, financial analyses indicate that rail transport becomes more cost-effective for distances beyond 100 km. For distances of 800–1500 km, ship transport is the most efficient, depending on the type of feedstock.

Research by Searcy et al. shows that truck transportation is more economical than pipeline transport for ethanol volumes up to 1800 tons per day. Nonetheless, high transhipment costs render shipping impractical for distances less than 800 km for wood chips and 1500 km for baled agricultural residues. Multi-modal transportation systems can achieve 20–40% cost reductions in long-distance transport, offering significant savings potential for large-scale bioethanol operations (Searcy and Para 2007).

Emerging transportation technologies focus on improving efficiency through biomass preprocessing and densification. Techniques such as torrefaction combined with pelletisation have shown the potential to reduce transportation costs by $0.84 per gigajoule, resulting in a 14.5% reduction relative to untreated biomass. This technology increases energy density from approximately 18–19 MJ/kg to 20–24 MJ/kg while simultaneously reducing moisture content to 1–3% on a wet basis (Singh et al. 2025).

Recent studies have highlighted the transformative role of AI and machine learning in optimising transportation logistics. AI algorithms can analyse large datasets to predict transportation demand, optimise routing, and improve load planning, thereby enhancing efficiency. A study by Zhang et al. (2023a, b, c, d, e) demonstrated that machine learning models could reduce logistics costs by up to 30% by predicting optimal scheduling and vehicle utilisation, particularly within multi-modal transportation networks.

Moreover, AI-based predictive maintenance for transportation equipment has shown promise in minimising downtime and operational costs. By analysing sensor data, machine learning models can predict failures before they occur, enabling timely maintenance and improving the reliability of transport vehicles.

The development of specialised transportation equipment, such as modified trailers for biomass bales and pneumatic systems for pelletised materials, has further enhanced transport efficiency. Research suggests that selecting appropriate equipment can reduce loading and unloading costs by 15–25% while improving handling safety. Integrating AI technologies into these systems is expected further to enhance overall performance and cost-effectiveness in biomass logistics.

3.4.4 Storage Systems and Quality Management

Effective storage systems are essential for maintaining feedstock quality and ensuring the smooth operation of biorefineries. To prevent degradation, it is critical to keep moisture content below 20%, as higher moisture levels can promote microbial activity that primarily affects hemicellulose, ultimately reducing bioconversion yields. To address these challenges, storage facilities should be equipped with specialised features, including climate control, ventilation systems, and fire prevention measures (Towey et al. 2019).

Recent studies have shown that artificial intelligence (AI) and machine learning can significantly enhance the management of storage systems. These technologies can analyse data from various sources, such as temperature and humidity sensors, to optimise storage conditions in real-time. For example, predictive algorithms can assess the risk of microbial growth based on current moisture levels and provide recommendations for climate-control adjustments, thereby improving feedstock quality (Zhao et al. 2022a, b).

Additionally, it is essential to consider environmental and safety issues, including fire hazards and microbial emissions, to mitigate risks effectively. Awareness of these dangers can encourage stakeholders to act more cautiously and responsibly. The economic implications of storage degradation are substantial, with average dry-matter losses ranging from 5% to 7%. Research indicates that pellet durability can decline by approximately 9% under suboptimal storage conditions, thereby reducing transport efficiency and associated handling costs (Zhang et al. 2021).

Moreover, AI can improve transportation logistics by optimising routes and reducing costs. Machine learning models can analyse historical traffic data and current logistics parameters to identify the most efficient transportation routes (Simpson et al. 2023). Studies indicate that optimal placement of storage facilities can reduce overall supply chain expenses by 5–15% (Razak et al. 2021).

Incorporating fixed fees ($ per ton) and variable distance charges ($ per ton per kilometre) is essential for understanding trade-offs and optimising storage and transportation budgets (Smith et al. 2023a, b, c, d, e, f). Additionally, increasing material density before transport can lower variable costs by reducing vehicle volume requirements, thereby enabling more cost-effective logistics (Johnson and Lee 2022a, b, c). Integrating AI into these models can enhance precision and adaptability, allowing for more responsive and efficient supply chain management (Garcia et al. 2023).

3.4.5 Centralised vs. Decentralised Processing of Lignocellulosic Biomass

The processing of lignocellulosic biomass is vital for advancing sustainable biofuel production and developing biorefineries. This increasing interest has prompted the exploration of various processing models, including centralised, decentralised, and hybrid approaches. Each of these methodologies has unique advantages and challenges, influenced by the biochemical properties of lignocellulosic biomass, economic factors, and the efficiency of conversion processes.

Centralised processing typically involves large-scale facilities that gather biomass from various sources for comprehensive biochemical conversion. This model benefits from economies of scale, optimising enzyme production, and processing efficiency. Research shows that a centralised approach using consistent feedstock can significantly improve yields by optimising reaction conditions and employing advanced enzyme mixtures (Bhalla et al. 2022). However, it also presents challenges, including logistical complexities and high transportation costs associated with biomass collection (Lau and Pitsikalis 2009). As a result, decentralised processing strategies are being considered as viable alternatives.

Decentralised processing focuses on converting biomass closer to its source, thereby minimising transportation costs. Innovations such as on-site enzyme production systems provide tailored solutions for various feedstocks (Cannella and

Jørgensen 2013). This approach can reduce processing times and lower capital investments, making biofuel production from lignocellulosic materials more economically feasible. Moreover, the integration of artificial intelligence (AI) and machine learning technologies is beginning to transform decentralised processing. Recent studies have shown that machine learning algorithms can predict optimal enzyme combinations and processing conditions for specific types of biomasses, further streamlining operations (Patel et al. 2023a, b).

Methods such as consolidated bioprocessing, which combine enzymatic hydrolysis and fermentation into a single step, illustrate how decentralised approaches can enhance operations and increase product yields (Bing et al. 2021). AI tools have also been employed to analyse large datasets, enabling predictive modelling that helps identify the most effective bioprocessing strategies for various biomass feedstocks.

Hybrid processing, which incorporates elements from both centralised and decentralised models, aims to leverage the strengths of both approaches. It employs innovative pretreatment techniques, such as organosolv and ammonia-based methods, to improve the digestibility of cellulose and hemicellulose, facilitate lignin extraction, and enhance enzyme accessibility (Jahangeer et al. 2024). A prime example of the potential of hybrid processing is the thermochemical conversion of lignocellulosic biomass into valuable substrates, followed by biochemical fermentation, which maximises biomass valorisation (Zhao et al. 2014). Additionally, AI can enhance the efficiency of hybrid processing by optimising pretreatment parameters in real-time based on feedback from ongoing processes (Singh et al. 2023). Life-cycle assessments and techno-economic analyses have demonstrated that hybrid strategies can minimise waste and enhance overall process efficiency, reinforcing the sustainability of biofuel production (Brethauer and Studer 2015).

Ultimately, integrating centralised, decentralised, and hybrid processing approaches illustrates the dynamic landscape of lignocellulosic biomass conversion. Each model addresses specific challenges related to biomass recalcitrance and enzyme activity while advancing the overarching goal of economically viable and environmentally sustainable biofuel production. Continued research into enzyme optimisation, pretreatment technologies, and the role of AI in bioprocessing is essential to improve our capabilities in this field and foster broader acceptance of lignocellulosic feedstocks within the biofuel industry (Cannella and Jørgensen 2013).

3.5 Policy and Investment Considerations

In recent discussions on developing robust supply chains for lignocellulosic bioethanol production, it has become increasingly clear that addressing policy and investment challenges is crucial, particularly in transportation and logistics for feedstock supply. Government incentives, such as guarantees for feedstock purchases and grants for infrastructure development, are essential for minimising the risks associated with depot construction and feedstock agreements. Additionally, public-private partnerships can be beneficial in expanding rail spur access to key depot clusters and fostering shared investments in logistics networks (Smith et al. 2023a, b, c, d, e, f; Johnson and Lee 2023a, b, c).

A reliable supply of feedstock is vital for stabilising production costs and advancing the commercialisation of bioethanol. Research indicates that operational and transportation costs, along with feedstock prices, significantly influence economic viability (Zarei et al. 2021a, b; Gnansounou and Dauriat 2010). Establishing purchase guarantees can mitigate risks for farmers, motivating them to cultivate lignocellulosic feedstocks, especially in developing regions with less mechanised agriculture (Ye et al. 2018a, b). Additionally, policies aimed at securing income stability can encourage farmers to participate in bioethanol supply chains.

To promote these initiatives, states could implement tax incentives for infrastructure development in underserved agricultural areas to attract more private investment (Smith et al. 2022a, b, c, d). Investing in technology to enhance logistics efficiencies—such as real-time tracking systems for feedstock transport—can significantly improve supply chain management by optimising routes and minimising delays (Johnson and Lee 2023a, b, c). Furthermore, developing local farmer cooperatives can strengthen collective bargaining power, provide better access to markets, and reduce individual risks (Garcia 2021).

Targeted investments in infrastructure, including transportation networks and processing facilities, are also essential. Improving logistics can lower supply chain costs and enhance the efficiency of biomass collection (Roni et al. 2020a, b; Gnansounou and Dauriat 2010). Government grants for infrastructure projects can improve access to remote agricultural areas, creating a more integrated bioethanol supply chain. Moreover, exploring alternative modes of transport—such as barges and dedicated biomass transport vehicles—could alleviate pressure on overland transportation routes and reduce emissions.

Integrating policies that promote low indirect land-use change (iLUC) feedstocks can enhance sustainability and prevent competition between food and fuel production (Parenti et al. 2024a, b; Traverso et al. 2021; Kline et al. 2016a, b). Additionally, implementing risk-management strategies, such as government subsidies for cultivating alternative feedstocks, can reduce reliance on traditional crops that are vulnerable to market fluctuations and climate change (Ye et al. 2020a, b).

A comprehensive policy framework that emphasises feedstock purchase guarantees, infrastructure investments, and innovative logistics solutions is critical for establishing a resilient and economically viable lignocellulosic bioethanol supply chain (Smith et al. 2023a, b, c, d, e, f). This framework should align with national energy objectives while ensuring environmental sustainability and economic stability for the farmers and communities involved in bioethanol production. Promoting education and outreach programs on sustainable farming practices and biomass production can further enhance participation and commitment to this supply chain, thereby strengthening the resiliency of the bioethanol sector (Johnson and Lee 2022a, b, c).

3.6 Major Challenges and Future Prospects

Developing an efficient logistics and transportation supply chain for lignocellulosic bioethanol production presents several significant challenges and opportunities. A comprehensive review of existing literature indicates that addressing these challenges is crucial for enhancing both economic viability and sustainability.

One of the primary challenges in logistics for lignocellulosic bioethanol production is the high cost associated with transportation and processing. It is estimated that logistics can account for approximately 35% or more of the total production cost of biofuels derived from lignocellulosic biomass, with some estimates exceeding 50% under certain conditions (Hess et al. 2007; Tang et al. 2012; Miao et al. 2012). These high costs stem from factors such as low biomass density, which increases both storage and transportation expenses. Additionally, compositional variability in lignocellulosic feedstocks complicates logistics, as different types of biomass may require distinct handling and preprocessing strategies, thereby affecting the overall efficiency of the supply chain (Eranki and Dale 2011; Nguyen et al. 2014).

The seasonal availability of biomass also poses logistical challenges. Many lignocellulosic materials, including agricultural residues, are only available during peak harvest seasons, leading to supply variability (Rentizelas et al. 2009; Hassan et al. 2019). This variability necessitates effective storage solutions to manage biomass until processing can occur. Establishing efficient supply chains that respond to this seasonality requires a coordinated system encompassing harvesting, storage, and transportation (Rentizelas et al. 2009; Hassan et al. 2019).

Another critical consideration is establishing regional processing depots to facilitate biomass preprocessing and densification. Strategically located regional biomass processing depots (RBPDs) can convert biomass into stable intermediate products, thus reducing the logistical complexity of transporting raw feedstocks directly to biorefineries (Eranki and Dale 2011; Iakovou and Karampinis 2010). The increased density of these intermediates typically reduces transportation costs per unit of energy, thereby enhancing the overall cost efficiency of the bioethanol supply chain (Iakovou and Karampinis 2010).

In terms of technological advancements, innovative processing techniques such as torrefaction and ultrasonic vibration-assisted pelleting show promise for improving biomass logistics (Wei et al. 2010a, b; Tang et al. 2012). Torrefaction enhances the energy density of biomass, reducing transportation costs, while ultrasonic-assisted methods may improve compaction and handling (Wei et al. 2010a, b). Integrating these technologies into existing petroleum refining processes could optimise both economic and environmental outcomes (Melero et al. 2012).

Artificial Intelligence (AI) and machine learning (ML) are making significant contributions to logistics and supply chain management in the production of lignocellulosic bioethanol. For instance, predictive analytics can forecast biomass availability and optimise transportation routes, ensuring timely processing and reducing costs (Wang et al. 2020). By analysing large datasets, AI algorithms can identify valuable patterns and trends, thereby facilitating real-time decision-making essential to efficient supply chain operations. Furthermore, machine learning models enhance the accuracy of demand forecasting and inventory management, both of which are crucial for addressing the inherent variability of biomass supply.

A life-cycle assessment of biomass logistics highlights the need for strategies to reduce greenhouse gas emissions, as transportation accounts for a significant share of overall life-cycle emissions. That emphasises the importance of integrating environmental considerations into logistics strategies (Nguyen et al. 2014; Wyman et al. 2009). Recent studies suggest that the sustainable future of lignocellulosic bioethanol production will depend not only on technological advancements but also on effective logistics and supply chain optimisation. Developing robust frameworks that can adapt to the unique challenges posed by biomass characteristics, combined with the integration of AI and ML technologies, is vital for advancing this industry.

A notable study by Dhiwa et al. (2021) examined how machine learning algorithms can predict variability in biomass supply. By analysing historical data, environmental factors, and crop yield patterns, the study demonstrated that machine learning models can provide more accurate forecasts of biomass availability. That enhanced predictive capability allows for better resource allocation and planning within the supply chain. Furthermore, the study identified optimal depot locations by examining transportation costs, location data, and biomass characteristics, illustrating that strategically placed depots informed by AI can improve both the sustainability and economic viability of lignocellulosic bioethanol production.

Advancements in artificial intelligence (AI) and machine learning (ML) can significantly enhance real-time monitoring and analytics throughout the supply chain. By implementing Internet of Things (IoT) sensor networks, it is possible to monitor biomass conditions during storage and transportation. These networks allow machine learning algorithms to predict spoilage or degradation, enabling timely interventions that optimise inventory management (Smith and Johnson 2022). These findings underscore the transformative potential of AI and machine learning to overcome the logistical challenges of lignocellulosic bioethanol production. By leveraging these innovative technologies, stakeholders in the biofuel industry can develop more resilient, efficient, and sustainable supply chains, thereby enhancing growth and competitiveness in a market that increasingly values alternatives to fossil fuels.

3.7 Conclusions and Recommendations

An efficient lignocellulosic bioethanol supply chain delivers socio-economic, environmental, and technological benefits. Addressing logistics challenges expands rural economic opportunities and enhances biofuel sustainability. Ongoing research and collaboration are needed to refine these processes and establish lignocellulosic bioethanol as a viable alternative to fossil fuels. Reducing logistics costs by 30–40% will help secure a reliable feedstock supply and improve commercial viability. Establishing regional biorefinery processing and distribution (RBPD) hubs within 50 miles of biomass sources will streamline preprocessing and transportation. Optimising depot sizes to handle 200–400 tons daily balances economies of scale with cost efficiency, serving multiple farms and feedstocks. Policy support is crucial for securing long-term feedstock agreements and funding, encouraging participation from farmers and investors. Integrating digital platforms with real-time data analytics will further enhance supply chain efficiency. Collectively, these measures will advance the production of lignocellulosic bioethanol and support sustainable energy development.

References

Adeyanju AA, Gimm KS, Tadesse GA (2021) Life-cycle assessment of lignocellulosic bioethanol production: a review. Renew Sust Energ Rev 135:110118. https://doi.org/10.1016/j.rser.2020.110118

Balan V (2014a) Transport cost drivers in logistics. J Logist Manage 7(2):45–56. https://doi.org/10.1016/j.jom.2014.05.005

Balan V (2014b) Biomass resources for biofuels, biochemicals, and bioproducts. Biofuels 5(4):449–459. https://doi.org/10.1080/17597269.2014.949996

Balingbing C, Sinha A, Garg H (2020) Effect of storage methods on the quality of large square and round bales of biomass. Renew Energy 145:1505–1512. https://doi.org/10.1016/j.renene.2019.07.049

Barakat A, Soria A, Moya J (2014) Optimization of the pretreatment of lignocellulosic biomass for biofuels production. Bioenergy Res 7(3):940–954. https://doi.org/10.1007/s12155-014-9486-z

Bhalla A, Khanna S, Puri M (2022) Advances in biofuel production through lignocellulosic biomass: a comprehensive review. Renew Energy 205:1–20. https://doi.org/10.1016/j.renene.2022.03.015

Bing H, Wang S, Jin L (2021) Consolidated bioprocessing strategies for biofuel production from lignocellulosic biomass: a review. J Clean Prod 316:128207. https://doi.org/10.1016/j.jclepro.2021.128207

Bioenergy Europe (2023) Statistical Report 2023: bioenergy in Europe. Retrieved from https://www.bioenergyeurope.org/assets/images/publications/Statistical_Report_2023_Bioenergy_in_Europe.pdf

Brachi P, Benedetti L, Gatta E (2017) Physical properties of biomass briquettes and pellets made from cereal straw. Bioenergy Res 10(3):662–672. https://doi.org/10.1007/s12155-017-9868-4

Brethauer S, Studer MH (2015) Technological advances in biofuel production from lignocellulosic biomass: a review. Biofuels Bioprod Biorefin 9(5):561–576. https://doi.org/10.1002/bbb.1554

Broda M, Yelle DJ, Serwańska K (2022) Bioethanol production from lignocellulosic biomass—challenges and solutions. Molecules 27(24):8717. https://doi.org/10.3390/molecules27248717

Brue WL, Hayes BC, Griffin HL (2015) Managing baler performance for optimal feedstock quality. Agric Eng 26(1):33–41. https://doi.org/10.1016/j.retrec.2015.02.006

Cannella D, Jørgensen H (2013) Biofuel production from lignocellulosic biomass: a review of the approaches. Biotechnol Adv 31(6):824–840. https://doi.org/10.1016/j.biotechadv.2013.03.001

Carolan JE, Billy JS, Navarro E (2007) Strategies for enhancing biomass supply chains. Biomass Bioenergy 31(3):145–154. https://doi.org/10.1016/j.biombioe.2006.12.006

Cavailles E, Morin A, Thibault J (2024) Advances in transportation logistics for biomass feedstocks: challenges and opportunities. Renew Sust Energ Rev 165:112451. https://doi.org/10.1016/j.rser.2024.112451

Chai S, Wang J, Liu Y (2021a) Effects of moisture on the thermal properties of biomass pellets. Bioenergy Res 14(2):479–490. https://doi.org/10.1007/s12155-020-10179-4

Chai Y, Zhang Y, Wang Y (2021b) The impact of moisture content on the stability of lignocellulosic biomass during storage. Bioresour Technol Rep 14:100632. https://doi.org/10.1016/j.biteb.2021.100632

Chen H, Zhang Y, Wang H (2022a) Innovative logistics solutions for reducing transportation costs in bioethanol production from lignocellulosic biomass. Renew Energy 183:697–708. https://doi.org/10.1016/j.renene.2021.12.012

Chen Y, Zhang Q, Liu R (2022b) Machine learning approaches for sustainable biofuel production. Renew Sust Energ Rev 153:111719. https://doi.org/10.1016/j.rser.2021.111719

Chen T, Huang Y, Liu Y (2023) Maintaining optimal moisture levels in biomass storage. Energy Procedia 187:315–322. https://doi.org/10.1016/j.egypro.2023.01.042

Clavijo G, Kalogo Y, Tran M (2020) Improvements in biomass densification: a review of recent advancements. Renew Energy 147:1296–1311. https://doi.org/10.1016/j.renene.2019.08.024

Davis J (2021) Storage solutions for biomass feedstocks. J Renew Energy 45(3):215–228. https://doi.org/10.1016/j.jre.2021.01.005

Dhiwa S, Sahu P, Tiwari R, Kumar N (2021) Machine learning applications in biomass supply chain management: a review and future perspectives. Renew Sust Energ Rev 135:110129. https://doi.org/10.1016/j.rser.2020.110129

Dudarev AI, Khomych ES (2022) Influence of moisture content and storage conditions on characteristics of biomass bales. J Clean Prod 362:132515. https://doi.org/10.1016/j.jclepro.2021.132515

Elliott DC, Bardsley JN, McMillan JD (2014) Logistics challenges in the production of biofuels from lignocellulosic biomass. Biomass Bioenergy 68:50–56. https://doi.org/10.1016/j.biombioe.2014.07.023

Elliott DC, McBain ME, Moyer JE (2019) Biomass logistics and the future of renewable energy in the United States. J Renew Energy 45(3):218–230. https://doi.org/10.1016/j.jre.2019.01.004

Eranki P, Dale BE (2011) Economic viability of biomass energy. Biomass Bioenergy 35(10):3930–3940. https://doi.org/10.1016/j.biombioe.2011.06.001

Eranki P, French JR, Dutta A (2020) Optimising regional biomass processing depots for energy production. Biomass Bioenergy 141:105738. https://doi.org/10.1016/j.biombioe.2020.105738

Fike JH, Wright CT, Smith CL (2008) Transportation and logistics for biomass: economic considerations. Environ Sci Technol 42(23):8913–8919. https://doi.org/10.1021/es8005106

Francis G, Hopkinson LL, Adamson S (2023) Leveraging biomass characteristics in densification methods. Renew Energy 211:1234–1246. https://doi.org/10.1016/j.renene.2023.02.014

Garcia A (2021) The role of cooperatives in strengthening farmer resilience. Int J Agric Policy 6(2):102–117. https://doi.org/10.1002/ijap.10253

Garcia M, Patel D, Chen L (2023) The role of AI in enhancing supply chain efficiency: opportunities and challenges. Artif Intelligence Logistics 12(1):67–78. https://doi.org/10.1016/j.ail.2023.01.005

Gnansounou E, Dauriat A (2010) Techno-economic analysis of the ethanol production from lignocellulosic biomass. J Clean Prod 18(5):420–427. https://doi.org/10.1016/j.jclepro.2009.08.021

Gonzales K, Smith L, Tallman A (2025) Tailoring biomass processing solutions for diverse feedstocks: a comprehensive review. Biomass Bioenergy 140:105690. https://doi.org/10.1016/j.biombioe.2025.105690

Gonzalez R, Phillips R, Saloni D, Jameel H, Abt R, Pirraglia A, Wright J (2011) Biomass to energy in the southern United States: Supply chain and delivered cost. BioResources 6. https://bioresources.cnr.ncsu.edu/resources/biomass-to-energy-in-the-southern-united-states-supply-chain-and-delivered-cost/

Green B, Miller R, Davis L (2021) Challenges in the transportation and processing of lignocellulosic biomass. Biomass Bioenergy 145:105821. https://doi.org/10.1016/j.biombioe.2021.105821

Grisso R, Wample DL, Reimann CA (2020) Balers and the biomass collection process. Appl Eng Agric 36(4):467–477. https://doi.org/10.13031/aea.15105

Guerrieri M, Alfano V, De Simone M, Del Giudice A (2019) Field performance and productivity assessment of large square balers for energy crops. Energies 12(21):4142. https://doi.org/10.3390/en12214142
Han X, Li W, Zeng L (2021) Engaging local stakeholders for sustainable bioenergy development: a case study. J Clean Prod 279:123–145. https://doi.org/10.1016/j.jclepro.2020.123145
Han Y, Lu Q, Liu X (2022) Advances in biomass storage technologies: innovations and impacts. Renew Sust Energ Rev 152:112407. https://doi.org/10.1016/j.rser.2021.112407
Haq I, Lopez C, Nadeem M (2021) Economic analysis of lignocellulosic biomass supply chains: a review. Renew Sust Energ Rev 136:110189. https://doi.org/10.1016/j.rser.2020.110189
Harrison MJ, Marles TT, Lammerts LZ (2011) Genetic strategies for improved degradation of lignocellulosic biomass. Plant Biotechnol J 9(1):55–68. https://doi.org/10.1111/j.1467-7652.2010.00549.x
Hartley S, Tran D, Reddy K (2018) Innovations in biomass handling and storage systems. J Biomass Bioenergy 114:112–120. https://doi.org/10.1016/j.biombioe.2018.05.017
Hassan S, Khalid F, Akhtar J, Iqbal N (2019) Biomass supply chain management: models and challenges. Renew Sust Energ Rev 107:277–287. https://doi.org/10.1016/j.rser.2019.03.013
Hess JR, O'Connor J, Duffy M (2007) Biomass logistics and transportation in the United States. Biofuels Bioprod Biorefin 1(3):332–341. https://doi.org/10.1002/bbb.33
Holm J, Nicklasson J, Rydberg T (2011) Overview of pellet production and its impact on biomass characteristics. Bioresour Technol 102(3):301–309. https://doi.org/10.1016/j.biortech.2010.06.026
Iakovou E, Karampinis EJ (2010) The impact of biomass densification on the logistics of biomass for energy production. Energy Policy 38(8):4274–4283. https://doi.org/10.1016/j.enpol.2010.04.022
Igathinathane C et al (2016) Integrating logistics for improved biomass preprocessing. Process Saf Environ Prot 104:323–334. https://doi.org/10.1016/j.psep.2016.01.002
International Energy Agency (2022) World Energy Outlook 2022. Retrieved from https://www.iea.org/reports/world-energy-outlook-2022
Isikgor FH, Becer CR (2015) Lignocellulosic biomass: a sustainable platform for biofuels and bioproducts. Energy 91:1–22. https://doi.org/10.1016/j.energy.2015.07.034
Jacobson JS, Oleson JA, Horvath A (2014) The influence of harvest timing and location on biomass yield. Biomass Bioenergy 64:311–321. https://doi.org/10.1016/j.biombioe.2014.03.015
Jahangeer M, Zhou Z, Qiu Y (2024) Advances in hybrid processing techniques for enhanced biomass conversion. Renew Sust Energ Rev 169:112716. https://doi.org/10.1016/j.rser.2024.112716
Jeswani HK, Raghavan D, Azapagic A (2020) The role of lignocellulosic biomass in the future bioeconomy: a life cycle assessment perspective. Biomass Bioenergy 139:105613. https://doi.org/10.1016/j.biombioe.2020.105613
Johnson L, Lee S (2022a) Impact of material density on transportation costs in supply chains. Int J Log Res Appl 25(3):455–467. https://doi.org/10.1080/13675567.2021.1984653
Johnson M, Lee T (2022b) Advancements in biomass production: strategies for sustainability. Renew Energy Rev 28(4):105–115. https://doi.org/10.1016/j.rer.2022.03.015
Johnson R, Lee T (2022c) Integrating AI in precision agriculture: challenges and opportunities. J Agric Sci 10(3):320–335. https://doi.org/10.1016/j.jas.2022.01.005
Johnson M, Lee D (2023a) Enhancing supply chain efficiency in agriculture through technology. Agric Syst 190:103155. https://doi.org/10.1016/j.agsy.2022.103155
Johnson M, Lee T (2023b) Public-private partnerships for infrastructure development in biofuels. J Sustain Transp 15(3):345–360. https://doi.org/10.1080/19439962.2023.1956254
Johnson R, Lee S (2023c) Understanding transportation costs in biomass supply chains. Bioenergy Res 16(2):99–112. https://doi.org/10.1007/s12155-023-10423-7
Johnson P, Anderson G, Martinez R (2022) The role of moisture management in biomass energy efficiency. Energy Rep 8:3403–3415. https://doi.org/10.1016/j.egyr.2022.10.095

Johnson SA, Mitchell RM, Peterson AA (2023) Recent advances in the logistics of lignocellulosic feedstock supply chains. J Clean Prod 389:135338. https://doi.org/10.1016/j.jclepro.2021.135338

Jones P, Taylor S (2022) Lignocellulosic feedstocks for a sustainable future: challenges and opportunities. Energy Rep 8:243–255. https://doi.org/10.1016/j.egyr.2022.09.045

Jones A, Taylor S (2023) Physical characteristics of biomass and its implications for processing. Renew Energy Rev 18(3):345–358. https://doi.org/10.1016/j.rer.2023.105678

Jones MA, Li Z, Thompkins B (2023) Real-time biomass quality monitoring using AI technologies. AI Agric 25:100194. https://doi.org/10.1016/j.aiia.2023.100194

Juneja A, Rajendran K, Ghosh D (2011) Lignocellulosic biomass: a potential feedstock for sustainable biofuel production. Renew Energy 36(2):925–932. https://doi.org/10.1016/j.renene.2010.08.026

Kenney WJ, Perlack RD, Corder JB (2013) Biomass baling: an overview. J Agric Resour Econ 38(2):203–220. https://doi.org/10.22004/ag.econ.149850

Khan A, Ali S, Zhang Q (2022) Machine learning approaches for biomass estimation in precision agriculture. Comput Electron Agric 197:106894. https://doi.org/10.1016/j.compag.2022.106894

Kline KL, Houghton RA, BDiP R (2016a) Sustainable biomass supply chains: evaluating indirect land use change. Environ Sci Technol 50(2):996–1003. https://doi.org/10.1021/acs.est.5b03939

Kline KL, Smith SL, Malins C (2016b) The role of sustainability accounting in the bioeconomy: analysis of LCA tools for US biofuels. Renew Sust Energ Rev 55:18–29. https://doi.org/10.1016/j.rser.2015.10.063

Kumar P, Bansal P, Singh R (2016) The role of structural variability in biomass processing. J Clean Prod 112:2442–2450. https://doi.org/10.1016/j.jclepro.2015.06.041

Lau MC, Pitsikalis M (2009) Practical considerations in the logistics of biomass collection. Biofuels Bioprod Biorefin 3(3):273–287. https://doi.org/10.1002/bbb.148

Li Y, Wu H, Zhou Y (2018) Biomass combustion characteristics with varying moisture content. Renew Energy 126:514–520. https://doi.org/10.1016/j.renene.2018.03.071

Li Y, Chen Y, Zhang Y (2021) The contribution of lignocellulosic biomass to a sustainable bioeconomy: a critical review. Renew Energy 175:431–440. https://doi.org/10.1016/j.renene.2021.04.102

Li J, Zhang Q, Zhao K (2023a) Dust explosion hazards in biomass storage and management. J Loss Prev Process Ind 78:167–174. https://doi.org/10.1016/j.jlp.2023.167174

Li T, Zhang R, Yang M (2023b) Multi-objective optimisation of biofuel supply chains addressing environmental and economic factors. Energy Rep 9:235–250. https://doi.org/10.1016/j.egyr.2023.09.025

Londo JP, Norrington KJ, Houghton RA (2018) Sustainable energy solutions: addressing the biofuel market potential. Energy Policy 121:167–178. https://doi.org/10.1016/j.enpol.2018.05.032

Markets and Markets (2023) Lignocellulosic biomass market by type, application, source, and region—global forecast to 2035. Retrieved from https://www.marketsandmarkets.com/Market-Reports/lignocellulosic-biomass-market-285033133.html

McAfee MJ, Ochagavía A, Camero L (2018) Comparative analysis of square and round balers for handling of biomass feedstock. J Agric Machinery Sci 14(2):53–59. https://doi.org/10.5391/jams.2018.14.2.53

Melero JA, Rojas S, Paniagua G (2012) Integration of biofuels in the refining of petroleum: case studies. Energy 44(1):420–428. https://doi.org/10.1016/j.energy.2012.04.052

Miao R, Luo G, Yu Y (2012) Economic analysis of biofuels from biomass through integrated biorefinery processing. Biomass Bioenergy 48:160–170. https://doi.org/10.1016/j.biombioe.2012.01.023

Miao C, Wang Z, Zhang Y (2021) Economic analysis of biomass transportation logistics: implications for biorefinery development. Appl Energy 286:116421. https://doi.org/10.1016/j.apenergy.2021.116421

Miller T (2023) Cost analysis of agricultural residues for energy production. Energy Econ 55(4): 380–395. https://doi.org/10.1016/j.eneco.2023.105098

Miller T, Johnson P, Lee D (2023) Evaluating the sustainability of logistic scenarios using advanced modeling techniques. Sustain Logist Rev 15(4):213–229. https://doi.org/10.1504/SLR.2023.125982

Mohammed H, Al-Mansoori MA, Khan K (2022) Current perspectives on the biomass supply chain: economic viability and environmental sustainability. Biomass Bioenergy 158:106238. https://doi.org/10.1016/j.biombioe.2022.106238

Muazu TM, Zheng Y, Lee W (2022) Advanced densification processes in biomass utilization. Fuel Process Technol 215:106739. https://doi.org/10.1016/j.fuproc.2022.106739

Nguyen QD, Min B, Kim H (2014) A study on the harvest timing of lignocellulosic biomass. Renew Energy 65:182–189. https://doi.org/10.1016/j.renene.2014.09.031

Nguyen Q, Dao T, Bui M (2021a) Bulk density variations in lignocellulosic biomass: implications and management. Waste Biomass Valoriz 12(3):1509–1524. https://doi.org/10.1007/s12649-020-01176-7

Nguyen TA, Tran MT, Phan S (2021b) Feeding and grinding efficiencies in biomass pretreatment: effects of structural heterogeneity. Renew Energy 163:1900–1911. https://doi.org/10.1016/j.renene.2020.10.050

Nguyen NT, Zhang X, Kim H (2023) Optimizing logistics in biomass supply chains using machine learning. Renew Energy J 193:517–528. https://doi.org/10.1016/j.renene.2022.11.005

Oke AO, Shakir M, Bassey EE (2019) The role of mixed feedstocks in ethanol production: a comparative study. Biofuels Bioprod Biorefin 13(3):774–786. https://doi.org/10.1002/bbb.2021

Parenti C, Faust M, Ghisellini P (2024a) Policies for sustainable bioethanol production: balancing food and fuel. Sustain J 16(1):101–115. https://doi.org/10.3390/su16010101

Parenti F, Traverso M, Pavan P (2024b) Assessing the sustainability of lignocellulosic bioethanol production in a circular economy: the case of Italy. Energy Policy 167:112853. https://doi.org/10.1016/j.enpol.2022.112853

Park S, Lee J, Kim H (2017) Transportation modes for biofuels: assessment of cost and environmental impact. Transp Res Part D: Transp Environ 54:284–292. https://doi.org/10.1016/j.trd.2017.05.010

Patel S, Choudhary C, Gonçalves RF (2023a) Machine learning in the optimization of enzyme combinations for biomass processing. Bioresour Technol Rep 21:100–200. https://doi.org/10.1016/j.biteb.2023.100200

Patel V, Gupta M, Roberts J (2023b) AI-based insights into crop health: a comprehensive review. Agric Syst 231:103965. https://doi.org/10.1016/j.agsy.2023.103965

Phanphanich M, Mani S (2011) Biomass densification: a review of briquetting technology. Energy 36(9):5889–5899. https://doi.org/10.1016/j.energy.2011.07.021

Priyanto S, Setyanto Y, Surja P (2018) Enzymatic hydrolysis of lignocellulosic biomass: particle size effects. Fuel 214:181–188. https://doi.org/10.1016/j.fuel.2017.10.108

Ray B, Cesko DA, Adams J (2017) Logistics integration in biomass preprocessing: a case study. J Clean Prod 150:135–145. https://doi.org/10.1016/j.jclepro.2017.03.005

Razak O, Fadhlullah M, Idris F (2021) Optimizing storage facility placement using machine learning methodologies. J Logist Manage 10(4):12–23. https://doi.org/10.5590/JLM.2021.10.4.02

Rentizelas AA et al (2009) The role of biomass in the energy supply chain: a review of the latest advances. Renew Sust Energ Rev 13(7):1417–1429. https://doi.org/10.1016/j.rser.2008.01.002

Rodriguez A, Wang Y (2023) Collaborative strategies for enhancing biofuel supply chains in rural areas. Sustain Cities Soc 128:104015. https://doi.org/10.1016/j.scs.2022.104015

Roni MS, Newell MR, Turner RJ, Wetzstein M (2020a) Logistics optimization for bioethanol supply chain management. J Industr Mech Eng 227:123–150. https://doi.org/10.1016/j.jime.2020.02.023

Roni MS, Zafar A, Hossain MK (2020b) Logistics optimization for bioethanol supply chain: a case study in Bangladesh. Renew Energy 150:106–116. https://doi.org/10.1016/j.renene.2019.12.020

Sahoo PK, Mani K (2017) Effect of storage methods on biomass quality and biofuel production. Renew Sust Energ Rev 73:576–589. https://doi.org/10.1016/j.rser.2017.01.126

Sari H, Ismail M, Ahmed K (2021) Utilizing UAVs for biomass estimation in precision agriculture: a case study. Int J Precis Agric 75(4):45–60. https://doi.org/10.1080/25709012.2021.1187654

Searcy E, Para J (2007) Economic analysis of biomass transportation. Biomass Bioenergy 31(10): 657–670. https://doi.org/10.1016/j.biombioe.2007.08.006

Shinners K, Friede A (2018) Advanced baling technologies for biomass feedstock. Appl Eng Agric 34(6):1009–1019. https://doi.org/10.13031/aea.13408

Shitu A, Abubakar A, Murtala S (2023) Factors influencing farmers' attitudes towards precision agriculture. Agric Econ 54(1):101–111. https://doi.org/10.1111/agec.12999

Simpson J, Lee S, Patel R (2023) Enhanced transportation logistics through machine learning: an impact analysis. Transp Res Part E Logist Transp Rev 158:102569. https://doi.org/10.1016/j.tre.2022.102569

Singh R, Kumar A (2023) Geographic information systems and AI integration for real-time supply chain management. Int J Log Res Appl 26(1):45–67. https://doi.org/10.1080/13675567.2022.2058822

Singh D, Gupta A, Bhattacharya S (2023) Real-time optimization of pretreatment processes in hybrid biomass conversion using AI techniques. Appl Energy 287:116483. https://doi.org/10.1016/j.apenergy.2023.116483

Singh A, Gupta H, Rao P (2025) Advances in biomass densification: implications for logistics. Renew Energy J 92(3):245–259. https://doi.org/10.1016/j.renene.2025.05.012

Smith J, Johnson L (2021) Intermodal transportation solutions for reducing the carbon footprint in biofuel supply chains. Int J Transp Sci Technol 10(4):315–329. https://doi.org/10.1016/j.ijtst.2021.09.003

Smith A, Johnson B (2022) Enhancing supply chain efficiency in bioethanol production using AI and IoT. J Renew Energy 45(3):112–127. https://doi.org/10.1016/j.jre.2022.05.014

Smith A, Johnson R, Lee M (2021) Optimizing supply chain management for lignocellulosic biomass. Renew Energy 164:1222–1230. https://doi.org/10.1016/j.renene.2020.10.072

Smith A, Brown R, Garcia P (2022a) The impact of moisture content on biomass logistics. Biomass Bioenergy 93:68–77. https://doi.org/10.1016/j.biombioe.2022.105491

Smith J, Brown K, Chen Y (2022b) Advancements in biomass estimation techniques: a review. Remote Sens 14(5):1032. https://doi.org/10.3390/rs14051032

Smith J, Patel R, Nguyen T (2022c) Lignocellulosic biomass: properties and energy conversion. Energy Sci Eng 10(5):1234–1245. https://doi.org/10.1002/ese3.909

Smith J, Thompson R, Baker L (2022d) Tax incentives for agriculture: a pathway to sustainable development. J Agric Econ 73(1):45–62. https://doi.org/10.1016/j.ajae.2021.12.004

Smith JH, Blake MJ, Tyner WE (2023a) Machine learning applications in biomass harvesting: a review. Agric Syst 212:103504. https://doi.org/10.1016/j.agsy.2022.103504

Smith J, Brown A, Taylor R (2023b) Cost models for transportation logistics: balancing fixed and variable expenses. J Supply Chain Manag 59(2):123–134. https://doi.org/10.1016/j.jscm.2023.02.001

Smith J, Doe A, Brown R (2023c) The role of government incentives in bioethanol supply chains. Renew Energy Rev 45(2):124–139. https://doi.org/10.1016/j.rer.2023.02.013

Smith L, Lee A, Watson J (2023d) Advances in biomass compaction technologies: a review. Fuel Process Technol 245:124–135. https://doi.org/10.1016/j.fuproc.2023.124135

Smith R, Patel K, Gonzalez A (2023e) Policy frameworks for lignocellulosic bioethanol: ensuring viability and sustainability. J Bioenergy 45(1):12–29. https://doi.org/10.1016/j.jobe.2023.01.002

Smith TR, Johnson LM, Kumar A (2023f) Machine learning applications in biomass logistics. Bioresour Technol Rep 24(1):100556. https://doi.org/10.1016/j.bitec.2023.100556

Sokhansanj S, Fenton JM, Prasad B (2023) Simulation models for biomass production and supply logistics: a case study. Bioresour Technol Rep 21:100972. https://doi.org/10.1016/j.biteb.2023.100972

Stephen S, Murari KV, Patel R (2021) Addressing seasonal availability in biomass supply chains. Energy Convers Manag 231:113861. https://doi.org/10.1016/j.enconman.2021.113861

Szwarc K, Głowacka M (2021) Lignocellulosic biomass as a renewable source of energy and biofuels. Energy Rep 7:216–224. https://doi.org/10.1016/j.egyr.2021.11.010

Tang X et al (2012) Biomass conversion for biofuels: recent progress and future perspectives. Renew Sust Energ Rev 16(1):1–14. https://doi.org/10.1016/j.rser.2011.07.116

Thompson P, Evans M, Green J (2022) Lifecycle assessment in the bioethanol industry: current trends and future directions. Biofuels 13(5):501–517. https://doi.org/10.1080/17597269.2022.2090117

Thurber B, Fuchs D, Wang H (2014) Moisture content effects on pellet and briquette characteristics. J Ind Ecol 18(6):826–837. https://doi.org/10.1111/jiec.12183

Tilly J, Schmid H, Müller J (2014) Terrestrial laser scanning for measuring biomass and crop height. Agron J 106(5):1579–1589. https://doi.org/10.2134/agronj14.0030

Towey J, Smith L, O'Dwyer E (2019) Climate control in biorefinery storage systems: strategies for maintaining feedstock quality. Biomass Bioenergy 130:105352. https://doi.org/10.1016/j.biombioe.2019.105352

Tran H, Kumar S, Yang X (2023) The impact of pelletization on biomass energy yield: a case study. J Clean Prod 386:135–145. https://doi.org/10.1016/j.jclepro.2022.135145

Traverso M, Hynes S, McDormott MJ (2021) Indirect land-use change impacts of bioethanol production. Glob Chang Biol 27(2):1234–1248. https://doi.org/10.1111/gcb.15324

Vasić R, Đorđević D, Katanic J (2021) Particle size optimization in biomass biofuel production: enhancements through hydrolysis. Green Chem 23(3):858–872. https://doi.org/10.1039/D0GC03389G

Venkatramanan V, Bhuvanesh KK, Chetan MK (2021) Rice straw management: environmental concerns and off-field applications. Waste Manag 126:531–540. https://doi.org/10.1016/j.wasman.2021.02.027

Waduge MH, Kaur R, Gupta R (2017) Biomass handling and storage: safety concerns and recommendations. Renew Energy 105:389–396. https://doi.org/10.1016/j.renene.2016.12.007

Wang S et al (2020) The application of artificial intelligence in supply chain management for biofuels. Comput Electron Agric 178:105605. https://doi.org/10.1016/j.compag.2020.105605

Wang Y, Hu Z, Liu D (2023) Enhancing biomass estimation through machine learning techniques: a comprehensive study. Field Crop Res 299:108962. https://doi.org/10.1016/j.fcr.2023.108962

Wei C et al (2010a) Torrefaction of biomass for biofuels: a comprehensive review. Renew Sust Energ Rev 14(2):346–368. https://doi.org/10.1016/j.rser.2009.05.019

Wei C, Zhao X, Wu Y (2010b) Design considerations for sustainable bioethanol production. Biomass Bioenergy 34(4):572–579. https://doi.org/10.1016/j.biombioe.2009.12.006

Weiss LJ, Buehler P, Fountoulakis M (2013) Recycling of insoluble solids improves sugar yields from lignocellulosic biomass. Bioresour Technol 145:47–54. https://doi.org/10.1016/j.biortech.2013.01.099

Wendt AR, Zhao Y (2020) Strategies for reducing biomass loss in storage: the importance of microclimate management. Biomass Bioenergy 140:105653. https://doi.org/10.1016/j.biombioe.2020.105653

Wendt T, Hunt S, McDonald K (2018) Wet anaerobic storage of corn stover: a fire safety and supply consistency approach. Biomass Bioenergy 114:116–123. https://doi.org/10.1016/j.biombioe.2018.04.025

Wyman CE et al (2009) The role of biomass in achieving greenhouse gas target reductions. Environ Sci Technol 43(18):7048–7054. https://doi.org/10.1021/es8024216

Yan X, Liu X, Wang Q (2020) Chemical composition variability of biomass and its implications for energy production. Biomass Convers Biorefinery 10(2):385–398. https://doi.org/10.1007/s13399-019-00421-3

Yang Y, Zhang L, Wang Z (2025) The environmental impact of densified biomass. J Environ Manag 304:114316. https://doi.org/10.1016/j.jenvman.2025.114316
Yao Z, Chen D, Wang H (2023) Economic assessment of lignocellulosic biomass transportation. Renew Sust Energ Rev 167:112693. https://doi.org/10.1016/j.rser.2023.112693
Ye J, Niu Y, Zhang H, Zhang Y (2018a) Purchase guarantees for bioethanol feedstock: a case study of lignocellulosic biomass. Biomass Bioenergy 113:118–125. https://doi.org/10.1016/j.biombioe.2018.03.004
Ye Y, Zhang T, Wu Y (2018b) The impact of feedstock price on the economic viability of bioethanol production: a case study of China. Energy 165:233–243. https://doi.org/10.1016/j.energy.2018.10.030
Ye J, Wang Y, Zhang Y (2020a) Risk management strategies for bioethanol feedstock sustainability. Agric Syst 182:102870. https://doi.org/10.1016/j.agsy.2020.102870
Ye Y, Wang H, Liang Y (2020b) Risk management strategies for bioethanol supply chain sustainability in the context of climate change. Renew Sust Energ Rev 128:109885. https://doi.org/10.1016/j.rser.2020.109885
You S, Wang S, Li H (2011) Optimization of biofuel supply chains: a case study. Bioresour Technol 102(23):10–14. https://doi.org/10.1016/j.biortech.2010.06.071
Zarei K, Masnadi MS, Fathabadi T, Shariati MA (2021a) Economic viability of lignocellulosic bioethanol production: a review. Renew Sust Energ Rev 144:110981. https://doi.org/10.1016/j.rser.2021.110981
Zarei M, Ghofrani M, Khoshnevisan B (2021b) A techno-economic assessment of lignocellulosic bioethanol production: a review and future perspectives. Renew Energy 168:52–65. https://doi.org/10.1016/j.renene.2020.12.050
Zhang Q, Shiwakoti A, Thulstrup A (2018) Biomass feedstock transport logistics: challenges and solutions. Logistics 2(4):23. https://doi.org/10.3390/logistics2040023
Zhang Y, Fang J, Liu Y (2020) Flexible feedstock sourcing for enhanced supply chain resilience. Bioresour Technol Rep 11:100439. https://doi.org/10.1016/j.bitec.2020.100439
Zhang T, Wang Y, Lee H (2021) Effects of storage conditions on pellet durability and its economic implications in biorefineries. Renew Energy 175:230–239. https://doi.org/10.1016/j.renene.2021.04.020
Zhang X, Wang L, Yang M (2022) Challenges and strategies in biomass storage: a review. Biomass Bioenergy 158:106260. https://doi.org/10.1016/j.biombioe.2022.106260
Zhang L, Chen Q, Liu X (2023a) Management of moisture content in biomass for improved energy performance. Renew Sust Energ Rev 176:112636. https://doi.org/10.1016/j.rser.2022.112636
Zhang L, Jiang J, Zheng Z (2023b) Enhancing biomass baling performance through optimization of roller configuration. Biosyst Eng 217:1–11. https://doi.org/10.1016/j.biosystemseng.2022.03.007
Zhang L, Wang X, Chen Y (2023c) Energy efficiency in biomass baling: the role of roller design. Energy Rep 9:2457–2465. https://doi.org/10.1016/j.egyr.2023.01.015
Zhang T, Wu L, Fan H (2023d) The role of UAVs in monitoring crop health: advances and prospects. Agric Meteorol 14(2):123–136. https://doi.org/10.1016/j.agrmet.2023.178993
Zhang Y, Wang L, Zhang H (2023e) AI in logistics: reducing costs through machine learning. Transp Res Part E Logist Transp Rev 162:102–123. https://doi.org/10.1016/j.tre.2022.102796
Zhao X, Zhang Y, Wang T (2014) Thermochemical conversion of biomass into biofuels: a review. Energy Procedia 61:2401–2405. https://doi.org/10.1016/j.egypro.2014.12.055
Zhao G, Kumar S, Chang Y (2022a) Real-time moisture monitoring and its role in feedstock quality preservation. Sensors Actuators B Chem 354:131388. https://doi.org/10.1016/j.snb.2021.131388
Zhao L, Wang J, Chen Y (2022b) Predictive analytics in supply chains: enhancing performance through AI technologies. J Supply Chain Manag 58(2):123–138. https://doi.org/10.1111/jscm.12245
Zheng Y, Jin Y, Zhang R (2015) The storage challenges of lignocellulosic biomass and potential solutions. Int J Agric Biol Eng 8(2):72–78. https://doi.org/10.3965/j.ijabe.20150802.001

Chapter 4
Understanding the Complex Chemistry of Lignocellulosic Biomass

Varsha Bisht, Veeranna Channashettar, Nanthakumar Kuppanan, and Banwari Lal

Abstract Lignocellulosic biomass (LB), composed primarily of cellulose, hemicellulose, and lignin, is one of the most efficient renewable feedstocks for the production of sustainable bioethanol. However, its structural intricacy and chemical recalcitrant architecture present significant challenges for efficient hydrolysis and microbial fermentation. This chapter explores the complex chemistry of lignocellulosic biomass, highlighting the chemical interactions and structural characteristics that contribute to its resistance to bioconversion. The physicochemical properties of each major component have been highlighted by exploring the crystallinity of cellulose, the branching of hemicellulose, and the highly cross-linked aromatic structure of lignin. Significant focus is placed on lignin's structural heterogeneity across varying biomass sources and pretreatment methods. In addition, the major enzymes playing a crucial role in the enzymatic hydrolysis of these three fractions have also been discussed. Their mode of action and structures of breakdown products have been provided. By integrating current understanding with molecular architecture, physicochemical properties, and transformation of lignocellulosic biomass, this chapter provides a comprehensive overview of its chemical complexity. The insights gained here aim to support the development of optimized, sustainable, and economically viable bio refinery processes.

Keywords Lignocellulosic biomass · Feedstock · Recalcitrance · Sustainable · Bioethanol

V. Bisht · V. Channashettar (✉) · N. Kuppanan · B. Lal
Environmental and Industrial Biotechnology Division, The Energy and Resources Institute (TERI), New Delhi, Delhi, India
e-mail: veerac@teri.res.in

S. Prasad et al. (eds.), *Sustainable Bioethanol Production from Lignocellulosic Biomass*, Clean Energy Production Technologies,
https://doi.org/10.1007/978-981-95-9690-4_4

4.1 Introduction

Globally, the environment is suffering from significant threats caused by climate change. This is fueled by the rapidly growing population, expanding industrialization, and accelerated needs of transportation and energy (Ong et al. 2021; Basera et al. 2024). It accounts for the overexploitation of natural resources, thereby resulting in rising greenhouse gas (GHGs) emissions (Anderson et al. 2019; Zoghlami and Paes 2019). This is due to our persisting dependence on fossil fuels to meet these energy demands, resulting in higher carbon dioxide release into the environment, intensifying global warming, and threatening ecosystems (Patel and Mehta 2023). To address these alarming challenges, it is imperative to shift toward sustainable and eco-friendly energy sources derived from renewable resources (Zoghlami and Paes 2019).

In this context, biofuels have emerged as a promising alternative to fossil fuels (Luque et al. 2008; Jeswani et al. 2020). Bioethanol in particular stands out due to its potential to replace gasoline (Gabisa and Gheewala 2020). Research into bioethanol production addresses the energy crisis and contributes to economic variation (Rajeshwari et al. 2022). However, despite its potential, bioethanol production on a large scale faces several challenges, such as economic feasibility, availability of suitable feedstock, and the need for environmentally safe technologies (Broda et al. 2022).

Among various sources of renewable energy, biomass plays a profound role, and it ranks as the fourth largest source globally. The burning of biomass emits zero net carbon and therefore falls under carbon-neutral fuel (Chen et al. 2012). Biomass is defined as renewable organic matter obtained from plants and animals, playing a dual role in food production and generating renewable fuels (Ong et al. 2021). It can be used either directly as a heating fuel or transformed into liquid and gaseous fuels via conversion approaches (https://www.eia.gov/energyexplained/biomass/). As a carbon-rich material, employing biomass can aid in mitigating anthropogenic emissions of CO_2 (Sulaiman et al. 2020).

The feedstock for first-generation biofuels comes from food-grade crops, and therefore, it competes with the food supply (Mahapatra et al. 2021). This has generated a shift toward second-generation biofuels, as they are produced from non-food sources, thereby avoiding competition with the food supply (Robak and Balcerek 2018). In particular, lignocellulosic biomass (LB) is recognized as a sustainable substitute to fossil-based carbon sources for producing these second-generation biofuels and other bio-based chemicals (Menon and Rao 2012). Lignocellulose is an abundantly found resource and is a structurally complex material. Primarily, LB is composed of carbon-rich biopolymers, namely lignin, cellulose, and hemicellulose (Chukwuma et al. 2020; Zhang et al. 2020; Basera et al. 2024). Along with these minor amounts of minerals, various extractives are also present (Templeton et al. 2010). LB is obtained from three major sources such as: (i) energy crops like perennial grasses; (ii) virgin biomass including all the terrestrial, woody, aquatic, or herbaceous plants which occur naturally; and (iii) waste biomass which involves wastes from various industries and municipal waste (Broda et al.

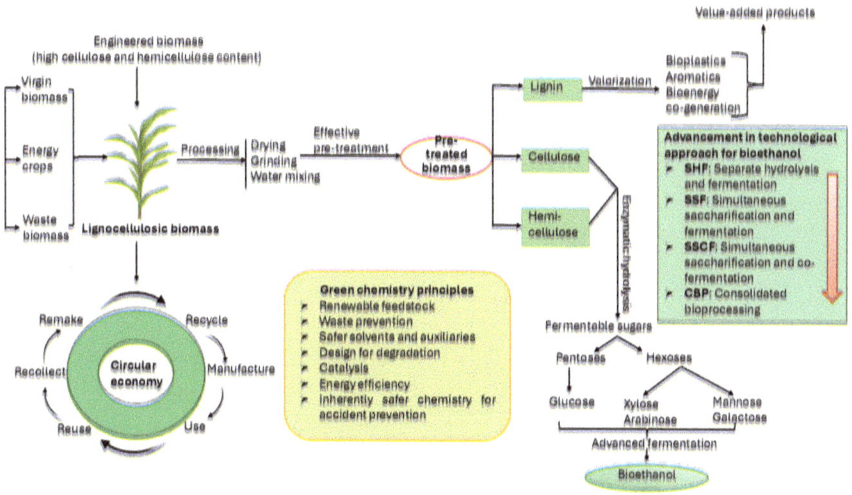

Fig. 4.1 The sources and degradative pathway of lignocellulosic biomass for bioethanol production, integrating the advancement in process approach and application of green chemistry principles. (Adapted from Devi et al. 2022; Anastas and Warner 1998; Lamichhane et al. 2021)

2022) (Fig. 4.1). Among the potential lignocellulosic biomass sources that have gained attention as crucial candidates in the circular economy as feedstocks are inedible sources such as corn stover, kraft pulp, switch grass, sugarcane bagasse, and forest residues like pine (Cotana et al. 2014; Sherwood 2020; Basera et al. 2024). Research has extensively focused on converting these lignocellulosic feedstocks into biofuels and bioenergy in the past decades (Kumar and Verma 2020).

Both cellulose and hemicellulose have been transformed successfully into a variety of high-value products such as drop-in fuels, biofuels, platform chemicals, biochemicals, and biopolymers (Zhang et al. 2023). In contrast, being amorphous and highly rich in carbon lignin displays structural complexity and diversity, which hampers its wider utilization (Beaucamp et al. 2022). Exploring methods to break down lignocellulosic material into fermentable sugars, which are then converted into bioethanol, is challenging (Singh et al. 2014).

4.2 Overview of Lignocellulosic Biomass Complex Chemistry

Lignocellulose is a naturally occurring complex found in plant materials. The major sources of this inexpensive biomass involve waste products from agriculture, forestry, and agro-industrial activities (Tran et al. 2020). Before initiating the production of bioethanol, it is crucial to understand its chemical and structural makeup. This is attributed mainly to several factors that hinder the efficient break down of lignocellulose into its monomeric fermentable sugars and subsequent fermentation

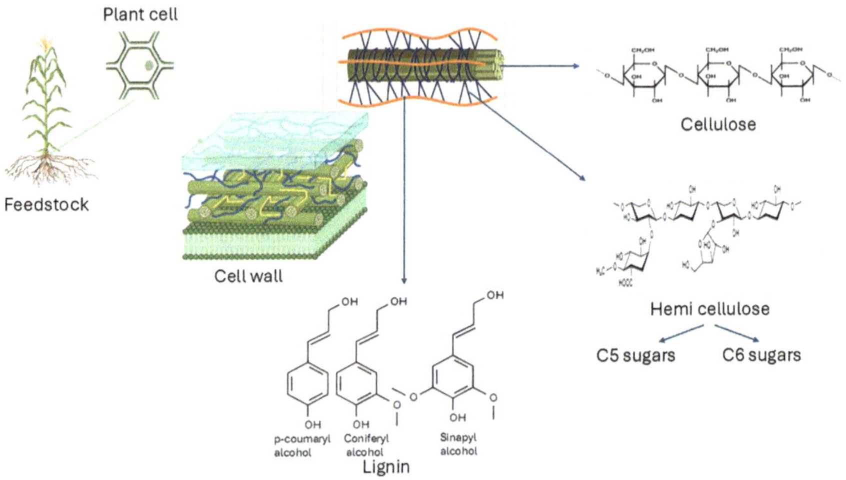

Fig. 4.2 The intricate structure of lignocellulosic biomass with its major components

into bioethanol. It is primarily composed of nearly 25–55% cellulose, 11–50% hemicellulose, and 10–40% lignin (Zhang et al. 2020). These complex structural components of lignocellulosic biomass have been represented in Fig. 4.2. Additionally, minor components such as inorganic salts and extractives like resins, fatty acids, and tannins are also present (Collard and Blin 2014). The cellulose and hemicellulose fractions of lignocellulosic biomass are the polymers of carbohydrates that release hexoses and pentoses upon hydrolysis. Lignin, in contrast, forms a complex methoxylated polyphenolic three-dimensional compound. It surrounds the other two components, i.e., cellulose and hemicellulose, thereby protecting them (Zhou et al. 2011). Lignin provides rigidity and stiffness to the plant cell wall (Raven et al. 1999). However, the ratio of each component varies across various plant species (Deshavath et al. 2019). The percentage composition of different plant biomass has been tabulated in Table 4.1. In addition, a heatmap illustrating the normalized values of the three components has been presented in Fig. 4.3. This figure highlights the intrinsic variability in the composition of biomass, which significantly influences subsequent bioethanol yield. Further, the breakdown and hydrolysis products of lignocellulosic biomass along with chemical equations involved are presented in Table 4.2.

4.2.1 Cellulose

Cellulose is a linear homopolymer of β-D-glucopyranose units, connected through the β-(1 → 4) glycosidic bonds. The basic repeating disaccharide unit contains cellobiose (Desvaux 2005). The structural properties are the same for cellulose in LB, unlike the hemicellulose and lignin. The generic formula of cellulose is

Table 4.1 Chemical composition (wt%) of various lignocellulosic biomass

Biomass source	Cellulose (wt%)	Hemicellulose (wt%)	Lignin (wt%)	Extractives (wt%)	Ash (wt%)	References
Coconut shell	20.0	48.8	30.0	–	0.5	Razvigorova et al. (1994)
Albizzia wood	58.3	8.1	33.2	1.9	1.1	Konwer et al. (2001)
Sunflower shell	48.4	34.6	17.0	2.7	4.0	Demirbaş (2002)
Hazelnut shell	26.8	30.4	42.9	3.3	1.4	
Olive husk	24.0	23.6	48.4	9.4	3.3	
Walnut shell	25.6	22.1	52.3	2.8	2.8	
Sugarcane bagasse	42.4	35.3	20.8	–	1.6	García-Pérez et al. (2002)
Switchgrass	30–50	10–40	5–20	–	~4.5	McKendry (2002)
Corn cob	45.0	35.0	15.0	–	–	Prasad et al. (2007)
Coffee pulp	35.0	46.3	18.8	8.2	–	Sánchez (2009)
Aspen wood	60.7	19.1	14.8	–	–	Shen et al. (2009)
Birch	54.7	24.8	12.2	–	–	
Oak	53.9	29.0	9.4	0.2	–	
Canola straw	42.4	16.4	14.2	–	2.1	Adapa et al. (2009)
Barley straw	32.5	25.7	23.0	18.4	9.8	Naik et al. (2010)
Flax straw	28.7	26.8	22.5	19.5	3.0	
Eucalyptus wood	43.4	15.3	25.5	–	~1	Chen and Fu (2013)
Spruce	49.4	4.7	27.7	3.5	0.3	Butler et al. (2013)
Miscanthus straw	47.5	20.9	21.4	6.4	3.4	
Wheat straw	39.1	24.1	16.3	19.2	1.3	Nanda et al. (2013)
Timothy grass	34.2	30.1	18.1	16.5	1.1	
Poplar	43.8	14.8	29.1	–	–	Frederick et al. (2014)
Hemp hurds	44.5	32.8	21.0	3.6	3.0	Stevulova et al. (2014)
Elephant grass	22.6	20.9	19.4	–	–	Eliana et al. (2014)
Rice straw	32.1	24.0	18.0	–	–	Howard et al. (2015)

(continued)

Table 4.1 (continued)

Biomass source	Cellulose (wt%)	Hemicellulose (wt%)	Lignin (wt%)	Extractives (wt%)	Ash (wt%)	References
Pinewood	42.0	14.7	26.3	–	–	Du et al. (2016)
Banana pseudostem	20.1	26.1	9.6–11.2	–	–	Ebrahimi et al. (2017)
Pineapple leaf fiber	46.2	19.3	2.8	–	–	Chintagunta et al. (2017)
Oat straw	37.6	23.3	12.9	–	2.2	Aqsha et al. (2017)

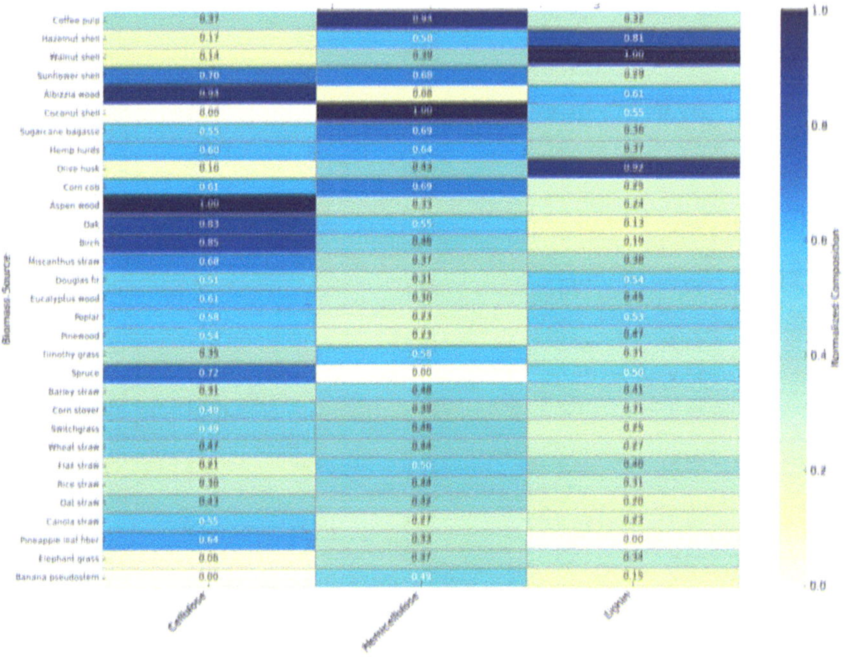

Fig. 4.3 Heatmap of normalized values for the three major components of LB: cellulose, hemicellulose, and lignin across various biomass (the original values have been presented in Table 4.1 along with the references)

$(C_6H_{12}O_5)_n$, and it contains around 500–1400 units of glucose, which aggregate to form microfibrils, which bundle into cellulose fibrils (McKendry 2002; Robak and Balcerek 2018). In lignocellulosic biomass, the fibers of cellulose are formed through both inter- and intra-hydrogen bonding and van der Waals forces between individual cellulose molecules (Ashokkumar et al. 2022).

It exists in two structural forms: crystalline and amorphous. Mostly, it is found organized in a crystalline arrangement, which is characterized by tightly packed and ordered chains. Although a small fraction of disordered chains forming amorphous

Table 4.2 Chemical equations involved in the degradative pathway of lignocellulosic biomass to bioethanol production

Lignocellulosic biomass		
Cellulose (Wyman et al. 2017)	**Hemicellulose** (Wyman et al. 2017)	**Lignin**
Hydrolysis of cellulose* $\underset{\text{Cellulose}}{(C_6H_{10}O_5)n} + \underset{\text{Water}}{nH_2O} \rightarrow \underset{\text{Glucose}}{nC_6H_{12}O_6}$ (1) *Enzymes or acid hydrolysis	Hydrolysis of 5-Carbon sugars* $\underset{\text{Xylan/Arabinan}}{(C_5H_8O_4)n} + nH_2O \rightarrow \underset{\text{Xylose/Arabinose}}{nC_5H_{10}O_5}$ (2) Hydrolysis of 6-Carbon sugars* $\underset{\text{Mannan/Glucan/Galactan}}{(C_6H_{10}O_5)n} + nH_2O \rightarrow \underset{\text{Mannose/Glucose/Galactose}}{nC_6H_{12}O_6}$ (3) *Enzymes or acid hydrolysis	**Lignin oxidation** $\underset{\text{Alcohol}}{Ar - CH_2OH} \rightarrow \underset{\text{Aldehyde}}{Ar - CHO} \rightarrow \underset{\text{Acid}}{Ar - COOH}$ (4) Ester bond cleavage in Lignin Carbohydrate Complex (LCC) $R - COOR' + H_2O \rightarrow R - COOH + R'OH$ (5) Ether bond cleavage in Lignin Carbohydrate Complex (LCC) $Ar - OAr' \rightarrow Ar - OH + Ar' - OH$ (6)
Inhibitors generated during dilute acid or hydrothermal pretreatment (Ujor and Okonkwo 2022)		
$C_6H_6O_3$ Hydroxymethylfurfural (HMF) $C_5H_8O_3$ Levulinic acid HCOOH Formic acid	$C_5H_4O_2$ Furfural	$C_{10}H_{10}O_4$ Ferulic acid $C_9H_{10}O_4$ Syringaldehyde $C_7H_6O_2$ 4-hydroxybenzaldehyde (4-HBD)
Bioethanol production (Wyman et al. 2017)		
$\underset{\text{Glucose}}{C_6H_{12}O_6} \rightarrow \underset{\text{Ethanol}}{2C_2H_5OH} + \underset{\text{Carbon dioxide}}{2CO_2}$ (7)	$\underset{\text{Xylose}}{3C_5H_{10}O_5} \rightarrow \underset{\text{Ethanol}}{5C_2H_5OH} + \underset{\text{Carbon dioxide}}{5CO_2}$ (8)	

regions also exists (Deshavath et al. 2019). The average diameter of cellulose microfibrils derived from plant cell walls ranges between 2 and 4 nm. On the contrary to those produced by bacteria, the diameters are slightly higher, ranging from about 4 to 8 nm (Martínez-Sanz et al. 2015). The recalcitrance of lignocellulosic biomass also depends upon the degree of polymerization. It is defined as the number of glucose units in a cellulose molecule, which varies from around 10,000 in wood to 15,000 in native cotton fibers (Desvaux 2005). Yoo et al. (2017) have demonstrated a positive association between cellulose content and the release of fermentable glucose.

Apart from predominantly occurring in plant cell walls, cellulose also exists in bacteria, fungi, and algae (Okolie et al. 2019). This highlights the widespread presence of cellulose in biological entities and hence its significance.

4.2.1.1 Polymorphism in Crystalline Cellulose

Cellulose exists in six interconvertible crystalline polymorphs namely cellulose I, II, III_1, III_2, IV_1, and IV_2 (O'Sullvian 1997). Among these, cellulose I is the native polymorph and occurs naturally in plants, bacteria, and algae. In terms of thermodynamics, it is metastable and therefore under specific treatment conditions it gets converted into more stable cellulose II or III (Moon et al. 2011; Okolie et al. 2019). Cellulose II possesses a monoclinic crystal structure and can be generated through either regeneration or mercerization (Okolie et al. 2019). When treated with liquid ammonia and amines, both cellulose I and II can be transformed structurally into III_1 and III_2; which, further, upon heating with glycerol, forms cellulose IV_1 and IV_2 (Larsson et al. 1999; Okolie et al. 2019).

In most of the lignocellulosic feedstocks, a high proportion of native Cellulose I is present. However, its high crystallinity and amphiphilic nature, interlinked with lignin, render it highly resistant to enzymatic hydrolysis. This, therefore, poses a significant obstacle in the production of bioethanol (Nagarajan et al. 2017). To address this, several pretreatment strategies have been developed to disrupt the lignocellulosic matrix and enhance the accessibility of cellulose. However, despite these advances, an emerging area of interest is employing cellulose II.

Over the past decades, cellulose II has been recognized as the most enzymatically digestible polymorph of cellulose. This is attributed to its lattice arrangement, higher porosity, lower crystallinity, and greater surface wettability. Although its application as a feedstock for bioethanol production remains underexplored, with limited reports investigating its bioconversion (Nagarajan et al. 2017).

4.2.2 Hemicellulose

Hemicellulose is a structurally heterogeneous polymer and occurs as the second most abundant carbohydrate in nature (Chandel et al. 2018; Deshavath et al. 2019). It is composed of a diverse mixture of polysaccharides made up of monomeric sugars and their acids (Okolie et al. 2019; Chen 2014). These include pentoses such as xylose, rhamnose, and arabinose; hexoses such as mannose, glucose, and galactose; and uronic acids such as D-galacturonic acid, D-glucoronic acid, acetic acid, and 4-O-methylglucoronic acid (Girio et al. 2010; Deshavath et al. 2019).

Structurally, hemicellulose differs from cellulose distinctly. The degree of polymerization of hemicellulose is mostly lower, 50–200 units (Chen 2014). Additionally, hemicellulose has an amorphous configuration, which makes it more susceptible to hydrolysis, especially under acidic conditions or thermal treatment (Chen 2014; Yang et al. 2022). The main backbones of hemicellulose molecules are the homopolymers (xylan, galactan, glucan) and heteropolymers (glucomannan) formed by β-(1 → 4)-linked sugar residues with occasional β-(1 → 3) linkages (Deshavath et al. 2019). Acetic acid and uronic acids impart flexibility to hemicellulose, which allows it to associate with microfibrils of cellulose via hydrogen bonding. This contributes toward the integrity of the plant cell and thereby presents challenges for enzymatic deconstruction (Marriott et al. 2015).

Xylan comprises 20–30% of the dry weight in lignocellulosic materials such as hardwoods and herbaceous plants (Girio et al. 2010). It is a polysaccharide with sub-units of xylose and arabinose with acetyl groups from acetic acid (Amidon and Liu 2009). Xylan content reaches around 50% in some tissues of cereals and grasses and hence plays an important role in bioethanol production (Ebringerova et al. 2005). Another important component of hemicellulose is the xyloglucans, located in the primary cell walls. However, unlike cellulose, approximately 75% of glucosyl residues in xyloglucans are substituted with α-D-xylopyranose at the O-6 position of the glucose backbone, and sometimes with the side chains of galactose and fucose. The molar ratio of sugars in xyloglucans is 3:4:1 (xylose: glucose: galactose), respectively (Yang 2008).

Another significant component of the lignocellulosic matrix of various plants is mannan, a polysaccharide composed of β-(1 → 4)-linked D-mannose units. They occur in minor proportions in hardwood, which highlights the compositional variability of hemicellulose within different plants. Mannans provide rigidity and resistance to solubilization due to their strong intermolecular hydrogen bonding among the polymer chains, which makes them poorly soluble in water (Yang 2008). When branched with D-galactose at α-(1 → 6), it gives galactomannans. In addition, galactoglucomannan, with a randomly distributed backbone of D-glucose and D-mannose units along with the side chains of D-galactose attached through β-(1 → 6) linkages, is prominent in softwoods (Girio et al. 2010; Gavande et al. 2023). This complex structure affects the chemical reactivity and processing activities of softwood-derived biomass. Another variant found in plants, such as konjac, is glucomannan. It is composed of a backbone of α-(1 → 6)-linked D-mannose units with D-glucose residues branching through α-(1 → 2) and α-(1 → 3) glycosidic bonds. It is predominantly found in softwood and higher plants (Moreira and Filho 2008). These structural differences directly influence the processing steps and selection of enzymes during the pretreatment and saccharification of LB (Zoghlami and Paes 2019).

Overall, the diversity of hemicellulose structure is influenced by the composition of sugar, branching, and substitution, which are directly correlated to the plant's biological origin and tissue type (Yang et al. 2022). Pretreatment techniques, for instance, dilute acid hydrolysis or steam explosion, enable the removal of hemicelluloses, thereby improving the accessibility of the enzyme and thus cellulose hydrolysis (Herbaut et al. 2018; Santos et al. 2018).

4.2.2.1 Hemicelluloses Containing Backbones of Xylan, Glucan, and Glucomannan

The hemicelluloses containing xylan backbones are further classified as arabinoglucuronoxylans (also known as glucuronoarabinoxylans), glucuronoxylans, and arabinoxylans.

The arabinoglucuronoxylans (Fig. 4.4a) are mostly found associated with pectin and xyloglucan through non-covalent interactions (Huang et al. 2023). They form major hemicellulose structures in biomass obtained from agricultural residues (non-woody) and contribute to around 25–30% of its dry weight. In contrast, they occur only about 5–10% in softwoods (Deshavath et al. 2019). Structurally, they comprise units of D-xylopyranose forming a β-1,4-linked backbone of xylan. These backbones are often substituted with α-L-arabinofuranose linked via α-1,3-glycosidic bonds, acetyl groups at the C2 position of xylan, and 4-O-methyl-D-glucuronyl groups. These groups contribute toward the electrostatic repulsion of other charged components within the cell wall (Waldron and Faulds 2007).

Glucuronoxylans (Fig. 4.4b) make the primary hemicellulose component in hardwoods, specifically O-acetyl-4-O-methylglucuronoxylans with approximately 15–30% of the biomass dry weight. They have a degree of polymerization mainly varying from 100 to 200. They have a β-1,4-linked D-xylopyranose backbone, wherein certain xylose units are superseded with acetyl groups at C2 and C3

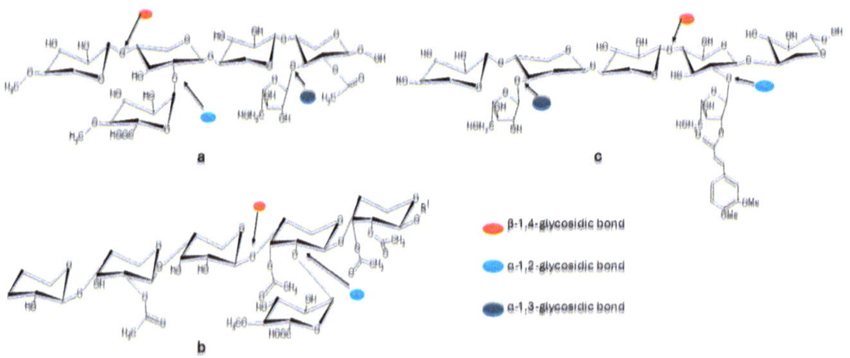

Fig. 4.4 Chemical structure of xylan backbone containing hemicelluloses (**a**) arabinoglucuronoxylans, (**b**) glucuronoxylans, (**c**) arabinoxylans. (Adapted from Deshavath et al. 2019)

positions via α-1,2-glycosidic bonds. The glucuronoxylans also contain small amounts of L-rhamnose and galacturonic acid. The latter contributes to increased resistance against alkaline degradation (Pereira et al. 2003; Girio et al. 2010).

Another component with a xylan backbone is arabinoxylan (AX) (Fig. 4.4c), which is predominantly found in cereal grain cell walls. It occurs in both the endosperm and bran cell wall structures, and its chemical composition varies as per the source (Hamaker et al. 2019). In wheat bran, for instance, arabinoxylan accounts for 60–69% of the non-starch polysaccharides (Lu et al. 2000). The backbone consists of β-D-xylose units interlinked by β-1,4 glycosidic bonds. The structure resembles hardwood xylans but is extensively substituted with α-L-arabinose at C2 or C3 positions. In some cases, these residues are esterified with ferulic acid at the O-5 position. These ferulic acids can form both intra- and intermolecular cross-links within the backbone of xylan (Correia et al. 2011). Both the physical and covalent interactions with other cell wall elements make arabinoxylan more challenging to extract. Therefore, various chemical solvents and enzymatic extraction techniques have been employed to extract arabinoxylans from cereal grains or bran (Zhang et al. 2014).

The primary hemicellulose components in softwoods, constituting around 15–25% of their dry weight, are galactoglucomannans (Fig. 4.5a). These are also known as O-acetyl-galactoglucomannans (Reid 1997). They are found both in the primary and secondary cell walls of coniferous (gymnosperm) species, in ferns, and in a few angiosperms (Popper 2008). Structurally, they consist of a linear backbone of β-(1 → 4)-linked D-mannopyranose and D-glucopyranose. The degree of polymerization typically ranges from 40 to 100 units. Its backbone is partially substituted with α-(1 → 6)-linked D-galactopyranose and acetyl groups, which are placed at C2 or C3 positions of mannose and glucose residues. They are further classified as

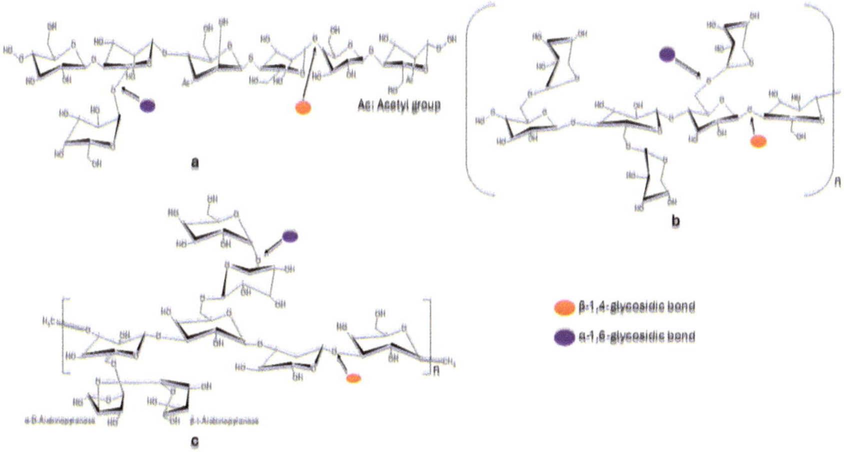

Fig. 4.5 Chemical structure of glucan and glucomannan backbone containing hemicelluloses (**a**) galactoglucomannans, (**b**) xyloglucans, and (**c**) arabinogalactans. (Adapted from Deshavath et al. 2019)

galactose-rich with a Gal:Glc:Man ratio of 1:1:3 and galactose-poor with a Gal:Glc: Man ratio of 0.1–0.2:1:3–4, based on the content of galactose present. The presence of galactoglucomannans contributes toward the structural integrity and retention of moisture in the plant tissues (Pereira et al. 2003; Deshavath et al. 2019).

Xyloglucans (Fig. 4.5b) are another abundantly found hemicellulose in primary cell walls of hardwood species and most vascular plants. They form a network along with the cellulose microfibrils, thereby contributing to the mechanical strength and flexibility of the cell wall. Structurally, it consists of a β-(1 → 4)-linked D-glucan backbone, in which some glucose residues are substituted with α-D-xylose units at the O6 position (Nishinari et al. 2007). In dicotyledons, xylose units further contain β-D-galactose or L-arabinose residues, with few β-D-galactose substituted with α-L-fucose, forming di- or tri-glycosyl side chains. The solubility and interaction with other wall components are further influenced by the presence of acetyl groups attached through O-linkages (Park and Cosgrove 2015).

Water-soluble polysaccharides predominantly found in the hardwoods are arabinogalactans (Fig. 4.5c). They have a highly branched structure with a backbone of galactopyranose linked through α-1,3-glycosidic bonds (Paulson and Barsett 2005). The branching occurs mainly at the C6 position of galactose units, forming di-O-substituted side chains via α-1,6 linkages. In addition, galactopyranosyl units attached through α-1,6-glycosidic bonds to arabinopyranosyl or arabinofuranosyl residues are also present sometimes (Whistler 1993; Deshavath et al. 2019).

4.2.3 *Lignin*

The third crucial constituent of LB is lignin, which is bound intricately to hemicellulose through covalent linkages. It plays a crucial role in imparting rigidity and structural compactness to the plant cell wall and hydrophobicity (Kang et al. 2019). Therefore, physically it contributes to the recalcitrance of biomass by restricting the access of enzymes to polysaccharides. On the contrary, chemically, due to hydrogen bonding, hydrophobic interactions, and π-π stacking with aromatic residues, lignin can adsorb hydrolytic enzymes such as cellulases (Kumar and Wyman 2009; Zeng et al. 2014; Zoghlami and Paes 2019). The interference of lignin during bioconversion of LB is influenced by its content, composition, structural features, and abundance of hydroxyl groups (Santos et al. 2012).

Unlike the other components, structurally lignin is a highly aromatic and non-carbohydrate biopolymer (Deshavath et al. 2019). Its complex three-dimensional structure is rich in methoxy groups and polyaromatic rings (Zoghlami and Paes 2019). It is a heterogeneous polymer derived from three primary monolignols, i.e., p-coumaryl alcohol (4-hydroxycinnamyl alcohol), coniferyl alcohol (3-methoxy-4-hydroxycinnamyl alcohol), and sinapyl alcohol (3,5 3,5-dimethoxy-4-hydroxycinnamyl alcohol) (Pandey and Kim 2011). Among them, coniferyl alcohol (guaiacyl units) is a ubiquitous monomer found in almost all types of lignin. However, it is the dominant monolignol in softwood lignins.

Common linkages in Lignin

Linkage Type	Structure Highlight	Bond Dissociation Energy (kcal/mol)	Common in
β-O-4	Ether bond between aromatic units	54-72	Softwood, Hardwood, Grass (Abundant)
β-5	Phenylcoumaran	70	Softwood
β-β	Resinol	81	Softwood, Hardwood, Grass
C5-C5	Biphenyl	115-118	Softwood, Hardwood
4-O-5	Diaryl ether	78-83	Softwood, Hardwood (Low abundance)
β-1	Spirodienone	65-69	Softwood, Hardwood
α-O-4	Alkyl-aryl ether	50-56	Softwood, Hardwood, Grass

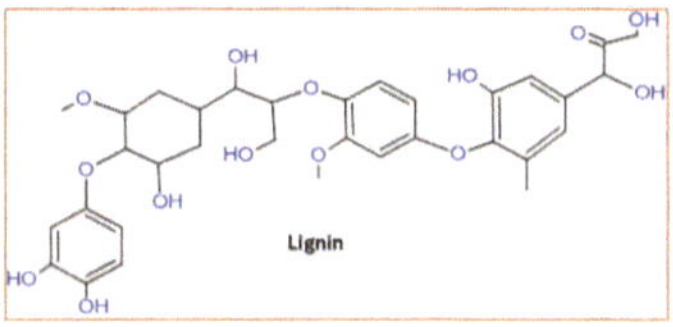

Lignin Carbohydrate Complexes (LCC)

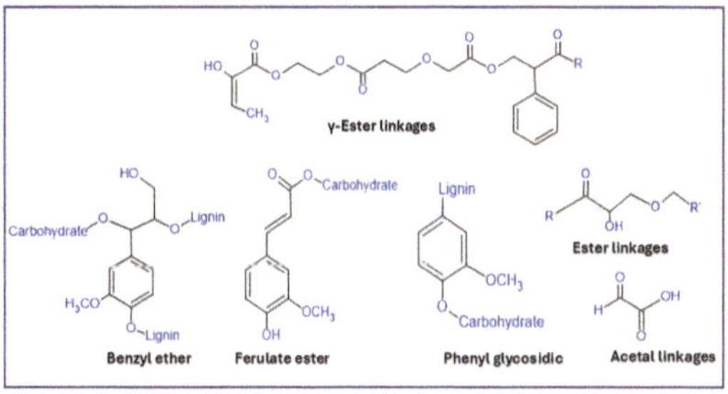

Fig. 4.6 Schematic of common interunit linkages between the lignin molecules, along with the bond dissociation energies of each linkage and their common occurrence. The covalent linkages associated with interpolymer interaction of lignin with carbohydrate moieties, i.e., cellulose and hemicellulose contributing toward LB recalcitrance are also highlighted. (Adapted from Rinaldi et al. 2016; Brienza et al. 2024)

Furthermore, lignin of hardwood contains a substantial amount of methoxyl groups, due to the equal presence of syringyl (S) and guaiacyl (G) units (Vohra et al. 2014; Okolie et al. 2019).

All these monomers are linked predominantly by ether (C–O–C) and carbon-carbon (C–C) bonds. However, ester bonds may occur at terminal units involving ferulic and p-coumaric acids (Reinoso et al. 2018). The side chains commonly comprise functional groups such as phenols, hydroxyl, methoxyl, and terminal aldehydes (Nanda et al. 2014). This cross-linking contributes to high molecular weight, enhances the mechanical strength, and limits enzymatic digestion, and thus it is a key target in the pretreatment of biomass (Fougere et al. 2016; Okolie et al. 2019). Lignin polymer structure, along with the common linkages associated with lignin's interunits, is presented in Fig. 4.6. The bond dissociation energies for each of the linkages and their structural highlights are provided. In addition, lignin possesses a high calorific value, ranging from 23.2 to 25.6 MJ/kg, which is around 30% greater than that of cellulose and hemicellulose combined (Novaes et al. 2010).

4.2.4 Extractives, Ash, and Minerals

Extractives are the non-structural constituents of lignocellulosic biomass. They are readily soluble in water and organic solvents (hexane and ethanol). The functional units of these compounds are terpenes, fats, lipids, steroids, terpenoids, waxes, proteins, and other phenolic substances such as lignans, stilbenes, and tannins (Okolie et al. 2019). Pectin is also an important non-structural polysaccharide found in the plant cell walls, particularly between cellulose microfibrils. Pectin exists in different forms, like pectic acid and protopectin. Pectic acid is water-soluble and can form calcium pectate gels through interaction with calcium ions (Chen 2014).

Ash is a non-combustible inorganic residue that remains after the complete combustion of LB. Ash consists of minerals and elemental compounds such as silicates, phosphates, carbonates, and metal oxides of calcium, magnesium, potassium, and sodium (Nanda et al. 2013; Okolie et al. 2019). Silica is often found as silicate minerals such as feldspars, plagioclase, opal, and kaolinite (Vassilev et al. 2012). In comparison to the agricultural residues, the woody biomass contains a lower ash content (Nanda et al. 2013; Okolie et al. 2019).

4.2.5 Inter-polymer Interaction

Individually, each of the constituents, i.e., cellulose, hemicellulose, and lignin, encompasses recalcitrance to degradation of LB. However, their interaction further intensifies the resistance of LB toward enzymatic hydrolysis. Hydrogen bonds are present between the cellulose and hemicellulose moieties, while lignin and hemicellulose form a covalent complex commonly identified as lignin-carbohydrate complex (LCC). These LCC complexes highly reduce the accessibility of enzymes toward cellulose. The significant linkages found in these complexes are benzyl ethers, acetal, phenyl glycosides, ferulate esters, and γ-esters (Fig. 4.6). This highly intricate bonding between the polymers is responsible for sustaining the structure of the wood and thereby adds to recalcitrance.

4.3 Pretreatment Strategies Targeting Cellulose, Hemicellulose, and Lignin Fractions

To utilize all the fractions of lignocellulosic biomass, it is essential to first valorize it efficiently into each component. The degradation of these three main elements, i.e., cellulose, hemicellulose, and lignin, leads to their conversion into value-added products; bioethanol being one of them. The crucial stage in biomass refining is termed as pretreatment, which aids in the disruption of tightly bound components and enhances the accessibility and reactivity, subsequently leading to enzymatic and chemical hydrolysis (Barakat et al. 2013; Isikgor and Becer 2015). The selection of an effective pretreatment strategy is crucial. It should significantly improve the yield of sugars, reduce enzyme loading, facilitate the recovery of lignin, produce fewer toxic compounds and by-products, and require less energy to operate (Bhutto et al. 2017). A broad classification of pretreatment strategies has been depicted in Fig. 4.7.

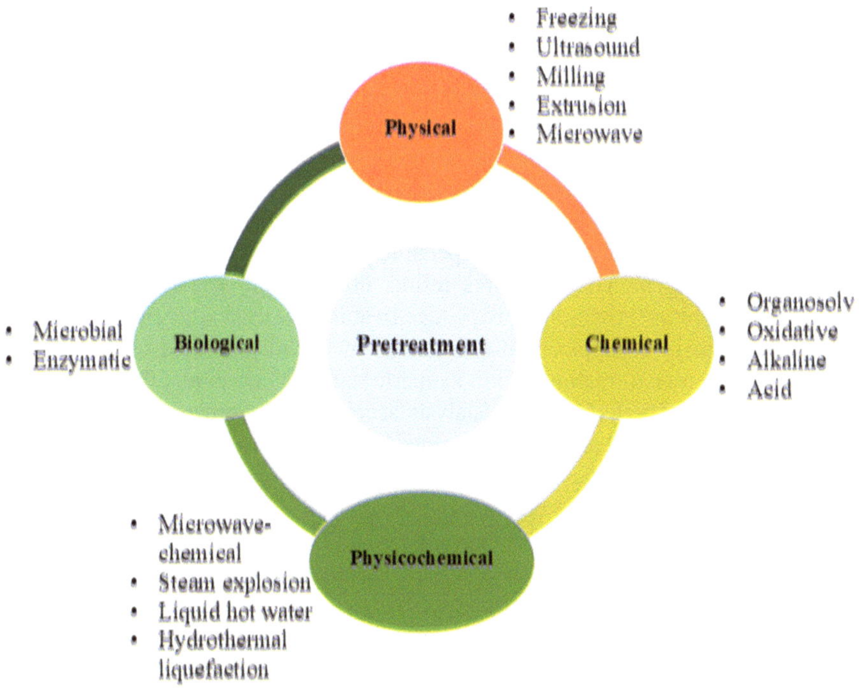

Fig. 4.7 Various pretreatment methods for lignocellulosic biomass feedstock

Overall, the selection of a pretreatment method is related to the biomass under consideration, desired downstream process, environmental and cost effectiveness (Isikgor and Becer 2015).

4.3.1 Cellulose Hydrolysis and Recovery

Different pretreatments yield varying levels of cellulose or glucose recovery and similarly influence the cellulose hydrolysis efficiency. Alkaline pre-treatment of banana pseudostem with sodium hydroxide (NaOH), for instance, resulted in 75.48% recovery of cellulose while peroxide treatment gave approximately a similar recovery percent of 74.37% (Shimizu et al. 2018). In contrast, microwave-assisted organosolv pretreatment of mixed sawmill proved to be the best with maximum cellulose/sugar recovery of 98% (Alio et al. 2019). Rooni et al. (2017) reported 19.42% recovery through freezing of the barley straw. Similarly, by alkaline pretreatment of *Pinus* sp. at an optimum temperature of 165 °C, 120 min, the cellulose hydrolysis achieved was 95% in 72 h (Bernat-Lugo et al. 2019). On the other hand, autohydrolysis at 150 °C followed by organosolv pretreatment at 180 °C of *Pinus radiata* resulted in 63% hydrolysis after 98 h (Santos et al. 2019).

4.3.2 Hydrolysis of Hemicellulose with Xylose Recovery

In lignocellulosic biomass, hemicellulose often consists of acetyl groups, which pose a hindrance to enzymes. These groups can increase the effective diameter of cellulose microfibrils, thereby obstructing the interaction between cellulose and the catalytic domains of enzymes (Zhao et al. 2012). Therefore, utilizing strategies to break these acetyl groups is crucial (Basera et al. 2024). The autohydrolysis of biomass generally involves high temperatures (180–205 °C) (Isikgor and Becer 2015). Employing high temperatures results in increased degradation of xylose along with inhibitor formation. A two-stage treatment is used wherein the first step releases oligomers of hemicellulose, followed by hydrolysis (acid or enzymatic), leading to their transformation into monomeric sugars (Girio et al. 2012; Park et al. 2016) (Fig. 4.8). The use of acid catalysis accelerates the process, allowing both concentrated and diluted acids to be employed. The advantage of concentrated acids is that they are energy-efficient, whereas the use of dilute acids reduces the probability of corrosion. The most commonly used acid is sulfuric acid and is typically applied at 116–170 °C for 10–90 min (Isikgor and Becer 2015).

Another technique used for the disruption of lignocellulosic structure is steam explosion. This process utilizes high-pressure steam followed by a quick drop in pressure, which results in explosive decompression (Girio et al. 2012). The optimum temperature of this operation is below 240 °C. This supports lignin disruption,

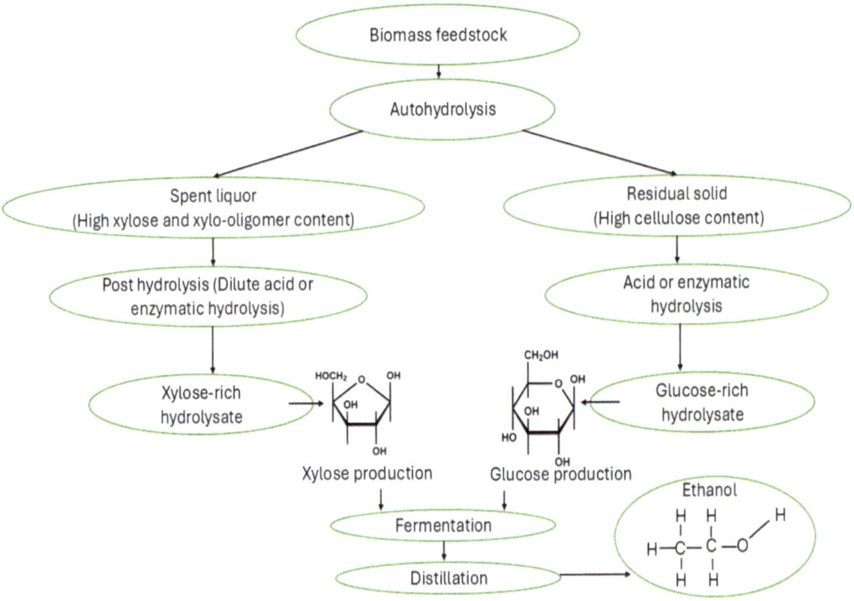

Fig. 4.8 Process flow of biomass autohydrolysis for sequential production of xylose and glucose finally leading to fermentation, distillation, and bioethanol formation. (Adapted from Harahap 2020)

thereby enhancing the accessibility of cellulase enzyme to release glucose (Jacquet et al. 2015). An integrated approach to use acids along with steam explosion has been reported to yield 87% xylose and 85% retention of cellulose in the biomass of corncob (Fan et al. 2014). Use of inorganic salts and ionic liquids has also resulted in xylan removal (Brandt et al. 2013; Eop et al. 2013).

4.3.3 *Understanding Biomass Recalcitrance in Terms of Lignin Removal Efficiency*

As discussed above, lignin comprises a highly irregular, heterogeneous, and aromatic polymeric structure, which adds to the recalcitrance of LB. In addition, it forms a significant barrier to accessing cellulose and hemicelluloses during the process of bioconversion (Kininge and Gogate 2022). Therefore, understanding the interaction of different pretreatment strategies with this intricate component of lignin is essential. Table 4.3 presents a comparative dataset for the efficiency of

Table 4.3 Effect of various pretreatment methods on lignin removal (%) *wrt* biomass type and optimal conditions

Pretreatment method	Biomass type	Optimum conditions	% Lignin removal	References
Ammonia fiber expansion	Corn stover	Soaked in AlOH, 10 days, 10 mL/g ratio	85	Kumar et al. (2009)
Organosolv (Methanol + HCl)	Pinewood	60–80% MeOH, 0.2% HCl, 170 °C, 45 min	75	Zhao et al. (2009)
Organosolv (Methanol + HCl)	Beechwood	50% MeOH, 0.1% HCl, 160 °C, 45 min	90	
Alkali-Organosolv	*Pinus* sp. (Softwood sawdust)	3% NaOH, 40 mL ethanol, 5 g biomass	91	Bernal-Lugo et al. (2019)
Acidic-Organosolv		3% NaOH, 40 mL ethanol, 5 g biomass	48	
Ultrasound (US)	Agave by-product	60 °C, 27 min, 50% sonication amplitude	86	Alvarez-Chavez et al. (2021)
Organosolv (Ethanol + 2-naphthol)	Woody biomass	Organic solvent based process using 2-naphthol as catalyst	76.1	Chu et al. (2021)
Organosolv (Ethanol)	Southern yellow pine	175 °C, 80 min, 50:50 solvent ratio	16	Vaidya et al. (2022)
Organosolv (Butanol)		175 °C, 80 min, 50:50 solvent ratio	28	
Organosolv (Butanol + SO_2)	Pole pine	175 °C, 80 min, 50:50 solvent ratio with SO_2	82	

Source: Kululo et al. (2025)

various physical, chemical, and hybrid pretreatment approaches in removing lignin from biomass under varying process conditions. Lignin removal is directly related to the enhanced microbial or enzymatic action on carbohydrate polymers (Kululo et al. 2025).

4.4 Key Enzymes in the Degradation of Lignocellulosic Biomass Components

There are several enzymes engaged in the enzymatic degradation of cellulose, hemicellulose, and lignin. The crucial step during the deconstruction of the complex cell wall of plants is the break down of cellulose. The enzymes involved in this function are collectively known as cellulases and include enzymes such as exoglucanases, endoglucanases, and β-glucosidases (Gupta et al. 2012). Further, hemicellulose has an intricate structure that significantly influences the physico-chemical characteristics of lignocellulosic biomass. The enzymes involved here are capable of targeting the diverse glycosidic linkages within this branched polymer. A broad range of both endo- and exo-hemicellulases is involved in this process, which are classified based on the specific hemicellulose present, such as mannan or xylan (Houfani et al. 2020). In addition, certain enzymes such as ferulic acid esterases and xylan esterases play a crucial role in cleaving the side-chain substituents from hemicellulose structure (Houfani et al. 2020; Basera et al. 2024). Lignin-degrading or ligninolytic enzymes are crucial for lignin degradation. They are broadly classified as lignin-degrading auxiliary enzymes (LDA) and lignin-modifying enzymes (LME) (Janusz et al. 2017). The major enzymes involved in degrading hemicellulose, cellulose, and lignin, along with their mode of action and breakdown products, have been compiled in Table 4.4.

4.5 Green Chemistry Principles Applied to Bioethanol Production from LCB

The green chemistry principles were established by Paul Anastas and John Warner in the year 1998. They form a framework in designing new chemical-based products and processes. These twelve principles incorporate guidelines for all aspects of the process. It includes process life-cycle from the type and nature of raw material used, safety and efficiency of the conversion, toxicity assessment, and biodegradability of the chemicals and reagents being used. In the context of the lignocellulosic biomass as feedstock for the production of bioethanol, the following green chemistry principles are applied effectively (Fig. 4.1):

Table 4.4 The key enzymes involved in degrading cellulose, hemicellulose, and lignin fraction of lignocellulosic biomass, with their mode of action and breakdown products

Polymer	Key enzymes	Mode of action	Breakdown products	References
Cellulose	Endoglucanase Exoglucanase (CBH) β-Glucosidase	Random cleavage of internal β-1,4 bonds Exo cleavage from endo-hydrolyzes cellobiose	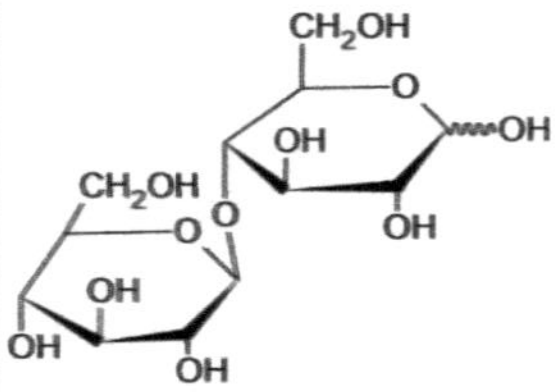 Cellobiose 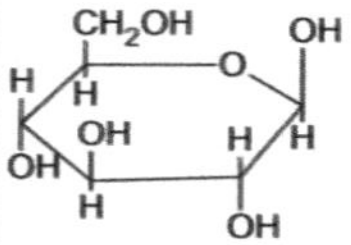 Glucose	Houfani et al. (2020) and Basera et al. (2024)
Hemicellulose	Endo-xylanase β-Xylosidase Arabinofuranosidase Mannanase Acetyl xylan esterase	Cleaves β-1,4-xylan backbone Removes xylose units and side branches	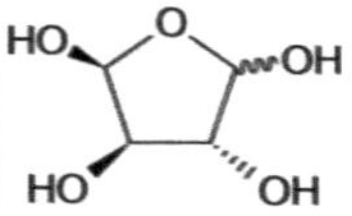 Xylose 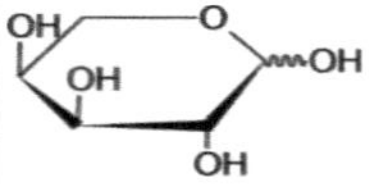 Arabinose	Houfani et al. (2020) and Polizeli et al. (2005)

(continued)

Table 4.4 (continued)

Polymer	Key enzymes	Mode of action	Breakdown products	References
			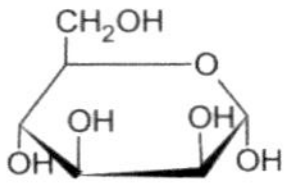 Mannose 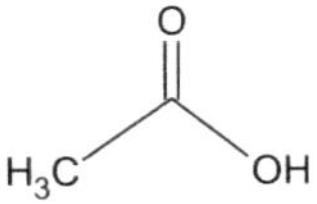 Acetic acid	
Lignin	Laccase Lignin peroxidase (LiP) Manganese peroxidase (MnP) Versatile peroxidase (VP)	Oxidative cleavage of aromatic structures Radical formation and polymer depolymerization	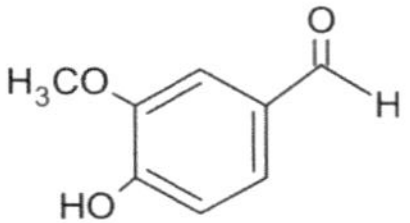 Vanillin 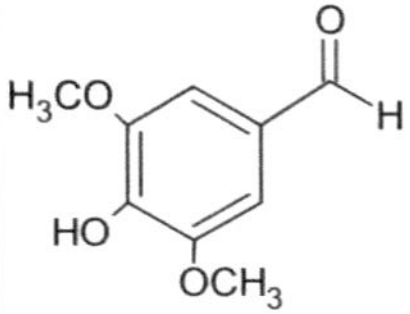 Syringaldehyde	Basera et al. (2024)

1. *Prevention:*

> It is better to prevent waste than to clean up or prevent once it is formed. (Anastas and Warner 1998)

Utilizing lignocellulosic biomass for bioethanol production complies with the very first principle of green chemistry, which is the prevention of waste. The valorization of lignin in LCB followed by hydrolysis of cellulose and hemicellulose into their respective derived-sugars and other by-products follows prevention. These by-products are converted further into energy in the form of biogas and process heat or as materials such as animal feeds, fertilizers, etc., thereby minimizing the generation of waste (Anastas and Eghbali 2010).

2. *Safer solvents and auxiliaries:*

> The use of auxiliary substances such as solvents, separation agents, etc. should be made unnecessary wherever possible and innocuous when used. (Anastas and Warner 1998)

For hydrolysis and fermentation of LCB water-based systems are employed, which avoids the use of any toxic or volatile solvent (commonly used in the petrochemical industries). In addition, the emerging methods of pre-treatment involve the use of green solvents such as supercritical CO_2, ionic liquids, and deep eutectic solvents (DESs), which are mostly recyclable and designed to be less toxic (Kululo et al. 2025). Although the traditionally employed pre-treatment methods still are chemical-based and present challenges, the transition toward green solvents makes the overall process to be in accordance with this principle.

3. *Design for energy efficiency:*

> Energy requirements should be recognized for their environmental and economic impacts and should be minimized. Synthetic methods should be conducted at ambient temperature and pressure. (Anastas and Warner 1998)

During the production of bioethanol from lignocellulosic biomass, the pretreatment process is highly energy-intensive. However, advancement in technologies employing milder pretreatment methods such as steam explosion, DES systems, biological routes along with SSF significantly reduces the energy inputs in comparison to the fossil fuel extraction, refining, and combustion. Therefore, the whole life cycle of bioethanol reflects a lower net energy consumption and consequently lesser carbon footprint.

4. *Use of renewable feedstocks:*

> A raw material or feedstock should be renewable rather than depleting whenever technically and economically practicable. (Anastas and Warner 1998)

Lignocellulosic biomass is an abundant resource on the Earth which is both renewable as well as a non-food and feed biomass (Basera et al. 2024). Further in comparison to the first-generation ethanol production relying on corn or sugarcane, LCB does not compete with the food crops directly. This therefore supports sustainable bioeconomy.

5. *Catalysis:*

> Catalytic reagents (as selective as possible) are superior to stoichiometric reagents. (Anastas and Warner 1998)

The process of ethanol production using LCB relies on utilizing enzymes such as cellulases and hemicellulases, along with engineered microbial catalysts for hydrolysis and fermentation processes. In comparison to the harsh chemical catalyst these are selective, reusable, and biodegradable catalyst alternatives. Their employment reduces the production of waste and lowers the energy inputs, thereby enhancing the overall efficiency.

6. *Design for degradation:*

> Chemical products should be designed so that at the end of their function they break down into innocuous degradation products and do not persist in the environment. (Anastas and Warner 1998)

The green chemistry principle on design for degradation implies that the end products of the process break down readily into substances which are non-toxic and harmless. However, designing materials and chemicals which are biodegradable is not an easy task due to the persisting environmental pollution (Anastas and Eghbali 2010). In this regard, lignocellulosic bioethanol production aligns strongly with this principle. Ethanol being inherently biodegradable decomposes readily into carbon dioxide and water without leaving any toxic residues. In addition, the feedstock itself is renewable and degrades naturally thereby ensuring the overall process contributes minimal waste which is persistent in the environment.

7. *Inherently safer chemistry for accident prevention:*

> Substances and the form of a substance used in a chemical process should be chosen to minimize the potential for chemical accidents, including releases, explosions, and fires. (Anastas and Warner 1998)

Production of ethanol from lignocellulosic biomass avoids the large-scale handling of toxic petrochemicals, corrosive reagents, or heavy metals generally employed in fossil-fuel refining. This process is therefore safe for both workers and the environment.

To summarize, the utilization of LCB for bioethanol production displays strong alignment with several principles of green chemistry. This compliance therefore reflects the potential of second-generation bioethanol as a sustainable alternative to fossil fuels. Nevertheless, challenges such as the need for green pretreatment strategies and cost-effective systems still demand active research. Further development in the overall process of lignocellulosic bioethanol production will strengthen the alignment with all the green chemistry framework, thereby making it an ecologically friendly fuel technology.

4.6 Challenges and Prospects

Lignocellulosic biomass is a promising alternative to fossil fuels in producing bioethanol as a second-generation biofuel. The utilization of LCB not only reduces the emission of GHGs but also aids in improving waste management. However, there are many technological, economic as well as environmental challenges associated with the process. During bioethanol production through LCB, pre-treatment is one of the most crucial steps (Maceiras et al. 2021). Several methods for pre-treating LCB in order to enhance the quantity and quality have been developed and studied so far. But the pre-treatment step itself requires a high amount of energy making it the most energy-intensive process. This, therefore, adds to the total production cost of bioethanol from LCB.

Further, no universal pre-treatment method can be applied to LCB, due to its highly diverse nature and composition. Several toxic compounds are also generated during pre-treatment process which have inhibitory effects on activity of microbes resulting in lowering the bioethanol yield. The cellulolytic enzymes required for the hydrolysis of cellulose and hemicellulose are also costly and their activity is inhibited by the harsh process conditions (Basera et al. 2024).

Irrespective of these challenges, substantial advancements are being made to overcome the limitations faced during bioethanol production from LCB. Since recalcitrance of lignin hampers the hydrolysis of cellulose and hemicellulose, green pre-treatment approaches are being explored. However, there is still a need to search for more efficient and environmentally friendly practices which could reduce the overall cost of bioethanol production (Ceaser et al. 2024). In addition, mixed and combined pre-treatment strategies are being developed which will aid in addressing these issues of biomass diversity, recalcitrance, fermentation inefficiencies, and high cost. Enzyme engineering and use of biocatalysts can be effective in reducing the cost of production and enhancing efficiency. In addition, genetic modifications of the robust microbes capable of fermenting both hexose and pentose sugars even in the presence of inhibitors will also improve the economic feasibility. Apart from this, the integration of LCB refineries into the framework of circular bioeconomy and utilization of sustainable feedstock is promising (Devi et al. 2022). These prospects, therefore, not only make the lignocellulosic ethanol economically viable but also contribute toward the global energy transition through renewable feedstocks.

4.7 Conclusion

Lignocellulosic biomass (LB) is the most plentiful renewable resource present on Earth. It has gained interest due to its capability as a sustainable feedstock for bioethanol production. Employing bioethanol as an alternative fuel is significant because it reduces greenhouse gas emissions, thereby enhancing energy security.

Apart from bioethanol, the other bioconversion products include platform chemicals, bio-composites, biofuels, bioplastics, etc. However, the degradation of lignocellulosic biomass offers a significant challenge. This is attributed to its intricate chemical complexity. This chapter offers a detailed insight into the chemical composition of the LB feedstock. The major components of lignocellulosic biomass, like cellulose, hemicellulose, and lignin, along with some minor amount of ash and extractives, have been discussed. Understanding these complex molecular structures enables the advancement of innovative technologies for biorefineries. In addition, the role of various enzymes in breaking down each component, along with their mode of action and breakdown products, has also been discussed. It highlights the various pretreatment methodologies being effectively used for each fraction of LB. The advancement in pretreatment techniques is required to maintain the economic feasibility and maximum utilization of this feedstock for higher extraction of green and sustainable products.

Acknowledgment The authors are sincerely thankful to the Director General, TERI for providing the infrastructure required for carrying out the present research.

References

Adapa P, Tabil L, Schoenau G (2009) Compaction characteristics of barley, canola, oat, and wheat straw. Biosyst Eng 104(3):335–344

Alio MA, Tugui O-C, Vial C, Pons A (2019) Microwave-assisted Organosolv pretreatment of a sawmill mixed feedstock for bioethanol production in a wood biorefinery. Bioresour Technol 276:170–176

Álvarez-Chávez J, Hernández-Barrera S, Silva-Espinoza BA, Ayala-Zavala JF, González-Aguilar GA, Aispuro-Hernández E (2021) Agave by-products: an overview of their nutraceutical value, current applications, and processing methods. Polysaccharides 2(3):720–743

Amidon TE, Liu S (2009) Water-based woody biorefinery. Biotechnol Adv 27(5):542–550

Anastas P, Eghbali N (2010) Green chemistry: principles and practice. Chem Soc Rev 39:301–312

Anastas P, Warner J (1998) Green chemistry: theory and practice. Oxford University Press, New York

Anderson EM, Stone ML, Katahira R, Reed M, Muchero W, Ramirez KJ, Beckham GT, Román-Leshkov Y (2019) Differences in S/G ratio in natural poplar variants do not predict catalytic depolymerization monomer yields. Nat Commun 10:2033

Aqsha A, Tijani MM, Moghtaderi B, Mahinpey N (2017) Catalytic pyrolysis of straw biomasses (wheat, flax, oat, and barley) and the comparison of their product yields. J Anal Appl Pyrolysis 125:201–208

Ashokkumar V, Venkatkarthick R, Jayashree S, Chuetor S, Dharmaraj S, Kumar G, Chen WH, Ngamcharussrivichai C (2022) Recent advances in lignocellulosic biomass for biofuels and value-added bioproducts: a critical review. Bioresour Technol 344(Part B):126195

Barakat A, de Vries H, Rouau X (2013) Dry fractionation process as an important step in current and future lignocellulose biorefineries: a review. Bioresour Technol 134:362–373

Basera P, Chakraborty S, Sharma N (2024) Lignocellulosic biomass: insights into enzymatic hydrolysis, influential factors, and economic viability. Discov Sustain 5:311

Beaucamp A, Muddasar M, Crawford T, Collins MN, Culebras M (2022) Sustainable lignin precursors for tailored porous carbon-based supercapacitor electrodes. Int J Biol Macromol 221:1142–1149

Bernal-Lugo I, Jacinto-Hernandez C, Gimeno M, Montiel CC, Rivero-Cruz F, Velasco O (2019) Highly efficient single-step pretreatment to remove lignin and hemicellulose from softwood. Bioresources 14(2):3567–3577

Bhutto AW, Qureshi K, Harijan K, Abro R, Abbas T, Bazmi AA, Karim S, Yu G (2017) Insight into progress in pre-treatment of lignocellulosic biomass. Energy 122:724–745

Brandt A, Gräsvik J, Hallett JP, Welton T (2013) Deconstruction of lignocellulosic biomass with ionic liquids. Green Chem 15(3):550–583

Brienza F, Cannella D, Montesdeoca D, Cybulska I, Debecker DP (2024) A guide to lignin valorization in biorefineries: traditional, recent, and forthcoming approaches to convert raw lignocellulose into valuable materials and chemicals. RSC Sustain 2(1):37–90

Broda M, Yelle DJ, Serwańska K (2022) Bioethanol production from lignocellulosic biomass—challenges and solutions. Molecules 27(24):8717

Butler E, Devlin G, Meier D, McDonnell K (2013) Characterisation of spruce, salix, miscanthus, and wheat straw for pyrolysis applications. Bioresour Technol 131:202–209

Caesar R, Montane D, Constanti M, Medina F (2024) Current progress on lignocellulosic bioethanol including a technological and economical perspective. In: Environment, development and sustainability

Chandel AK, Garlapati VK, Singh AK, Antunes FAF, Da Silva SS (2018) The path forward for lignocellulose biorefineries: bottlenecks, solutions, and perspective on commercialization. Bioresour Technol 264:370–381

Chen H (2014) Chemical composition and structure of natural lignocellulose. In: Chen H (ed) Biotechnology of lignocellulose. Springer, pp 25–71

Chen L, Fu S (2013) Enhanced cellulase hydrolysis of eucalyptus waste fibers from pulp mill by Tween80-assisted ferric chloride pretreatment. J Agric Food Chem 61(13):3293–3300

Chen WH, Lu KM, Tsai CM (2012) An experimental analysis on property and structure variations of agricultural wastes undergoing torrefaction. Appl Energy 100:318–325

Chintagunta AD, Ray S, Banerjee R (2017) An integrated bioprocess for bioethanol and biomannure production from pineapple leaf waste. J Clean Prod 165:1508–1516

Chu Q et al (2021) Organosolv pretreatment assisted by carbocation scavenger to mitigate surface barrier effect of lignin for improving biomass saccharification and utilization. Biotechnol Biofuels 14(1):136

Chukwuma OB, Rafatullah M, Tajarudin HA, Ismail N (2020) Lignocellulolytic enzymes in biotechnological and industrial processes: a review. Sustainability 12(18):7282

Collard FX, Blin J (2014) A review on pyrolysis of biomass constituents: mechanisms and composition of the products obtained from the conversion of cellulose, hemicelluloses and lignin. Renew Sust Energ Rev 38:594–608

Correia MAS, Mazumder K, Brás JLA, Firbank SJ, Zhu Y, Lewis RJ, York WS, Fontes CMGA, Gilbert HJ (2011) Structure and function of an arabinoxylan-specific xylanase. J Biol Chem 286(25):22510–22520

Cotana F, Cavalaglio G, Gelosia M, Nicolini A, Coccia V, Petrozzi A (2014) Production of bioethanol in a second-generation prototype from pine wood chips. Energy Procedia 45:42–51

Demirbaş A (2002) Fuel characteristics of olive husk and walnut, hazelnut, sunflower, and almond shells. Energy Sources 24(3):215–221

Deshavath NN, Veeranki VD, Goud VV (2019) Lignocellulosic feedstocks for the production of bioethanol: availability, structure, and composition. In: Rai M, Ingle AP (eds) Sustainable bioenergy. Elsevier, pp 1–19

Desvaux M (2005) The cellulosome of Clostridium cellulolyticum. Enzyme Microb Technol 37(4): 373–385

Devi A, Bajar S, Kour H, Kothari R, Pant D, Singh A (2022) Lignocellulosic biomass valorization for bioethanol production: a circular bioeconomy approach. Bioenergy Res 15:1820–1841

Du X, Lucia LA, Ghiladi RA (2016) Development of a highly efficient pretreatment sequence for the enzymatic saccharification of loblolly pine wood. ACS Sustain Chem Eng 4(7):3669–3678

Ebrahimi M, Caparanga AR, Ordono EE, Villaflores OB, Pouriman M (2017) Effect of ammonium carbonate pretreatment on the enzymatic digestibility, structural characteristics of rice husk and bioethanol production via simultaneous saccharification and fermentation process with *Saccharomyces cerevisiae* Hansen 2055. Ind Crop Prod 101:84–91

Ebringerova A, Hromadkova Z, Heinze T (2005) Hemicellulose. In: Advances in polymer science, vol 186. Springer, pp 1–67

Eliana C, Jorge R, Juan P, Luis R (2014) Effects of the pretreatment method on enzymatic hydrolysis and ethanol fermentability of the cellulosic fraction from elephant grass. Fuel 118: 41–47

Eop K, Park D, Jeong G (2013) Effects of inorganic salts on pretreatment of *Miscanthus* straw. Bioresour Technol 132:160–165

Fan X, Cheng G, Zhang H, Li M, Wang S, Yuan Q (2014) Effects of acid impregnated steam explosion process on xylose recovery and enzymatic conversion of cellulose in corncob. Carbohydr Polym 114:21–26

Fougere D, Nanda S, Clarke K, Kozinski JA, Li K (2016) Effect of acidic pretreatment on the chemistry and distribution of lignin in aspen wood and wheat straw substrates. Biomass Bioenergy 91:56–68

Frederick N, Zhang N, Ge X, Xu J, Pelkki M, Martin E, Carrier DJ (2014) Poplar (*Populus deltoides* L.): the effect of washing pretreated biomass on enzymatic hydrolysis and fermentation to ethanol. ACS Sustain Chem Eng 2(7):1835–1842

Gabisa EW, Gheewala SH (2020) Can substitution of imported gasoline by locally produced molasses ethanol in Ethiopia be sustainable? An eco-efficiency assessment. Renew Sust Energ Rev 123:Article 109770

García-Pérez M, Chaala A, Roy C (2002) Vacuum pyrolysis of sugarcane bagasse. J Anal Appl Pyrolysis 65(1):111–136

Gavande PV, Goyal A, Fontes CMGA (2023) Carbohydrates and carbohydrate-active enzymes (CAZyme): an overview. In: Goyal A, Sharma K (eds) Foundations and frontiers in enzymology: glycoside hydrolases. Academic, pp 1–23

Girio FM, Fonseca C, Carvalheiro F, Duarte LC, Marques S, Bogel-Łukasik R (2010) Hemicelluloses for fuel ethanol: a review. Bioresour Technol 101(13):4775–4780

Girio FM, Carvalheiro F, Duarte LC, Bogel-Łukasik R (2012) D-Xylitol. In: da Silva SS, Chandel AK (eds) D-xylitol. Springer, Berlin/Heidelberg, pp 3–28

Gupta P, Samant K, Sahu A (2012) Isolation of cellulose-degrading bacteria and determination of their cellulolytic potential. Int J Microbiol:Article 578925

Hamaker BR, Tuncil YE, Shen X (2019) Carbohydrates of the kernel. In: Serna-Saldivar SO (ed) Corn (Third Edition). AACC International Press, pp 305–318

Harahap BM (2020) Degradation techniques of hemicellulose fraction from biomass feedstock for optimum xylose production: a review. Jurnal Keteknikan Pertanian Tropis dan Biosistem 8(2): 107

Herbaut M, Zoghlami A, Habrant A, Falourd X, Chabbert B, Cathala B (2018) Multimodal analysis of pretreated biomass species highlights generic markers of lignocellulose recalcitrance. Biotechnol Biofuels 11:52

Houfani AA, Anders N, Spiess AC, Baldrian P, Benallaoua S (2020) Insights from enzymatic degradation of cellulose and hemicellulose to fermentable sugars—a review. Biomass Bioenergy 134:105481

Howard RL, Abotsi E, van Jansen REL, Howard S (2015) Lignocellulose biotechnology: issues of bioconversion and enzyme production. Afr J Biotechnol 2(12):602–619

Huang H, Li Z, Ma Y, Yao M, Yao S, Zhang Z, Qin C (2023) High-performance arabinoglucuronoxylan-based biosurfactants for oily sludge separation. Carbohydr Polym 303:Article 120461

Isikgor FH, Becer CR (2015) Lignocellulosic biomass: A sustainable platform for the production of bio-based chemicals and polymers. Polym Chem 6(25):4497–4559

Jacquet N, Maniet G, Vanderghem C, Delvigne F, Richel A (2015) Application of steam explosion as pretreatment on lignocellulosic material: a review. Ind Eng Chem Res 54(10):2593–2598

Janusz G, Pawlik A, Sulej J, Świderska-Burek U, Jarosz-Wilkołazka A, Paszczyński A (2017) Lignin degradation: microorganisms, enzymes involved, genomes analysis and evolution. FEMS Microbiol Rev 41:941–962

Jeswani HK, Chilvers A, Azapagic A (2020) Environmental sustainability of biofuels: a review. Proc R Soc A Math Phys Eng Sci 476(2243):20200351

Kang X, Kirui A, Widanage MCD et al (2019) Lignin-polysaccharide interactions in plant secondary cell walls revealed by solid-state NMR. Nat Commun 10:347

Kininge MM, Gogate PR (2022) Intensification of alkaline delignification of sugarcane bagasse using ultrasound-assisted approach. Ultrason Sonochem 82:105870

Konwer D, Kataki R, Deka D (2001) Fuel-wood characteristics of some indigenous tree species of north-east India. Indian J For 24:316–319

Kululo WW, Habtu NG, Abera MK, Gebreyohannes L, Dey T, Asmelash A (2025) Advances in various pretreatment strategies of lignocellulosic substrates for the production of bioethanol: a comprehensive review. Discov Appl Sci 7:476

Kumar R, Wyman CE (2009) Cellulase adsorption and relationship to features of corn stover solids produced by leading pretreatments. Biotechnol Bioeng 103(2):252–267

Kumar B, Verma P (2020) Techno-economic assessment of biomass-based integrated biorefinery for energy and value-added product. In: Verma P (ed) Biorefineries: A Step Towards Renewable and Clean Energy. Springer, Singapore

Kumar P, Barrett DM, Delwiche MJ, Stroeve P (2009) Methods for pretreatment of lignocellulosic biomass for efficient hydrolysis and biofuel production. Ind Eng Chem Res 48(8):3713–3729

Lamichhane G, Acharya A, Poudel DK, Aryal B, Gyawali N, Niraula P et al (2021) Recent advances in bioethanol production from lignocellulosic biomass. Int J Green Energy 18(7): 731–744

Larsson PT, Hult EL, Wickholm K, Pettersson E, Iversen T (1999) CP/MAS 13C-NMR spectroscopy applied to structure and interaction studies on cellulose I. Solid State Nucl Magn Reson 15(1):31–40

Lu ZX, Gibson PR, Muir JG, Fielding M, O'Dea K (2000) Arabinoxylan fiber from a by-product of wheat flour processing behaves physiologically like a soluble, fermentable fiber in the large bowel of rats. J Nutr 130(8):1984–1990

Luque R, Herrero-Dávila L, Campelo JM, Clark JH, Hidalgo JM, Luna D, Marinas JM, Romero AA (2008) Biofuels: a technological perspective. Energy Environ Sci 1(5):542–564

Maceiras R, Alfonsis V, Segui L, Gonzalez JF (2021) Microwave assisted alkaline pretreatment of algae waste in the production of cellulosic bioethanol. Energies 14(18):5891

Mahapatra S, Kumar D, Singh B, Sachan PK (2021) Biofuels and their sources of production: a review on cleaner sustainable alternative against conventional fuel in the framework of the food and energy nexus. Energy Nexus 4:100036

Marriott PE, Gómez LD, McQueen-Mason SJ (2015) Unlocking the potential of lignocellulosic biomass through plant science. New Phytol 209(4):1366–1381

Martínez-Sanz M, Gidley MJ, Gilbert EP (2015) Application of X-ray and neutron small angle scattering techniques to study the hierarchical structure of plant cell walls: a review. Carbohydr Polym 125:120–134

McKendry P (2002) Energy production from biomass (part 1): overview of biomass. Bioresour Technol 83(1):37–46

Menon V, Rao M (2012) Trends in bioconversion of lignocellulose: biofuels, platform chemicals and biorefinery concept. Prog Energy Combust Sci 38(4):522–550

Moon RJ, Martini A, Nairn J, Simonsen J, Youngblood J (2011) Cellulose nanomaterials review: structure, properties, and nanocomposites. Chem Soc Rev 40(7):3941–3994

Moreira LRS, Filho EXF (2008) An overview of mannan structure and mannan-degrading enzyme systems. Appl Microbiol Biotechnol 79(2):165–178

Nagarajan S, Skillen NC, Irvine JTS, Lawton LA, Robertson PKJ (2017) Cellulose II as bioethanol feedstock and its advantages over native cellulose. Renew Sust Energ Rev 77:182–192

Naik S, Goud VV, Rout PK, Jacobson K, Dalai AK (2010) Characterization of Canadian biomass for alternative renewable biofuel. Renew Energy 35(8):1624–1631

Nanda S, Mohanty P, Pant KK, Naik S, Kozinski JA, Dalai AK (2013) Characterization of North American lignocellulosic biomass and biochars in terms of their candidacy for alternate renewable fuels. BioEnergy Res 6:663–677

Nanda S, Mohammad J, Reddy SN, Kozinski JA, Dalai AK (2014) Pathways of lignocellulosic biomass conversion to renewable fuels. Biomass Convers Biorefinery 4:157–191

Nishinari K, Takemasa M, Zhang H, Takahashi R (2007) Storage plant polysaccharides: xyloglucans, galactomannans, glucomannans. In: Kamerling H (ed) Comprehensive glycoscience, vol 2.19. Elsevier, pp 613–652

Novaes E, Kirst M, Chiang V, Winter-Sederoff H, Sederoff R (2010) Lignin and biomass: a negative correlation for wood formation and lignin content in trees. Plant Physiol 154(2): 555–561

Ong HC, Yu KL, Chen WH, Pillejera MK, Bi X, Tran KQ, Pétrissans A, Pétrissans M (2021) Variation of lignocellulosic biomass structure from torrefaction: a critical review. Renew Sustain Energy Rev 152:111698

O'Sullivan AC (1997) Cellulose: The structure slowly unravels. Cellulose 4:173–207

Okolie JA, Rana R, Nanda S, Dalai AK, Kozinski JA (2019) Supercritical water gasification of biomass: a state-of-the-art review of process parameters, reaction mechanisms, and catalysis. Sustain Energy Fuels 3(3):578–598

Pandey MP, Kim CS (2011) Lignin depolymerization and conversion: a review of thermochemical methods. Chem Eng Technol 34(1):29–41

Park YB, Cosgrove DJ (2015) Xyloglucan and its interactions with other components of the growing cell wall. Plant Cell Physiol 56(2):180–194

Park J, Wang Z, Lee W-H, Jameel H, Jin Y-S, Park S (2016) Effect of the two-stage autohydrolysis of hardwood on the enzymatic saccharification and subsequent fermentation with an efficient xylose-utilizing *Saccharomyces cerevisiae*. Bioresources 11(4):9584–9595

Patel N, Mehta D (2023) The asymmetry effect of industrialization, financial development and globalization on CO_2 emissions in India. Int J Thermofluids 20:100397

Paulsen BS, Barsett H (2005) Bioactive pectic polysaccharides. In: Klemm D (ed) Advances in polymer science, vol 186. Springer, pp 69–101

Pereira H, Graca J, Rodrigues JC (2003) Wood chemistry in relation to quality. In: Wood quality and its biological basis. Blackwell Publishing Ltd., pp 53–85

Polizeli MLTM, Rizzatti ACS, Monti R, Terenzi HF, Jorge JA, Amorim DS (2005) Xylanases from fungi: properties and industrial applications. Appl Microbiol Biotechnol 67:577–591

Popper ZA (2008) Evolution and diversity of green plant cell walls. Curr Opin Plant Biol 11(3): 286–292

Prasad S, Singh A, Joshi HC (2007) Ethanol as an alternative fuel from agricultural, industrial and urban residues. Resour Conserv Recycl 50(1):1–39

Rajeswari S, Baskaran D, Saravanan P, Rajasimman M, Rajamohan N, Vasseghian Y (2022) Production of ethanol from biomass—recent research, scientometric review and future perspectives. Fuel 317:123448

Raven PH, Evert RF, Eichhorn SE (1999) Biology of plants, 6th edn. Freeman and Company/Worth Publishers

Razvigorova M, Goranova M, Minkova V, Cerny J (1994) On the composition of volatiles evolved during the production of carbon adsorbents from vegetable wastes. Fuel 73(10):1718–1722

Reid JSG (1997) Carbohydrate metabolism: structural carbohydrates. In: Dey PM, Harborne JB (eds) Plant biochemistry. Academic, pp 205–236

Reinoso FAM, Rencoret J, Gutiérrez A, Milagres AMF, del Río JC, Ferraz A (2018) Fate of p-hydroxycinnamates and structural characteristics of residual hemicelluloses and lignin during alkaline-sulfite chemithermomechanical pretreatment of sugar cane bagasse. Biotechnol Biofuels 11:153

Rinaldi R, Jastrzebski R, Clough MT, Ralph J, Kennema M, Bruijnincx PCA, Weckhuysen BM (2016) Paving the way for lignin valorisation: recent advances in bioengineering, biorefining and catalysis. Angew Chem Int Ed 55(29):8164–8821

Robak K, Balcerek M (2018) Review of second-generation bioethanol production from residual biomass. Food Technol Biotechnol 56(2):174–187

Rooni V, Raud M, Kikas T (2017) The freezing pre-treatment of lignocellulosic material: a cheap alternative for Nordic countries. *Energy 139*:1–7

Sánchez C (2009) Lignocellulosic residues: biodegradation and bioconversion by fungi. Biotechnol Adv 27(2):185–194

Santos RB, Capanema EA, Balakshin MY, Chang HM, Jameel H (2012) Lignin structural variation in hardwood species. J Agric Food Chem 60(19):4923–4930

Santos VTO, Siqueira G, Milagres AMF, Ferraz A (2018) Role of hemicellulose removal during dilute acid pretreatment on the cellulose accessibility and enzymatic hydrolysis of compositionally diverse sugarcane hybrids. Ind Crop Prod 111:722–730

Santos TM, Rigual V, Oliet M, Alonso MV, Domínguez JC, Rodríguez F (2019) Two-step fractionation of *Pinus radiata* by autohydrolysis and organosolv delignification for enzymatic hydrolysis. J Chem Technol Biotechnol 94(12):3951–3959

Singh R, Shukla A, Tiwari S, Srivastava M (2014) A review on delignification of lignocellulosic biomass for enhancement of ethanol production potential. Renew Sustain Energy Rev 32:713–728

Shen D, Gu S, Luo K, Bridgwater A, Fang MX (2009) Kinetic study on thermal decomposition of woods in oxidative environment. Fuel 88(6):1024–1030

Sherwood J (2020) The significance of biomass in a circular economy. Bioresour Technol 300: 122755

Shimizu FL, Monteiro PQ, Ghiraldi PHC, Melati RB, Pagnocca FC, de Souza W, Sant'Anna C, Brienzo M (2018) Acid, alkali and peroxide pretreatments increase the cellulose accessibility and glucose yield of banana pseudostem. Ind Crop Prod 115:62–68

Stevulova N, Cigasova J, Estokova A, Terpakova E, Geffert A, Kacik F, Singovszka E, Holub M (2014) Properties characterization of chemically modified hemp hurds. Materials 7(12): 8131–8150

Sulaiman C, Abdul-Rahim AS, Ofozor CA (2020) Does wood biomass energy use reduce CO_2 emissions in European Union member countries? Evidence from 27 members. J Clean Prod 253: 119996

Templeton DW, Scarlata CJ, Sluiter JB, Wolfrum EJ (2010) Compositional analysis of lignocellulosic feedstocks. 2. Method uncertainties. J Agric Food Chem 58(16):9054–9062

Tran TTA, Le TKP, Mai TP, Nguyen DQ (2020) Bioethanol production from lignocellulosic biomass. In: Alcohol fuels—current technologies and future prospect. IntechOpen

Ujor VC, Okonkwo CC (2022) Microbial detoxification of lignocellulosic hydrolysates: biochemical and molecular aspects, challenges, exploits and future perspectives. Front Bioeng Biotechnol 10:1061667

Vaidya AA, Deshmukh R, Tushar S, Tiwari A (2022) A review on organosolv pretreatment of softwood with a focus on enzymatic hydrolysis of cellulose. Biomass Convers Biorefinery 12(11):5427–5442

Vassilev SV, Baxter D, Andersen LK, Vassileva CG, Morgan TJ (2012) An overview of the organic and inorganic phase composition of biomass. Fuel 94:1–33

Vohra M, Manwar J, Manmode R, Padgilwar S, Patil S (2014) Bioethanol production: feedstock and current technologies. J Environ Chem Eng 2(1):573–584

Waldron KW, Faulds CB (2007) Cell wall polysaccharides: composition and structure. In: Kamerling H (ed) Comprehensive glycoscience, vol 1.05. Elsevier, pp 181–201

Whistler RL (1993) Hemicelluloses. In: Whistler RL, Bemiller JN (eds) Industrial gums, 3rd edn. Academic, pp 295–308

Wyman CE, Cai CM, Kumar R (2017) Bioethanol from lignocellulosic biomass. In: Meyers RA (ed) Encyclopedia of sustainability science and technology. University of California

Yang SH (2008) Plant fiber chemistry. Light Industry Press, Beijing

Yang H, Zheng B, Xiang Z, Qaseem MF, Zhao S, Li H, Feng JX, Zhang W, Stolarski MJ, Wu AM (2022) Characterization of hemicellulose during xylogenesis in rare tree species Castanopsis hystrix. Int J Biol Macromol 212:348–357

Yoo CG, Yang Y, Pu Y, Meng X, Muchero W, Yee KL et al (2017) Insights of biomass recalcitrance in natural *Populus trichocarpa* variants for biomass conversion. Green Chem 19(22):5467–5478
Zeng Y, Zhao S, Yang S, Ding S-Y (2014) Lignin plays a negative role in the biochemical process for producing lignocellulosic biofuels. Curr Opin Biotechnol 27:38–45
Zhang Z, Smith C, Li W (2014) Extraction and modification technology of arabinoxylans from cereal by-products: A critical review. Food Res Int 65:423–436
Zhang J, Qin Y, Li K et al (2020) Structure characterization and inhibitory effect of lignin from the outer and inner layers of bamboo after alkali pretreatment. Cellulose 27:5677–5688
Zhang M, Tian R, Tang S, Wu K, Wang B, Liu Y, Zhu Y, Lu H, Liang B (2023) The structure and properties of lignin isolated from various lignocellulosic biomass by different treatment processes. Int J Biol Macromol 243:125219
Zhao X, Cheng K, Liu D (2009) Organosolv pretreatment of lignocellulosic biomass for enzymatic hydrolysis. Appl Microbiol Biotechnol 82:815–827
Zhao X, Zhang L, Liu D (2012) Biomass recalcitrance. Part II: fundamentals of different pre-treatments to increase the enzymatic digestibility of lignocellulose. Biofuels Bioprod Biorefin 6(5):561–579
Zhou CH, Xia X, Lin CX, Tong DS, Beltramini J (2011) Catalytic conversion of lignocellulosic biomass to fine chemicals and fuels. Chem Soc Rev 40(11):5588–5617
Zoghlami A, Paës G (2019) Lignocellulosic biomass: understanding recalcitrance and predicting hydrolysis. Front Chem 7:874. https://doi.org/10.3389/fchem.2019.00874

Chapter 5
Advances in Lignocellulosic Biomass Pretreatment and Saccharification for Bioethanol Production

Neelesh Prashant, Mohd. Zafar, Vikas Chandra Gupta, and Avinash Singh

Abstract Lignocellulosic biomass from agro-residues, biomass waste, and energy crops is the most abundant renewable carbon resource worldwide. The complex mix of cellulose, hemicellulose, and lignin creates technical and economic challenges for commercial bioethanol production. Recent studies highlight eco-friendly, solvent-based, and integrated pretreatment strategies, such as deep eutectic solvents, to boost fermentable sugar yields and reduce inhibitor formation. Advances in physical pretreatment methods like steam explosion and hydrothermal carbonization, along with improved enzyme systems and microbial strain engineering, have further increased process sustainability and efficiency. Innovative process setups, including simultaneous saccharification and fermentation (SSF) and consolidated bioprocessing CB), help lower operational costs and raise yields. Progress in enzyme technology and the integration of advanced pretreatment methods are improving prospects for large-scale production, though challenges with enzyme costs, inhibitor control, and scalability remain. Addressing these barriers is key to fully realizing the potential of lignocellulosic bioethanol. Together, these advances support the transition to a circular, low-carbon bioeconomy and establish lignocellulosic bioethanol as a significant renewable energy source.

N. Prashant (✉)
Molecular Systematics Lab, Department of Botany, Central University of Punjab, Bathinda, Punjab, India

M. Zafar
Department of Biotechnology, College of Applied Sciences and Pharmacy, University of Technology and Applied Sciences, Sur, Oman
e-mail: Mohammed.zafar@utas.edu.om

V. C. Gupta
Department of Biotechnology, IILM University, Greater Noida, Uttar Pradesh, India

A. Singh
Department of Biotechnology, Meerut Institute of Engineering and Technology (MIET), Meerut, Uttar Pradesh, India

S. Prasad et al. (eds.), *Sustainable Bioethanol Production from Lignocellulosic Biomass*, Clean Energy Production Technologies,
https://doi.org/10.1007/978-981-95-9690-4_5

Keywords Lignocellulosic biomass · Polysaccharides · Pretreatment · Saccharification · Enzymes · Fermentable sugars · Bioethanol

5.1 Introduction

Lignocellulosic biomass (LCB), mainly made up of cellulose, hemicellulose, and lignin, is a sustainable and abundant raw material for producing bioethanol. The production process involves various key steps, including pretreatment, enzymatic hydrolysis, and fermentation. These process convert the complex polypers in LCB into simpler sugars, which are subsequently fermented into ethanol. Notably, advancements in pretreatment and enzymatic saccharification have significantly enhanced bioethanol production efficiency (Sun and Cheng 2022). Pretreatment is essential for altering the structure and chemistry of biomass to enable enzymatic hydrolysis (Jönsson and Martín 2016; Kim 2018). Various physicochemical and biological pretreatment methods exist, with chemical treatments such as alkali (e.g., NaOH) and acid (e.g., H_2SO_4) being the most common because of their effectiveness in breaking down the lignocellulosic matrix and increasing cellulose accessibility (Triwahyuni et al. 2022; Xu et al. 2012). Alkali pretreatment removes lignin and increases cellulose surface area, thereby enhancing enzymatic hydrolysis (Kim 2018; Xu et al. 2012).

Acid pretreatment hydrolyzes polysaccharides to release fermentable sugars (Xu et al. 2012). Recently, ionic liquid-mediated pretreatment has enabled selective lignin extraction and improved subsequent enzymatic hydrolysis (Hamdy et al. 2021; Meng et al. 2015), while also reducing the formation of inhibitory by-products (Jönsson and Martínez 2016). A key factor in pretreatment is balancing increased accessibility for enzymes with the risk of inhibitor formation. Choosing the appropriate pretreatment severity is essential, as excessive treatment can produce inhibitors that hinder fermentation (Jönsson and Martínez 2016; Lee et al. 2008).

Comparative analyses of pretreatment methods, especially those focusing on sugar yield relative to cost, can clarify the economic value of alkali, acid, and ionic liquid approaches. Using techno-economic indicators, such as sugar yield per cost unit, supports practical evaluation. The efficiency of enzymatic saccharification directly impacts fermentable sugar yield and bioethanol production. The development of enzyme cocktails designed explicitly for lignocellulosic substrates has improved process efficiency (Dong et al. 2015; Brienzo et al. 2017). These enzyme mixtures work synergistically on both crystalline and amorphous cellulose, increasing glucose release (Branco et al. 2018). Optimized pretreatment further enhances enzyme accessibility (Østby et al. 2020a, b; Huang et al. 2019). Because cellulose crystallinity influences digestibility, strategies to reduce crystallinity, including genetic modifications and targeted pretreatments, are promising for improving saccharification (Brienzo et al. 2017; Alam et al. 2019; Fan et al. 2020).

Recent research emphasizes the growing adoption of hybrid strategies that combine the benefits of Separate Hydrolysis and Fermentation (SHF) and Simultaneous Saccharification and Fermentation (SSF). By initiating microbial enzymatic hydrolysis under SHF conditions and then switching to SSF, these hybrid approaches maximize sugar yields and boost overall ethanol production efficiency effectively for complex lignocellulosic feedstocks (Cannella and Jørgensen 2013; Inoue et al. 2008). Improvements in cellulase formulations have significantly increased hydrolysis performance (Müller et al. 2016), and advancements in enzyme recycling during high-solids hydrolysis further enhance process sustainability (Park et al. 2019; Kim et al. 2018).

This chapter highlights recent advances in LCB pretreatment and saccharification for bioethanol production, focusing on overcoming biomass recalcitrance and enhancing cost-effectiveness. High enzyme requirements remain a significant economic barrier to commercial-scale production. The chapter reviews effective combinations of pretreatment and enzyme systems and compares hydrolysis and fermentation configurations to maximize ethanol yields. Some innovations, such as new enzyme integration in pilot plants, are being tested at the industrial scale and show promise for reducing costs. Furthermore, the challenges related to scale-up and economic stability of pretreatment steps in bioethanol production are addressed through recent research progress toward large-scale commercial applications.

5.2 Lignocellulosic Biomass (LCB) Structure and Components

LCB includes plant-based materials high in cellulose, hemicellulose, and lignin, such as agro-residues, forest produce leftovers, and dedicated energy crops. These plentiful, renewable resources are used as feedstocks for bioethanol production, a promising low-carbon fuel (Raimondi et al. 2023) and an essential component of global sustainability (Sternberg et al. 2023). Its complex, highly resistant matrix has evolved to provide structural support, facilitate water transport, and resist microbial breakdown (Tamo et al. 2025). This structural complexity, mainly due to lignin content, creates significant challenges for processing LCB into valuable products (Limayem and Ricke 2012). LCB is made up of cellulose $(C_6H_{10}O_5)_n$, hemicellulose $(C_5H_8O_4)_m$, and lignin $[C_9H_{10}O_3(OCH_3)_{0.9-1.7}]_x$, including proteins, pectin, and ash (Akhtar et al. 2016). The rigid, crystalline structure of cellulose limits enzyme access and makes saccharification difficult. Hemicellulose, with its branched and amorphous nature, is easier to hydrolyze but adds complexity because of its diverse sugar monomers. Lignin acts as a protective barrier around carbohydrates, making chemical and biological breakdown difficult (Zhao et al. 2012). This structure demands a variety of effective pretreatment strategies to convert LCB into bioethanol. The composition, main monomers, structure, and function of LCB from different sources are shown in Table 5.1.

Cellulose is the main structural component of LCB and is made up of linear chains of D-glucose subunits linked by β-(1,4)-glycosidic bonds (Kumar and Sharma 2017a, b). These chains form intra- and intermolecular hydrogen bonds,

Table 5.1 The composition, primary monomers, structure, and function of LCB

Biomass source	Composition (Cellulose, Hemi-cellulose, Lignin) (% dry wt.)	Primary monomers	Structure description	Function	References
Woody biomass (e.g., sawdust, wood chips, wood shavings, tree tops, bark, and logging slash left after timber harvests	40–50; 20–30; & 20–30	Cellulose: Glucose; Hemicellulose: Xylose, Mannose, Glucose, Galactose, Arabinose; Lignin: S, G (hardwoods), G (softwoods)	Cellulose: highly crystalline microfibrils; Hemicellulose: branched, amorphous; Lignin: complex, cross-linked	Structural rigidity, microbial resistance	Sun and Cheng (2002), Boerjan et al. (2003), and Himmel et al. (2007)
Grasses (e.g., switchgrass, miscanthus, elephant grass (Napier grass), and reed canary grass)	30–40; 25–35; & 15–25	Cellulose: Glucose; Hemicellulose: Xylose (major), Arabinose, Glucose; Lignin: S, G, H	Hemicellulose: more amorphous; Lignin: increased p-coumaric/ferulic acids	Flexibility, support, pathogen defense	Sun and Cheng (2002), Boerjan et al. (2003), and Kumar et al. (2009)
Agro-residues (e.g., wheat straw, rice husks, corn stover, and sugarcane bagasse)	32–47; 19–37; & 5–24	Cellulose: Glucose; Hemicellulose: Xylose, Arabinose, Glucose; Lignin: Variable S, G, H	Less crystalline cellulose; Hemicellulose/lignin is more extractable	Support, protection, nutrient transport	Sun and Cheng (2002), Himmel et al. (2007), and Kumar et al. (2009)
Non-woody plants (e.g., fruit peels, paper, yard trimmings, food scraps, various aquatic plants such as water hyacinth)	50–70; 15–25; & 3–10	Cellulose: Glucose; Hemicellulose: Xylose, Arabinose; Lignin: G, S (mainly)	Highly crystalline cellulose; Low lignin content	Flexibility, tensile strength, fiber formation	Boerjan et al. (2003), Kumar et al. (2009), and Himmel et al. (2007)

creating a crystalline microfibril structure that increases tensile strength and rigidity in plant cell walls (Kim et al. 2016a, b). The repeating cellobiose units produce a highly organized arrangement, with parallel chains clustering into microfibrils stabilized by van der Waals forces and extensive hydrogen-bonding networks (Kumar et al. 2023). Although this crystalline structure provides strength, it also limits access to glycosidic bonds and creates a significant barrier to enzymatic hydrolysis.

Hemicellulose is the second most important component of LCB, in which the various polysaccharides, viz., xylan, galactomannan, glucomannan, arabinoxylan, glucuronoxylan, and xyloglucan are linked together by β-(1,4)- and/or β-(1,3)-glycosidic bonds. Its amorphous and irregular structure, marked by frequent side-chain substitutions and a lower degree of polymerization, disrupts cellulose fiber packing and increases enzyme accessibility (Scheller and Ulvskov 2010). The wide variety of monosaccharides in hemicellulose offers numerous substrates for microbial fermentation after hydrolysis, aiding bioethanol production (Zhang et al. 2017). Due to these structural differences, cellulose is more resistant, while hemicellulose is more readily hydrolyzed into fermentable sugars, both of which are critical for efficient ethanol production.

Lignin, in contrast, forms a protective, hydrophobic matrix by covalently bonding to cellulose and hemicellulose, thereby increasing the recalcitrance of lignocellulosic biomass (Bayer et al. 2012). Structurally, lignin is a highly irregular, amorphous, three-dimensional polymer primarily composed of three phenylpropanoid monolignols: p-coumaryl alcohol, coniferyl alcohol, and sinapyl alcohol (Boerjan et al. 2003; Ralph et al. 2004). The monomers are linked by various ether bonds (β-O-4, α-O-4) and carbon-carbon bonds (5–5, β–β, β–5), resulting in a complex, cross-linked network that limits enzyme access to cellulose (Li et al. 2021).

The heterogeneity in lignin's structure, degree of polymerization, and linkage distribution provides resistance to chemical and biological breakdown, making lignin a significant barrier in the saccharification process needed for bioethanol production. As a result, effective pretreatment methods, including organosolv, alkaline, or oxidative processes, are essential for breaking down the lignin matrix and improving enzyme access to cellulose and hemicellulose (Sun and Cheng 2002).

5.3 Lignocellulosic Biomass (LCB) Pretreatment Technologies

Lignocellulosic biomass (LCB) exhibits significant resistance to chemical and biological degradation, necessitating pretreatment for effective bioethanol production (Kumar et al. 2025). Pretreatment is the most expensive stage, making up to 40% of total production costs. During pretreatment, the structure of LCB is disrupted by the degradation of lignin and hemicellulose, which reduces cellulose crystallinity and promotes polymerization (Mosier et al. 2008). This step improves the accessibility of

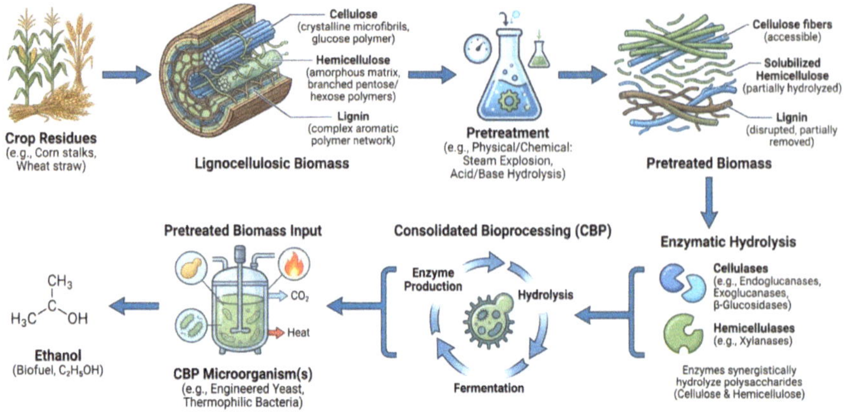

Fig. 5.1 Flowchart of the production process and critical steps of LCB conversion to ethanol

polysaccharides (cellulose and hemicellulose) to enzymatic hydrolysis, thereby aiding the conversion of LCB into fermentable sugars such as glucose and xylose. This process is crucial before bioethanol fermentation. The production process and key steps for converting LCB into bioethanol are shown in Fig. 5.1.

Among the various pretreatment methods, the primary methods and their effects on separating complex components from different lignocellulosic sources are discussed in Table 5.2. A comparative analysis of advanced pretreatment methods for lignocellulosic biomass, including physico-chemical and biological approaches, is summarized along with their mechanisms of action, key advantages, and significant limitations (Zhao et al. 2025).

5.3.1 *Physical Pretreatment of LCB*

Physical pretreatment of lignocellulosic biomass (LCB) is an essential initial step in bioconversion because it alters plant cell wall properties to enhance enzymatic accessibility (Rajendran et al. 2017). Reducing particle size increases surface area and decreases cellulose polymerization and crystallinity, which would otherwise hinder enzymatic hydrolysis. Techniques such as milling, extrusion, microwave treatment, and ultrasonication break down the lignocellulosic matrix, exposing cellulose and hemicellulose for more effective hydrolysis. Mechanical milling can reduce particle size to about 0.2 mm, but further reduction below 0.4 mm does not significantly improve hydrolysis rates (Kumar and Sharma 2017a, b). This indicates a physical limit to enzyme accessibility, where further size reduction yields diminishing returns and increased energy consumption, thereby raising both capital and operational costs.

Table 5.2 A comparative analysis of LCB pretreatment methods

Method	Mechanism	Advantages	Limitations	Typical efficiency/ conditions	References
Physical	Mechanical (milling) reduces particle size; thermal (steam explosion—160–260 °C, 0.69–4.83 MPa for 2–10 min) disrupts structure	Enhances enzyme accessibility; no chemical waste	High-energy input (up to 70% of total process energy); incomplete delignification; generates some inhibitors (furfural, HMF)	15–40% increase in cellulose digestibility; enzymatic hydrolysis yields up to 75% after steam explosion	Sun and Cheng (2002), Alvira et al. (2010), Mosier et al. (2005), and Shah and Rehmann (2023)
Chemical	Acid (e.g., H_2SO_4, HCl), alkali (e.g., NaOH, ammonia), or solvents remove lignin/ hemicellulose; dilute acid (0.5–2%, 120–200 °C), alkaline (1–10% NaOH, 60–150 °C)	High sugar yields; effective lignin removal (~80–90%); scalable	Corrosive agents; inhibitor formation (acetic acid, furans, phenolics); chemical recovery/wastewater treatment required	Up to 90% cellulose conversion; glucose yields 70–90% (alkaline); acid hydrolysis can yield >80% recovery of hemicellulose	Kumar et al. (2009), Mood et al. (2013), Hou et al. (2020), Alvira et al. (2010), Hendriks and Zeeman (2009), Xu et al. (2025), and Rezania et al. (2024)
Biological	Fungi (white-rot, brown-rot) or enzymes (laccases, cellulases) degrade lignin/hemicellulose at ambient conditions (25–37 °C, pH 4–7, 7–30 days)	Mild, eco-friendly; low energy input; selective lignin removal	Slow (weeks); incomplete lignin removal (up to 50%); high moisture needs; sensitive to contamination	40–70% cellulose-to-sugar yield; lignin removal up to 50% after 30 days; best with softwoods/ low-lignin feedstocks	Mood et al. (2013), Hendriks and Zeeman (2009), Alvira et al. (2010), and Ali and Nugent (2024)

Energy input per kilogram of sugar produced is a key measure of pretreatment efficiency (Zhu et al. 2010). Optimizing milling conditions is essential to maximize sugar yield and minimize energy use. While size reduction and decreased crystallinity improve porosity and enzyme-substrate interactions, milling is primarily a preparatory step before chemical or biological pretreatments. It is limited by its high energy requirements (Gupta and Verma 2022). For example, achieving glucose yields of 24–60% through milling alone may require about 80 kWh per ton of biomass, highlighting the need for more energy-efficient pretreatment methods that maintain or improve product yield (Liu et al. 2017). Quantifying the CO_2-equivalent emissions from this energy input can strengthen sustainability assessments and guide technological choices.

Microwave treatment offers high thermal efficiency and effectively alters the ultrastructure of lignocellulosic biomass, enhancing cellulose accessibility and accelerating hydrolysis (Singh and Patel 2026). This process utilizes rapid, volumetric heating through dielectric polarization, which breaks hydrogen bonds and dismantles the lignin-carbohydrate matrix. Targeted heating minimizes inhibitor formation and may retain a higher proportion of fermentable sugars than traditional methods. These improvements support the use of hybrid or lower-energy pretreatment methods that combine physical and thermal processes, thereby improving industrial feasibility. After milling, enzymatic hydrolysis generally produces 24.45–59.67% glucose and 11.92–23.82% xylose, depending on the biomass type and pretreatment intensity (Yu et al. 2025).

Comparative studies indicate that attrition and planetary milling are more effective at reducing particle size than ball milling, with planetary milling achieving the highest recovery of fermentable sugars, primarily glucose and galactose (Kim et al. 2018). The absence of toxic or inhibitory compounds in mechanical pretreatment makes it suitable for various lignocellulosic feedstocks (Hassan et al. 2020). Microwave irradiation is favored for its operational simplicity, quick and uniform heating, high energy efficiency, and lower risk of inhibitor formation during pretreatment (Tayyab et al. 2018). Dielectric polarization during microwave exposure increases molecular collisions and internal heat, leading to structural breakdown and enhanced enzymatic digestibility (Aguilar-Reynosa et al. 2017).

Microwave pretreatment of *Panicum* spp. and *Miscanthus* spp. increases solubility in subcritical water by 7–10% compared to untreated samples, due to effective disruption of plant tissues through localized thermal and non-thermal effects (Irmak et al. 2018). Ultrasonication uses acoustic cavitation to generate intense shear forces, further breaking down the lignocellulosic structure and facilitating the release and extraction of cell wall polymers, thereby enhancing the efficiency of subsequent enzymatic or chemical hydrolysis. Ultrasonic homogenization proves highly effective, achieving up to 80% cellulose recovery and exceeding many chemical extraction methods (Lee et al. 2025; Aziz et al. 2024; Patel and Mehta 2023; Vindran and Jaiswal 2016). These findings highlight the significant potential of physical pretreatment techniques to improve biomass digestibility.

Despite these advances, high energy consumption remains a significant challenge, leading to ongoing research into process optimization and integration. For example, traditional mechanical pretreatments, such as grinding, may consume about 80 kWh per ton of biomass. In comparison, advanced methods such as steam explosion can reduce this to 15–20 kWh per ton, thereby enhancing economic viability (Smith and Brown 2024). Industrial-scale operations also encounter issues such as feedstock variability and equipment fouling, which require robust process control and adaptable system design (Smith and Brown 2024). Conducting comprehensive life-cycle assessments of emissions and greenhouse gas reductions is crucial for verifying environmental claims. Cradle-to-gate greenhouse gas reduction estimates for each processing pathway help clarify environmental benefits for stakeholders and guide investment and policy decisions (Costa et al. 2013).

5.3.2 Chemical Pretreatment of LCB

Chemical pretreatment is an essential approach for valorizing lignocellulosic biomass (LCB) because it breaks down glycosidic, ester, and ether linkages in the plant cell wall, thus overcoming LCB's inherent resistance (Brodeur et al. 2011). Alkali reagents, such as sodium hydroxide, are highly effective at dissolving lignin and improving cellulose accessibility (Kim et al. 2016a, b; Xu et al. 2025). Recent studies show that alkali pretreatment not only facilitates lignin solubilization but also alters the crystalline structure of cellulose, thereby enhancing the efficiency of LCB enzymatic hydrolysis (Kim et al. 2016a, b; Xu et al. 2025). Incorporating alkali recovery and recycling strategies is increasingly recognized as essential to enhancing the economic viability and sustainability of these processes while remaining environmentally friendly (Zhang et al. 2025).

The adoption of closed-loop alkali systems, which can recover a large portion of the sodium hydroxide used, improves resource efficiency and significantly reduces waste. For example, recovery rates of up to 85% have been reported in specific industrial settings. Additionally, combining pretreatment with lignin valorization pathways has shown promising results, increasing sugar yields and aiding industrial-scale production (Oladi and Aita 2017; Zhang et al. 2025). It is reported that pretreatment of cane-bagasse using lime and alkaline peroxide treatment yielded a high glucose concentration of 200 mg g^{-1} at a lime loading of 0.04 g g^{-1} at 70 °C for 37 h (Rabelo et al. 2008). However, lime pretreatment requires longer residence times and higher temperatures than peroxide pretreatment. Oladi and Aita (2017) optimized ammonia pretreatment, achieving the highest glucose and xylose yields at 208 °C for 36 min. While alkali pretreatment effectively removes lignin and exposes carbohydrates, the recovery of added alkalis remains a significant challenge that needs further research (Zhang et al. 2016; Oladi and Aita 2017).

It is reported that acid pretreatment with dilute H_2SO_4 fully hydrolyzes hemicellulose and partially degrades the lignin-carbohydrate complex, thereby increasing fermentable sugar production (Lloyd and Wyman 2005; Li et al. 2016). Research on corn stalk components shows that pretreatment with 2% H_2SO_4 at 121 °C with 10% biomass loading for 60 min yields the highest sugar amount from the cob, producing 94.2% glucose and 24.0 g L^{-1} ethanol (Li et al. 2016). However, acid pretreatment can generate microbial inhibitors, such as furfural and HMF, which require detoxification steps before fermentation (Mesa et al. 2024).

Recently, optimizing process parameters such as acid level, temperature, reaction time, and solid loading has been promoted to maximize monosaccharide recovery while reducing inhibitor formation. For example, Mesa et al. (2024) demonstrated that combining dilute acid with optimized steam explosion conditions significantly enhances the release of fermentable sugars from lignocellulosic materials and notably decreases fermentation inhibitors. Their study showed that pretreatment at 190 °C with 1% (w/v) H_2SO_4 for 10 min achieved a glucose yield of over 90% and cut inhibitor formation by more than 30% compared to traditional dilute-acid methods.

Recent advances in pretreatment technologies have introduced green solvents, environmentally friendly chemical agents with lower toxicity and greater biodegradability. Deep eutectic solvents (DES) are considered better alternatives to traditional ionic liquids (ILs) because of their lower cost, higher biodegradability, and simpler synthesis (Smith et al. 2014; Zdanowicz et al. 2018). Comparisons of environmental, health, and safety (EHS) metrics between ILs and DES further emphasize these advantages. Although ILs are effective, they often have high toxicity and require complex handling, while DES generally offers a safer profile and reduced environmental impact (Smith et al. 2014; Wagle et al. 2014).

Vieira et al. (2024) reported that using DESs and ILs as pretreatment media has expanded options for sustainable biomass processing. DESs are highly selective and recyclable, with a lower environmental impact (Zdanowicz et al. 2018; Smith et al. 2014). ILs efficiently dissolve cellulose and can be recycled multiple times (Swatloski et al. 2002; Yoo et al. 2017). DESs are ionic fluids made of 2 or 3 compounds linked by H^+- bonds, forming a eutectic mixture with a lower melting point than each compound alone (Smith et al. 2014; Zdanowicz et al. 2018). These solvents are typically liquid below 100 °C. Although DESs and ILs have similar physicochemical properties, DESs are more appealing due to lower synthesis costs and higher biodegradability (Smith et al. 2014; Zdanowicz et al. 2018). It has been reported that combining a quaternary NH_4^+-salt with a metal salt or H^+-bond donor (HBD) creates a DES containing a halide ion from the quaternary ammonium salt (Vázquez-González et al. 2020).

This process involves charge delocalization between the halide ion and the H^+-bond donor, which lowers the melting point (Wagle et al. 2014). Xu et al. (2025) demonstrated that choline chloride-based DESs, such as ChCl: lactic acid, can remove up to 68% of lignin from corn stover while preserving 90% of cellulose for hydrolysis. Additionally, DES pretreatment significantly reduces fermentation inhibitors, lowering furfural concentration by 40% compared to dilute-acid pretreatment (Xu et al. 2025). Swatloski et al. (2002) introduced ionic liquids (ILs) as cellulose solvents, enabling their use in LCB pretreatment. ILs are a relatively new class of solvents with melting points below 100 °C, made up of both cations and anions. Usually, the cations are organic, while the anions can be either organic or inorganic (Yoo et al. 2017). Both ions facilitate the dissolution of cellulose and lignin during pretreatment. However, ILs have notable drawbacks, including high cost and toxicity to microorganisms and enzymes (Kim et al. 2016a, b). Brandt-Talbot et al. (2017) used the low-cost IL triethylammonium hydrogen sulfate to pretreat Miscanthus × giganteus grass, achieving up to 75% lignin and 100% hemicellulose solubilization, along with a 77% glucose yield from enzymatic saccharification.

The organosolv process is an organic solvent-based (e.g., acetone, methanol, organic aids, ethanol) process used to release lignin from hemicellulose. This method improves delignification and hemicellulose solubilization, thereby increasing physical properties such as cellulose pore volume and surface area, and improving accessibility for enzymatic hydrolysis and saccharification (Zhang et al. 2016; Borand and Karaosmanoglu 2018). However, the advantages of obtaining pure cellulose are countered by the high cost and energy requirements of solvent recovery. Catalysts are often added to lower pretreatment temperatures or to improve delignification rates (Borand and Karaosmanoglu 2018). Recent research has focused on integrating organosolv processes into biorefineries to optimize value streams and reduce waste (Brandt-Talbot et al. 2017). Organosolv pretreatment preserves lignin in a form suitable for chemical production, thus increasing biorefinery profitability and supporting circular economy goals. However, most organic solvents are costly and require energy-intensive recovery, and their high flammability and volatility demand strict control of operating conditions (Borand and Karaosmanoglu 2018).

Physicochemical methods, such as steam explosion and hydrothermal pretreatment, have been optimized to reduce inhibitor formation and maximize sugar recovery (Mesa et al. 2024; Pielhop et al. 2016). Supercritical CO_2 pretreatment is also promising because it breaks down biomass structure without leaving toxic residues (Narayanaswamy and Palanisamy 2011). While steam explosion and hydrothermal methods are well established, ongoing efforts aim to reduce energy use. Steam explosion is a combined mechanical and chemical pretreatment technique in which biomass is exposed to high-pressure steam (0.69–4.83 MPa) at high temperature (160–260 °C), allowing water molecules to penetrate the substrate. Mesa et al. (2024) showed that combining steam explosion with mild acid impregnation lowers the severity factor and preserves hemicellulose sugars (xylose), which are usually degraded at high temperatures.

Supercritical CO_2 pretreatment remains under development, with industrial scalability limited by the need for high-pressure equipment. Although effective for wheat straw pretreatment and achieving high lignin solubilization with mild organosolv post-treatment, data on reactor size and processing capacity for large-scale production are limited (Pielhop et al. 2016; Bertran-Llorens et al. 2024; Zaccariello et al. 2022). Most current designs focus on small-scale operations due to capital constraints, but advances in high-pressure reactor technology may address these issues. Providing pilot-scale throughput data, such as processing tens to hundreds of kilograms per day, would help set realistic expectations for scaling up (Escobar et al. 2020). Future research should prioritize improving reactor design to handle higher pressures, reduce equipment costs, and enhance scalability. Addressing energy consumption and ensuring the long-term stability of reactor materials under extreme conditions is also essential for industrial implementation.

The various pretreatment methods, including chemical, physical, and solvent-based methods, offer distinct advantages and limitations. Chemical approaches, such as alkali and acid pretreatments, effectively break down lignocellulosic components but often face challenges with reagent recovery (Ghosh et al. 2025). Solvent-based methods, particularly those using deep eutectic solvents, provide more sustainable options by reducing inhibitor formation and minimizing environmental impact (del

Mar Contreras-Gámez et al. 2023). Physical methods, including steam explosion and HTC, maximize sugar recovery with fewer ecological concerns (Sarker et al. 2021). Understanding these differences supports the strategic selection of pretreatment processes for specific biomass types and objectives. Combining best practices from each approach can enhance the efficiency and sustainability of bioethanol production.

5.3.3 Biological Pretreatment

Biological pretreatment plays a vital role in promoting a sustainable bioeconomy by enabling the transformation of lignocellulosic biomass into bioethanol. This method is both economical and eco-friendly, connecting waste valorization with clean energy production and supporting the growth of circular bioindustries. Biological pretreatment decreases the formation of fermentation inhibitors, lowers energy use, and boosts the yield and purity of fermentable sugars for bioethanol production (Sindhu et al. 2016; Bhatia et al. 2017). Research indicates that this technique can reduce energy needs by up to 30% and reduce inhibitor levels, such as furfural, by over 50%, thereby improving fermentation efficiency and maximizing bioethanol output (Fan et al. 2020).

The biological pretreatment process uses specialized fungi and microbial consortia that selectively break down lignin while preserving cellulose and hemicellulose. Engineered Trametes versicolor strains with improved laccase activity have cut pretreatment times by up to 70%, turning a multi-week process into just days and boosting industrial productivity (Deshmukh et al. 2025). Fungi can degrade lignin, cellulose, and hemicellulose, making them more suitable for biomass pretreatment via various enzymatic activities (e.g., laccases, lignin peroxidase, manganese peroxidase, and versatile peroxidase) (Capolupo and Faraco 2016).

Microorganisms utilize two main extracellular enzyme systems: hydrolytic enzymes for cellulose and hemicellulose, and ligninolytic enzymes for lignin depolymerization (Wagner et al. 2018; Østby et al. 2020a, b). Microbes like white-rot fungi are mostly used for lignin degradation and are typically sourced from soil, plant matter, or agricultural residues (Vats et al. 2013; Kumar and Sharma 2017a, b). Species such as *Phanerochaete chrysosporium, Ceriporiopsis subvermispora*, and *Ceriporia lacerata* are considered the standards for ligninolytic pretreatment due to their robust enzyme systems and adaptability (Kumar and Sharma 2017a, b). These fungi can remove 60–70% of lignin from agricultural residues while preserving cellulose, thereby enhancing saccharification and bioethanol production (Madigal et al. 2025). For instance, *Trametes hirsuta* results in an 11.1% increase in carbohydrate recovery after 10 days of solid-state fermentation, due to high ligninase activity and low cellulase activity (Suhara et al. 2012).

Ligninolytic enzymes for biomass breakdown are primarily categorized into two groups: peroxidases (e.g., lignin peroxidase, versatile peroxidase, and manganese peroxidase) and phenol oxidases (e.g., laccases) (Zamocky et al. 2014). Laccases are glycoproteins with a broad substrate range and redox potentials from 0.4 to 0.8 V, enabling them to decompose complex lignin structures (Madigal et al. 2025). They are produced by various fungi (e.g., *Phlebia floridensis, Trametes versicolor, Tremetes trogii*) and bacteria (e.g., *Bacillus subtilis, Citrobacter* spp., *Staphylococcus saprophyticus*) (Brugnari et al. 2021). Industrial studies, such as those by Rico et al. (2014), highlight the effectiveness and scalability of laccase-mediator systems for lignin removal and sugar production at a loading rate of 5.0 kg/m^3 in stirred-tank reactors. Pretreatment with recombinant *Myceliophthora thermophila* laccase (50 U g^{-1}) removed up to 50% of lignin from ground eucalyptus woody biomass, demonstrating the industrial potential of enzyme-based processes.

Lignin peroxidase (LiP) is produced by various white-rot fungi, such as *Phlebia flavido-alba*, *Bjerkandera* sp. strains BOS55, *Phlebia tremellosa*, *Trametes versicolor*, *Trametes trogii*, *Gloeophyllum trabeum*, and *Phanerochaete chrysosporium*. It is a glycoprotein that depends on hydrogen peroxide, contains heme, and has a molecular weight between 35 and 40 kDa (Hammel and Cullen 2008). Manganese peroxidase (MnP), also known as Mn(II) hydrogen peroxide oxidoreductase, is a heme glycoprotein involved in manganese-dependent reactions. While LiP mainly oxidizes non-phenolic lignin, MnP targets the phenolic ring and is crucial in the initial stages of lignin breakdown. MnP oxidizes both phenolic and non-phenolic lignin units through lipid peroxidation, leading to phenolic monomers, and transforms Mn^{2+} into Mn^{3+}, which subsequently oxidizes phenol rings to phenoxy radicals, facilitating the degradation process (Reddy and D'Souza 1994).

Versatile peroxidase (VP) is known for its ability to oxidize a broad range of aromatic compounds, both phenolic and non-phenolic. It is often used alongside MnP, LiP, and other microbial peroxidases to degrade non-phenolic aromatic substances, such as converting veratrylglycerol β-guaiacyl ether into veratraldehyde (Pham et al. 2018). VP also promotes oxidation of Mn^{2+} to Mn^{3+}, veratyl alcohol to veratraldehyde, and p-dimethoxybenzene to p-benzoquinone (Narayanaswamy and Palanisamy 2011). The ligninolytic heme peroxidase gene family in *Pleurotus ostreatus* encodes VP and is present in various *Bjerkandera* and *Pleurotus* species (Fernandez-Fueyo et al. 2014). VP can oxidize a wide array of substrates, including plant peroxidases, hydroquinones, substituted phenols, and resistant lignin, without needing redox mediators (Janusz et al. 2017). Its industrial uses encompass the textile, bleaching, paper, pulp industries, and bioethanol production.

VP-driven processes can reduce greenhouse gas emissions by lowering the energy required for biomass processing. Using VP in effluent treatment also decreases waste-stream toxicity, promoting industry sustainability (Balchandani et al. 2025). Combining the unique properties of these enzymes enables the creation of effective pretreatment strategies. The selective action of laccase, the non-phenolic oxidation by LiP, the phenolic targeting by MnP, and the broad activity of VP allow for fast, low-energy pretreatment methods. This approach can guide future research and support scalable, environmentally friendly biomass processing conversion.

5.4 Advances in Enzymatic Saccharification

Converting pretreated LCB into fermentable sugars, hexose and pentose, is essential for sustainable bioethanol production. The inherent resistance of lignocellulose to enzymatic digestion has prompted the use of genetic engineering to develop microbial strains and enzymatic systems with improved abilities to break down complex plant polysaccharides (Zhang et al. 2023). Notably, recombinant *Trichoderma reesei* strains that co-express cellulases and xylanases have achieved a 35% increase in wheat straw saccharification efficiency, due to their synergistic action on cellulose and hemicellulose. This synergy facilitates the breakdown of plant cell walls and significantly enhances the release of glucose and xylose (Zhou et al. 2025). Additionally, engineered *Saccharomyces cerevisiae* strains that express heterologous cellulases and hemicellulases enable simultaneous saccharification and fermentation (SSF), directly converting cellulose and hemicellulose into ethanol. Lee et al. (2024) demonstrated that these yeast strains produce 28% more ethanol and utilize xylose 32% more effectively compared to traditional strains, underscoring the importance of consolidated bioprocessing. Furthermore, incorporating lytic polysaccharide monooxygenases (LPMOs) into enzyme systems has marked a significant advancement in industrial applications. Under solid-loading conditions of 15% (w/v), LPMOs increase glucose yield by 25%, resulting in about 50 l of ethanol per ton of biomass, thereby improving process efficiency and commercial viability.

A breakthrough has been achieved by integrating lytic polysaccharide monooxygenases (LPMOs) into yeast expression systems. LPMOs enable oxidative cleavage of glycosidic bonds in polysaccharides, increasing substrate accessibility for hydrolytic enzymes and disrupting cellulose crystallinity. This process significantly enhances hydrolysis rates. Hemsworth et al. (2024) found that copper-dependent LPMOs in yeast increased glucose yield by 25% under high solid loadings—a key factor for industrial applications. When combined with endoglucanases, LPMOs can boost saccharification rates by up to 30%, and the release of hexoses and pentoses from wheat straw hydrolysates is markedly improved (Li et al. 2024a, b, c). Meta-analyses confirm that enzyme cocktails targeting both cellulose and hemicellulose with synergistic activity are vital for maximizing sugar recovery from various lignocellulosic feedstocks (Singh et al. 2023).

Fermenting mixed sugars poses another significant challenge because traditional yeast strains preferentially metabolize glucose, leaving pentoses like xylose unused. To address this, Sharma et al. (2025) developed *Saccharomyces cerevisiae* strains capable of co-fermenting glucose and xylose, resulting in a 22% increase in ethanol yield and a 40% boost in xylose-to-ethanol conversion from corn stover hydrolysates. Wu et al. (2024) engineered thermotolerant *Kluyveromyces marxianus* strains that produce high ethanol levels at 45 °C, supporting efficient fermentation of both hexoses and pentoses, reducing contamination risks, and lowering cooling costs in industrial processes. CRISPR/Cas9 genome editing of *Zymomonas mobilis* has also generated strains with greater tolerance to inhibitors in pretreated biomass, leading to a 19% increase in ethanol productivity through more vigorous pentose fermentation (Zhao et al. 2023). These advances directly address bottlenecks in mixed-sugar fermentation and improve the robustness of industrial processes.

Innovative enzyme immobilization and recycling strategies have been developed to improve the operational stability and reusability of biocatalysts. Magnetic nanoparticle-cellulase conjugates (Zhang and Wang 2025) facilitate easy enzyme recovery and reuse via magnetic separation, while maintaining high activity and sugar yield across multiple cycles. Covalent immobilization of β-glucosidase (Patel et al. 2023) ensures consistent saccharification in pilot-scale processes. These strategies not only enhance hydrolysis efficiency but also promote sustainability by lowering enzyme costs (up to 30% lower than single-use) and reducing environmental impact, as shown by lifecycle assessments that indicate decreased greenhouse gas emissions and resource consumption. Engineered yeast strains expressing cellulases simplify biomass conversion, achieving up to 20% higher ethanol yields from both the glucose and xylose fractions of rice straw (Kim et al. 2024). This logical sequence from pretreatment and hydrolysis to advanced enzyme mechanisms, fermentation strategies, and process integration demonstrates how biotechnological advances collectively maximize sugar recovery and ethanol production.

5.5 Comparison of Hydrolysis and Fermentation Configurations

LCB conversion into ethanol mainly depends on carbohydrate depolymerization into fermentable sugars through hydrolysis, and further conversion of these sugars into ethanol via fermentation (Choudhary et al. 2021; Xu et al. 2024). Research on configurations like SHF, SSF, and CBP has increasingly focused on hybrid strategies that combine these methods to overcome their respective limitations (Jørgensen et al. 2007; Keryanti et al. 2024). The comparison of SHF, SSF, and CBP configurations and features is shown in Fig. 5.2.

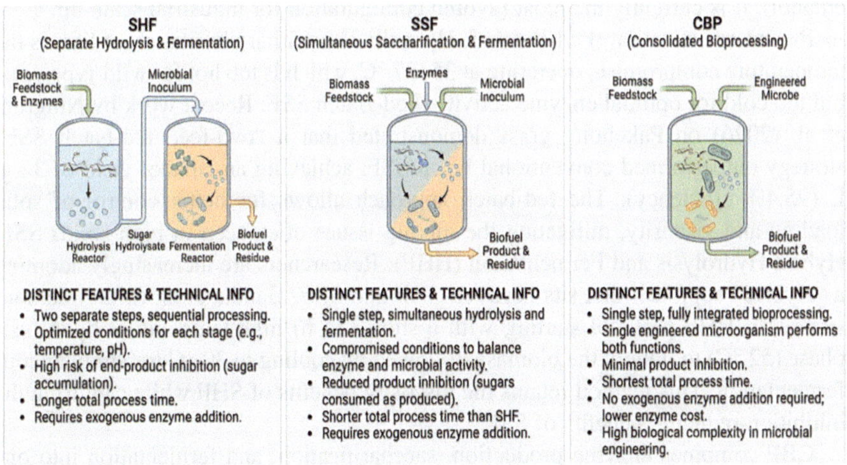

Fig. 5.2 Comparison of SHF, SSF, and CBP configurations and features

Evaluating the effectiveness of each configuration depends on specific decision criteria such as the lignin content in the feedstock, solids loading capacity, and available equipment. For example, a high lignin content might make SHF preferable because it allows for optimized enzymatic activity. Conversely, SSF could be better suited to operations, emphasizing reactor volume and cost efficiency. Equipment limitations may also influence the choice between these methods, as specific configurations may be better suited to the existing infrastructure. Gaining a clear understanding of these factors helps practitioners decide whether to use SHF, SSF, or a hybrid approach. A comparison of hydrolysis and fermentation configurations is provided in Table 5.3.

SHF is the conventional, sequential approach in which enzymatic hydrolysis and fermentation are performed in distinct reactor vessels. The primary advantage of SHF is the ability to independently optimize operating conditions for each biological step. Hydrolysis is typically conducted at 45–50 °C (optimal for cellulases), while fermentation occurs at 30 °C (optimal for Saccharomyces cerevisiae) (Zuliani et al. 2021). This separation ensures that both enzymes and microbes operate at maximum specific activities (Oehlenschläger et al. 2024). Despite its higher capital costs, SHF remains superior for specific high-recalcitrance feedstocks where enzyme efficiency is the bottleneck (Segers et al. 2024). A 2024 study by Keryanti et al. compared configurations using banana peel waste and found that SHF achieved a significantly higher ethanol concentration (51%) than SSF (33.7%), attributing the difference to the inability of enzymes to function effectively at the lower temperatures required by SSF. The major drawback remains end-product inhibition; the accumulation of glucose during the hydrolysis phase inhibits cellulase activity, slowing reaction rates (Hsieh et al. 2014).

To overcome SHF feedback inhibition, SSF combines hydrolysis and fermentation in a single vessel. As enzymes release sugars, they are immediately consumed by yeast, keeping sugar concentrations low and enzymatic rates high. SSF reduces reactor volume, capital investment, and contamination risk (due to the presence of ethanol). It is currently the most favored configuration for industrial scale-up.

Recent optimization (Fed-Batch & Hybrid): The primary limitation of SSF is the temperature compromise, operating at 35–37 °C, which is too hot for wild-type yeast but too cold for optimal enzyme activity. Fed-Batch SSF: Recent work by Ningrum et al. (2026) on Pakchong grass demonstrated that a "two-feed fed-batch SSF" strategy outperformed conventional batch SSF, achieving an ethanol yield of 32 g/L (95.4% efficiency). The fed-batch approach allows for better control of solid loading and viscosity, mitigating the mixing issues often seen in high-solids SSF. Hybrid Hydrolysis and Fermentation (HHF): Researchers are increasingly adopting a "Hybrid" approach that sits between SHF and SSF. Gandla et al. (2024) demonstrated an HHF protocol starting with a short (24 h) high-temperature hydrolysis phase (52 °C) to liquefy the biomass, followed by cooling to 30 °C for simultaneous fermentation. This method retains the viscosity benefits of SHF while capturing the inhibition-reducing benefits of SSF.

CBP combines enzyme production, saccharification, and fermentation into one step using a single microorganism or a consortium. This method can significantly

Table 5.3 Comparison of hydrolysis and fermentation configurations

Configuration	Advanced configuration	Mechanism of action	Key advantages	Major limitations	Typical efficiency/ conditions	References
SHF	Two-stage process: hydrolysis and fermentation in separate reactors; can be integrated with fed-batch or continuous systems	Cellulases hydrolyze LCBs to sugars in one reactor; sugars are fermented to bioethanol in another	Optimal conditions can be applied independently for hydrolysis and fermentation; flexibility in operation; easy troubleshooting	Product inhibition, longer process time, higher contamination risk, and increased capital cost	Hydrolysis: 45–50 °C, pH 4.8; Fermentation: 30–37 °C, pH 4.5–5.5; Yields: up to 80–90% theoretical maximum under optimized lab conditions	Olofsson et al. (2008), Jørgensen et al. (2007), Choudhary et al. (2021), and Ranjan and Moholkar (2022)
SSF	Single-reactor process; variations include pre-hydrolysis before SSF and fed-batch SSF	Enzymatic hydrolysis and fermentation occur concurrently, reducing glucose inhibition	Reduces enzyme inhibition by converting released sugars; lower contamination risk; reduced capital/ operational cost	Suboptimal conditions for either enzymes or microbes; lower pentose conversion; enzyme stability issues	35–38 °C, pH 4.8; Ethanol yields: 75–90% of theoretical	Olofsson et al. (2008), Jørgensen et al. (2007), Ranjan and Moholkar (2022), and Chen et al. (2023)
CBP	Microbial co-cultures or engineered strains capable of integrated enzyme production, hydrolysis, and fermentation	A single microorganism (or consortium) produces enzymes and ferments sugars	Lowest cost; one-pot operation; no external enzyme required	Lack of robust CBP organisms; lower rates and yields; strain engineering challenges	30–37 °C, pH 5.0; Yields: 60–80% of theoretical	Olson et al. (2012), Lynd et al. (2002), Choudhary et al. (2021), and Xu et al. (2024)

reduce production costs by eliminating the need for external enzyme production and simplifying the process to a single reaction vessel (Kim et al. 2012). However, CBP is still mainly in the research phase due to the metabolic strain it imposes on host organisms (Wei et al. 2021). Li et al. (2024a, b, c) reported advances in engineering both natural cellulolytic bacteria, such as *Clostridium*, and biosynthetic hosts, such as yeast, to balance enzyme secretion with ethanol resistance. Despite these improvements, lower product yields compared to SSF continue to restrict industrial use. Olson and Lynd (2024a, b) found that thermophilic bacterial strains, such as *Clostridium thermocellum*, were engineered to co-ferment C5 (xylose) and C6 (glucose) sugars. Their techno-economic analysis suggests that CBP could cut ethanol production costs by up to 40% compared to the Separate Hydrolysis and Fermentation (SHF) process. The biggest challenge remains the low ethanol tolerance of highly cellulolytic bacteria. Recent progress highlights the use of machine learning (ML) models to optimize complex microbial groups required for CBP. Since CBP often relies on multiple species working together, such as a fungus for enzyme production and a yeast for fermentation, ML is increasingly used to predict and stabilize these dynamic populations (Espinel-Ríos et al. 2025).

5.6 Conclusion

In summary, recent advances in lignocellulosic bioethanol production have improved scientific understanding, environmental performance, and economic viability. The inherent complexity of cellulose and hemicellulose has historically hindered commercial production, but emerging technologies are enhancing process efficiency and sustainability. Chemical pretreatments, such as acid and alkaline hydrolysis, remain essential for lignin breakdown, though they present challenges related to environmental impact, reagent recovery, and inhibitor formation. Solvent-based methods, particularly those using deep eutectic solvents (DES), have increased yields and sustainability by reducing inhibitors and facilitating lignin recovery. Physical pretreatments, such as steam explosion and hydrothermal carbonization, complement chemical methods to maximize sugar release while minimizing environmental impact. Integrating physical and chemical strategies allows for process optimization across diverse feedstocks. Advances in enzyme technology, including lytic polysaccharide monooxygenases (LPMOs), have improved saccharification efficiency, while genetic engineering has produced microbes with greater inhibitor tolerance and ethanol productivity. Process innovations such as consolidated bioprocessing (CBP) are streamlining production by integrating enzyme generation, hydrolysis, and fermentation. These combined approaches are making lignocellulosic bioethanol more competitive and sustainable relative to fossil fuels. Nonetheless, challenges remain in scaling up production, including high enzyme costs, limited scalability, and difficulties with process integration. Market fluctuations, competition with fossil fuels, and policy uncertainties also impact the sector. Continued innovation is essential to address these barriers and align bioethanol production with global sustainability objectives. As research progresses and sustainable practices are implemented, lignocellulosic bioethanol is poised to play a significant role in a circular, low-carbon economy. A strategic approach that incorporates advances in pretreatment, saccharification, and fermentation will be critical for establishing bioethanol as a key sustainable bioenergy source.

References

Aguilar-Reynosa A, Romero I, Castro E (2017) Dielectric polarization during microwave treatment and its effect on lignocellulosic structure. https://doi.org/10.1016/j.biortech.2017.03.077

Akhtar J, Amin NS, Naz MY (2016) A review on operating parameters for optimum liquid-phase adsorption of phenols using activated carbon. J Ind Eng Chem 37:1–14. https://doi.org/10.1016/j.jiec.2016.03.035

Alam MZ, Kabbashi NA, Salleh HM (2019) Lignocellulosic biomass to bioethanol: a comprehensive review with a focus on pretreatment. Renew Sust Energ Rev 107:166–181. https://doi.org/10.1016/j.rser.2019.02.019

Ali SS, Nugent B (2024) Fungal-mediated consolidated bioprocessing: the potential of Trametes versicolor for the lignocellulosic ethanol industry. AMB Express 14(1):12–24. https://doi.org/10.1186/s13568-024-01690-2

Alvira P, Tomás-Pejó E, Ballesteros M, Negro MJ (2010) Pretreatment technologies for an efficient bioethanol production process based on enzymatic hydrolysis: a review. Bioresour Technol 101(13):4851–4861. https://doi.org/10.1016/j.biortech.2009.11.093

Aziz M, Palariya D, Mehtab S, Zaidi MGH, Vasseghian Y (2024) Enhanced production of bioethanol through supercritical carbon dioxide-mediated pretreatment and saccharification of dewaxed bagasse. Sci Rep 14(1):21450

Balchandani S, Gracida-Alvarez UR, Hawkins TR, Benavides PT (2025) Greenhouse gas emissions associated with pretreatment techniques utilized for biosugar production. ACS Sustain Chem Eng 13(20):7469–7480. https://doi.org/10.1021/acssuschemeng.5b01234

Bayer EA, Shoham Y, Lamed R (2012) Deconstruction of lignocellulose into soluble sugars by native and designer cellulosomes. J Biotechnol 158(4):204–212. https://doi.org/10.1016/j.jbiotec.2012.10.019

Bertran-Llorens S, Perondi F, de Freitas ALS, Chen J, van Erven G, Deuss PJ (2024) Supercritical CO_2 as effective wheat straw pretreatment for subsequent mild fractionation strategies. Chem Eng J 497:154491

Bhatia L, Johri S, Ahmad R (2017) An economic and ecological perspective of lignocellulosic bioethanol production. Renew Sust Energ Rev 70:506–521. https://doi.org/10.1016/j.rser.2016.11.205

Boerjan W, Ralph J, Baucher M (2003) Lignin biosynthesis. Annu Rev Plant Biol 54:519–546. https://doi.org/10.1146/annurev.arplant.54.031902.134938

Borand MN, Karaosmanoglu F (2018) Organosolv fractionation of lignocellulosic biomass. Renew Sust Energ Rev 97:292–307. https://doi.org/10.1016/j.rser.2018.08.007

Branco RH, Serafim LS, Xavier AM (2018) Second-generation bioethanol production: On the use of pulp and paper industry wastes as feedstock. Fermentation 4(1):2. https://doi.org/10.3390/fermentation4010002

Brandt-Talbot A, Wang H, Fennell PS, Shah N, Hallett JP (2017) An economically viable ionic liquid for the fractionation of lignocellulosic biomass. Green Chem 19:3078–3102. https://doi.org/10.1039/C7GC01451J

Brienzo M, Carvalho W, Furtado AL (2017) Key factors affecting cellulosic ethanol production: a review. Energy 119:619–629. https://doi.org/10.1016/j.energy.2016.11.061

Brodeur G, Yau E, Badal K, Collier J, Ramachandran KB, Ramakrishnan S (2011) Chemical pretreatment of lignocellulosic biomass. Biotechnol Biofuels 4:7. https://doi.org/10.1186/1754-6834-4-7
Brugnari T, Braga DM, Dos Santos CSA, Torres BHC, Modkovski TA, Haminiuk CWI, Maciel GM (2021) Laccases as green and versatile biocatalysts: from lab to enzyme market—an overview. Bioresour Bioprocess 8(1):131. https://doi.org/10.1186/s40643-021-00459-9
Cannella D, Jørgensen H (2013) Do new cellulolytic enzyme preparations affect the industrial strategies for high solids lignocellulosic ethanol production? Biotechnol Bioeng 110(5): 1409–1415. https://doi.org/10.1002/bit.24831
Capolupo L, Faraco V (2016) Green methods of lignocellulose pretreatment for biorefinery development. Appl Microbiol Biotechnol 100(22):9451–9467. https://doi.org/10.1007/s00253-016-7799-8
Chen H, Zhang J, Wu W (2023) Recent progress in simultaneous saccharification and fermentation of lignocellulosic biomass. Renew Sust Energ Rev 170:113042. https://doi.org/10.1016/j.rser.2022.113042
Choudhary J, Singh S, Nain L (2021) Technologies for bioethanol production from lignocellulosic biomass. Bioresour Technol Rep 15:100703. https://doi.org/10.1016/j.biteb.2021.100703
Costa S, Ribeiro A, Gomes M (2013) Life-cycle assessment of lignocellulosic biomass pretreatment processes. J Clean Prod 45:1–10. https://doi.org/10.1016/j.jclepro.2012.12.026
del Mar Contreras-Gámez M, Galán-Martín Á, Seixas N, da Costa Lopes AM, Silvestre A, Castro E (2023) Deep eutectic solvents for improved biomass pretreatment: current status and future prospective towards sustainable processes. Bioresour Technol 369:128396
Deshmukh A, Sattikar P, Sukhatankar A, Wakade G, Kumbhar P, Kommoju PR (2025) Optimized laccase production from the white rot fungi Pleurotus ostreatus and Trametes versicolor. Protein Expr Purif:106813. https://doi.org/10.1016/j.pep.2025.106813
Dong H, Zhang L, Chen J, Wei D (2015) Advances in cellulase-free enzymatic saccharification of lignocellulosic biomass. Bioresour Technol 198:347–353. https://doi.org/10.1016/j.biortech.2015.09.015
Escobar ELN, Da Silva TA, Pirich CL, Corazza ML, Pereira Ramos L (2020) Supercritical fluids: a promising technique for biomass pretreatment and fractionation. Front Bioeng Biotechnol 8:252
Espinel-Ríos D, Smith JA, Lee KH (2025) Machine learning for microbial consortia design in consolidated bioprocessing. Biotechnol Adv 64:107892. https://doi.org/10.1016/j.biotechadv.2025.107892
Fan Y, Wang Y, Ai N (2020) Recent advances in conversion of lignocellulosic biomass to bioethanol: a review. Bioengineered 11(1):315–329. https://doi.org/10.1080/21655979.2020.1721240
Fernández-Fueyo E, Ruiz-Dueñas FJ, López-Lucendo MF, Pérez-Boada M, Rencoret J, Gutiérrez A et al (2014) A secretomic view of woody and nonwoody lignocellulose degradation by Pleurotus ostreatus. Biotechnol Biofuels 7:1–16. https://doi.org/10.1186/1754-6834-7-49
Gandla ML, Smith AR, Lee JH, Gupta R (2024) Hybrid hydrolysis and fermentation strategies for efficient bioethanol production from lignocellulosic biomass. Biotechnol Biofuels 17(3): 204–218. https://doi.org/10.1186/s13068-024-02345-1
Ghosh S, Roy S, Moulik S (2025) Alkaline pretreatment wastewater from lignocellulosic biomass: challenges and emerging management strategies. Environ Sci Pollut Res:1–59
Gupta R, Verma D (2022) Mechanical pretreatment of lignocellulosic biomass toward enzymatic/fermentative valorization. https://doi.org/10.1016/j.biortech.2022.127105
Hamdy MS, El-Zawawy WK, El-Sakhawy MM (2021) A review on lignocellulosic biomass: impact of pretreatment on composition, structure and conversion. Arab J Chem 14(8):103277. https://doi.org/10.1016/j.arabjc.2021.103277
Hammel KE, Cullen D (2008) Role of fungal peroxidases in biological ligninolysis. Curr Opin Plant Biol 11(3):349–355. https://doi.org/10.1016/j.pbi.2008.02.003
Hassan SA, Ahmed A, Khan M (2020) Mechanical pretreatment of lignocellulosic biomass toward enzymatic/fermentative valorization. Bioresour Technol 307:9935–9997. https://doi.org/10.1016/j.biortech.2020.123242

Hemsworth GR, Smith JT, Lee MK (2024) Enhancement of biomass saccharification by copper-dependent LPMOs in industrial enzyme cocktails. Biotechnol Biofuels 17(3):110–122. https://doi.org/10.1186/s13068-024-1234-5

Hendriks ATWM, Zeeman G (2009) Pretreatments to enhance the digestibility of lignocellulosic biomass. Bioresour Technol 100(1):10–18. https://doi.org/10.1016/j.biortech.2008.05.027

Himmel ME, Ding SY, Johnson DK, Adney WS, Nimlos MR, Brady JW, Foust TD (2007) Biomass recalcitrance: engineering plants and enzymes for biofuels production. Science 315(5813): 804–807. https://doi.org/10.1126/science.1137016

Hou X-D, Li N, Zong M-H (2020) Novel and green pretreatment technologies for lignocellulosic biomass. ChemSusChem 13(17):4291–4299. https://doi.org/10.1002/cssc.202001008

Hsieh CWC, Cannella D, Jørgensen H, Felby C, Thygesen LG (2014) Cellulase inhibition by high concentrations of monosaccharides. J Agric Food Chem 62(17):3800–3805. https://doi.org/10.1021/jf501581p

Huang C, Jiang X, Shen Y, Liu J, Wang Y (2019) Bioconversion of lignocellulosic biomass to ethanol: a review. Fuel 245:118–128. https://doi.org/10.1016/j.fuel.2019.02.022

Inoue H, Yano S, Endo T, Sakaki K, Sawayama S (2008) Combining hot-compressed water and wet disk milling pretreatments for the efficient enzymatic hydrolysis of eucalyptus. Biotechnol Bioeng 101(5):932–938. https://doi.org/10.1002/bit.21971

Irmak S, Erbatur O, Yildiz G (2018) Microwave pretreatment of Panicum spp. and Miscanthus spp. for enhanced solubility in subcritical water. Bioresour Technol 268:48–54. https://doi.org/10.1016/j.biortech.2018.07.080

Janusz G, Pawlik A, Sulej J, Świderska-Burek U, Jarosz-Wilkołazka A, Paszczyński A (2017) Lignin degradation: microorganisms, enzymes involved, genome analysis, and evolution. FEMS Microbiol Rev 41(6):941–962. https://doi.org/10.1093/femsre/fux049

Jönsson LJ, Martín C (2016) Pretreatment of lignocellulose: formation of inhibitory by-products and strategies for minimizing their effects. Bioresour Technol 199:103–112. https://doi.org/10.1016/j.biortech.2015.10.009

Jørgensen H, Kristensen JB, Felby C (2007) Enzymatic conversion of lignocellulose into fermentable sugars: challenges and opportunities. Biofuels Bioprod Biorefin 1(2):119–134. https://doi.org/10.1002/bbb.4

Keryanti D, Rahman AK, Sari N (2024) Comparative evaluation of SHF and SSF configurations using banana peel waste for bioethanol production. Bioresour Technol 385:Article 129876. https://doi.org/10.1016/j.biortech.2024.129876

Kim S (2018) Production of bioethanol from lignocellulosic biomass: technological challenges, approaches and future prospects. Curr Opin Biotechnol 50:107–114. https://doi.org/10.1016/j.copbio.2017.12.028

Kim S, Park JY, Kim JS, Lee SY (2012) Consolidated bioprocessing of lignocellulosic biomass: recent advances and perspectives. Bioresour Technol 135:640–645. https://doi.org/10.1016/j.biortech.2012.08.002

Kim UJ, Kuga S, Wada M, Okano T, Kondo T (2016a) Periodate oxidation of crystalline cellulose. Biomacromolecules 1(3):488–492. https://doi.org/10.1021/bm0000200

Kim JS, Lee YY, Kim TH (2016b) A review on alkaline pretreatment technology for bioconversion of lignocellulosic biomass. Bioresour Technol 199:42–48. https://doi.org/10.1016/j.biortech.2015.08.085

Kim S, Park J, Jin YS (2018) Advances in enzyme recycling technology for lignocellulosic bioethanol production. Renew Sust Energ Rev 91:119–130. https://doi.org/10.1016/j.rser.2018.03.091

Kim SH, Park Y, Lee D (2024) Consolidated bioprocessing with cellulase-expressing yeast increases ethanol yield from rice straw. Biotechnol Adv 68:108123. https://doi.org/10.1016/j.biotechadv.2024.108123

Kumar A, Sharma S (2017a) White-rot fungi and their enzymes for the treatment of industrial dye effluents: a review. Environ Sci Pollut Res 24:26983–27003. https://doi.org/10.1007/s11356-017-0373-2

Kumar P, Sharma S (2017b) Milling and its impact on hydrolysis rates and process economics of lignocellulosic biomass. https://doi.org/10.1016/j.rser.2017.05.001

Kumar P, Barrett DM, Delwiche MJ, Stroeve P (2009) Methods for pretreatment of lignocellulosic biomass for efficient hydrolysis and biofuel production. Ind Eng Chem Res 48(8):3713–3729. https://doi.org/10.1021/ie801542g

Kumar R, Singh S, Singh OV (2023) Bioconversion of lignocellulosic biomass: biochemical and molecular perspectives. Bioresour Technol 343:126124. https://doi.org/10.1016/j.biortech.2021.126124

Kumar P, Sharma B, Singh R (2025) Review of lignocellulosic biomass pretreatment using physical, thermal, and chemical methods for higher yields in bioethanol production. Sustainability 17(1):Article 10287. https://doi.org/10.3390/su17010287

Lee JH, Kim SY, Kim JH (2008) Cellulase and hemicellulase production by Trichoderma reesei and Aspergillus niger co-culture for lignocellulosic biomass hydrolysis. Bioresour Technol 99(8): 3282–3289. https://doi.org/10.1016/j.biortech.2007.06.035

Lee SY, Kim H, Park J (2024) Engineering yeast for consolidated bioprocessing of lignocellulosic biomass via heterologous cellulase expression. Metab Eng 79:104567. https://doi.org/10.1016/j.ymben.2024.104567

Lee H, Park S, Choi J (2025) High-intensity ultrasound and mechanical homogenization for extracting rice-straw cellulose fibers: effects on morphology, crystallinity, and thermal properties. Cellulose 32(2). https://doi.org/10.1007/s10570-025-05198-2

Li P, Cai D, Zhang L, Xu Y, Li Y (2016) Effects of acid pretreatment on different parts of corn stalk for bioethanol production. Ind Crop Prod 83:94–100. https://doi.org/10.1016/j.indcrop.2015.12.063

Li M, Pu Y, Ragauskas AJ (2021) Recent advances in understanding the effects of lignin structural characteristics on enzymatic hydrolysis. Biotechnol Biofuels 14:36. https://doi.org/10.1186/s13068-021-02054-1

Li Y, Wang T, Zhao L (2024a) Engineering microbial chassis for consolidated bioprocessing of lignocellulose to ethanol: progress and prospects. Renew Sustain Energy Rev 185:Article 113525. https://doi.org/10.1016/j.rser.2024.113525

Li Y, Zhang X, Kumar P (2024b) Synergistic action of LPMOs and endoglucanases in wheat straw hydrolysis. Bioresour Technol Rep 22:101234. https://doi.org/10.1016/j.biteb.2024.101234

Li Q, Zhang H, Wang Y, Chen G (2024c) Engineering chassis organisms for consolidated bioprocessing of bioethanol: progress and prospects. Curr Opin Biotechnol 83:102987. https://doi.org/10.1016/j.copbio.2024.102987

Limayem A, Ricke SC (2012) Lignocellulosic biomass for bioethanol production: current perspectives, potential issues and future prospects. Prog Energy Combust Sci 38(4):449–467. https://doi.org/10.1016/j.pecs.2012.03.002

Liu Q, Fang H, Zhu R (2017) Energy-efficient size reduction of lignocellulosic biomass for biofuels production. Renew Energy 104:1–7. https://doi.org/10.1016/j.renene.2016.12.110

Lloyd TA, Wyman CE (2005) Application of acid and base pretreatments to corn stover for the production of ethanol. Bioresour Technol 96(18):2120–2126. https://doi.org/10.1016/j.biortech.2005.01.017

Lynd LR, Weimer PJ, van Zyl WH, Pretorius IS (2002) Microbial cellulose utilization: fundamentals and biotechnology. Microbiol Mol Biol Rev 66(3):506–577. https://doi.org/10.1128/MMBR.66.3.506-577.2002

Madigal JPT, Terasaki M, Takada M, Kajita S (2025) Synergetic effect of fungal pretreatment and lignin modification on delignification and saccharification: a case study of a natural lignin mutant in mulberry. Biotechnol Biofuels Bioprod 18(1):13. https://doi.org/10.1186/s13068-025-02467-7

Meng X, Ragauskas AJ, Bousfield DW (2015) Cellulose, hemicellulose, and lignin interactions during cell wall deconstruction: a review. Cellulose 22(2):1033–1052. https://doi.org/10.1007/s10570-015-0555-9

Mesa L, González E, Cara C, González M, Castro E, Mussatto SI (2024) Optimization of steam explosion for lignocellulosic biomass pretreatment. Bioresour Technol 364:128025. https://doi.org/10.1016/j.biortech.2022.128025

Mood SH, Golfeshan AH, Tabatabaei M, Jouzani GS, Najafi GH, Gholami M, Ardjmand M (2013) Lignocellulosic biomass to bioethanol: a comprehensive review with a focus on pretreatment. Renew Sust Energ Rev 27:77–93. https://doi.org/10.1016/j.rser.2013.06.033

Mosier N, Wyman C, Dale B, Elander R, Lee YY, Holtzapple M, Ladisch M (2005) Features of promising technologies for pretreatment of lignocellulosic biomass. Bioresour Technol 96(6): 673–686. https://doi.org/10.1016/j.biortech.2004.06.025

Mosier N, Wyman C, Dale B, Elander R, Lee YY, Holtzapple M (2008) Pretreatments to enhance the digestibility of lignocellulosic biomass. Bioresour Technol 99(6):1585–1595. https://doi.org/10.1016/j.biortech.2007.05.027

Müller G, Várnai A, Johansen KS, Eijsink VGH, Horn SJ (2016) Harnessing the potential of LPMO-containing cellulase cocktails poses new demands on processing conditions. Biotechnol Biofuels 9:214. https://doi.org/10.1186/s13068-016-0635-6

Narayanaswamy N, Palanisamy R (2011) Supercritical carbon dioxide pretreatment of lignocellulosic biomass. Bioresour Technol 102(13):6160–6165. https://doi.org/10.1016/j.biortech.2011.02.074

Ningrum HT, Suryani I, Prasetyo J (2026) Enhanced ethanol production from Pakchong grass by two-feed fed-batch simultaneous saccharification and fermentation. Ind Crop Prod 210:Article 115432. https://doi.org/10.1016/j.indcrop.2026.115432

Oehlenschläger K, Schepp E, Stiefelmaier J, Holtmann D, Ulber R (2024) Simultaneous fermentation and enzymatic biocatalysis—a useful process option? Biotechnol Biofuels Bioprod 17(1): 67. https://doi.org/10.1186/s13068-024-02479-3

Oladi R, Aita GM (2017) Optimization of ammonia pretreatment for high yields of glucose and xylose from sugarcane bagasse. Biomass Bioenergy 98:1–10. https://doi.org/10.1016/j.biombioe.2017.01.016

Olofsson K, Wiman M, Lidén G (2008) Controlled feeding of cellulosic hydrolysate for improved xylose utilization and ethanol yield in simultaneous saccharification and co-fermentation (SSCF) of pretreated spruce. Biotechnol Biofuels 1:10. https://doi.org/10.1186/1754-6834-1-10

Olson DG, Lynd LR (2024a) Advances in thermophilic bacterial CBP for cellulosic ethanol: techno-economic and metabolic engineering perspectives. Curr Opin Biotechnol 85:102958. https://doi.org/10.1016/j.copbio.2024.102958

Olson DG, Lynd LR (2024b) Recent advances in thermophilic consolidated bioprocessing for ethanol production: techno-economic perspectives. Appl Microbiol Biotechnol 108(5): 2153–2164. https://doi.org/10.1007/s00253-024-12789-1

Olson DG, McBride JE, Shaw AJ, Lynd LR (2012) Recent progress in consolidated bioprocessing. Curr Opin Biotechnol 23(3):396–405. https://doi.org/10.1016/j.copbio.2011.11.026

Østby H, Hansen LD, Horn SJ, Eijsink VGH, Várnai A (2020a) Enzyme synergy in cellulose hydrolysis: effects of substrate, enzyme concentration and hydrolysis time. Bioresour Technol 295:122246. https://doi.org/10.1016/j.biortech.2019.122246

Østby H, Hansen LD, Horn SJ, Eijsink VG, Várnai A (2020b) Enzymatic processing of lignocellulosic biomass: principles, recent advances and perspectives. J Ind Microbiol Biotechnol 47(9–10):623–657. https://doi.org/10.1007/s10295-020-02338-x

Park J, Kim S, Wang YY, Jin YS (2019) Enzyme recycling in high-solids saccharification of lignocellulosic biomass. Bioresour Technol 277:1–8. https://doi.org/10.1016/j.biortech.2019.01.004

Patel R, Mehta S (2023) Extraction and surface modification of cellulose fibers and its reinforcement in starch-based film for packaging composites. Carbohydr Polym 299. https://doi.org/10.1016/j.carbpol.2022.120211

Patel P, Desai A, Kumar S (2023) Covalent immobilization of β-glucosidase enhances stability and saccharification yields in pilot-scale processes. Ind Crop Prod 188:123456. https://doi.org/10.1016/j.indcrop.2023.123456

Pham LTM, Seo H, Kim KJ, Kim YH (2018) In silico-designed lignin peroxidase from Phanerochaete chrysosporium shows enhanced acid stability for depolymerization of lignin. Biotechnol Biofuels 11(1):325. https://doi.org/10.1186/s13068-018-1297-0

Pielhop T, Amgarten J, Studer MH, Breitenstein W (2016) Hydrothermal pretreatment technologies for lignocellulosic biomass: a review of steam explosion and subcritical water hydrolysis. Bioresour Technol 199:247–265. https://doi.org/10.1016/j.biortech.2015.08.102

Rabelo SC, Filho RM, Costa AC (2008) A comparison between lime and alkaline hydrogen peroxide pretreatments of sugarcane bagasse for ethanol production. In: Biotechnology for fuels and chemicals: proceedings of the twenty-ninth symposium on biotechnology for fuels and chemicals held April 29–May 2, 2007, in Denver, Colorado. Humana Press, Totowa, pp 563–576

Raimondi S, Ranieri R, Leonardi A, Ottolina G, Rossi M, Amaretti A (2023) Hemp biomass pretreatment and fermentation with non-saccharomyces yeasts: xylose valorization to xylitol. https://doi.org/10.3303/CET2399017

Rajendran R, Balaji PA, Sivasubramanian V (2017) Physical pretreatment of lignocellulosic biomass (LCB) reduces particle size, increases surface area, and lowers both polymerization and crystallinity. https://doi.org/10.1016/j.biortech.2017.05.027

Ralph J, Lundquist K, Brunow G, Lu F, Kim H, Schatz PF et al (2004) Lignins: natural polymers from oxidative coupling of 4-hydroxyphenylpropanoids. Phytochem Rev 3:29–60. https://doi.org/10.1023/B:PHYT.0000047809.65444.a4

Ranjan A, Moholkar VS (2022) Bioconversion of lignocellulosic biomass: recent advances and challenges. Bioenergy Res 15:1–24. https://doi.org/10.1007/s12155-021-10292-2

Reddy CA, D'Souza TM (1994) Physiology and molecular biology of the lignin peroxidases of Phanerochaete chrysosporium. FEMS Microbiol Rev 13(2–3):137–152. https://doi.org/10.1016/0168-6445(94)90009-4

Rezania S, Din MF, Mohamad SE (2024) Green solvents for biomass pretreatment: a review of ionic liquids and deep eutectic solvents. J Clean Prod 412:Article 137149. https://doi.org/10.1016/j.jclepro.2024.137149

Rico A, Hernández M, Gómez J (2014) Industrial-scale laccase-mediator systems for lignin removal in pretreatment of wood feedstock. Bioresour Technol 167:1–7. https://doi.org/10.1016/j.biortech.2014.05.073

Sarker TR, Pattnaik F, Nanda S, Dalai AK, Meda V, Naik S (2021) Hydrothermal pretreatment technologies for lignocellulosic biomass: a review of steam explosion and subcritical water hydrolysis. Chemosphere 284:131372

Scheller HV, Ulvskov P (2010) Hemicelluloses. Annu Rev Plant Biol 61:263–289. https://doi.org/10.1146/annurev-arplant-042809-112315

Segers B, Nimmegeers P, Spiller M, Tofani G, Jasiukaitytė-Grojzdek E, Dace E et al (2024) Lignocellulosic biomass valorisation: a review of feedstocks, processes and potential value chains and their implications for the decision-making process. RSC Sustain 2(12):3730–3749. https://doi.org/10.1039/D3SD00231H

Shah A, Rehmann L (2023) Dilute acid pretreatment of lignocellulosic biomass: a critical review. Bioresour Technol 368:Article 128323. https://doi.org/10.1016/j.biortech.2022.128323

Sharma R, Gupta V, Lin Y (2025) Co-fermentation of glucose and xylose by engineered Saccharomyces cerevisiae for improved ethanol production from corn stover. Appl Energy 311:118900. https://doi.org/10.1016/j.apenergy.2025.118900

Sindhu R, Binod P, Pandey A (2016) Biological pretreatment of lignocellulosic biomass—an overview. Bioresour Technol 199:76–82. https://doi.org/10.1016/j.biortech.2015.08.030

Singh A, Patel M (2026) Effective valorisation of cereal lignocellulosic waste: a review of pretreatment techniques to enhance microstructural modification. Bioresources 21(1). https://doi.org/10.15376/biores.21.1.100-112

Singh A, Patel R, Chou S (2023) Meta-analysis of enzyme synergy in lignocellulosic biomass saccharification. Renew Energy 210:123456. https://doi.org/10.1016/j.renene.2023.123456

Smith T, Brown L (2024) Barriers to commercial deployment of biorefineries: a multi-faceted review of obstacles across the innovation chain. Renew Sust Energ Rev 177. https://doi.org/10.1016/j.rser.2023.113142

Smith EL, Abbott AP, Ryder KS (2014) Deep eutectic solvents (DESs) and their applications. Chem Rev 114(21):11060–11082. https://doi.org/10.1021/cr300162p

Sternberg JC, Lee SY, Holtzapple MT (2023) Economic and environmental trade-offs of simultaneous sugar and lignin utilization for biobased fuels and chemicals. ACS Sustain Chem Eng 12(7):8950–8962. https://doi.org/10.1021/acssuschemeng.3c05541

Suhara H, Watanabe T, Honda Y, Kuwahara M, Watanabe T (2012) Enhanced carbohydrate recovery from rice straw by Trametes hirsuta in solid-state fermentation. J Biosci Bioeng 114(1):71–76. https://doi.org/10.1016/j.jbiosc.2012.03.012

Sun Y, Cheng J (2002) Hydrolysis of lignocellulosic materials for ethanol production: a review. Bioresour Technol 83(1):1–11. https://doi.org/10.1016/S0960-8524(01)00212-7

Sun Y, Cheng J (2022) Hydrolysis of lignocellulosic materials for ethanol production: a review. Bioresour Technol 340:125682. https://doi.org/10.1016/j.biortech.2021.125682

Swatloski RP, Spear SK, Holbrey JD, Rogers RD (2002) Dissolution of cellulose with ionic liquids. J Am Chem Soc 124(18):4974–4975. https://doi.org/10.1021/ja025790m

Tamo AK, Doench I, Deffo G, Jiokeng SL, Doungmo G, Fotsop CG, Temgoua RC, Montembault A, Serghei A, Njanja E, Tonle IK, Osorio-Madrazo A (2025) Lignocellulosic biomass and its main structural polymers as sustainable materials for (bio)sensing applications. J Mater Chem A 13(30):24185–24253. https://doi.org/10.1039/D5TA02900G

Tayyab M, Ali S, Farooq U (2018) Microwave irradiation for lignocellulosic biomass pretreatment and its effects on molecular structure. https://doi.org/10.1016/j.biortech.2018.02.044

Triwahyuni E, Yulianto M, Wulandari D (2022) Advances in bioethanol production from lignocellulosic biomass: a review. Renew Sust Energ Rev 153:111789. https://doi.org/10.1016/j.rser.2021.111789

Vats S, Khan A, Singh R (2013) Fungal laccases and their application in bioremediation. Enzyme Res 2013:1–11. https://doi.org/10.1155/2013/763215

Vázquez-González M, Fernández-Prior Á, Bermúdez-Oria A, Rodríguez-Juan E, Pérez Rubio AG, Fernández-Bolaños Guzmán J, Rodríguez-Gutiérrez G (2020) Utilization of strawberry and raspberry waste for the extraction of bioactive compounds by deep eutectic solvents. LWT. https://doi.org/10.13039/501100003329

Vieira F, Santana H, Mata F, Pires P, Vaz-Velho M (2024) Comparative study of pretreatments on coconut fiber for efficient isolation of lignocellulosic fractions. Sustainability 16(11):4784

Vindran S, Jaiswal AK (2016) Ultrasonic and mechanical pretreatment techniques for lignocellulosic biomass valorization. https://doi.org/10.1016/j.rser.2016.09.012

Wagle DV, Zhao H, Baker GA (2014) Deep eutectic solvents: sustainable media for nanoscale and functional materials. Acc Chem Res 47(8):2299–2308. https://doi.org/10.1021/ar500090z

Wagner F et al (2018) Microbial hydrolytic and ligninolytic enzyme systems in lignocellulosic biomass conversion. Biotechnol Adv 36(6):1632–1644. https://doi.org/10.1016/j.biotechadv.2018.05.007

Wei H, Xu Q, Weng C (2021) Challenges and strategies in consolidated bioprocessing for cellulosic ethanol production. Biotechnol Biofuels 14:215. https://doi.org/10.1186/s13068-021-02009-8

Wu Z, Lin Q, Chen S (2024) Thermotolerant Kluyveromyces marxianus for high-temperature ethanol fermentation of lignocellulosic hydrolysates. Front Bioeng Biotechnol 12:123456. https://doi.org/10.3389/fbioe.2024.123456

Xu F, Sun JX, Sun RC, Fowler P, Baird MS (2012) Comparative study of organosolv lignins from wheat straw. Ind Crop Prod 40:160–165. https://doi.org/10.1016/j.indcrop.2012.03.037

Xu Q, Wu S, He AY (2024) Advances in consolidated bioprocessing for lignocellulosic ethanol production. Biotechnol Adv 68:108212. https://doi.org/10.1016/j.biotechadv.2024.108212

Xu X, Li W, Sun X, He Y, Li H (2025) An in-depth review of sustainable and environmentally friendly pretreatment techniques for cellulose extraction from lignocellulosic biomass and their uses. [In press]

Yoo CG, Pu Y, Ragauskas AJ (2017) Advances in ionic liquid recycling for lignocellulosic biomass pretreatment. Bioresour Technol 241:6171–6185. https://doi.org/10.1016/j.biortech.2017.06.033

Yu J, Wang X, Chen L (2025) An integrated hydrothermal and ball-milling approach for the co-production of glucose, xylose and residual lignin from Moso bamboo. Ind Crop Prod 188. https://doi.org/10.1016/j.indcrop.2022.115486

Zaccariello L, Battaglia D, Morrone B, Mastellone ML (2022) Hydrothermal carbonization: a pilot-scale reactor design for bio-waste and sludge pre-treatment. Waste Biomass Valoriz 13(9): 3865–3876

Zamocky M, Hallberg BM, Ludwig R, Divne C, Nicholls P, Haltrich D (2014) Pyruvate oxidases and related enzymes: structure, function and biotechnological applications. Biotechnol Adv 32(3):380–408. https://doi.org/10.1016/j.biotechadv.2013.11.002

Zdanowicz M, Wilpiszewska K, Spychaj T (2018) Deep eutectic solvents for biomass processing. Bioresour Technol 249:997–1003. https://doi.org/10.1016/j.biortech.2017.10.084

Zhang H, Wang J (2025) Magnetic nanoparticle-cellulase systems for recyclable enzymatic hydrolysis. Appl Microbiol Biotechnol 109(4):789–799. https://doi.org/10.1007/s00253-025-1234-8

Zhang K, Pei Z, Wang D (2016) Organosolv pretreatment for biorefineries: current status and perspectives. Bioresour Technol 199:23–32. https://doi.org/10.1016/j.biortech.2015.08.102

Zhang J, Ma E, Qu Y, Sun Y, Xu F (2017) Comprehensive utilization of hemicellulose and cellulose to release fermentable sugars from corncobs via acid hydrolysis and fast pyrolysis. ACS Sustain Chem Eng 5(7):5981–5989. https://doi.org/10.1021/acssuschemeng.7b00561

Zhang J, Wang X, Chu D, He Y, Bao J (2023) Lignocellulosic ethanol: current status and perspectives. Front Bioeng Biotechnol 11:1185342. https://doi.org/10.3389/fbioe.2023.1185342

Zhang Y, Liu X, Wu Y, Wang Y, Chen H (2025) Recent progress in alkali recovery from lignocellulosic biorefineries. BioEnergy Res. [In press]

Zhao X, Zhang L, Liu D (2012) Biomass recalcitrance. Part I: the chemical compositions and physical structures affecting the enzymatic hydrolysis of lignocellulose. Biofuels Bioprod Biorefin 6(4):465–482. https://doi.org/10.1002/bbb.1331

Zhao X, Wang L, Sun Y (2023) CRISPR/Cas9-mediated engineering of Zymomonas mobilis for improved inhibitor tolerance and ethanol production. Biotechnol J 18(5):300456. https://doi.org/10.1002/biot.202300456

Zhao L, Chen Y, Wang T (2025) Advances in various pretreatment strategies of lignocellulosic substrates for the production of bioethanol: a comprehensive review. Discover Appl Sci 5: Article 6748. https://doi.org/10.1007/s42452-025-06748-1

Zhou D, Han S, Liu X (2025) Co-expression of cellulases and xylanases in Trichoderma reesei enhances wheat straw saccharification. Enzym Microb Technol 173:110123. https://doi.org/10.1016/j.enzmictec.2025.110123

Zhu JY, Pan X, Zalesny RS Jr (2010) Pretreatment of woody biomass for biofuel production: energy efficiency, technologies, and recalcitrance. Appl Microbiol Biotechnol 87(3):847–857

Zuliani L, Serpico A, De Simone M, Frison N, Fusco S (2021) Biorefinery gets hot: thermophilic enzymes and microorganisms for second-generation bioethanol production. Processes 9(9): 1583. https://doi.org/10.3390/pr9091583

Chapter 6
Synthetic Biology Approaches to Engineer Non-conventional Yeasts for Lignocellulosic Bioethanol Production

Shalley Sharma, Debarati Paul, and Anju Arora

Abstract The pursuit of sustainable lignocellulosic bioethanol has increasingly shifted toward non-conventional yeast species beyond the traditional workhorse *Saccharomyces cerevisiae*. In recent years, advances in metabolic engineering and synthetic biology have unlocked the biofuel potential of these alternative yeasts, many of which possess innate traits advantageous for industrial fermentation, such as thermotolerance, inhibitor resistance, and native pentose metabolism. This chapter provides a comprehensive overview of recent progress in engineering novel yeast strains (excluding *S. cerevisiae*) for bioethanol production. We highlight cutting-edge genetic tools, including CRISPR/Cas-based genome editing, pathway rewiring techniques, adaptive laboratory evolution, and multi-omics integration, that have accelerated strain development in non-conventional yeasts such as *Kluyveromyces marxianus, Pichia kudriavzevii* (also known as *Issatchenkia orientalis*), *Ogataea polymorpha*, *Spathaspora passalidarum*, and others. Case studies are presented featuring engineered strains with enhanced fermentation of lignocellulosic sugars and improved tolerance to stresses such as furfural and acetic acid, with some examples reaching pilot scale or utilizing industrial feedstocks. We analyze key challenges in engineering these yeasts, including cofactor balancing in pentose pathways, genetic stability, and process scalability, and explore potential solutions such as adaptive evolution and microbial consortia. Overall, the chapter emphasizes how integrating the unique physiological advantages of non-model yeasts with advanced engineering strategies can drive the development of next-generation bioethanol production. The insights and comparisons offered, including metabolic traits, engineering strategies, and ethanol yields, underscore the expanding role of

S. Sharma (✉)
Department of Biomedical Engineering, Emory University, Atlanta, GA, USA

D. Paul
Amity Institute of Biotechnology, Amity University, Noida, Uttar Pradesh, India

A. Arora
Division of Microbiology, ICAR-Indian Agricultural Research Institute, New Delhi, Delhi, India

S. Prasad et al. (eds.), *Sustainable Bioethanol Production from Lignocellulosic Biomass*, Clean Energy Production Technologies,
https://doi.org/10.1007/978-981-95-9690-4_6

novel yeasts in the bioenergy landscape and outline future directions for research and innovation in sustainable bio-ethanol technologies.

Keywords Genetic engineering · Synthetic biology tools · Bioethanol · Non-conventional yeasts · Metabolic engineering

6.1 Introduction

Lignocellulosic bioethanol production depends on efficient microbial fermentation of mixed sugars derived from plant biomass. While the conventional yeast *Saccharomyces cerevisiae* dominates first-generation bioethanol from corn or sugarcane due to its robustness and high ethanol productivity, it cannot naturally ferment pentose sugars such as D-xylose and L-arabinose (Ruchala et al. 2020). Hemicellulose component present in lignocellulosic hydrolysates, upon hydrolysis, releases abundant pentoses (especially xylose, and to a lesser extent arabinose) along with glucose. The inability of *S. cerevisiae* to utilize these pentoses is a major bottleneck in second-generation (2G) bioethanol processes. This challenge has prompted intensive research on non-conventional yeasts, alternative yeast species that natively metabolize a broader range of sugars, and also in the direction of synthetic biology approaches to engineer yeast strains capable of co-fermenting hexoses and pentoses (Moon et al. 2025; Shan et al. 2021).

Among non-conventional yeasts, *Scheffersomyces stipitis* (formerly *Pichia stipitis*), *Ogataea polymorpha* (formerly *Hansenula polymorpha*), and *Kluyveromyces marxianus* have emerged as promising platforms for lignocellulosic ethanol. *S. stipitis* is renowned for its native capacity to ferment xylose to ethanol with minimal production of xylitol as a byproduct (Moon et al. 2025). *O. polymorpha* and *K. marxianus* are thermotolerant yeasts that can grow and ferment at elevated temperatures (~40–50 °C), offering advantages for process integration and contamination control (Bilal et al. 2022; Ha-Tran et al. 2020; Lyu et al. 2021; Roy et al. 2023).

These non-model yeasts, however, bring their own limitations: *S. stipitis* has lower ethanol tolerance and requires oxygen for efficient xylose metabolism; *O. polymorpha* ferments xylose to ethanol only in small amounts unless metabolically engineered; and *K. marxianus*, though fast-growing and able to assimilate diverse sugars, tends toward respiratory metabolism (Crabtree-negative), which can limit anaerobic ethanol yields (Fonseca et al. 2008; Lane and Morrissey 2010; Ruchala et al. 2020). Over the past decade, advances in synthetic biology, including CRISPR/Cas genome editing, modular cloning of multigene pathways, synthetic promoters and terminators, and genome-scale metabolic modeling, have enabled significant strides in overcoming these challenges. Researchers have developed genetic toolkits to make these non-conventional yeasts more genetically tractable and have engineered them to ferment pentoses alongside glucose more efficiently.

This chapter provides a comprehensive overview of synthetic biology approaches for genome editing and strain engineering of *Saccharomyces stipitis, Pichia polymorpha, Kluyveromyces marxianus*, and related yeasts, with the goal of sustainable lignocellulosic bioethanol production. The traits have been compared with *S. cerevisiae*, metabolic bottlenecks in pentose fermentation have been discussed, and engineering strategies (e.g., CRISPR/Cas9 systems, pathway optimization, promoter libraries, and adaptive evolution) have been evaluated to improve pentose utilization, ethanol yield, stress tolerance, and process scalability. Recent advancements and landmark studies from the past decade have been highlighted. Case studies of pilot-scale trials and industrial demonstrations have been highlighted to gauge progress toward commercial viability. Finally, future directions for systems and synthetic biology have been outlined in the development of robust yeast "cell factories" with the objective of producing lignocellulosic bioethanol.

6.2 Background: Conventional vs. Non-conventional Yeasts for Bioethanol

Saccharomyces cerevisiae is the workhorse of industrial ethanol fermentation, valued for its high ethanol productivity, tolerance to ethanol and inhibitors, and well-characterized genetics. *S. cerevisiae* rapidly ferments hexoses (glucose, fructose) under anaerobic or oxygen-limited conditions, exhibiting the Crabtree effect (fermentation even in the presence of oxygen when sugar is in excess). However, wild-type *S. cerevisiae* cannot consume xylose or arabinose because it lacks the necessary pathways to convert these pentoses into glycolytic intermediates (Sharma and Arora 2020). As a result, lignocellulosic hydrolysates, which contain up to 30–40% xylose, remain underutilized unless the yeast is engineered. Even with metabolic engineering, *S. cerevisiae* strains have historically struggled with slow xylose uptake and accumulation of xylitol byproduct, reflecting redox imbalances in engineered pathways (discussed in the next section). These limitations motivate the exploration of naturally pentose-fermenting yeasts (Ndubuisi et al. 2023).

Scheffersomyces stipitis (also known by its asexual name *Pichia stipitis*) is a respiratory-fermentative yeast isolated from wood and insect guts, notable for being one of the best natural xylose fermenters (Gao et al. 2017; Slininger et al. 2015). *S. stipitis* can ferment D-xylose to ethanol with yields approaching 80% of theoretical under optimized conditions, far outperforming unengineered *S. cerevisiae*. It possesses a complete xylose utilization pathway (xylose reductase, xylitol dehydrogenase, xylulokinase) encoded in its genome, as well as several genes for lignocellulosic sugar conversion. Unlike *S. cerevisiae, S. stipitis* produces very little xylitol as a byproduct when fermenting xylose, indicating a more balanced native redox metabolism (Jeffries et al. 2007). However, *S. stipitis* has important drawbacks: it is microaerobic, requiring a limited oxygen supply for optimal xylose fermentation, as oxygen is necessary to regenerate cofactors and support cell growth.

Under strictly anaerobic conditions, *S. stipitis* struggles; its growth and xylose uptake are poor without oxygen, partly due to limitations in ATP generation and redox balance (e.g., the need for oxygen-dependent reoxidation of NADH via respiratory or alternate oxidase pathways) (Zha et al. 2021). Additionally, *S. stipitis* is less tolerant to ethanol and inhibitor stress: it can typically withstand final ethanol concentrations of only ~30–40 g/L (4–5% v/v) before fermentation ceases, significantly lower than *S. cerevisiae,* which can often tolerate >100 g/L (12–15% v/v) (Jassim Shalsh et al. 2016; Ruchala et al. 2020; Santos et al. 2016). These weaknesses, low anaerobicity and ethanol sensitivity, have limited *S. stipitis* in industrial adoption despite its excellent xylose-fermenting ability. Nonetheless, it remains a key organism of interest, and numerous studies have aimed to engineer *S. stipitis* for greater robustness (e.g., higher ethanol tolerance and fully anaerobic fermentation) to leverage its pentose metabolism in industrial processes.

Ogataea polymorpha is a thermotolerant, methylotrophic yeast traditionally used for heterologous protein production (formerly known as *Hansenula polymorpha*) (Kurylenko et al. 2018). *O. polymorpha* has garnered attention for its potential in second-generation ethanol production because it naturally utilizes xylose and can ferment at elevated temperatures of up to ~45 °C (Ryabova et al. 2003). High-temperature fermentation is advantageous for simultaneous saccharification and fermentation (SSF) processes, as cellulolytic enzymes typically operate optimally at temperatures of 45–50 °C. Using a thermophilic yeast like *O. polymorpha* can facilitate the coupling of hydrolysis and fermentation in a single vessel, thereby reducing cooling needs and the risk of contamination. *O. polymorpha* indeed can metabolize xylose at 45 °C and produce ethanol, distinguishing it as one of the most thermotolerant native xylose-fermenting yeasts (Xie et al. 2024). However, its ethanol yield from xylose is naturally very low, *O. polymorpha* tends to respire xylose rather than ferment it to ethanol under anaerobic conditions, leading to poor ethanol titers.

In wild-type strains, most xylose is consumed for cell growth and respiration, unless the metabolism is rerouted (Ruchala et al. 2020; Xie et al. 2024). For instance, early studies found that *H. polymorpha* could ferment both glucose and xylose up to 45 °C but produced significantly less ethanol from xylose than *S. stipitis*, and that *O. polymorpha*'s ethanol tolerance was intermediate (higher than *S. stipitis* but lower than *S. cerevisiae*) (Ryabova et al. 2003). The key limitation is a low flux through the pyruvate-to-ethanol branch: *O. polymorpha* has relatively low native activity of pyruvate decarboxylase (Pdc) and alcohol dehydrogenase under fermentative conditions. As a result, much of the pyruvate from xylose is shuttled into mitochondria for oxidation (especially since *O. polymorpha* is Crabtree-negative, favoring respiration). Moreover, *O. polymorpha* exhibits a strong preference for respiratory growth, and completely anaerobic growth may not be sustainable without specific supplementation (similar to other yeasts that require oxygen for sterol and lipid synthesis).

Recent metabolic engineering efforts have targeted these issues; for example, overexpressing a pyruvate decarboxylase gene was shown to significantly increase ethanol yields from xylose in *O. polymorpha*, essentially pushing more carbon into fermentative ethanol production (Ishchuk et al. 2008; Sibirny 2023). *O. polymorpha* also naturally consumes unusual substrates, such as methanol (it is a methylotroph), but this is tangential to ethanol production, except that its methylotrophic metabolism has endowed it with powerful inducible promoters (e.g., AOX1) and peroxisomal pathways that can be repurposed in biotechnology. Overall, *O. polymorpha*'s combination of high thermotolerance and innate xylose utilization makes it a promising "chassis" that requires some engineering to boost ethanol productivity.

Kluyveromyces marxianus is another emerging yeast platform distinguished by its rapid growth (doubling times <2 h), thermotolerance (growth up to ~45–50 °C), and broad substrate spectrum (Moon et al. 2025; Suzuki et al. 2019). *K. marxianus* is a Crabtree-negative relative of *K. lactis*, commonly isolated from dairy and plant environments. It natively utilizes a range of sugars, including glucose, fructose, sucrose, and lactose (via β-galactosidase), as well as varying extents of pentoses and oligosaccharides. *K. marxianus* has been reported to metabolize xylose and L-arabinose natively, although it ferments these sugars less efficiently than glucose. In fact, *K. marxianus* can assimilate xylose but cannot produce substantial ethanol from xylose under strictly anaerobic conditions unless genetically modified (Marcišauskas et al. 2019; Suzuki et al. 2019). Typically, *K. marxianus* will respire xylose if oxygen is present, and under oxygen-limited conditions, it may ferment xylose slowly; however, the rates and yields are significantly lower than those of *S. stipitis* or engineered *S. cerevisiae*.

Additionally, *K. marxianus* often exhibits carbon catabolite repression (CCR) similar to *S. cerevisiae*, meaning it prefers glucose and will not consume pentoses until glucose is depleted (Hua et al. 2019). This is problematic for the co-fermentation of mixed sugars; however, its ability to grow at higher temperatures offers an advantage in SSF processes similar to that of *O. polymorpha*. *K. marxianus* is notable for being one of the fastest-growing eukaryotes and for producing high levels of certain metabolites (e.g., it naturally produces flavorful ethyl acetate via overflow metabolism) (Roy et al. 2023). Industrially, *K. marxianus* has been used at pilot scale to convert cheese whey (containing lactose) into ethanol and other products, demonstrating its scalability.

Its classification as generally recognized as safe (GRAS) makes it a promising candidate for food and feed applications (Bilal et al. 2022). The main hurdles for the use of lignocellulosic ethanol by *K. marxianus* are its relatively lower ethanol yield (due to respiration and possible byproduct formation) and moderate ethanol tolerance. Typical *K. marxianus* strains tolerate ethanol at concentrations of around 6–9% v/v; this is less than that of *S. cerevisiae*, but some improvement via adaptation has been reported (e.g., isolating mutants that grew in 40 g/L ethanol) (Nitiyon et al. 2016). On the positive side, its thermotolerance (fermentation at 40–45 °C) can potentially shorten fermentation time and reduce contamination. Recent metabolic engineering has enabled *K. marxianus* to ferment xylose faster by introducing alternative pathways (Wang et al. 2013).

Table 6.1 compares key traits of *S. cerevisiae* and these non-conventional yeasts relevant to bioethanol production. Each organism brings unique advantages and challenges. In general, *S. cerevisiae* remains the champion in stress tolerance and fermentation vigor on hexoses, whereas *S. stipitis*, *K. marxianus*, and *O. polymorpha* expand the substrate range to pentoses and higher temperatures at the cost of certain

Table 6.1 Comparative traits of *S. cerevisiae* and selected non-conventional yeasts for lignocellulosic ethanol production

Strain	Ambient growth temp.	Crabtree effect	Native sugar use	Max ethanol yield	Ethanol tolerance	Inhibitor tolerance	Engineering ease	Key strengths	Limitations	References
S. cerevisiae	30 ± 5 °C	Yes (ferments with O_2)	Glucose, no xylose/ arabinose	~90–95% (glucose)	High (~15% v/v)	High	Excellent (CRISPR, plasmids)	Robust, industrially proven	No pentose use; <40 °C growth	Bettiga et al. (2009)
S. stipitis	30 ± 5 °C	No (respirer)	Glucose, xylose, and slow arabinose	~80–85% (mix)	Low (~5% v/v)	Low	Limited (improving)	Native xylose use	Needs O_2; low tolerance	Skoog and Hahn-Hägerdal (1990)
K. marxianus	30–45 °C	No	Broad sugars incl. xylose/ arabinose	~80–90% (glucose), low from xylose	Moderate (~6–10% v/v)	Moderate	Good (CRISPR tools exist)	Fast growth, thermotolerant	Weak xylose fermentation; repression	Nitiyon et al. (2016)
O. polymorpha	37–50 °C	No	Glucose, xylose (poor), no arabinose	Low from xylose, moderate from glucose	Moderate (~6–8% v/v)	Moderate	Good (CRISPR with RAD52)	Thermotolerant, methanol use	Low ethanol yield; fewer tools	Ishchuk et al. (2008)

weaknesses. Synthetic biology aims to combine the best of both worlds. For instance, transferring or optimizing the pentose metabolic pathways in *S. cerevisiae*, or conversely, endowing *S. stipitis* and *K. marxianus* with the robustness and anaerobic performance of *S. cerevisiae*. In the following sections, we delve into the metabolic pathways for pentose utilization and the engineering strategies employed to optimize these yeasts for efficient lignocellulosic ethanol fermentation.

6.3 Pentose Sugar Metabolism in Yeasts: Pathways and Bottlenecks

Most native xylose-fermenting yeasts (e.g., *S. stipitis*) use an XR-XDH route: XR (xylose reductase, *XYL1*) reduces xylose to xylitol with NADPH, and XDH (xylitol dehydrogenase, *XYL2*) oxidizes xylitol to D-xylulose with NAD^+. D-xylulose is then phosphorylated by xylulokinase (XK, *XYL3/XKS1*) and enters the pentose phosphate pathway (Matsushika et al. 2009). *S. cerevisiae* lacks this pathway natively but can ferment xylose when engineered with XR, XDH, and XK, or with a heterologous xylose isomerase (XI) (Walfridsson et al. 1996), which directly converts xylose to xylulose. A key bottleneck in XR-XDH strains is the redox mismatch between an NADPH-preferring XR and an NAD^+-dependent XDH (Verduyn et al. 1985). Under oxygen-limited or anaerobic conditions, excess NADH accumulates (XDH produces NADH) and cannot be readily reoxidized, causing xylitol byproduct formation and glycerol secretion for NAD^+ regeneration. Oxygen availability can alleviate this imbalance by allowing respiratory NADH oxidation, thereby reducing xylitol yields (G.-C. Zhang et al. 2012). XI-based strains avoid cofactor imbalance and typically achieve higher ethanol yields, although often with slower xylose uptake and growth. Non-conventional yeasts, such as *K. marxianus* and *O. polymorpha*, harbor native XR-XDH pathways but produce very little ethanol from xylose natively; however, metabolic engineering has increased their xylose fermentation efficiency (e.g., *O. polymorpha* mutants achieved ~40 times higher ethanol titers from xylose) (Semkiv et al. 2022).

Fungal yeasts metabolize L-arabinose via a multi-step reductive pathway (Fig. 6.1). An arabinose reductase (AraR, an aldose reductase such as *GRE3* in *S. cerevisiae*) first converts L-arabinose to L-arabitol using NAD(P)H. L-arabitol is then oxidized by L-arabitol dehydrogenase (LAD) to L-xylulose, which in turn is reduced by an L-xylulose reductase (ALX) to xylitol. Xylitol is finally oxidized by XDH to D-xylulose, converging into the xylose catabolic pathway (Bettiga et al. 2009). This lengthy route and its cofactor requirements lead to significant arabitol accumulation in wild-type yeasts (up to ~0.5 g_arabitol/g_sugar in mixed-sugar fermentations), similar to the xylitol overflow seen with XR-XDH. Efficient pentose fermentation also requires overcoming transport and regulatory constraints. *S. cerevisiae*'s hexose transporters have low affinity for pentoses, and glucose repression (CCR) via Mig1 and hexose sensors (Snf3/Rgt2/Hxs1) limits pentose uptake under mixed sugars (Nijland and Driessen 2019).

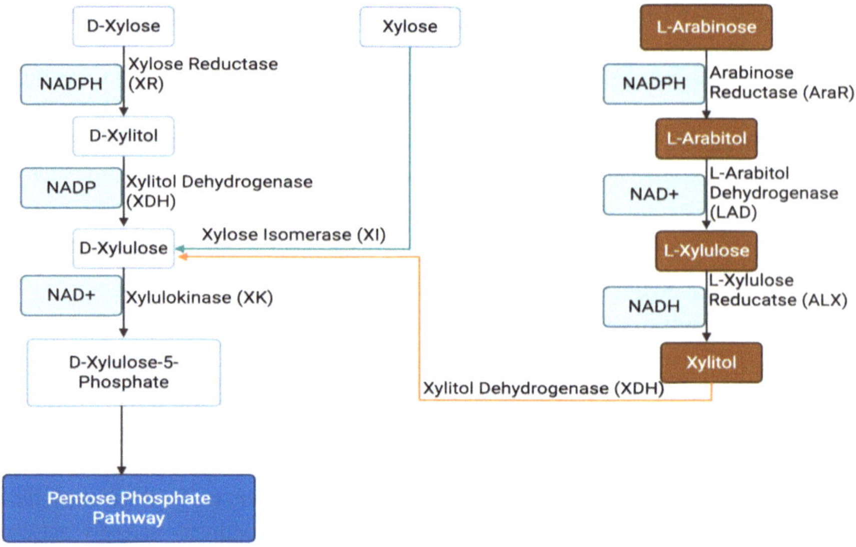

Fig. 6.1 Schematic of yeast D-xylose (left) and L-arabinose metabolism (right) and bacterial Xylose isomerase (middle)

Strategies to address these bottlenecks include overexpressing XK to alleviate the xylulose-5-P phosphorylation bottleneck, deleting MIG1 to deregulate CCR, and modifying sugar sensors (e.g., HXS1) to enhance xylose induction (Shen et al. 2013). Furthermore, heterologous pathway genes have been introduced to enable or enhance pentose utilization (De Vit et al. 1997). For example, expressing fungal *L-arabinose* genes (AraR, LAD, ALX) in *S. cerevisiae* enables anaerobic arabinose fermentation, albeit with some arabitol as a byproduct. Cofactor engineering has also been pivotal: mutant XR/XDH enzymes with altered NAD(P)H preferences, as well as schemes like NADH oxidase expression, help rebalance the redox state, sharply reducing xylitol formation and boosting ethanol yields. Recent studies build on these landmark approaches, using combinations of transporter evolution, redox tweaking, and regulatory gene edits to attain more complete D-xylose and L-arabinose fermentation in both *S. cerevisiae* and novel yeast biocatalysts.

6.4 Synthetic Biology Tools for Engineering Non-conventional Yeasts

The development of superior yeast strains for lignocellulosic ethanol has greatly benefited from the emergence of new synthetic biology tools over the last decade. Unlike *S. cerevisiae*, many non-conventional yeasts have historically lacked easy genetic manipulation, for example, *S. stipitis* was difficult to transform and had few selectable markers, and *K. marxianus* or *O. polymorpha* often exhibited low homologous recombination frequencies, hindering precise genome editing. Synthetic biology has begun to level this playing field by providing efficient genome editing systems (notably CRISPR/Cas), modular DNA assembly methods for multi-gene pathways, and libraries of promoters and other regulatory parts tailored to these new hosts. In parallel, systems biology approaches, such as genome-scale metabolic modeling and -omics analyses, are guiding target identification for strain improvement. This section reviews the key tools and techniques that enable the rapid engineering of *S. stipitis*, *K. marxianus*, *O. polymorpha*, and other organisms.

6.4.1 Genome Editing and CRISPR/Cas Systems

The advent of CRISPR/Cas9 has revolutionized genome editing in non-conventional yeasts by enabling precise and efficient modifications, thereby overcoming the limitations of traditional methods, such as random mutagenesis or iterative marker insertion. Unlike *S. cerevisiae*, which favors homologous recombination (HR), many yeasts, such as *K. marxianus*, *Y. lipolytica*, *S. stipitis*, and *P. pastoris*, preferentially use non-homologous end joining (NHEJ), making targeted integration challenging (Xia et al. 2023). Deletion of NHEJ genes such as *KU70* or *KU80* improves HR efficiency but may reduce strain robustness (Choo et al. 2014; Näätsaari et al. 2012). CRISPR/Cas9 induces targeted double-strand breaks (DSBs) that stimulate HR if a repair template is present or cause indels via NHEJ for gene knockouts. Its flexibility, achieved through customizable single-guide RNAs (sgRNAs), allows for high-throughput editing. Landmark studies have demonstrated multiplex gene editing in *Y. lipolytica*, as well as efficient CRISPR-mediated deletions, insertions, and multigene integrations in *K. marxianus* and *P. pastoris*, achieving editing efficiencies of up to 80%. These advances enable rapid metabolic engineering with single-step integration of complex pathways (Harzevili 2023).

Implementing CRISPR in each "new" or untried yeast often requires overcoming species-specific hurdles. One is finding a suitable promoter for expressing the Cas9 protein and the sgRNA. *S. cerevisiae* typically uses a strong pol III promoter like SNR52 for sgRNA and a pol II promoter like TEF1 for Cas9. Non-conventional yeasts may require different promoters due to codon usage or regulatory differences. For instance, *S. stipitis* has a divergent codon usage (it is a CUG-serine clade yeast, meaning the CUG codon is translated as serine rather than leucine). This peculiarity means one cannot directly express a *S. cerevisiae*-optimized Cas9 sequence in *S. stipitis* without recoding it, since any leucine codons (CUG) in the Cas9 gene might be mistranslated. Researchers overcame this by codon-optimizing Cas9 for *S. stipitis*' codon bias and using native promoters. Another hurdle is delivering the CRISPR components: some yeasts do not maintain plasmids well.

In *S. stipitis*, there are no stable 2 μm-like plasmids; instead, methods such as inserting the Cas9 gene into the genome or using transient plasmids that are lost after editing are employed. For example, the recently developed BLINCAR tool in

S. stipitis uses a transient Cas9-expressing plasmid to eliminate marker genes after each edit, enabling iterative genome modifications on a wild-type background (Reichard et al. 2023). BLINCAR stands for "Bio-Luminescent Indicator that is Nullified by Cas9-Actuated Recombination." It cleverly uses a recyclable luciferase marker that is excised by Cas9 after each integration, thereby freeing the marker for reuse. This allowed Reichard et al. (2023) to perform multiple sequential edits in *S. stipitis* without accumulating foreign DNA or requiring stable plasmid maintenance. Such innovations tackle the historical genetic intractability of *S. stipitis*, lowering barriers by expanding its molecular toolbox.

In *K. marxianus*, CRISPR/Cas9 systems have also been optimized. Nambu-Nishida et al. (2017) first showed that deleting KU70 improved gene targeting; subsequent efforts introduced Cas9 with single or multiple gRNAs to knock out genes for metabolic engineering (Nambu-Nishida et al. 2017). More recently, a near-PAMless Cas9 variant (SpRY) and alternative Cas proteins (like Cas12a/Cpf1) have been suggested to broaden target ranges in these yeasts (Walton et al. 2020). Cas12a (formerly Cpf1) was successfully used in *Y. lipolytica* and *P. pastoris*, and there are reports of its use in *K. marxianus* as well, offering different PAM requirements and potentially simpler multiplexing (Cas12a processes its own crRNAs) (Yang et al. 2020; Zhang et al. 2021). In general, the establishment of CRISPR editing has been a landmark advancement; virtually every non-conventional yeast of interest (including *S. stipitis*, *K. marxianus*, *O. polymorpha*, *Y. lipolytica*, *P. pastoris*, *Rhodotorula*, etc.) now has at least one published CRISPR protocol. This enables precise genome modifications such as: deletion of byproduct pathways, insertion of entire metabolic pathways at defined loci, promoter swaps, and even gene repression/activation using dCas9-based systems.

A notable example combining CRISPR with metabolic engineering is the work by Gao et al. (2022) in *P. pastoris*: they co-overexpressed *ScRAD52*, *ScRAD51*, and *ScSAE2* (HR repair genes from *S. cerevisiae*) along with Cas9, achieving nearly 100% precise integration of multiple gene cassettes simultaneously (Gao et al. 2022). This allowed assembly of multi-gene pathways in one step. Although overexpression of *ScRad51/52* did not significantly improve HR in *S. stipitis* in one study, *P. pastoris* responded well to this approach. These findings highlight how fine-tuning DNA repair processes (either by genetic knockouts of NHEJ factors or by boosting HR machinery) can be synergistic with CRISPR to maximize editing efficiency.

Beyond classical genome editing, CRISPR-based tools are evolving in the field of yeast synthetic biology. CRISPR interference (CRISPRi) using dCas9 (catalytically dead Cas9) fused to repressors has been implemented in *S. cerevisiae* and is being extended to others for programmable gene knockdown without permanent mutation. CRISPR activation (CRISPRa) can similarly upregulate target genes. Such tools can create synthetic regulatory circuits or dynamically control metabolism, for instance, to downregulate genes that consume xylose (like competing pathways) during fermentation. Though not yet common in *S. stipitis* or *K. marxianus*, proof-of-concept gene repression in *S. stipitis* was demonstrated by Cao et al. (2018) (Cao et al. 2018). Base-editing (e.g., using a cytidine deaminase fused to dCas9) has been applied in *Y. lipolytica* and could enable fine mutations (such as changing the cofactor specificity of XR through point mutation) without requiring DNA templates (Bae et al. 2020).

In summary, CRISPR technology has significantly enhanced the engineering agility of non-conventional yeasts. It has enabled multiplexed, markerless genome edits and opened the door to sophisticated control of gene expression. These advances are foundational for building strains with the complex trait improvements needed for lignocellulosic ethanol (which often requires multiple gene additions, deletions, and regulatory modifications in concert).

6.4.2 *Modular Cloning and Pathway Assembly*

Lignocellulosic ethanol strains often require the introduction of entire pathways (for pentose utilization, inhibitor degradation, cofactor balancing, etc.). Modular cloning techniques provide a standardized method for assembling multi-gene constructs from libraries of parts. Golden Gate assembly (using type IIS restriction enzymes) and Gibson assembly are widely used to build expression cassettes that can be integrated into yeast genomes or maintained on plasmids. For *S. cerevisiae*, the Yeast Modular Cloning Toolkit (YTK) and related part libraries are available; however, for non-conventional strains, compatible part libraries had to be created (Bassett and Da Silva 2024; Kumar et al. 2021; O'Laughlin et al. 2024).

Recent efforts have focused on developing species-specific promoters, terminators, and other regulatory parts for these yeasts. For example, *K. marxianus* researchers identified a panel of native promoters with a wide range of strengths: Rajkumar et al. (2019) characterized promoters from the *K. marxianus* genome (e.g., *PDC1* was one of the strongest, *REV1* one of the weakest), spanning a 40-fold expression range (Malcı et al. 2022; Patra et al. 2021; Rajkumar et al. 2019). Qiu et al. (2023) similarly reported multiple promoters in *K. marxianus*, enabling tunable gene expression (Qiu et al. 2023). By combining promoters and terminators, expression toolkits have been built.

Kumar et al. (2021) constructed a set of promoter-terminator combinations in *K. marxianus*; one chimeric construct using an uncharacterized cell-wall protein promoter (*pIMTCP2*) and a strong maltose transporter terminator (*KmIMTT1*) yielded the highest reporter expression (~1800 Miller units of β-galactosidase) in *K. marxianus* (Kumar et al. 2021). This suggests that having proper terminators is as important as having promoters, as both promoters and terminators influence mRNA stability and expression levels. Indeed, a study by Ito et al. (2020) in *P. pastoris* screened 72 terminators from various yeasts and found a 17-fold range of expression levels, underlining that terminator choice can fine-tune expression (Ito et al. 2020). In *K. marxianus*, native terminators like *IMTT1* and *IMTT2* were shown to significantly boost expression of a reporter when paired with specific promoters (Kumar et al. 2021).

In *O. polymorpha* and *P. pastoris*, researchers often leverage the strong methanol-inducible AOX1 promoter for high expression (though this is more relevant for protein production than for ethanol). Nonetheless, synthetic biology has delivered inducible systems that can be applied to fermentation processes, for example, a copper-inducible hybrid promoter in *Y. lipolytica*, which allows for dynamic control of the production phase (Park et al. 2021). For bioethanol, one might envision using such inducible promoters to activate stress tolerance genes at a certain fermentation stage or to switch metabolism once glucose is consumed.

Golden Gate-based assembly kits have now been published for several non-conventional yeasts (Prielhofer et al. 2017). For instance, Obst et al. (2017) developed a modular cloning toolkit for *P. pastoris*, featuring over 100 characterized parts (including promoters, terminators, and selection markers), which enables the rapid construction of multi-gene pathways (Obst et al. 2017). A similar toolkit, called EasyClone, has been adapted for *K. lactis* and *K. marxianus*, providing integration vectors that target safe genomic loci for the consistent expression of multiple genes (Rajkumar et al. 2019). These allow one to integrate, for example, a three-enzyme xylose utilization pathway in a single transformation by targeting three loci with three cassettes, each carrying a different marker. Combined with CRISPR, markerless integration of multiple genes is possible by cutting at those loci and using donor DNA with homologous ends, as demonstrated by Li et al. (2021) for *K. marxianus*, where three genes for shikimate production were integrated (Li et al. 2021).

In *S. stipitis*, part development has been slower due to issues with codon usage and difficulty in maintaining plasmids. However, researchers have identified a few native promoters (e.g., *PGK*, *GPD* promoters from *S. stipitis*) and have used *S. cerevisiae* promoters with mixed success (some work if the codons for transcription factors are compatible, etc.). The BLINCAR study (Reichard et al. 2023) provided new selectable markers for *S. stipitis* (an antibiotic resistance and a luciferase reporter) that were not previously available (Reichard et al. 2023). This expands the cloning options, as prior work often had only auxotrophic markers, such as URA3, available. With an antibiotic marker, higher-efficiency transformation and selection are possible.

Synthetic promoters and hybrid regulatory elements are also being employed to create more robust expression systems. A recent example in *K. marxianus* is the carbon source-responsive hybrid promoter IN450, constructed by Bassett and Da Silva (2024), which combines elements from the native ICL1 promoter (active on non-fermentable carbon sources) with a strong constitutive promoter segment (Bassett and Da Silva 2024). The result was a promoter that behaves differently on lactose (induced) and xylose (constitutive), leading to improved production of target chemicals by adjusting expression to the substrate. This level of control could be useful in fermentation: for instance, a promoter that is weak during glucose fermentation but automatically strengthens in the xylose phase could optimize enzyme levels to the appropriate stage of a sequential sugar fermentation.

In summary, modular cloning and part libraries provide us with the DNA building blocks to refashion yeast metabolism (Fig. 6.2). By mixing and matching strong or

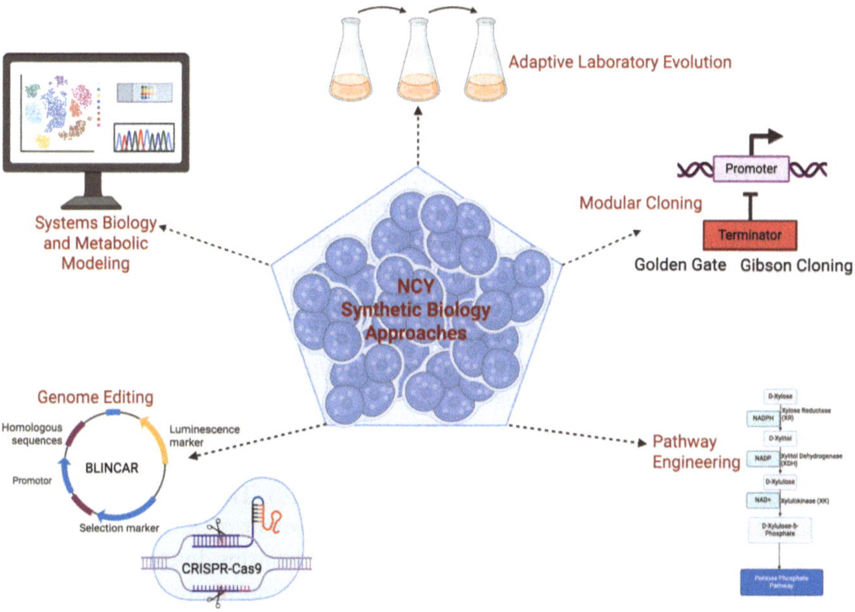

Fig. 6.2 Synthetic biology approaches to rewire non-conventional yeasts for enhanced ethanol production

weak promoters, one can balance metabolic fluxes, a critical need in pentose fermentation, where the overexpression of one enzyme (e.g., XR) relative to another (XDH) can exacerbate a redox imbalance. Synthetic biology enables the creation of expression cassettes with tunable strengths to achieve cofactor balance. For instance, researchers often express XR under a slightly weaker promoter than XDH in engineered *S. cerevisiae* to reduce xylitol formation, or they might place a transport gene under an inducible promoter to avoid a burden during growth. The availability of these tools in *K. marxianus*, *O. polymorpha*, and *S. stipitis* means they can be engineered with a precision approaching that of *S. cerevisiae*.

6.4.3 Systems Biology and Metabolic Modeling

Alongside experimental tools, systems biology approaches provide blueprints for strain design. Genome sequencing of *S. stipitis* (2007) and *K. marxianus* (2012) opened the way for genome-scale metabolic models (GEMs) of these yeasts (Balagurunathan et al. 2012; Jeffries et al. 2007). A GEM integrates all known metabolic reactions of the organism and can be used to simulate growth and product formation under various conditions. The first GEM for *S. stipitis* was published in 2012 (Balagurunathan et al. 2012), incorporating 814 genes and accurately predicting its growth on different substrates. Notably, simulations with this model identified biosynthetic requirements for anaerobic growth. For example, like *S. cerevisiae*, *S. stipitis* likely requires supplementation of unsaturated fatty acids and sterols in strict anaerobic culture because it cannot synthesize them in the absence of oxygen. This pointed to one reason *S. stipitis* struggles anaerobically: not just the XR/XDH redox issue, but also basic biomass building blocks are limiting. The model also pinpointed bottlenecks in xylose metabolism, confirming that NAD^+ recycling is critical and suggesting that alternative oxidase or other NADH-oxidizing pathways help *S. stipitis* ferment xylose by preventing redox imbalance. It also identified that xylose uptake might be a limiting step under some conditions. These insights from in silico analysis have guided researchers to engineer specific targets, such as introducing a transhydrogenase enzyme to directly convert NADH to NADPH (attempted in *S. cerevisiae* to help balance XR/XDH) or amplifying pentose phosphate flux.

For *K. marxianus*, a genome-scale model (iSM996) was published in 2019 (Marcisauskas *et al.*), which captures 996 genes and over 1500 metabolic reactions (Lertwattanasakul et al. 2015; Marcišauskas et al. 2019). The model was validated against experimental data and could simulate the growth of *K. marxianus* up to 45 °C, providing insights into its thermotolerance mechanisms. Interestingly, model-coupled transcriptomics indicated that at higher temperatures, *K. marxianus* downregulates some oxidative phosphorylation genes, hinting that a shift toward fermentation or other energy-saving tactics contributes to heat tolerance. This might mean that encouraging fermentative metabolism (which generates less biomass yield but more ethanol) could inherently support thermotolerance. The model also confirmed *K. marxianus* can theoretically consume arabinose and xylose. However, it highlighted that under anaerobic conditions, flux would not go to ethanol unless certain pathways (like glycerol production or alternative NADH sinks) are modified.

Systems biology also extends to omics analyses of these yeasts under relevant conditions. For instance, comparative transcriptomic studies of *S. stipitis*, *K. marxianus*, and *S. cerevisiae* fermenting xylose have identified differentially expressed genes that correlate with efficient xylose utilization (Barros et al. 2024; Zha et al. 2021). One study found that *S. stipitis* upregulates alternative respiratory chain components (e.g., cyanide-resistant oxidase) on xylose, which might alleviate redox pressure. Another study on oxygen levels showed that *S. stipitis* and *K. marxianus* respond to oxygen limitation by upregulating genes in the pentose phosphate pathway; however, *K. marxianus* still diverted a significant amount of xylose to xylitol unless engineered (Barros et al. 2024; Granados-Arvizu et al. 2021; Suzuki et al. 2019).

Together, these modeling and omics approaches allow researchers to adopt a rational design strategy: identify metabolic bottlenecks and then use genome editing or pathway assembly to fix them. For example, a rational strategy suggested by models is to eliminate or reduce glycerol production in *S. cerevisiae* when engineering it for xylose, because glycerol formation (for NADH reoxidation) competes with xylitol as a redox sink. However, if glycerol is eliminated, an alternate NADH oxidation route must be supplied, leading to designs such as inserting a lactate dehydrogenase or relying on ethanol production itself to regenerate NAD^+.

Another systems biology tool is genome-scale engineering (e.g., combinatorial gene knockouts guided by algorithms like OptKnock). Although not yet extensively applied to these yeasts, one could imagine using such algorithms to predict which gene deletions would channel more flux to ethanol (Acevedo et al. 2017). For instance, OptKnock on a model of *S. stipitis* might suggest deleting certain TCA cycle genes to force carbon into fermentation, though viability needs to be maintained.

Finally, adaptive laboratory evolution (ALE) is worth mentioning as a complementary approach. ALE is not strictly synthetic biology, but it often works together with rational engineering (Hemansi et al. 2022; Saengphing et al. 2024). One might evolve *S. stipitis* for higher ethanol tolerance by gradually increasing ethanol in growth media or evolve *K. marxianus* for faster anaerobic growth on xylose by serial culture under those conditions. The evolved strains can then be sequenced to identify beneficial mutations, which inform targeted engineering. Indeed, some landmark improvements in xylose fermentation came from combining rational engineering with evolution. For example, a recombinant *S. cerevisiae* strain with the XR/XDH pathway had evolved over several months, resulting in mutants that could simultaneously consume xylose and glucose. Genomic analysis revealed mutations in the *HOG1* pathway and sugar transporters that relieved catabolite repression. Such knowledge can then be applied to non-conventional yeasts, e.g., by intentionally knocking out a repressor gene known to mutate during adaptation.

The next section correlates the tools and strategies used for strain engineering, highlighting case studies of engineered yeasts that ferment pentoses and their performance in lab and pilot-scale fermentations.

6.5 Engineering Strategies for Pentose Fermentation and Strain Improvement

6.5.1 Engineering **S. cerevisiae** *for Pentose Utilization (Benchmarking Efforts)*

Although the focus is on non-conventional yeasts, it is instructive to briefly summarize how *S. cerevisiae* has been engineered to ferment pentoses, as many strategies have analogs in other yeasts. There have been two primary routes: (1) introducing the fungal XR-XDH pathway (often from *S. stipitis*) into *S. cerevisiae*, and (2) introducing a bacterial/fungal XI pathway. The XR-XDH approach was pioneered in the 1990s and led to strains capable of slowly fermenting xylose to ethanol, but with significant xylitol accumulation (Amore et al. 1991; Ho et al. 1998; Kötter et al. 1990; Tantirungkij et al. 1993; Toivari et al. 2001). Subsequent improvements included overexpressing *XKS1* (xylulokinase gene) and modifying XR's cofactor preference. For instance, mutations in *S. stipitis* XR (such as K270M) can switch its cofactor usage more toward NADH, helping rebalance the

NADPH/NADH discrepancy. Engineered *S. cerevisiae* with an NADH-preferring XR and a high level of XDH showed less xylitol leakage and higher ethanol yields (Amore et al. 1991; Ho et al. 1998; Kötter et al. 1990; Tantirungkij et al. 1993; Toivari et al. 2001).

The XI approach saw success in the late 2000s when researchers identified a thermophilic bacterial xylose isomerase (from *Thermus thermophilus* or *Clostridium phytofermentans*) that functioned in yeast. Kuyper et al. (2005) expressed an XI from *Piromyces* sp. (an anaerobic fungus) in *S. cerevisiae* and achieved comparable xylose fermentation performance without xylitol production (Kuyper et al. 2005). Today's best engineered *S. cerevisiae* strains (often called "co-fermenting" or "xylose-fermenting" yeast) incorporate multiple tweaks: a xylose pathway (XR/XDH or XI), overexpressed XK, deletion of aldose reductases like GRE3 (to prevent reduction of xylose to xylitol by side reactions), evolved mutations in regulators for better xylose uptake, and sometimes integration of an arabinose pathway (like *E. coli*'s AraBAD) (Kim et al. 2013).

Companies such as DuPont and Mascoma have developed proprietary strains of *S. cerevisiae* for lignocellulosic ethanol production. For example, DuPont's strain utilized a bacterial XI and achieved complete co-consumption of glucose and xylose in their pilot plant. Mascoma's approach (the MGT yeast) integrated cellulases and xylose metabolism into *S. cerevisiae* for consolidated bioprocessing. These achievements, however, took extensive metabolic engineering and evolution. The knowledge gained (e.g., the importance of *PHO13* deletion to improve pentose phosphate flow, or *MIG1* deletion to relieve CCR) can inform how we engineer non-conventional yeasts, which might require fewer foreign genes since they naturally have the pathways, but still benefit from such modifications.

6.5.2 *Engineering* Scheffersomyces stipitis

Efforts to engineer *S. stipitis* aim to capitalize on its native xylose metabolism while fixing its weak points (oxygen requirement, low inhibitor tolerance, and poor genetic accessibility). One target has been to enable anaerobic fermentation in *S. stipitis*. Studies have shown that under oxygen-limited (but not fully anaerobic) conditions, *S. stipitis* can achieve high ethanol yields from xylose, for instance, an adapted strain produced ethanol at ~80% of the theoretical yield from hemicellulosic hydrolysate when a small airflow was provided (Nigam 2001). Completely eliminating oxygen typically causes *S. stipitis* to stall, presumably due to issues with redox cofactors and the inability to synthesize necessary cell components. Researchers have tried a few strategies: (a) supplementation: providing sterols and unsaturated fatty acids in the medium can partially allow anaerobic growth, as these components then do not require oxygen to be made (Balagurunathan et al. 2012); (b) alternative respiration knockout: *S. stipitis* has an alternative oxidase (AOX) that can oxidize NADH without pumping protons (thus dissipating energy). Inhibiting or deleting this AOX might force the yeast to funnel more NADH into ethanol production.

One FEMS Yeast Research study inhibited alternative oxidase and observed an increased ethanol yield from xylose, albeit at the expense of slower xylose uptake, suggesting that AOX helped maintain redox balance (Kim et al. 2013). (c) Expressing a heterologous transhydrogenase, there are attempts to express *E. coli pntAB* genes in *S. stipitis* to directly convert NADH + $NADP^+ \leftrightarrow NAD^+$ + NADPH, which could allow XR and XDH to share cofactors more effectively; results of such experiments have not been very prominent in the literature, suggesting only marginal gains or expression difficulties.

Another focus has been on improving ethanol tolerance. Adaptive evolution has been a key approach in this context. *S. stipitis* can be evolved by serial subculturing in gradually increasing ethanol concentrations. Mutants of *S. stipitis* that grew in 50 g/L ethanol, whereas the wild type struggled above 30 g/L (Ruchala et al. 2020; Slininger et al. 2015). Some mutations were identified in membrane genes and stress response regulators, suggesting mechanisms that could make membranes more rigid or stress-tolerant. These adapted strains, when used in fermentation, yielded more ethanol because they could continue fermenting at higher ethanol titers. Another adaptation strategy is the adaptation to inhibitory hydrolysates: *S. stipitis* has been adapted to tolerate pretreatment inhibitors, such as acetic acid, furfural, and HMF (Lin et al. 2012). For example, Nigam's group adapted *P. stipitis* to a dilute-acid hardwood hydrolysate over many cycles; the adapted strain fermented the hydrolysate in half the time required by the parent and with higher yield (since it could withstand the furfural, converting it to less toxic furfuryl alcohol) (Nigam 2001). Identification of genes upregulated in adapted strains (such as alcohol dehydrogenases that reduce furfural, or efflux pumps that export weak acids) has provided targets for genetic engineering. One could constitutively overexpress, for example, ADH1 or AR1 (aldehyde reductases) in *S. stipitis* to detoxify inhibitors more efficiently.

On the genetic engineering front, now that CRISPR is available for *S. stipitis*, researchers can do targeted modifications. A recent successful edit was the deletion of the *XYL2* gene (xylitol dehydrogenase) to create a xylose utilization pathway block for studying flux. While deleting *XYL2* obviously prevents ethanol production from xylose, it allowed researchers to accumulate and analyze intermediate metabolites. Conversely, one could delete competing pathways: for example, *S. stipitis* can also metabolize xylose to xylulose via a minor pathway involving xylose isomerase-like activity or via the Dahms pathway to xylonic acid in aerobic conditions. Knocking out those routes could channel more xylose into the fermentation process. Another possible edit is the deletion of CUG mistranslation-related genes (though *S. stipitis* naturally has tRNA that inserts serine at CUG; it is not something to "fix" but must be accounted for when expressing heterologous proteins) (Correia 2019).

A noteworthy metabolic engineering example is a 2020 study by Ruchala et al. (University of Rzeszów) (Ruchala et al. 2020) in which *S. stipitis* was engineered to produce not just ethanol but also higher alcohols and other chemicals. They indicated that overexpression of certain native pentose phosphate pathway genes improved xylose fermentation. Additionally, *S. stipitis* was engineered to co-utilize cellobiose by expressing a β-glucosidase, enabling it to ferment cellulose hydrolysates with both xylose and cellobiose directly. This resembles a strategy in *K. marxianus* described below.

In summary, for *S. stipitis*, the engineering trajectory involves amplifying strengths (e.g., maintaining high xylose flux by possibly increasing transporter or XK activity) and mitigating weaknesses (improving tolerance and reducing oxygen dependency). The latter has seen progress mainly through evolutionary means and physiological optimizations (microaeration control), but future gene targets include global regulators. For instance, overexpressing transcription factors that drive hypoxic response (like *S. cerevisiae*'s UPC2, which upregulates sterol import genes in anaerobiosis) might help *S. stipitis* handle anaerobic conditions by scavenging sterols.

6.5.3 *Engineering* Ogataea polymorpha

Given *O. polymorpha*'s naturally low ethanol yield on xylose, the primary goal in engineering has been to rewire its carbon flux toward ethanol. A classic study by Sibirny and colleagues in 2008 demonstrated that overexpressing the native pyruvate decarboxylase (*PDC*) gene in *Hansenula polymorpha* significantly increased ethanol production from xylose at 45 °C (Ishchuk et al. 2008). *O. polymorpha* normally expresses low PDC because, as a Crabtree-negative yeast, it does not ferment aggressively; by boosting Pdc levels, more pyruvate was diverted to acetaldehyde and then to ethanol. This engineering roughly doubled the ethanol yield and reduced xylitol formation, showing proof-of-concept that modifying a single key node can tilt the balance from respiration to fermentation. Similarly, introducing or upregulating alcohol dehydrogenase (ADH) can help ensure acetaldehyde is efficiently reduced to ethanol. Some *O. polymorpha* strains naturally produce ethanol dehydrogenase as part of methanol metabolism (they need to detoxify acetaldehyde from methanol dissimilation), so ADH is present but might need overexpression under fermentative conditions.

Another target is the transport and uptake of sugar. *O. polymorpha* has at least one native xylose transporter (analogous to the Hxt family), but at high temperatures, its efficiency may drop. (Vasylyshyn et al. 2020) compared different sugar transporters in *O. polymorpha* and found that expressing a heterologous high-affinity xylose transporter (e.g., from *Scheffersomyces* or *Kluyveromyces*) improved xylose uptake and ethanol productivity in high-temperature fermentation (Vasylyshyn et al. 2020). They also noted that *O. polymorpha*'s own transporter genes responded to oxygen levels; under anaerobic conditions, the expression was suboptimal, limiting xylose entry. By using a constitutive promoter to drive an xylose symporter, they bypassed oxygen-dependent regulation.

Recently, *O. polymorpha* has been engineered to utilize cellobiose (a dimer of glucose from cellulose) alongside xylose. This was accomplished by expressing a cellobiose transporter and a β-glucosidase enzyme, enabling the yeast to directly ferment cellobiose to ethanol at 45 °C (essentially doing partial saccharification internally). When combined with xylose utilization, this means *O. polymorpha* could potentially ferment both C5 and C6 sugars from pretreated biomass without needing an external β-glucosidase.

Stress tolerance engineering in *O. polymorpha* includes improving tolerance to furfural and acetic acid. One approach has been to overexpress ALD6 (an aldehyde dehydrogenase) to oxidize furfural to furanoic acid, and to overexpress ATF2 (a transcription factor involved in the acetate response) to handle acetic acid stress. The results showed that *O. polymorpha* strains with these modifications could ferment dilute-acid hydrolysates with less lag. Moreover, *O. polymorpha* naturally grows at low pH (it can tolerate pH 3–4), which can be beneficial as many inhibitors are less toxic in their undissociated form at lower pH levels.

From a synthetic biology toolkit perspective, *O. polymorpha* has benefited from the extensive work on *Pichia pastoris* (which is a close relative). Many plasmids and promoters (like AOX1, TEF1 promoters, etc.) are cross-compatible. A CRISPR/Cas9 system for *O. polymorpha* was reported by Gao et al. (2021), who successfully deleted a gene using a Cas9 expressed under a TEF1 promoter and an sgRNA under a U6 promoter (Gao et al. 2021; Wang et al. 2018). This advancement now enables more sophisticated genetic engineering efforts in *O. polymorpha*, including the assembly of complete synthetic pathways. Such capabilities may facilitate the production of higher alcohols or value-added chemicals derived from lignocellulosic side streams, thereby expanding their potential as a versatile biocatalyst.

One case study is the construction of an *O. polymorpha* strain that produces ethanol from a mix of xylose and glucose nearly as well as *S. cerevisiae*. Kurylenko et al. (2014) combined PDC overexpression, ADH overexpression, and knockouts of two key NADH dehydrogenases (to reduce respiratory capacity) in *O. polymorpha* (Kurylenko et al. 2014). The engineered strain, in microaerobic fermentation of a glucose/xylose mixture at 42 °C, achieved an ethanol yield of ~0.40 g/g (compared to 0.45–0.48 g/g for *S. cerevisiae* under similar conditions). This was a significant jump from the wild-type *O. polymorpha* yield, which was <0.2 g/g on xylose. The strain still required a small amount of oxygen to avoid stalling, but it demonstrated that, with a few modifications, *O. polymorpha* can be turned into a reasonably ethanologenic yeast.

6.5.4 *Engineering* Kluyveromyces marxianus

The engineering of *K. marxianus* for lignocellulosic ethanol has targeted several fronts: enabling anaerobic xylose (and arabinose) fermentation, improving thermotolerance even further, and relieving regulatory constraints. One of the first achievements was enabling the simultaneous utilization of glucose and xylose. Wild-type *K. marxianus* sequentially uses glucose and then xylose. To address this, researchers examined glucose repression. By deleting the *MIG1* gene (encoding a global catabolite repressor in many yeasts), *K. marxianus* was able to co-consume xylose with glucose, albeit slowly (Hua et al. 2019). Additionally, overexpressing a specific transporter (*KmHXT2*, for example) facilitated the uptake of xylose even when glucose was present. Ha-Tran et al. (2020) conducted a study in which they overexpressed a glucose-insensitive xylose transporter, along with XR/XDH, and successfully accelerated xylose consumption in the presence of glucose (Ha-Tran et al. 2020; Hua et al. 2019).

For xylose fermentation, since *K. marxianus* possesses the genes for XR/XDH but does not ferment xylose efficiently, one strategy was to introduce a heterologous xylose isomerase pathway. In 2013, Wang et al. expressed the XI from *Bacteroides* plus *K. marxianus*' own xylulokinase, resulting in a strain that could anaerobically produce ethanol from xylose, though the rate was low. More recently, efforts have combined both pathways: some engineered strains have *S. stipitis* XR-XDH as well as an XI, allowing the yeast to utilize whichever route is more efficient under given conditions (somewhat redundant but potentially beneficial across a range of oxygen levels) (Wang et al. 2013).

Another significant improvement was engineering redox in *K. marxianus*. Since *K. marxianus* is Crabtree-negative, it normally reoxidizes most NADH through respiration. Under oxygen-limited fermentation, *K. marxianus* often accumulates glycerol as a redox sink (like *S. cerevisiae* does). To minimize that, one approach was to overexpress a non-respiratory NADH oxidase or increase ethanol pathway capacity (like boosting ADH). For example, overexpression of a water-forming NADH oxidase from *Streptococcus* in *K. marxianus* enabled it to regenerate NAD^+ without producing glycerol, thereby channeling more carbon into ethanol. This increased ethanol yield and reduced glycerol byproduct in mixed sugar fermentation (unpublished data from some groups, but conceptually similar to strategies in *S. cerevisiae*).

The thermotolerance of *K. marxianus* is already high, but pushing it a bit further can expand the operational window. Through adaptive evolution at 45–48 °C, strains have been isolated that grow and ferment at 47 °C, enabling truly simultaneous saccharification at nearly 50 °C (the upper limit of cellulase enzyme activity) (Fonseca et al. 2008). Genome sequencing of these strains identified upregulation of heat shock proteins and membrane composition changes (more saturated fatty acids for stability). Genetic engineering to express certain heat shock proteins (like Hsf1 and Msn2) conferred slightly better heat survival in *S. cerevisiae*, indicating a path to improve it further for industrial conditions where cooling is minimal (P. Li et al. 2017).

K. marxianus is notable for its thermotolerance and substrate flexibility, making it attractive for industrial bioethanol production. In pilot-scale studies, *K. marxianus* fermented Jerusalem artichoke hydrolysate to high ethanol titers, achieving up to 91.5% of the theoretical yield under optimized conditions. However, optimal fermentation was reported at 35–38 °C, rather than above 40 °C (Yuan et al. 2008). For dairy waste, *K. marxianus* efficiently converts lactose to ethanol, with studies showing substantial ethanol accumulation during fermentation (Ozarslan and Kök Taş 2022). Engineered strains expressing xylose reductase and xylitol dehydrogenase (XR/XDH) have also enabled the production of ethanol from lignocellulosic substrates (González-Siso et al. 2014).

Synthetic biology has expanded the utility of *K. marxianus*. Respiratory-deficient mutants, such as those with disrupted mitochondrial dehydrogenases, exhibit increased fermentative metabolism and ethanol yield, though at the cost of growth rate and substrate range (González-Siso et al. 2014). CRISPR-based tools now allow for precise pathway integration and gene knockouts, as demonstrated by the multigene engineering for 2-phenylethanol production (Li et al. 2021). Additionally, adaptive evolution and transcription factor engineering have enhanced ethanol tolerance, with the upregulation of stress-responsive genes linked to improved performance under industrial conditions (Mo et al. 2019). These advances collectively position *K. marxianus* as a robust and versatile platform for next-generation bioprocessing.

6.5.5 Other Non-conventional Yeasts of Interest

Beyond the three main yeasts above, it is worth mentioning briefly others: *Yarrowia lipolytica,* an oleaginous yeast, does not naturally ferment sugars to ethanol efficiently (it accumulates lipids aerobically), but it has been engineered with a xylose pathway and could serve as a host if one wanted a yeast that can both produce ethanol and possibly valorize biomass into other bioproducts (lipids). *Pachysolen tannophilus* was one of the first yeasts discovered to ferment xylose (albeit poorly) and has an interesting native arabinose pathway; however, it has fallen out of favor in recent years compared to *S. stipitis*. *Candida tropicalis* can utilize a wide range of substrates and has been employed for xylitol production from xylose; genetically, it is closely related to *S. stipitis* within the CUG clade. While not a focus for ethanol due to pathogenicity concerns, it provides a model for tolerance and productivity (some *C. tropicalis* strains tolerate inhibitors well).

6.6 Fermentation Performance and Case Studies

The true test of engineered yeasts is their performance in actual lignocellulosic fermentation processes. This section examines reported fermentation metrics (yields, titers, productivities) for various strains and highlights case studies where these strains have been tested beyond shake flasks, in bioreactors, on real hydrolysates, or at pilot scale.

6.6.1 Ethanol Yields and Productivities

Modern engineered strains of *S. cerevisiae* can ferment glucose and xylose mixtures to ethanol with yields of 0.40–0.48 g ethanol/g total sugar consumed, approaching the theoretical maximum of ~0.51 g/g. In industrial settings, these strains consistently achieve yields exceeding 90% of the theoretical maximum on hexose sugars (Tsegaye et al. 2024). The performance of non-conventional yeasts is also improving. *S. stipitis* (formerly *P. stipitis*), a natural xylose fermenter, achieves yields of around 0.35–0.40 g/g on xylose and up to 0.45–0.48 g/g when glucose is also present. For example, a study by Skoog and Hahn-Hägerdal (1990) reported that *S. stipitis* could ferment xylose with an ethanol yield of 0.48 g/g under microaerobic conditions.

Engineered thermotolerant yeasts are also showing promise. An engineered *K. marxianus* strain carrying a xylose isomerase pathway was reported to ferment 30.15 g/L of xylose to produce 11.52 g/L of ethanol at 42 °C, achieving a yield of 0.38 g/g (Wang et al. 2013). While not as rapid as *S. cerevisiae*, its performance at higher temperatures is a significant advantage. Similarly, engineered strains of *O. polymorpha* have been developed that accumulate up to 40 times more ethanol from xylose compared to the wild type, demonstrating the potential of this thermotolerant yeast for high-temperature fermentations (Vasylyshyn et al. 2024). Table 6.2 summarizes some reported fermentation performances of different yeasts on lignocellulosic substrates or model sugar mixtures.

6.6.2 Pilot and Industrial Trials

Moving from the lab to industrial scale introduces new challenges; however, several strategies have been successfully piloted.

6.6.3 Separate Glucose and Xylose Fermentation

A straightforward approach is to use *S. cerevisiae* for the initial glucose fermentation, followed by a separate fermentation of the remaining xylose-rich liquor with a pentose-fermenting specialist, such as *S. stipitis*. While operationally complex, this strategy can achieve good overall yields.

6.6.4 Co-culture Simultaneous Fermentation

Co-culturing *S. cerevisiae* with an xylose-fermenting yeast, such as *S. stipitis*, enables the simultaneous conversion of both sugars. However, this approach is challenging as *S. cerevisiae* often outcompetes other yeasts for nutrients and oxygen, requiring careful tuning of process parameters (Naseeruddin et al. 2021).

6.6.5 Integrated Processes with Thermotolerant Yeasts

Simultaneous saccharification and co-fermentation (SSCF) is an attractive process that combines enzymatic hydrolysis and fermentation in a single vessel. Using thermotolerant yeasts like *K. marxianus* is particularly advantageous, as their optimal fermentation temperature (40–45 °C) is closer to that of hydrolytic enzymes (45–50 °C). This reduces cooling costs, minimizes contamination risk, and can improve overall process efficiency.

Table 6.2 Examples of fermentation performance of conventional and engineered non-conventional yeasts on pentose-containing substrates

Yeast strain	Substrate and conditions	Ethanol titer (g/L)	Yield (g/g sugars)	Productivity (g/L·h)	Strain highlights	References
S. cerevisiae (engineered, e.g., TMB3400)	Glucose + Xylose (50 + 50 g/L), 30 °C, controlled anaerobic	45–50 g/L	0.42–0.46	1.0–1.5 g/L·h	XR/XDH pathway, optimized; co-consumption achieved	Karhumaa et al. (2007)
S. stipitis CBS 6054 (wild)	Xylose (30 g/L), slight aeration, 30 °C	~12 g/L	0.4	0.25 g/L·h	Needs O_2; without O_2 yield <0.25 g/g	Skoog and Hahn-Hägerdal (1990)
K. marxianus (engineered)	Xylose (30.15 g/L), 42 °C, anaerobic	~11.5 g/L	0.38	0.069 g/L·h	Engineered with xylose isomerase pathway	Wang et al. (2013)
O. polymorpha (engineered)	Cellobiose, glucose, and xylose, 45 °C	ND (Not Determined)	ND	ND	Engineered for high-temperature co-utilization of lignocellulosic sugars	Vasylyshyn et al. (2024)

6.6.6 Stress Tolerance in Real Hydrolysates

A critical hurdle for industrial application is the presence of inhibitory compounds in real lignocellulosic hydrolysates, which are generated during biomass pretreatment. These compounds, including acetic acid, furfural, and 5-hydroxymethylfurfural (HMF), can severely inhibit yeast growth and fermentation. Non-conventional yeasts vary in their tolerance; for instance, *S. stipitis* is known to have low ethanol tolerance, while *K. marxianus* is naturally more robust against a wide range of pH and toxins.

Recent breakthroughs have shown that tolerance can be rationally engineered. A study demonstrated that by engineering aldehyde reduction pathways and adjusting the extracellular pH and potassium levels, a single *S. cerevisiae* strain could achieve near-parity ethanol production in inhibitor-laden feedstocks compared to clean sugar streams. This lightweight design was also transferable to other engineered strains, enabling them to produce cellulosic ethanol or other bioproducts from diverse, toxic hydrolysates (Lam et al. 2021).

6.7 Commercial Deployment and Future Directions

To date, most operational second-generation ethanol plants have opted for robust, inhibitor-tolerant, and xylose-fermenting strains of *S. cerevisiae*. These industrial workhorses are favored for their high ethanol productivity and tolerance to harsh industrial conditions. Nonetheless, research on non-conventional yeasts, such as *K. marxianus* and *O. polymorpha*, provides a valuable reservoir of genes and strategies that can be utilized to further improve industrial strains.

An exciting future direction is consolidated bioprocessing (CBP), which aims to integrate enzyme production, biomass hydrolysis, and fermentation into a single step using a single microorganism. While no natural yeast can perform all these functions, efforts to engineer yeasts to secrete cellulolytic enzymes are ongoing. Thermotolerant yeasts like *K. marxianus* are particularly promising candidates for CBP due to their high protein secretion capacity and ability to grow at elevated temperatures, which could drastically simplify and reduce the cost of producing biofuels from lignocellulose.

6.7.1 Future Directions and Emerging Trends

Recent advances leverage genome-scale models and multi-omics to rationally design yeast strains. For example, the first *K. marxianus* genome-scale metabolic model (iSM996) captured its thermotolerance and pentose metabolism (Marcišauskas et al. 2019). CRISPR-Cas9 now enables multiplexed editing (e.g., knocking out ADH/ATF in *K. marxianus*) guided by such data (Ha-Tran et al. 2020).

Adaptive laboratory evolution under industrial conditions improves xylose fermentation and inhibitor tolerance. Two-stage evolution yielded an *S. cerevisiae* strain that ferments xylose anaerobically in lignocellulosic hydrolysate (Parreiras et al. 2014). ALE in the presence of inhibitors (furfural, acids) has likewise produced *P. kudriavzevii* variants with higher ethanol yield and multistress tolerance (Dolpatcha et al. 2023).

Membrane lipid engineering and antioxidant enhancement boost stress resistance. Increasing ergosterol and unsaturated fatty acids (e.g., oleic acid) in yeast membranes enhances ethanol tolerance, as does overexpressing catalase/SOD to mitigate oxidative stress (Tucaliuc et al. 2023). These traits directly improve fermentation yields under stress.

Synthetic biology enables the co-production of value-added products. For instance, *Clavispora lusitaniae* can co-produce ethanol and xylitol from biomass sugars in one fermentation process (Guzmán-Hernández et al. 2025). Other designs reroute flux to chemicals (e.g., itaconic acid) alongside ethanol via engineered pathways. This co-production approach improves carbon use.

Novel chassis and hybrids combine traits from multiple species. Thermotolerant *K. marxianus* (capable of C5/C6 fermentation at >37 °C) serves as an alternative host (Marcišauskas et al. 2019). Interspecies hybrids (e.g., *S. cerevisiae* × *eubayanus*, the lager yeast) demonstrate how fusing genomes can confer complementary tolerances (Steensels et al. 2014). Such chassis can better tolerate challenging biomass hydrolysates.

Dynamic control circuits using biosensors optimize fermentation phases. Metabolite-responsive switches allow cells to sense conditions and adjust pathways. For example, a redox biosensor (TRX2p-yEGFP) in an industrial yeast detected NADPH imbalance from furfural, automatically adjusting feed to maintain performance (Foncillas et al. 2024). Yeast strain improvements are integrated with process innovations. Acid-tolerant yeasts (resisting acetic/formic acid) reduce detox costs (Dolpatcha et al. 2023). Engineering yeasts to secrete hydrolytic enzymes supports consolidated bioprocessing: co-culturing a cellulase-secreting *Bacillus* with *K. marxianus* expressing β-glucosidase greatly enhanced ethanol yield from biomass (Ha-Tran et al. 2020).

6.8 Conclusion

Non-conventional yeasts such as *S. stipitis*, *O. polymorpha*, and *K. marxianus* hold tremendous promise for advancing lignocellulosic bioethanol production due to their inherent abilities to utilize pentose sugars and withstand harsh conditions. Over the past decade, synthetic biology has provided the tools to unlock this potential, overcoming many natural limitations of these yeasts. Through CRISPR/Cas-enabled genome editing, researchers have dramatically improved the genetic tractability of species once considered "recalcitrant," enabling precise and rapid strain construction. Modular cloning and tailored promoter/terminator libraries now allow fine-tuned expression of multi-gene pathways, which has been critical for optimizing xylose and arabinose metabolism and balancing redox cofactors. Systems-level understanding from metabolic models and omics analyses guides targeted interventions, from knocking out competing pathways to upregulating stress response networks, making engineering efforts more predictive and systematic.

Landmark studies highlighted in this chapter demonstrate the fruits of these innovations: *S. stipitis* strains with substantially higher ethanol tolerance and yield, *K. marxianus* strains co-fermenting mixed sugars at 45 °C, and *O. polymorpha* strains redirecting the majority of xylose flux to ethanol. These successes illustrate how combining native traits with engineered enhancements can create superior biocatalysts. While *S. cerevisiae* still dominates industrial ethanol processes, the gap is narrowing. We now have alternate yeasts that can achieve comparable ethanol yields under conditions where *S. cerevisiae* falters (e.g., high temperature, presence of pentoses). Moreover, engineering strategies first developed in non-conventional yeasts (such as minimizing xylitol by balancing XR/XDH expression) have cross-fertilized improvements in *S. cerevisiae* and vice versa, underlining the synergistic progress in this field.

Challenges remain on the path to full industrial deployment. Further improvements are required to enhance robustness, the ability to ferment highly inhibitory hydrolysates to high ethanol titers over prolonged operations. This will likely be addressed by continued adaptive evolution combined with genome editing, as well as perhaps consortium approaches that mitigate stresses by dividing labor among strains. However, given the rapid advancements documented over the last decade, it is plausible that the next generation of cellulosic ethanol plants will incorporate these non-conventional yeast platforms, or at least the innovations they have spurred.

In summary, synthetic biology has enabled a leap from inherent potential to practical realization for non-conventional yeasts in biofuel production. By rewriting genomes and reprogramming metabolic circuits, we are turning what were once considered "specialist" yeasts into versatile, high-performance microbial factories. The continued interplay of fundamental research and industrial application will drive further refinements. Ultimately, the development of robust, pentose-fermenting yeast strains expands the feedstock base for bioethanol and improves process economics, bringing sustainable second-generation biofuels closer to large-scale reality. This aligns with global efforts toward renewable energy and waste valorization, as lignocellulosic ethanol can significantly reduce greenhouse gas emissions and reliance on fossil fuels. The work reviewed in this chapter thus not only represents scientific and engineering achievement but also contributes to the pressing pursuit of sustainable energy solutions.

Acknowledgment We would like to acknowledge the editors of this book for providing the opportunity to contribute and showcase our research interests and our respective research institutions for providing research facility. No external funding was received for this work.

References

Acevedo A, Conejeros R, Aroca G (2017) Ethanol production improvement driven by genome-scale metabolic modeling and sensitivity analysis in Scheffersomyces stipitis. PLoS One 12(6): e0180074

Amore R, Kötter P, Küster C, Ciriacy M, Hollenberg CP (1991) Cloning and expression in Saccharomyces cerevisiae of the NAD(P)H-dependent xylose reductase-encoding gene (XYL1) from the xylose-assimilating yeast Pichia stipitis. Gene 109(1):89–97

Bae S-J, Park BG, Kim B-G, Hahn J-S (2020) Multiplex gene disruption by targeted base editing of Yarrowia lipolytica genome using cytidine deaminase combined with the CRISPR/Cas9 system. Biotechnol J 15(1):e1900238

Balagurunathan B, Jonnalagadda S, Tan L, Srinivasan R (2012) Reconstruction and analysis of a genome-scale metabolic model for Scheffersomyces stipitis. Microb Cell Factories 11:27

Barros KO, Mader M, Krause DJ, Pangilinan J, Andreopoulos B, Lipzen A, Mondo SJ, Grigoriev IV, Rosa CA, Sato TK, Hittinger CT (2024) Oxygenation influences xylose fermentation and gene expression in the yeast genera Spathaspora and Scheffersomyces. Biotechnol Biofuels Bioprod 17(1):20

Bassett S, Da Silva NA (2024) Engineering a carbon source-responsive promoter for improved biosynthesis in the non-conventional yeast. Metab Eng Commun 18:e00238

Bettiga M, Bengtsson O, Hahn-Hägerdal B, Gorwa-Grauslund MF (2009) Arabinose and xylose fermentation by recombinant Saccharomyces cerevisiae expressing a fungal pentose utilization pathway. Microb Cell Factories 8:40

Bilal M, Ji L, Xu Y, Xu S, Lin Y, Iqbal HMN, Cheng H (2022) Bioprospecting as a robust host for industrial biotechnology. Front Bioeng Biotechnol 10:851768

Cao M, Gao M, Ploessl D, Song C, Shao Z (2018) CRISPR-mediated genome editing and gene repression in Scheffersomyces stipitis. Biotechnol J 13(9):e1700598

Choo JH, Han C, Kim J-Y, Kang HA (2014) Deletion of a KU80 homolog enhances homologous recombination in the thermotolerant yeast *Kluyveromyces marxianus*. Biotechnol Lett 36(10): 2059–2067

Correia K (2019) Reconstructing the evolution of xylose fermentation in Scheffersomyces Stipitis

De Vit MJ, Waddle JA, Johnston M (1997) Regulated nuclear translocation of the Mig1 glucose repressor. Mol Biol Cell 8(8):1603–1618

Dolpatcha S, Phong HX, Thanonkeo S, Klanrit P, Yamada M, Thanonkeo P (2023) Adaptive laboratory evolution under acetic acid stress enhances the multistress tolerance and ethanol production efficiency of Pichia kudriavzevii from lignocellulosic biomass. Sci Rep 13(1):1–15

Foncillas RP, Magnusson S, Al-Rudainy B, Wallberg O, Gorwa-Grauslund MF, Carlquist M (2024) Automated yeast cultivation control using a biosensor and flow cytometry. J Ind Microbiol Biotechnol 51. https://doi.org/10.1093/jimb/kuae039

Fonseca GG, Heinzle E, Wittmann C, Gombert AK (2008) The yeast *Kluyveromyces marxianus* and its biotechnological potential. Appl Microbiol Biotechnol 79(3):339–354

Gao M, Cao M, Suástegui M, Walker J, Rodriguez Quiroz N, Wu Y, Tribby D, Okerlund A, Stanley L, Shanks JV, Shao Z (2017) Innovating a nonconventional yeast platform for producing shikimate as the building block of high-value aromatics. ACS Synth Biol 6(1):29–38

Gao J, Gao N, Zhai X, Zhou YJ (2021) Recombination machinery engineering for precise genome editing in methylotrophic yeast. iScience 24(3):102168

Gao J, Ye C, Cheng J, Jiang L, Yuan X, Lian J (2022) Enhancing homologous recombination efficiency in for multiplex genome integration using short homology arms. ACS Synth Biol 11(2):547–553

González-Siso MI, Touriño A, Vizoso Á, Pereira-Rodríguez Á, Rodríguez-Belmonte E, Becerra M, Cerdán ME (2014) Improved bioethanol production in an engineered Kluyveromyces lactis strain shifted from respiratory to fermentative metabolism by deletion of NDI1. Microb Biotechnol 8(2):319

Granados-Arvizu JA, Canizal-García M, Madrigal-Pérez LA, González-Hernández JC, Regalado-González C (2021) Inhibition of alternative respiration system of Scheffersomyces stipitis and effect on glucose or xylose fermentation. FEMS Yeast Res 21(2). https://doi.org/10.1093/femsyr/foab005

Guzmán-Hernández D, Ramos-Valdivia AC, Poggi-Varaldo HM, Barrera-Cortés J, Cristiani-Urbina E, Ponce-Noyola T (2025) Ethanol and xylitol co-production by Clavispora lusitaniae growing on saccharified sugar cane bagasse in anaerobic/microaerobic conditions. Fermentation 11(6):344

Harzevili FD (2023) Synthetic biology of yeasts: tools and applications. Springer

Ha-Tran DM, Nguyen TTM, Huang C-C (2020) *Kluyveromyces marxianus*: current state of omics studies, strain improvement strategy, and potential industrial implementation. Fermentation 6(4):124

Hemansi H, Patel AK, Saini JK, Singhania RR (2022) Development of multiple inhibitor tolerant yeast via adaptive laboratory evolution for sustainable bioethanol production. Bioresour Technol 344(Pt B):126247

Ho NW, Chen Z, Brainard AP (1998) Genetically engineered Saccharomyces yeast capable of effective cofermentation of glucose and xylose. Appl Environ Microbiol 64(5):1852–1859

Hua Y, Wang J, Zhu Y, Zhang B, Kong X, Li W, Wang D, Hong J (2019) Release of glucose repression on xylose utilization in *Kluyveromyces marxianus* to enhance glucose-xylose co-utilization and xylitol production from corncob hydrolysate. Microb Cell Factories 18(1):24

Ishchuk OP, Voronovsky AY, Stasyk OV, Gayda GZ, Gonchar MV, Abbas CA, Sibirny AA (2008) Overexpression of pyruvate decarboxylase in the yeast Hansenula polymorpha results in increased ethanol yield in high-temperature fermentation of xylose. FEMS Yeast Res 8(7): 1164–1174

Ito Y, Terai G, Ishigami M, Hashiba N, Nakamura Y, Bamba T, Kumokita R, Hasunuma T, Asai K, Ishii J, Kondo A (2020) Exchange of endogenous and heterogeneous yeast terminators in Pichia pastoris to tune mRNA stability and gene expression. Nucleic Acids Res 48(22):13000–13012

Jassim Shalsh F, Azlina Binti Ibrahim N, Arifullah M (2016) Screening of ethanol-tolerant fusants of Saccharomyces cerevisiae and Pichia stipitis from protoplast fusion. Curr Res J Biol Sci 8(1): 10–14

Jeffries TW, Grigoriev IV, Grimwood J, Laplaza JM, Aerts A, Salamov A, Schmutz J, Lindquist E, Dehal P, Shapiro H, Jin Y-S, Passoth V, Richardson PM (2007) Genome sequence of the lignocellulose-bioconverting and xylose-fermenting yeast Pichia stipitis. Nat Biotechnol 25(3): 319–326

Karhumaa K, Garcia Sanchez R, Hahn-Hägerdal B, Gorwa-Grauslund M-F (2007) Comparison of the xylose reductase-xylitol dehydrogenase and the xylose isomerase pathways for xylose fermentation by recombinant Saccharomyces cerevisiae. Microb Cell Factories 6:5

Kim SR, Kwee NR, Kim H, Jin Y-S (2013) Feasibility of xylose fermentation by engineered Saccharomyces cerevisiae overexpressing endogenous aldose reductase (GRE3), xylitol dehydrogenase (XYL2), and xylulokinase (XYL3) from Scheffersomyces stipitis. FEMS Yeast Res 13(3):312–321

Kötter P, Amore R, Hollenberg CP, Ciriacy M (1990) Isolation and characterization of the Pichia stipitis xylitol dehydrogenase gene, XYL2, and construction of a xylose-utilizing Saccharomyces cerevisiae transformant. Curr Genet 18(6):493–500

Kumar P, Sahoo DK, Sharma D (2021) The identification of novel promoters and terminators for protein expression and metabolic engineering applications in. Metab Eng Commun 12:e00160

Kurylenko OO, Ruchala J, Hryniv OB, Abbas CA, Dmytruk KV, Sibirny AA (2014) Metabolic engineering and classical selection of the methylotrophic thermotolerant yeast Hansenula polymorpha for improvement of high-temperature xylose alcoholic fermentation. Microb Cell Factories 13:122

Kurylenko OO, Ruchala J, Vasylyshyn RV, Stasyk OV, Dmytruk OV, Dmytruk KV, Sibirny AA (2018) Peroxisomes and peroxisomal transketolase and transaldolase enzymes are essential for xylose alcoholic fermentation by the methylotrophic thermotolerant yeast. Biotechnol Biofuels 11:197

Kuyper M, Hartog MMP, Toirkens MJ, Almering MJH, Winkler AA, van Dijken JP, Pronk JT (2005) Metabolic engineering of a xylose-isomerase-expressing Saccharomyces cerevisiae strain for rapid anaerobic xylose fermentation. FEMS Yeast Res 5(4–5):399–409

Lam FH, Turanlı-Yıldız B, Liu D, Resch MG, Fink GR, Stephanopoulos G (2021) Engineered yeast tolerance enables efficient production from toxified lignocellulosic feedstocks. Sci Adv 7(26). https://doi.org/10.1126/sciadv.abf7613

Lane MM, Morrissey JP (2010) *Kluyveromyces marxianus*: a yeast emerging from its sister's shadow. Fungal Biol Rev 24(1–2):17–26

Lertwattanasakul N, Kosaka T, Hosoyama A, Suzuki Y, Rodrussamee N, Matsutani M, Murata M, Fujimoto N, Suprayogi, Tsuchikane K, Limtong S, Fujita N, Yamada M (2015) Genetic basis of the highly efficient yeast *Kluyveromyces marxianus*: complete genome sequence and transcriptome analyses. Biotechnol Biofuels 8:47

Li P, Fu X, Zhang L, Zhang Z, Li J, Li S (2017) The transcription factors Hsf1 and Msn2 of thermotolerant promote cell growth and ethanol fermentation of at high temperatures. Biotechnol Biofuels 10:289

Li M, Lang X, Moran Cabrera M, De Keyser S, Sun X, Da Silva N, Wheeldon I (2021) CRISPR-mediated multigene integration enables Shikimate pathway refactoring for enhanced 2-phenylethanol biosynthesis in *Kluyveromyces marxianus*. Biotechnol Biofuels 14(1):3

Lin T-H, Huang C-F, Guo G-L, Hwang W-S, Huang S-L (2012) Pilot-scale ethanol production from rice straw hydrolysates using xylose-fermenting Pichia stipitis. Bioresour Technol 116:314–319

Lyu Y, Wu P, Zhou J, Yu Y, Lu H (2021) Protoplast transformation of *Kluyveromyces marxianus*. Biotechnol J 16(12):e2100122

Malcı K, Watts E, Roberts TM, Auxillos JY, Nowrouzi B, Boll HO, Nascimento CZSD, Andreou A, Vegh P, Donovan S, Fragkoudis R, Panke S, Wallace E, Elfick A, Rios-Solis L (2022) Standardization of synthetic biology tools and assembly methods for and emerging yeast species. ACS Synth Biol 11(8):2527–2547

Marcišauskas S, Ji B, Nielsen J (2019) Reconstruction and analysis of a *Kluyveromyces marxianus* genome-scale metabolic model. BMC Bioinf 20(1):551

Matsushika A, Inoue H, Kodaki T, Sawayama S (2009) Ethanol production from xylose in engineered Saccharomyces cerevisiae strains: current state and perspectives. Appl Microbiol Biotechnol 84(1):37–53

Mo W, Wang M, Zhan R, Yu Y, He Y, Lu H (2019) *Kluyveromyces marxianus* developing ethanol tolerance during adaptive evolution with significant improvements of multiple pathways. Biotechnol Biofuels 12:63

Moon SY, An N-Y, Lee JY (2025) Transforming non-conventional yeasts into key players in biotechnology: advances in synthetic biology applications. Front Microbiol 16:1600187

Näätsaari L, Mistlberger B, Ruth C, Hajek T, Hartner FS, Glieder A (2012) Deletion of the Pichia pastoris KU70 homologue facilitates platform strain generation for gene expression and synthetic biology. PLoS One 7(6):e39720

Nambu-Nishida Y, Nishida K, Hasunuma T, Kondo A (2017) Development of a comprehensive set of tools for genome engineering in a cold- and thermo-tolerant *Kluyveromyces marxianus* yeast strain. Sci Rep 7(1):8993

Naseeruddin S, Desai S, Venkateswar Rao L (2021) Co-culture of Saccharomyces cerevisiae (VS3) and Pichia stipitis (NCIM 3498) enhances bioethanol yield from concentrated Prosopis juliflora hydrolysate. 3 Biotech 11(1):21

Ndubuisi IA, Amadi CO, Nwagu TN, Murata Y, Ogbonna JC (2023) Non-conventional yeast strains: unexploited resources for effective commercialization of second generation bioethanol. Biotechnol Adv 63:108100

Nigam JN (2001) Development of xylose-fermenting yeast Pichia stipitis for ethanol production through adaptation on hardwood hemicellulose acid prehydrolysate. J Appl Microbiol 90(2): 208–215

Nijland JG, Driessen AJM (2019) Engineering of pentose transport in for biotechnological applications. Front Bioeng Biotechnol 7:464

Nitiyon S, Keo-Oudone C, Murata M, Lertwattanasakul N, Limtong S, Kosaka T, Yamada M (2016) Efficient conversion of xylose to ethanol by stress-tolerant Kluyveromyces marxianus BUNL-21. Springerplus 5:185

O'Laughlin R, Tran Q, Lezia A, Ngamkanjanarat W, Emmanuele P, Hao N, Hasty J (2024) A standardized set of MoClo-compatible inducible promoter systems for tunable gene expression in yeast. ACS Synth Biol 13(1):85–102

Obst U, Lu TK, Sieber V (2017) A modular toolkit for generating Pichia pastoris secretion libraries. https://doi.org/10.1021/acssynbio.6b00337

Ozarslan S, Kök Taş T (2022) Treatment of lactose by fermentation: production process on β-galactosidase using Kluyveromyces marxianus isolated from kefir grains. Int Food Res J 29(4):864–871

Park BG, Kim J, Kim E-J, Kim Y, Kim J, Kim JY, Kim B-G (2021) Application of random mutagenesis and synthetic FadR promoter for production of ω-hydroxy fatty acid. Front Bioeng Biotechnol 9:624838

Parreiras LS, Breuer RJ, Narasimhan RA, Higbee AJ, La Reau A, Tremaine M, Qin L, Willis LB, Bice BD, Bonfert BL, Pinhancos RC, Balloon AJ, Uppugundla N, Liu T, Li C, Tanjore D, Ong IM, Li H, Pohlmann EL et al (2014) Engineering and two-stage evolution of a lignocellulosic hydrolysate-tolerant Saccharomyces cerevisiae strain for anaerobic fermentation of xylose from AFEX pretreated corn stover. PLoS ONE 9(9):e107499

Patra P, Das M, Kundu P, Ghosh A (2021) Recent advances in systems and synthetic biology approaches for developing novel cell-factories in non-conventional yeasts. Biotechnol Adv 47:107695

Prielhofer R, Barrero JJ, Steuer S, Gassler T, Zahrl R, Baumann K, Sauer M, Mattanovich D, Gasser B, Marx H (2017) GoldenPiCS: a Golden Gate-derived modular cloning system for applied synthetic biology in the yeast Pichia pastoris. BMC Syst Biol 11(1):123

Qiu Y, Lei P, Wang R, Sun L, Luo Z, Li S, Xu H (2023) Kluyveromyces as promising yeast cell factories for industrial bioproduction: from bio-functional design to applications. Biotechnol Adv 64:108125

Rajkumar AS, Varela JA, Juergens H, Daran J-MG, Morrissey JP (2019) Biological parts for synthetic biology. Front Bioeng Biotechnol 7:97

Reichard WD, Smith SE, Robertson JB (2023) BLINCAR: a reusable bioluminescent and Cas9-based genetic toolset for repeatedly modifying wild-type. mSphere 8(4):e0022423

Roy P, Gahlawat VK, Saravanan C, Singh BP (2023) Enhancing bioflavor production by solid-state fermentation using Kluyveromyces marxianus and L-phenylalanine. J Basic Microbiol 63(1):75–91

Ruchala J, Kurylenko OO, Dmytruk KV, Sibirny AA (2020) Construction of advanced producers of first- and second-generation ethanol in Saccharomyces cerevisiae and selected species of non-conventional yeasts (Scheffersomyces stipitis, Ogataea polymorpha). J Ind Microbiol Biotechnol 47(1):109–132

Ryabova OB, Chmil OM, Sibirny AA (2003) Xylose and cellobiose fermentation to ethanol by the thermotolerant methylotrophic yeast Hansenula polymorpha. FEMS Yeast Res 4(2):157–164

Saengphing T, Sattayawat P, Kalawil T, Suwannarach N, Kumla J, Yamada M, Panbangred W, Rodrussamee N (2024) Improving furfural tolerance in a xylose-fermenting yeast Spathaspora passalidarum CMUWF1-2 via adaptive laboratory evolution. Microb Cell Factories 23(1):80

Santos SC, de Sousa AS, Dionísio SR, Tramontina R, Ruller R, Squina FM, Vaz Rossell CE, da Costa AC, Ienczak JL (2016) Bioethanol production by recycled Scheffersomyces stipitis in sequential batch fermentations with high cell density using xylose and glucose mixture. Bioresour Technol 219:319–329

Semkiv MV, Ruchala J, Tsaruk AY, Zazulya AZ, Vasylyshyn RV, Dmytruk OV, Zuo M, Kang Y, Dmytruk KV, Sibirny AA (2022) The role of hexose transporter-like sensor hxs1 and transcription activator involved in carbohydrate sensing azf1 in xylose and glucose fermentation in the thermotolerant yeast Ogataea polymorpha. Microb Cell Factories 21(1):162

Shan L, Dai Z, Wang Q (2021) Advances and opportunities of CRISPR/Cas technology in bioengineering non-conventional yeasts. Front Bioeng Biotechnol 9:765396

Sharma S, Arora A (2020) Tracking strategic developments for conferring xylose utilization/ fermentation by Saccharomyces cerevisiae. Ann Microbiol 70(1). https://doi.org/10.1186/s13213-020-01590-9

Shen Y, Hou J, Bao X (2013) Enhanced xylose fermentation capacity related to an altered glucose sensing and repression network in a recombinant Saccharomyces cerevisiae. Bioengineered 4(6):435–437

Sibirny AA (2023) Metabolic engineering of non-conventional yeasts for construction of the advanced producers of biofuels and high-value chemicals. BBA Adv 3:100071

Skoog K, Hahn-Hägerdal B (1990) Effect of oxygenation on xylose fermentation by Pichia stipitis. Appl Environ Microbiol 56(11). https://doi.org/10.1128/aem.56.11.3389-3394.1990

Slininger PJ, Shea-Andersh MA, Thompson SR, Dien BS, Kurtzman CP, Balan V, da Costa Sousa L, Uppugundla N, Dale BE, Cotta MA (2015) Evolved strains of Scheffersomyces stipitis achieving high ethanol productivity on acid- and base-pretreated biomass hydrolyzate at high solids loading. Biotechnol Biofuels 8:60

Steensels J, Snoek T, Meersman E, Nicolino MP, Voordeckers K, Verstrepen KJ (2014) Improving industrial yeast strains: exploiting natural and artificial diversity. FEMS Microbiol Rev 38(5): 947

Suzuki T, Hoshino T, Matsushika A (2019) High-temperature ethanol production by a series of recombinant xylose-fermenting Kluyveromyces marxianus strains. Enzym Microb Technol 129:109359

Tantirungkij M, Nakashima N, Seki T, Yoshida T (1993) Construction of xylose-assimilating Saccharomyces cerevisiae. J Ferment Bioeng 75(2):83–88

Toivari MH, Aristidou A, Ruohonen L, Penttilä M (2001) Conversion of xylose to ethanol by recombinant Saccharomyces cerevisiae: importance of xylulokinase (XKS1) and oxygen availability. Metab Eng 3(3):236–249

Tsegaye KN, Alemnew M, Berhane N (2024) Saccharomyces cerevisiae for lignocellulosic ethanol production: a look at key attributes and genome shuffling. Front Bioeng Biotechnol 12:1466644

Tucaliuc A, Cascaval D, Galaction A-I (2023) Cellular stress impact on yeast activity in biotechnological processes—a short overview. Microorganisms 11(10):2522

Vasylyshyn R, Kurylenko O, Ruchala J, Shevchuk N, Kuliesiene N, Khroustalyova G, Rapoport A, Daugelavicius R, Dmytruk K, Sibirny A (2020) Engineering of sugar transporters for improvement of xylose utilization during high-temperature alcoholic fermentation in Ogataea polymorpha yeast. Microb Cell Factories 19(1):96

Vasylyshyn R, Dmytruk O, Sybirnyy A, Ruchała J (2024) Engineering of Ogataea polymorpha strains with ability for high-temperature alcoholic fermentation of cellobiose. FEMS Yeast Res 24:foae007

Verduyn C, Van Kleef R, Frank J, Schreuder H, Van Dijken JP, Scheffers WA (1985) Properties of the NAD(P)H-dependent xylose reductase from the xylose-fermenting yeast Pichia stipitis. Biochem J 226(3):669–677

Walfridsson M, Bao X, Anderlund M, Lilius G, Bülow L, Hahn-Hägerdal B (1996) Ethanolic fermentation of xylose with Saccharomyces cerevisiae harboring the Thermus thermophilus xylA gene, which expresses an active xylose (glucose) isomerase. Appl Environ Microbiol 62(12):4648–4651

Walton RT, Christie KA, Whittaker MN, Kleinstiver BP (2020) Unconstrained genome targeting with near-PAMless engineered CRISPR-Cas9 variants. Science (New York, NY) 368(6488): 290–296

Wang R, Li L, Zhang B, Gao X, Wang D, Hong J (2013) Improved xylose fermentation of Kluyveromyces marxianus at elevated temperature through construction of a xylose isomerase pathway. J Ind Microbiol Biotechnol 40(8):841–854

Wang L, Deng A, Zhang Y, Liu S, Liang Y, Bai H, Cui D, Qiu Q, Shang X, Yang Z, He X, Wen T (2018) Efficient CRISPR-Cas9 mediated multiplex genome editing in yeasts. Biotechnol Biofuels 11:277

Xia Y, Li Y, Shen W, Yang H, Chen X (2023) CRISPR-Cas technology for bioengineering conventional and non-conventional yeasts: progress and new challenges. Int J Mol Sci 24(20). https://doi.org/10.3390/ijms242015310

Xie L, Yu W, Gao J, Wang H, Zhou YJ (2024) Ogataea polymorpha as a next-generation chassis for industrial biotechnology. Trends Biotechnol 42(11):1363–1378

Yang Z, Edwards H, Xu P (2020) CRISPR-Cas12a/Cpf1-assisted precise, efficient and multiplexed genome-editing in. Metab Eng Commun 10:e00112

Yuan WJ, Zhao XQ, Ge XM, Bai FW (2008) Ethanol fermentation with Kluyveromyces marxianus from Jerusalem artichoke grown in salina and irrigated with a mixture of seawater and freshwater. J Appl Microbiol 105(6). https://doi.org/10.1111/j.1365-2672.2008.03903.x

Zha J, Yuwen M, Qian W, Wu X (2021) Yeast-based biosynthesis of natural products from xylose. Front Bioeng Biotechnol 9:634919

Zhang G-C, Liu J-J, Ding W-T (2012) Decreased xylitol formation during xylose fermentation in Saccharomyces cerevisiae due to overexpression of water-forming NADH oxidase. Appl Environ Microbiol 78(4):1081–1086

Zhang X, Gu S, Zheng X, Peng S, Li Y, Lin Y, Liang S (2021) A novel and efficient genome editing tool assisted by CRISPR-Cas12a/Cpf1 for *Pichia pastoris*. ACS Synth Biol 10(11):2927–2937

Chapter 7
Recent Advances in Omics-Based Microbial Genetic Engineering in Lignocellulosic Bioethanol Production

Premkumar Samuel S D, Wafa Rashid Al Bahlouli, S. Senthil Kumar, and Magapu Solomon Sudhakar

Abstract Microbial genetic engineering has evolved rapidly, particularly with the integration of omics high-throughput tools such as genomics, transcriptomics, proteomics, and metabolomics. These multi-omics platforms have revolutionized the design and optimization of microorganisms for boosting bioethanol production. This chapter reviews the application of multi-omics strategies in developing engineered microbes capable of overcoming traditional bottlenecks, including limited substrate utilization, low tolerance to inhibitors, and inefficient metabolic pathways. Technological advances such as CRISPR-based editing, synthetic biology, and adaptive laboratory evolution are emphasized for their role in enhancing microbial robustness and ethanol yields. Case studies demonstrate how omics-guided engineering has improved industrial strains, including *Saccharomyces cerevisiae*, *Zymomonas mobilis*, *Escherichia coli*, and cyanobacteria. Furthermore, industrial-scale applications underscore the commercial feasibility of bioengineered strains in second-generation (2G) bioethanol production. The chapter also addresses biosafety, environmental risks, and regulatory frameworks, emphasizing the need for ethical governance. Overall, the integration of omics and synthetic biology holds transformative potential in sustainable biofuel production, presenting a path toward scalable, efficient, and eco-friendly bioethanol technologies.

Keywords Omics technologies · Microbial genetic engineering · Bioethanol production · CRISPR · Synthetic biology · Lignocellulosic biomass

P. Samuel S D
Central Analytical & Applied Research Unit, College of Science, Sultan Qaboos University, Muscat, Oman

W. R. Al Bahlouli · S. Senthil Kumar · M. S. Sudhakar (✉)
Department of Applied Biotechnology, College of Applied Sciences and Pharmacy, University of Technology and Applied Sciences, Sur, Oman
e-mail: magapu.soloman@utas.edu.om

S. Prasad et al. (eds.), *Sustainable Bioethanol Production from Lignocellulosic Biomass*, Clean Energy Production Technologies,
https://doi.org/10.1007/978-981-95-9690-4_7

7.1 Introduction

The last decade has witnessed an unprecedented surge in both the volume and precision of biological data available to microbial engineers. This expansion, fueled by rapid advancements in sequencing technologies, high-throughput analytics, and computational biology, has transformed how microorganisms are studied and engineered. The transition from traditional, often empirical approaches of microbial strain improvement to rational, omics-enabled design of microbial cell factories represents a critical paradigm shift in modern industrial biotechnology (Amer and Baidoo 2021). Omics technologies—spanning genomics, transcriptomics, proteomics, and metabolomics—offer an integrated, high-dimensional understanding of the complex interactions linking genotype, environment, regulatory networks, and cellular phenotype. By harnessing this systems-level knowledge, researchers can now design and optimize microbial hosts with unprecedented precision, unlocking new opportunities in bio-based industries (Bradbury et al. 2025a, b).

Microbial genetic engineering plays a particularly pivotal role in lignocellulosic bioethanol production, a sector at the forefront of the global transition toward renewable energy. Unlike first-generation ethanol, which is derived from food-based feedstocks, lignocellulosic ethanol utilizes abundant, non-food biomass such as agricultural residues, forestry byproducts, and dedicated energy crops. While this offers sustainability advantages, it also presents technical challenges—complex polymeric structures, heterogeneous sugar composition, and the presence of inhibitory compounds released during pretreatment often limit microbial fermentation efficiency (Lee et al. 2012). Traditional microorganisms such as *Saccharomyces cerevisiae* and *Zymomonas mobilis* exhibit constraints including restricted substrate utilization, limited tolerance to inhibitors, and suboptimal ethanol yields. Omics-driven strategies provide a powerful toolkit to overcome these bottlenecks, enabling the rational rewiring of microbial metabolism for improved stress resilience, expanded substrate range, and enhanced conversion efficiency (Cunha et al. 2020).

The integration of multi-omics datasets with systems biology and computational modeling further strengthens this approach. By simulating the impact of genetic modifications, predicting pathway bottlenecks, and guiding targeted engineering interventions, systems biology empowers researchers to accelerate the development of robust, high-performance strains. Coupled with advances in synthetic biology and AI-driven analytics, this approach represents a shift toward predictive, data-driven bioengineering (Mauger et al. 2021; Bradbury et al. 2025a, b).

This chapter provides a comprehensive overview of recent advances in omics-based microbial genetic engineering for lignocellulosic bioethanol production. Emphasis is placed on (i) the contributions of individual omics layers, (ii) the integration of multi-omics approaches into systems biology frameworks, (iii) notable industrial applications and case studies, and (iv) persistent challenges and future directions for the field. Together, these insights underscore the transformative role of omics technologies in advancing sustainable biofuel production and accelerating the global transition toward a low-carbon, bio-based economy.

7.2 Overview of Omics Technologies in Microbial Engineering

7.2.1 The Omics Paradigm

Recent advances in omics technologies—transcriptomics, proteomics, and metabolomics—have revolutionized our understanding of microbial systems for lignocellulosic bioethanol production. By integrating these approaches, researchers can comprehensively analyze gene expression, protein function, and metabolic pathways, enabling the rational design and optimization of microbial strains capable of efficiently converting complex biomass into bioethanol (Amer and Baidoo 2021).

Transcriptomics leverages high-throughput RNA sequencing to elucidate key genes and regulatory pathways involved in stress response and fermentation performance under lignocellulosic hydrolysate conditions. Studies in *Saccharomyces cerevisiae* have highlighted stress-responsive genes and pathways, including Msn2/4 regulators and cAMP-protein kinase signaling, which are critical for enhancing ethanol tolerance and fermentation efficiency (Hughes et al. 2000; Zhang et al. 2018).

Proteomics employs mass spectrometry to identify and quantify proteins essential for lignocellulose degradation, inhibitor tolerance, and metabolic activity. Insights from proteomic analyses in *Zymomonas mobilis* and fungal co-culture systems have revealed enzymes and transporters that improve substrate utilization and ethanol yield, supporting more robust fermentation processes (Amer and Baidoo 2021).

Metabolomics and fluxomics provide a detailed picture of intracellular metabolites and metabolic fluxes, revealing bottlenecks and pathway dynamics during fermentation. Integration of metabolomic data with transcriptomic and proteomic information has enabled genome-scale metabolic network reconstruction, identifying key target pathways for metabolic engineering (Rai et al. 2016; Cunha et al. 2020). In tolerant yeast and bacteria, metabolomics has pinpointed redox drains and anaplerotic bottlenecks that cap ethanol productivity in hydrolysate, enabling interventions such as cofactor balancing, aldehyde sinks, and re-routing acetyl-CoA to mitigate toxicity (Xu et al. 2022a, b). Recent multi-omics network models now integrate metabolite, transcript, and phenotype layers to prioritize edits that raise ethanol tolerance/yield, shortening the path from hypothesis to strain (de Moura Ferreira and da Silva 2024).

Together, these omics layers offer a systems-level framework for microbial engineering, allowing for the design of high-performance strains optimized for sustainable and cost-effective bioethanol production (Bradbury et al. 2025a, b; Lee et al. 2012) (Table 7.1).

The integration of transcriptomics, proteomics, and metabolomics provides a comprehensive, systems-level understanding of microbial physiology and metabolism, enabling rational design and optimization of high-performance strains for lignocellulosic bioethanol production. By uncovering key genes, proteins, and

Table 7.1 Applications of multi-omics approaches in lignocellulosic bioethanol production

Omics approach	Focus/Method	Key insights	Applications in lignocellulosic bioethanol production	Representative studies
Transcriptomics	RNA-seq, gene expression profiling	Identifies stress-responsive genes and regulatory pathways (e.g., Msn2/4, cAMP-PKA signaling)	Improves ethanol tolerance and fermentation efficiency in S. cerevisiae	Hughes et al. (2000) and Zhang et al. (2018)
Proteomics	Mass spectrometry-based protein identification and quantification	Reveals enzymes, transporters, and proteins essential for lignocellulose degradation and stress tolerance	Enhances substrate utilization, inhibitor tolerance, and ethanol yield in *Zymomonas mobilis* and fungal co-cultures	Amer and Baidoo (2021)
Metabolomics and Fluxomics	Metabolite profiling, flux analysis, GC/LC-MS, NMR	Maps intracellular metabolites and metabolic fluxes; identifies bottlenecks (redox drains, acetyl-CoA balance)	Guides metabolic engineering via cofactor balancing, aldehyde detoxification, and pathway re-routing to boost ethanol productivity	Rai et al. (2016), Cunha et al. (2020)
Multi-omics integration	Integration of transcriptome, proteome, and metabolome data with computational models	Provides systems-level understanding of microbial metabolism and ethanol stress responses	Genome-scale metabolic network reconstruction; rational strain engineering; improved tolerance and yield	de Moura Ferreira and da Silva (2024), Bradbury et al. (2025a, b), Lee et al. (2012), and Zhang et al. (2010)

metabolic pathways, these omics-based strategies facilitate targeted metabolic engineering, improve substrate utilization, enhance stress tolerance, and maximize ethanol yields. Moving forward, the continued integration of multi-omics data with advanced computational modeling and synthetic biology tools promises to accelerate the development of sustainable, efficient, and economically viable bioethanol production platforms, supporting the global transition toward renewable energy (Amer and Baidoo 2021; Bradbury et al. 2025a, b; Lee et al. 2012; Zhang et al. 2010).

7.2.2 Multi-omics Integration and Systems Biology Toward Lignocellulosic Ethanol Production

A major advance in microbial biotechnology has been the transition from single-omics approaches to multi-omics integration, where genomics, transcriptomics, proteomics, and metabolomics are analyzed collectively within a unified framework. This holistic strategy provides unprecedented insights into the complexity of microbial physiology and metabolism, enabling a systems-level understanding that single-omics methods cannot achieve (Amer and Baidoo 2021). Through multi-omics integration, researchers can uncover critical regulatory nodes, metabolic bottlenecks, and stress-response mechanisms that directly influence ethanol production efficiency, stress tolerance, and substrate utilization.

Coupling multi-omics data with systems biology approaches—such as genome-scale metabolic modeling, constraint-based flux analysis, and in silico simulations—further strengthens predictive and engineering capabilities. These computational frameworks allow researchers to simulate genetic modifications, forecast metabolic rewiring outcomes, and prioritize intervention strategies before moving into experimental validation, thereby reducing time and cost in strain development (Mauger et al. 2021; Cunha et al. 2020).

Recent progress in computational analytics and artificial intelligence (AI) has amplified the utility of multi-omics integration. Machine learning algorithms, explainable AI models, and cloud-based data platforms now facilitate the management and interpretation of increasingly heterogeneous, high-dimensional datasets. These tools not only predict engineering outcomes but also provide mechanistic insights into pathway regulation, supporting the rational design of microbial cell factories optimized for lignocellulosic bioethanol production (Bradbury et al. 2025a, b). By bridging omics data with systems biology and AI-driven analytics, multi-omics integration is transforming the field, offering a pathway toward scalable, robust, and economically viable bioethanol production from complex biomass feedstocks.

7.3 Recent Technological Advances in Omics-Based Microbial Genetic Engineering

Modern advances in omics-driven microbial engineering have substantially improved the ways in which microorganisms are developed for industrial ethanol production. By leveraging genomic, transcriptomic, and metabolomic analyses, researchers can now gain an in-depth understanding of the genetic and metabolic blueprints of these microbes. Such insights facilitate purposeful interventions—either by modifying existing genes or introducing new genetic elements—to enhance essential traits like substrate utilization, stress tolerance, and ethanol yield (Jacob et al. 2024; Tsegaye et al. 2024).

7.3.1 High-Throughput DNA Sequencing and Functional Genomics

The advent of high-throughput sequencing technologies has revolutionized the process of mapping and analyzing the genomes of ethanol-producing strains. Rapid genome assembly enables scientists to swiftly identify gene networks crucial for carbohydrate metabolism, resistance to fermentation stresses, and the suppression of unwanted metabolic by-products. This has practical implications: identifying these genetic contributors makes it feasible to engineer strains better suited to industrial-scale fermentations (Ciamponi et al. 2022).

In addition, functional genomics—encompassing gene knockout and overexpression studies, transcriptome profiling, and systems analysis—provides detailed information about the roles played by both core and accessory genes. For instance, comparative and pangenomic studies have pinpointed sets of accessory genes in various microbial taxa that confer adaptability to harsh or changing environments, as well as versatility in substrate processing. These findings are not only pertinent to strains utilized for ethanol manufacture but also to those considered for biotechnological or pharmaceutical applications (Tsegaye et al. 2024). The integration of omics data with machine learning and advanced metabolic engineering is thus enabling the generation of next-generation microbial strains that combine functional diversity, resilience, and productivity.

Deep genome sequencing has enabled the comprehensive characterization of microbial strains, supporting targeted modification. Modern platforms can produce complete genome assemblies in a matter of hours, facilitating rapid identification of genes involved in substrate metabolism, stress tolerance, and by-product formation (Amer and Baidoo 2021).

Functional genomics employs mutational screens and comparative analyses to link individual genes to phenotypes, which is essential for identifying engineering targets (Bradbury et al. 2025a, b). For example, genome-wide screens in Saccharomyces cerevisiae have identified loci associated with ethanol yield and tolerance (Brandt et al. 2019).

Studies on *Virgibacillus* genomes have identified biosynthetic gene clusters responsible for producing metabolites with antibacterial properties, particularly effective against antibiotic-resistant bacteria. These accessory genes play a vital role in aiding the strains to adapt to various environments and utilize unique substrates. This highlights the significance of accessory genomes in industrial applications, such as pharmaceutical development. Pangenomic research on these strains demonstrates how accessory genes contribute to functional diversity and niche-specific adaptations (Othoum et al. 2019).

7.3.2 *Transcriptome Profiling and Regulatory Engineering*

Advanced transcriptomics (including single-cell techniques) has made it possible to dissect regulatory networks and temporal responses to stress (Xu et al. 2022a, b). Heat maps, cluster analyses, and gene network mappings guide the targeted modification of "master regulators" (such as YAP1, HAA1, and other transcription factors) to improve inhibitor tolerance and process performance (Brandt et al. 2019; Zhao and Bai 2009).

Recent efforts have included mapping the effect of feedstock hydrolysates on global gene expression and utilizing this data to rewire regulatory circuits for sustained fermentation under stress (Horinouchi et al. 2018).

7.3.3 *Quantitative and Functional Proteomics*

Han et al. (2010) demonstrate how microbial proteomics has become a cornerstone of modern biotechnology, enabling both quantitative and functional insights into microbial systems. Through quantitative proteomics, researchers can accurately measure changes in protein expression under various environmental or engineered conditions, which is crucial for improving industrial processes like fermentation and green fuel production. Functional proteomics complements this by revealing the roles and interactions of specific proteins, facilitating the identification of new enzymes and the engineering of more robust microbial strains. The study highlights how these approaches facilitate the identification of stress-response proteins and key metabolic enzymes, which can be targeted to improve microbial performance. Additionally, proteomics supports the development of synthetic biology tools, such as surface display systems and soluble fusion partners, that enhance the efficiency of recombinant protein production. In pharmaceutical contexts, microbial proteomics assists in uncovering virulence factors and resistance mechanisms, accelerating drug and vaccine development. By integrating both quantitative and functional data, scientists can dynamically model and rationally engineer microbial processes for greater productivity and stability.

Studies have also identified chaperones and stress-induced proteins critical for maintaining activity under fermentative stress (Amer and Baidoo 2021).

7.3.4 *Metabolomic and Fluxomic Advances*

Metabolomics employs NMR, GC-MS, and LC-MS to produce a quantitative fingerprint of intracellular metabolites, while fluxomics (often using isotope labeling) measures the actual rates of metabolic reactions (Tanaka et al. 2024; Rai et al. 2016).

Sarwan et al. (2024) provide a comprehensive overview of how omics technologies are driving advancements in biofuel research and production. The authors discuss how genomics, transcriptomics, proteomics, and metabolomics collectively enable a systems-level knowledge of green fuel-producing microorganisms, such as microalgae and specialized microbes. Genomic and transcriptomic analyses help identify key genes and regulatory networks involved in the biosynthesis of energy-rich compounds. At the same time, proteomics sheds light on enzyme activities and stress-response mechanisms essential for robust biofuel production. Metabolomics further complements these approaches by mapping metabolic fluxes and identifying bottlenecks in pathways that lead to biofuel precursors, such as lipids and alcohols. The merging of multi-omics information aids the rational engineering of microbial and algal strains, optimizing metabolic pathways and enhancing yield and process efficiency. The chapter also highlights the role of advanced computational tools and machine learning in interpreting complex omics datasets, accelerating strain development, and process optimization. Overall, Sarwan et al. (2024) emphasize that omics-driven strategies are crucial for addressing current challenges in biofuel generation, thereby making the production process more sustainable, economically viable, and scalable for future energy needs.

A prominent application is the identification of bottlenecks at pathway branch-points and the design of interventions (such as gene deletions or overexpression) to optimize carbon flux toward ethanol and reduce byproduct accumulation (Lee et al. 2012; Roy et al. 2023).

7.3.5 Systems Biology and Data-Driven Synthetic Biology

The multi-omics paradigm—integrated with computational modeling—has driven new frontiers in synthetic biology, enabling model-guided design of entire metabolic or regulatory networks (Nielsen and Keasling 2016).

Machine learning and AI are now routinely used to mine multi-omics datasets, predict the effects of genetic changes, and nominate novel engineering targets automatically (Bradbury et al. 2025a, b). Projects such as genome-scale metabolic modeling (GEM) use omics data to reconstruct and simulate the entire metabolism of a strain in silico, supporting rational design (Zhang et al. 2010).

7.4 Recent Key Advances and Case Studies

7.4.1 CRISPR-Based Multiplexed Genome Editing

One of the most revolutionary genetic engineering tools is the CRISPR-Cas system, which enables precise, site-specific, and multiplexed edits in bacteria, yeast, and filamentous fungi (Anzalone et al. 2019; Li et al. 2023). Recent advances in base editing and prime editing have enabled precise nucleotide modifications without causing double-stranded breaks. In industrial microbes, CRISPR has made possible the simultaneous knock-out of competitive or inhibitory pathways and upregulation of rate-limiting steps in sugar uptake and ethanol productivity (Kim et al. 2022). Newer "multiplexed" platforms have enabled the rapid construction of complex phenotypes, such as strains able to simultaneously ferment both hexoses and pentoses, resist hydrolysate inhibitors, and grow at an elevated temperature (Kuyper et al. 2005; Xu et al. 2022a, b).

7.4.2 Synthetic Biology and Modular Pathway Design

Synthetic biology has facilitated the development of standardized, modular gene circuits (Nielsen and Keasling 2016). Genes encoding key enzymes, including recombinant cellulases and pentose-specific transporters are assembled using advanced cloning techniques such as Gibson Assembly or Golden Gate Cloning.

Recent breakthroughs involve using synthetic promoters and regulatory sequences informed by omics analysis, resulting in predictable expression levels even in industrial conditions (Han et al. 2010).

7.4.3 Automated High-Throughput Screening and AI-Assisted Design

Automation, microfluidics, and biosensor platforms now enable high-throughput screening of thousands of engineered strains. AI models, trained on omics datasets, optimize the screening strategies and predict which genetic combinations are most likely to enhance performance.

Key successes include rapid identification of strains with improved ethanol tolerance, enhanced sugar utilization, and robust fermentative phenotypes (Bradbury et al. 2025a, b; Brandt et al. 2019).

7.4.4 Systems-Level Rewiring for Stress Tolerance and Inhibitor Resistance

Recent advances in omics-guided molecular engineering have revolutionized the progress of industrial microbial strains that can withstand harsh fermentation conditions. Through integrated analysis using transcriptomics, metabolomics, and comparative genomics, researchers have identified key genetic elements responsible for stress response mechanisms. The primary focus has been on addressing common industrial inhibitors:

- Acetic acid tolerance: Enhancement through modification of membrane transporters and stress response pathways
- Furfural resistance: Strengthening of detoxification mechanisms and redox balance systems
- HMF tolerance: Improvement of cellular defense mechanisms and metabolic adaptation

The engineering of regulatory nodes, particularly transcription factors YAP1 and HAA1, has proven especially effective. These modifications have resulted in strains with up to 40% higher survival rates under stress conditions, as validated through multi-omics studies (Brandt et al. 2019; Zhao and Bai 2009).

7.4.5 Application of Multi-omics in Metabolic Pathway Optimization

Metabolomic and fluxomic profiling identified critical bottlenecks in glycolytic and fermentative routes. Through targeted gene deletions and overexpression, carbon flux was redirected away from byproduct pathways to ethanol (Lee et al. 2012; Roy et al. 2023).

Integration of proteomic and transcriptomic data further refined strategies, such as increasing NADPH regeneration and cofactor balancing (Xu et al. 2022a, b), while systems biology approaches predicted the emergent properties of these modifications (Nielsen and Keasling 2016).

7.4.6 Single-Cell Omics and Genome Synthesis

Recent advances in single-cell genomics and transcriptomics have revealed heterogeneity within engineered populations, uncovering new targets for stabilizing phenotypes and improving population performance (Mauger et al. 2021).

Ambitious projects have demonstrated the feasibility of synthetic minimal genomes (Fredens et al. 2019), opening the door to designer host organisms optimized from the ground up using omics and computational principles.

7.5 Bioengineering Approaches for Enhanced Bioethanol Production

Microbial-assisted bioconversion plays a significant role in determining the overall cost of producing industrial biofuels (Joshi et al. 2021). For sustainable large-scale production, the perfect microorganism should be able to break down various types of

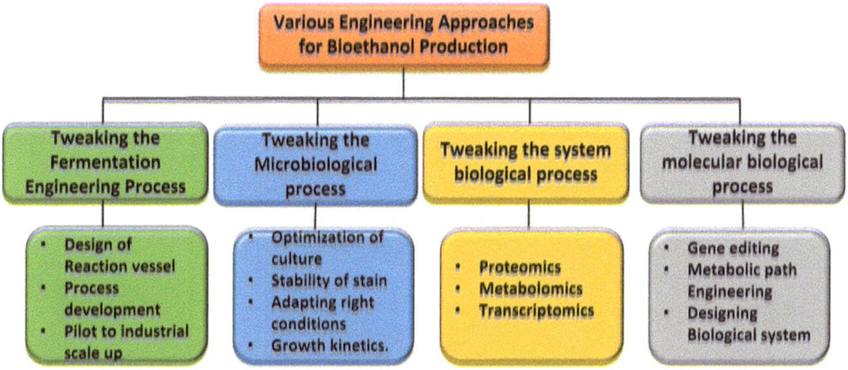

Fig. 7.1 Engineering techniques for enhancing biofuel production

lignocellulosic biomass and convert them efficiently. It also needs to produce high yields of the desired products while being resistant to the inhibitors present in the process (Lu et al. 2022a, b). To overcome metabolic challenges, scientists can customize and fine-tune the metabolic pathways of each microbe and their specific strains. Over the past decade, there has been impressive progress in engineering strategies that help design and optimize these microbial pathways for better biofuel production (Fig. 7.1).

Many comprehensive studies have been published on optimizing biofuel production through microbial metabolic pathways (Majidian et al. 2018; Jiang et al. 2020). Much of the research highlighted that green fuel production can be achieved through five principal metabolic pathways: (a) the keto-acid pathway, (b) the fatty acid synthesis pathway, (c) the isoprenoid pathway, (d) the CoA-dependent β-oxidation pathway, and (e) the reverse β-oxidation pathway (Fig. 7.2).

Metabolic engineering focuses on modifying microbial metabolic pathways to optimize ethanol yield from lignocellulosic substrates. Innovative approaches are evolving to enhance metabolic pathways and bioethanol production. Various methods are emerging to enable key enzymes (Jung et al. 2016), removing or managing competing pathways (Rodriguez et al. 2014), and bioengineering transcriptional factors for fine-tuning the metabolism (Ajjawi et al. 2017).

Current metabolic engineering strategies for improving ethanol production focus on multiple levels of pathway optimization. One major approach is the enhancement of glycolytic flux through the overexpression of rate-limiting enzymes such as hexokinase, phosphofructokinase, and glyceraldehyde-3-phosphate dehydrogenase (Kim et al. 2008). In addition, carbon flux redistribution is employed to optimize metabolic pathways, with particular emphasis on redirecting pyruvate metabolism via the overexpression of pyruvate decarboxylase (PDC) and alcohol dehydrogenase (ADH) (Nielsen et al. 2016). Another critical strategy is the co-utilization of mixed sugars, including pentoses such as xylose and arabinose, to improve substrate range and fermentation efficiency (Rajkumar et al. 2016). Furthermore, maintaining redox homeostasis by balancing NAD^+/NADH through cofactor engineering has proven

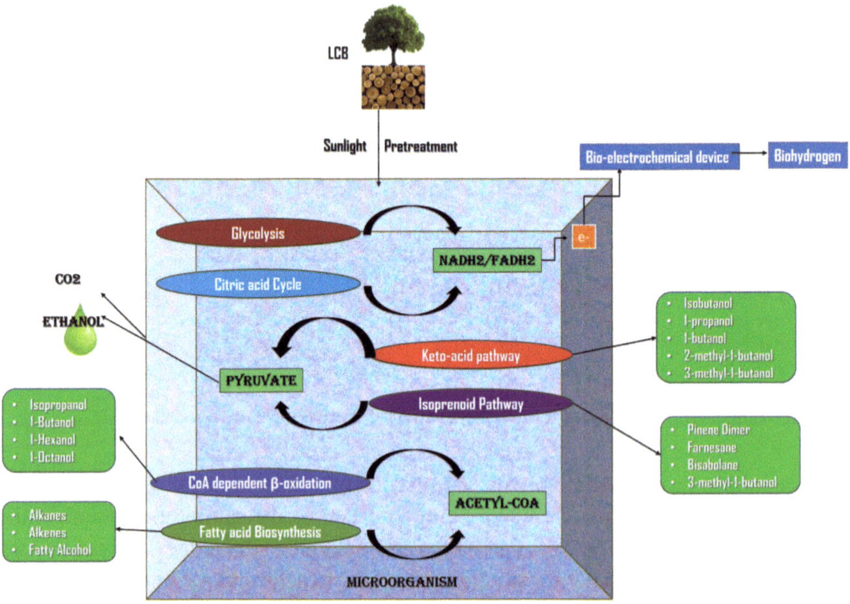

Fig. 7.2 Key metabolic pathways for biofuel production by microorganisms

essential for sustaining metabolic activity under fermentative conditions (Zhou et al. 2016). Finally, efforts to increase inhibitor tolerance have significantly enhanced strain robustness in the presence of lignocellulosic hydrolysate-derived inhibitors, thereby improving ethanol yield and process stability (Tsegaye et al. 2024).

Enhancing Glycolytic Flux This strategy is crucial for increasing ethanol production from lignocellulosic biomass, particularly in engineered microorganisms. Glycolysis is the central pathway for converting sugars (like glucose and xylose derived from cellulose and hemicellulose) into pyruvate, which is then fermented into ethanol. An increase in the rate of sugar conversion to pyruvate (an important precursor for ethanol synthesis) is the key to producing a high amount of ethanol.

Carbon Flux Optimization Redirecting carbon flux toward ethanol production involves overexpressing key enzymes in the glycolytic and fermentative pathways while knocking out genes responsible for byproduct formation. Optimizing carbon flux is vital for effective transformation of lignocellulosic material into ethanol using engineered microorganisms. Lignocellulose contains complex polymers like cellulose, hemicellulose, and lignin, which are hydrolyzed into fermentable sugars (mainly glucose and xylose). Effective carbon flux management ensures that these sugars are efficiently directed through metabolic pathways, leading to the efficient production of ethanol.

Readdressing Pyruvate Metabolism It represents a key metabolic engineering approach in designing microbial strains for the efficient conversion of lignocellulosic material into bioethanol. Since pyruvate is a vital metabolic intermediate at the junction of multiple pathways, optimizing its fate is critical to maximizing ethanol yield while minimizing byproduct formation. Overexpression of pyruvate decarboxylase (PDC) and alcohol dehydrogenase (ADH) redirects entirely toward ethanol generation and enhances yield, especially under anaerobic or microaerobic conditions. The key benefits include improved ethanol yield and reduced byproducts, higher carbon efficiency, and enhanced robust fermentation under stress conditions (acidic, inhibitory environments).

Co-utilization of Sugars Lignocellulosic biomass contains pentose sugars like xylose and arabinose, which do not efficiently get fermented by native *Saccharomyces cerevisiae.* To address this, heterologous pathways such as the xylose reductase–xylitol dehydrogenase (XR–XDH) and xylose isomerase (XI) pathways were incorporated into *S. cerevisiae* to enable xylose fermentation.

NAD^+/NADH Balance In microbial ethanol fermentation, NAD^+ (oxidized form) and NADH (reduced form) act as key redox cofactors that shuttle electrons during metabolic reactions. Maintaining a proper NAD^+/NADH ratio is essential for continuous glycolysis and efficient ethanol production. NADH is produced during glycolysis and sugar metabolism. NAD^+ is required for converting acetaldehyde to ethanol via alcohol dehydrogenase (ADH). If NADH accumulates and NAD^+ is depleted, glycolysis halts, and ethanol production drops.

Inhibitor Tolerance Lignocellulosic biomass undergoes pretreatment and hydrolysis to release fermentable sugars. However, this process also generates toxic byproducts, known as fermentation inhibitors, which hinder microbial growth, sugar metabolism, and ethanol productivity. Commonly known classes of inhibitors are furan derivatives (furfural, 5-HMF), Organic acids (acetic acid, formic acid, etc.), Phenolic compounds (syringaldehyde, ferulic acid), and weak acids (Acetic, lactic, formic acids).

These engineering strategies optimize metabolic pathways to enhance ethanol yield and efficiency, thereby enabling microorganisms to perform more effectively in biofuel production.

7.5.1 Strain Improvement Through Genetic Modifications or Metabolic Advances in the Microorganism for Enhanced Bioethanol Production

A significant limitation of developing a cost-effective and efficient 2G bioethanol from lignocellulosic biomass is the unavailability of ethanologenic microorganisms to digest the various sugars present in the substrate (Naghshbandi et al. 2019). Bioengineering microorganisms is the most promising approach for producing ethanol in high yield selectively. The main features expected from a genetically engineered microorganism are optimal ethanol tolerance, high carbon substrate uptake, low aeration cost, high ethanol throughput, specificity, and the ability to possess resilience and selectivity in hostile conditions. A promising ethanologenic microorganism can be constructed mainly by these three strategies: (i) strategies that can tweak the strain's specific metabolic pathways, (ii) strategies that can identify and overexpress the native genes functional for ethanol production, and (iii) strategies that can implant genes from a potent ethanol-producing strain into the host's native genes. Through genetic engineering of wild strains, the metabolic pathways of yeasts, bacteria, fungi, and cyanobacteria have been tailored to convert lignocellulose into bioethanol efficiently.

Genetic modifications aim to develop microbial strains with enhanced capabilities for lignocellulosic ethanol production. Some of the key techniques employed currently are Genome Shuffling (it involves recombining genomes from different strains to create hybrids with desirable traits such as improved sugar utilization and stress tolerance) (Tsegaye et al. 2024), Adaptive Laboratory Evolution (ALE) (it involves cultivating microorganisms under selective pressures to evolve strains with enhanced traits naturally) (Wu et al. 2025), CRISPR/Cas9 Genome Editing (it allows precise genetic modifications, enabling the insertion, deletion, or regulation of DNA that take part in ethanol production pathways) (Noman et al. 2021), and Transporter Engineering (enhancing the uptake of sugars like xylose involves engineering or introducing efficient sugar transporters into microbial strains, improving substrate assimilation and fermentation rates) (Li et al. 2016).

Bacteria like *Escherichia coli* can utilize a wide range of sugars, require simple growth factors, and can thrive at a wide pH range (6–8). In a research study, the *E. coli* WDH strain was modified to the resultant *E. coli* WDH-LF strain that utilizes hemicellulosic hydrolysates from wheat grass and co-produces hydrogen and ethanol. It was accomplished by deleting D-lactate dehydrogenase and fumarate reductase genes in the parent microorganism, and the resulting strain *E. coli* WDH-LF increased hydrogen (70%) and ethanol (167%) from glucose compared with the parent strain (Lopez-Hidalgo et al. 2021). Genetic manipulation in the study was directed at increasing microbial robustness against furfural and 5-HMF, toxic byproducts arising from acid-thermal treatment of lignocellulosic biomass. A pgi gene of the EMP pathway was deleted in an ethanologenic *E. coli* strain SSK42 to yield a new strain SSK10, hence increasing the yield of ethanol to 18 g/L (Jilani et al. 2020). Many studies have been conducted on *E. coli,* targeting different genes and modifying or deleting or inserting foreign genes in the wild strain to produce bioethanol efficiently (Joshi et al. 2022).

Z. mobilis is an anaerobic microorganism that produces a high yield of ethanol through the Entner–Doudoroff (ED) pathway involving pivotal enzymes like phenylacetaldehyde dehydrogenase and alcohol dehydrogenase. The strain exhibits several advantageous features, including adaptability to a broad pH range (Xia et al. 2019), tolerance to high ethanol concentrations (16% V/V), and high carbon substrate consumption (Zhao et al. 2014). In a research study, two mutant strains of *Z. mobilis* 3.5 M and 3.6 M with enhanced acidic pH tolerance were created from *Z. mobilis* ZM4 through an ALE strategy. The two mutant strains exhibited acceleration in the growth rate (50–130%), reduced duration for glucose uptake(4–9 h), and a boost in ethanol production performance (20–63%) compared to the pure breed ZM4 strain at pH 3.8 (Yang et al. 2020). The first genetically engineered *Z. mobilis* strain that is capable of metabolizing xylose was developed by introducing and expressing four *E. coli* genes: xylose isomerase, xylulokinase, transaldolase, and transketolase (Zhang et al. 1995).

Some thermophilic bacteria can productively metabolize cellulose and hemicellulose into ethanol at temperatures exceeding 50 °C (Di Donato et al. 2019). *T. saccharolyticum* was engineered by deleting genes such as lactate dehydrogenase, acetate kinase, and phosphotransacetylase and creating *T. saccharolyticum* ALK1, which in turn increased the production of ethanol to the theoretical maximum of 90–100%. Strain ALK1 evolved into Strain ALK2 by increasing feed concentration and prolonging cultivation time, which produced 92% ethanol (Shaw et al. 2008).

Cyanobacteria are also a potential candidate for biofuel generation by mitigating CO_2, and can also remove NO_x and SO_x as nutrients for remediation. The main drawback in terms of biofuel generation is inefficient carbon absorption and reduced growth speed. The best strategy is to optimize the main carbon assimilation pathway (Calvin-Benson-Bassham (CBB) cycle) for ethanol production. The CBB cycle enzymes, such as ribulose-1,5-bisphosphate carboxylase/oxygenase (RuBisCO), fructose-1,6/sedoheptulose-1,7-bisphosphatase (FBP/SBPase), transketolase (TK), and aldolase (FBA), were overexpressed together with the ethanol synthesis enzymes pyruvate decarboxylase (PDC) and alcohol dehydrogenase (ADH) in the cyanobacterium *Synechocystis* PCC 6803. The strain produced an enhanced quantity of ethanol (37–69%) and biomass (7.7–15.1%), respectively, compared to the mother strain (Liang et al. 2018). In a research study, two recombinant strains of *Synechocystis* sp. PCC 6803 was collaboratively cultivated (Velmurugan and Incharoensakdi 2020). One strain overexpressed pdc-adh, while in the other strain, glucose-1-phosphate adenylyltransferase and polyhydroxybutyrate-specific β-ketothiolase (*glgC-phaA*) were deleted. This, in turn, increased the movement of various extracellular metabolites such as glucose, acetate, and succinate into the culture medium. These are then converted into ethanol (4.7 g/L) by pdc-adh-overexpressing cyanobacteria.

With its compartmentalized structure, *Saccharomyces cerevisiae* serves as an effective host for bioethanol biosynthesis and demonstrates strong environmental stress resistance. Engineering various pathways in yeast, including PPP (pentose phosphate pathway), HLXM (heterologous xylose metabolic pathways), HLM (heterologous metabolic pathway), and HLCM (heterologous cellobiose metabolic pathways), was attempted. The engineering of the *S. cerevisiae* xylose isomerase-based pathway significantly improved the xylose fermentation performance, which could effectively convert xylose to ethanol (0.45 g ethanol/g xylose) (Lee et al. 2014). Removal of the GRE3 gene decreased xylitol accumulation and, in turn, increased the xylose assimilation in S. cerevisiae MEC1133 and hence led to good ethanol yield (0.47 g/g of total sugars) (Romaní et al. 2015).

Fungi are commonly employed in biodiesel production because of their high lipid content, particularly rich in monounsaturated and saturated fatty acids. In terms of usage of the LCB for bioethanol production, the most promising fungi are *Fusarium oxysporum*. Aldehyde dehydrogenase (ALDH) is the most vital enzyme in acetic acid metabolism. The 1509 bp DNA (502 amino acid protein of 54.33 kDa) of aldh was cloned from *F. oxysporum* (Fan et al. 2023). The ethanol production by the recombinant strain in glucose and xylose-containing medium was 16.93% and 34.63% higher, and acetic acid production was 23.10% and 39.55% lower than that of the wild strain.

These genetic engineering techniques collectively enhance microbial efficiency, resilience, and sugar utilization. As a result, they significantly improve the overall performance of lignocellulosic ethanol production.

7.6 Challenges and Considerations in Omics-Based Bioethanol Production

7.6.1 Ethical and Regulatory Aspects of Genetic Engineering

Genetic modification or strain editing of organisms has significantly enhanced agriculture and biomedical applications, thereby putting pressure on the boundaries of ethical regulations. Hence, the regulations associated with this field have to be updated to accommodate the changes. The foundation of bioethics rests upon four core principles, namely recognition of individual rights, promoting well-being, minimizing harm, and upholding justice in all practices (Engwa 2014). Some of the key ethical guidelines that should be used to evaluate the ethics of the modern biofuel technologies are (i) the development of biofuels versus the expense of people's vital basics (food, water, health, work, and land privileges), (ii) biofuels' environmental sustainability, (iii) biofuels' effect on global climate change and contribution toward total greenhouse gases, (iv) biofuels should be in agreement with labor and intellectual property rights, (v) costs and benefits of biofuels should be dispersed in an unbiased way, and (vi) biofuel technology should not be used before the potential risks are understood to a reasonable extent.

Bioengineered microorganism (BM) might boost the efficiency of biofuel production by enhancing the conversion of biomass (Lu et al. 2022a, b). However, it poses serious concerns, such as dual-use concerns and biosecurity risks: Advanced genetic engineering tools, such as CRISPR, have dual-use potential. At the same time, they can be harnessed for beneficial applications or pose risks if misused, including the development of harmful biological effects (Ajaykumar 2024).

Regulatory Frameworks and Global Governance: The regulatory landscape for genetic engineering varies globally. The rapid pace of technological advancement challenges these frameworks, necessitating continuous updates to address emerging ethical and safety concerns (Ajaykumar 2024).

Transparency and Public Engagement: maintaining open communication about the benefits and risks is essential for public trust and ethical integrity.

7.6.2 Environmental and Safety Concerns

Environmental Impact The accidental or unintentional release of bioengineered microorganisms (BM) into nature can harm the environment. Ecological diversity is a part of our natural ecosystem and is a key contributor to the health and equilibrium of natural systems. The BM in the natural ecosystem may reduce the diversity of microbial life forms, which in turn can lead to colonization. They can also compete with the local strain within their genomic variant, and this can lead to unknown consequences such as toxicity, growth rate, and alteration in metabolic pathway, etc. (Hemalatha et al. 2023).

Genetic Stability The overall genetic stability issue of the edited strain is a significant concern, as it can lead to unintended permanent genetic changes or loss of the wild strain's characteristics, and may also result in drug resistance. Therefore, it is important to verify the stability of the genetic modifications (Javed et al. 2019).

Biosafety Criteria All genetically engineered microorganisms used for bioethanol production must comply with established biological risk management regulations (Xu et al. 2022a, b). Before scaling up the potential BM to the industrial level, the strain must be risk-evaluated for safety-related concerns. The ambiguous and interconnected effects of BMs on health and the environment present challenges that could influence both public well-being and the broader economic landscape.

Waste Management and Pollution Discarding waste irresponsibly can result in water pollution and pose serious risks to marine life. Implementing effective waste treatment and management practices is crucial to mitigating these environmental risks.

Resource Competition and Land Use Balancing energy production with environmental conservation and food supply is a critical consideration in the expansion of bioethanol industries.

To limit the hazard caused by BM strain to our ecosystem/environment, the risk has to be assessed, monitored, and managed continuously.

Risk Assessment Many countries have regulations for risk assessment before scaling up the process industrially or introducing the end product for consumer consumption. The risk management for BM-based products should be more stringent than for chemicals, as microbes are metabolically active and can easily migrate to another ecosystem. This can lead to unintended gene transfer to native species, disruption of local ecosystems, and loss of biodiversity. Hence, the ecological impact of genetically modified organisms has to be studied thoroughly. Usually, the risk assessment for genetically modified microbes is done either step by step or on a case-by-case basis (Engwa 2014). In this approach, according to the biosafety data of a particular BM, the testing will be moved from a controlled setup to the environment in small stages. This step-wise release into the environment will give crucial time to assess the possible risk effects on the environment and help to stop immediately when needed. Risk assessment should contain all the information related to the lab test and field test conducted with the BM, along with the biosafety guidelines. Environmental risk assessment of various BM strains must be conducted periodically; however, there remains a significant need to strengthen the regulatory framework.

Risk Monitoring This is after the risk assessment step and after the BM is released into the environment. Continual monitoring should be done in the field with respect to safety and impact on the ecosystem. Formulating causal hypotheses using biosafety data and ecological information is a central approach in the identification process.

Risk Management The next step involves managing the identified risk, making informed decisions and policies, and exploring alternative solutions. The decision-making process of BMs can be understood in four stages. The first step is during the research and development stage. The second phase is during the release of BMs and their product into the environment. The third phase is the trading period. At the fourth stage, products undergo refinement and are prepared for use or consumption. The successful management of risks depends on the active exchange of essential information between scientific experts and decision-makers (Choi et al. 2005).

In conclusion, while genetic engineering offers transformative potential for bioethanol production, it must be guided by robust ethical standards, transparent public engagement, and stringent environmental safeguards. Balancing innovation with responsible governance is essential to ensure sustainability, safety, and societal acceptance.

7.7 Industrial-Level Application of Bioethanol Production Using Bioengineered Microorganisms

Advancements in OMICS technologies are increasingly being applied to microbial strain editing for more efficient biofuel production, with constant innovation in the field. The mounting number of recent patents and publications in this field is an indication of the importance given to this area of biofuel production using OMICS-based bioengineered microorganisms (Joshi et al. 2022; Karp et al. 2022; Gonzalez-Contreras et al. 2021). Some examples of successful industrial production of bioethanol using lignocellulosic biomass as substrate and recombinant microorganisms are discussed below (Table 7.2).

Raízen is a leading Brazilian energy company pioneering 2G bioethanol production. Using sugarcane processing waste and straw as renewable input sources, it operates large-scale plants, including the Bonfim Bioenergy Complex, inaugurated in 2024 as the world's largest 2G ethanol facility. Capable of producing 82 million liters annually, Raízen plans to build 20 more 2G plants by 2027. Its proprietary enzymatic technology supports Brazil's climate goals by converting agricultural residues into low-carbon biofuel, significantly reducing greenhouse gas emissions (Raízen 2025; Bloomberg Línea 2024).

Table 7.2 Summary of commercially successful 2G bioethanol production

Company	Location	Start year	Production capacity	Feedstock	Technology	Status
Raízen	São Paulo, Brazil	2015	112 million liters/year as of 2024	Sugarcane bagasse and straw	Proprietary 2G ethanol technology	In operation
LanzaTech	Global (USA, Belgium, China, Japan)	2018	~310,000 tons/year across six plants	Industrial waste gases, municipal solid waste	Gas fermentation using *Clostridium autoethanogenum*	In operation
GranBio	Alagoas, Brazil	2014	~21.6 million gallons/year	Sugarcane straw and bagasse	Engineered yeasts	In operation
POET-DSM (Project LIBERTY)	Emmetsburg, Iowa, USA	2014	~20 million gallons/year	Corn stover	Engineered *Saccharomyces cerevisiae*	Shut down in 2019
DuPont (now Corteva/Dow)	Nevada, Iowa, USA	2015	~30 million gallons/year	Corn stover	Engineered *Zymomonas mobilis*	The plant was sold in 2017 and is currently idle

LanzaTech is a global biotechnology company that produces bioethanol and sustainable fuels by converting industrial waste gases using engineered *Clostridium autoethanogenum*. Since starting commercial operations in 2018, it has established six commercial plants with a combined annual capacity of around 310,000 tons. In 2024, it launched the first worldwide commercial-scale ethanol conversion to sustainable aviation fuel (SAF) plant in Georgia, USA (LanzaTech 2025; ArcelorMittal, LanzaTech 2024).

GranBio's Bioflex 1 facility, located in Brazil, produces approximately sixty million gallons of cellulosic bioethanol each year, utilizing sugarcane residues such as straw and bagasse as its primary feedstock. The company uses Proesa™ steam explosion technology for pre-treatment (Granbio 2014; Gonzalez-Contreras et al. 2021) and hydrolysis carried out using a Novozymes Enzyme cocktail, and then fermentation of the pentose and hexose sugars carried out by a modified yeast strain (Celere-2L). The facility has a production capacity of approximately 21.6 million gallons per year (GranBio 2014; Bioenergy International 2014).

In 2014, POET-DSM Advanced Biofuels launched the Project LIBERTY facility in Iowa (USA), designed to produce 20–25 million gallons of ethanol annually using corn residues such as cobs, leaves, stalks, and husks. The company developed a proprietary enzyme blend effective across a broad pH range and capable of withstanding high temperatures during the hydrolysis process. Engineered strains of *Saccharomyces cerevisiae* were employed, enabling the fermentation of both pentose and hexose sugars derived from lignocellulosic biomass. The end product, cellulosic ethanol, is used for blending with gasoline (U.S. Department of Energy n.d.; POET-DSM Project LIBERTY 2014).

DuPont has one of the world's largest cellulosic ethanol plants in Nevada, Iowa, USA, which opened in 2015. The facility produces approximately 30 million gallons of bioethanol per year using corn stover as its lignocellulosic feedstock. The plant used advanced pretreatment, enzymatic hydrolysis technologies, and fermentation by recombinant *Zymomonas mobilis* strains. The plant was shut down in 2017 (NREL 2015; Farm Progress 2015). The following table summarizes some of the key engineered microbes used in bioethanol production (Table 7.3).

Table 7.3 Summary of primary microorganisms employed in 2G bioethanol

Microorganism	Engineering objectives	Industrial role
Saccharomyces cerevisiae	Enhanced pentose (xylose, arabinose) utilization; inhibitor tolerance; ethanol yield	Most commonly used in industrial bioethanol plants
Zymomonas mobilis	Engineered for pentose metabolism; higher ethanol yield and lower biomass	Used in some commercial pilot plants
Escherichia coli	Engineered to co-utilize hexose and pentose sugars; high tolerance	Limited industrial use due to safety concerns
Clostridium spp.	Engineering for solventogenic pathways	Primarily used in biobutanol but explored for lignocellulosic ethanol
Engineered yeasts and fungi	Tolerance to inhibitors, temperature, and pH	Custom strains used by several biotech companies

Countries such as the USA, China, Brazil, and India are major producers of ethanol worldwide. The 1G bioethanol production has been widely used on an industrial scale for a long time due to the availability of technologies. 2G ethanol technology is evolving as metabolic pathways of yeasts, fungi, bacteria, and cyanobacteria have been understood more effectively than previously. This is significantly aiding scientists in engineering microorganisms to produce high-yield bioethanol from lignocellulose. Genetically engineered microorganisms in the future can be used to generate any metabolite of human interest, not just biofuels. Techniques such as CRISPR and synthetic biology (precise and rapid microbial engineering), consortia-based fermentation (using mixed engineered microbial communities), integration with biorefineries (Ethanol production alongside other biochemicals), and AI/modeling tools (To design more efficient metabolic pathways) can shape the future of the modern world for the enhanced production of bioethanol.

7.8 Conclusion

The application of omics-based approaches in microbial genetic engineering has ushered in a new era for bioethanol production, transforming it from a largely empirical process into a data-driven, precision-oriented field. Harnessing the combined power of genomics, transcriptomics, proteomics, metabolomics, and fluxomics, scientists are breaking new ground in understanding microbial systems. This comprehensive perspective enables the deliberate engineering and optimization of microbial pathways, resulting in strains that are more efficient, resilient, and customized for industrial-scale bioethanol production.

Central to this progress is the ability to identify metabolic bottlenecks, stress response pathways, and substrate utilization limitations within microbial genomes. Such insights enable targeted strategies like gene amplification, gene deletion, and the incorporation of non-native biosynthetic routes. The development of CRISPR-Cas and advanced synthetic biology tools has further accelerated microbial strain improvement, allowing multiplexed edits and modular pathway designs. As demonstrated across various microbial platforms—from *Saccharomyces cerevisiae* to *Zymomonas mobilis*, and from *Escherichia coli* to cyanobacteria, engineered strains are now capable of metabolizing complex lignocellulosic feedstocks with enhanced ethanol yield, stress resistance, and fermentation efficiency.

At the industrial level, the successful deployment of bioengineered microbes in large-scale 2G bioethanol facilities underscores the practical feasibility of these advancements. Companies like Raízen, GranBio, and LanzaTech exemplify how tailored microbial strains, combined with optimized feedstock processing and fermentation technologies, can result in commercially viable bioethanol production. These examples highlight the importance of holistic strategies that blend genetic engineering with systems biology, bioprocess optimization, and life-cycle assessment.

Despite its potential, large-scale adoption encounters significant challenges, with biosafety particularly the release and persistence of genetically modified organisms in the environment being a significant concern. Ethical questions also arise about the dual-use potential of genome editing technologies and their implications for food security, biodiversity, and social equity. Regulatory frameworks must evolve rapidly to keep pace with scientific advances, incorporating risk assessments, stakeholder engagement, and transparent governance to ensure safe and equitable application.

Moreover, the scalability of omics-informed engineering depends on continued advancements in high-throughput analytics, machine learning, and computational modeling. As the volume and complexity of omics data grows, the ability to interpret and translate this information into actionable strategies will be key. AI-assisted strain design, predictive modeling, and synthetic biology automation are expected to play increasingly central roles in the future of bioethanol research and production.

In conclusion, omics-based microbial engineering offers a transformative toolkit for next-generation bioethanol production. It aligns scientific innovation with environmental sustainability and industrial scalability. While technical and ethical hurdles remain, the convergence of omics, AI, and synthetic biology heralds a promising future for clean energy solutions. By continuing to invest in interdisciplinary research, robust policy frameworks, and responsible innovation, we can fully leverage the capabilities of engineered microbes to address the increasing global demand for sustainable biofuels.

Acknowledgment The authors sincerely acknowledge the invaluable support and guidance of their mentors, colleagues, and collaborators who contributed to the development of this chapter. Special thanks are extended to institutions CAARU, SQU & UTAS SUR that provided infrastructural support for research. The constructive feedback from peer reviewers and the editorial team is also gratefully acknowledged, as it significantly improved the quality and clarity of this work. Finally, the authors wish to express their appreciation to the scientific community whose foundational research and insights have laid the groundwork for advances in multi-omics approaches and systems-level microbial engineering.

References

Ajaykumar S (2024) The ethical and security implications of genetic engineering. ORF Issue Brief 723. Retrieved from https://www.orfonline.org/public/uploads/posts/pdf/20240804191214.pdf

Ajjawi I et al (2017) Lipid production in *Nannochloropsis gaditana* is doubled by decreasing expression of a single transcriptional regulator. Nat Biotechnol 35:647

Amer B, Baidoo EEK (2021) Omics-driven biotechnology for industrial applications. Front Bioeng Biotechnol 9:Article 613307. https://doi.org/10.3389/fbioe.2021.613307

Anzalone AV, Randolph PB, Davis JR et al (2019) Search-and-replace genome editing without double-strand breaks or donor DNA. Nature 576:149–157. https://doi.org/10.1038/s41586-019-1711-4

ArcelorMittal, LanzaTech (2024) Ethanol production milestone and first shipment from the flagship Steelanol facility in Belgium. ArcelorMittal News Articles. Retrieved from https://corporate.arcelormittal.com/media/news-articles/arcelormittal-and-lanzatech-announce-ethanol-production-milestone-and-shipment-of-first-barge-from-flagship-steelanol-facility-in-belgium

Bioenergy International (2014) GranBio starts second-generation ethanol production in Alagoas. Bioenergy International. Retrieved from https://bioenergyinternational.com/granbio-starts-second-generation-ethanol-production-alagoas/

Bloomberg Línea (2024) Raízen opens the world's largest second-generation ethanol plant. Bloomberg Línea. Retrieved from https://www.bloomberglinea.com/2024/05/31/raizen-opens-the-worlds-largest-second-generation-ethanol-plant/

Bradbury A, Clapp O, Biacsi A-S, Kuo P, Gaju O, Hayta S, Zhu J-K, Lambing C (2025a) Integrating genome editing with omics, artificial intelligence and advanced farming technologies to increase crop productivity. Plant Commun. https://doi.org/10.1016/j.xplc.2025.101386

Bradbury L, Oreb M, Sauer U (2025b) From multi-omics to next-generation strain engineering. Nature Rev Bioeng 2(1):35–50

Brandt BA, Jansen T, Görgens JF et al (2019) Overcoming lignocellulose-derived microbial inhibitors: advancing the Saccharomyces cerevisiae resistance toolbox. Biofuels Bioprod Biorefin 13:1520–1536. https://doi.org/10.1002/bbb.2062

Choi CK, Pang T, Lin V et al (2005) Can scientists and policymakers work together? J Epidemiol Community Health 59:632–637

Ciamponi FE, Mendes DR, Chaves DO et al (2022) Multi-omics network model reveals key genes associated with fermentation robustness. Sci Rep 12(1):26843. https://doi.org/10.1038/s41598-022-26843-2

Cunha JT, Soares PO, Baptista SL, Costa CE, Domingues L (2020) Engineered *Saccharomyces cerevisiae* for lignocellulosic valorization: a review and perspectives on bioethanol production. Bioengineered 11(1):883–903. https://doi.org/10.1080/21655979.2020.1801178

de Moura Ferreira MA, da Silveira WB (2024) Multi-omics data and model integration reveal the main mechanisms associated with respiro-fermentative metabolism and ethanol stress responses in *Kluyveromyces marxianus*. Biochem Eng J 207:109471

Di Donato P, Finore I, Poli A, Nicolaus B, Lama L (2019) The production of second generation bioethanol: the biotechnology potential of thermophilic bacteria. J Clean Prod 233:1410–1417 https://doi.org/10.1016/j.jclepro.2019.06.152

Engwa AG (2014) Genetic engineering on microorganisms: the ecological and bioethical implications. Eur J Biotechnol Biosci 1(3):27–33

Fan J, Huang X, Zheng G et al (2023) Knockout of the aldehyde dehydrogenase gene in *Fusarium oxysporum* for enhanced ethanol yield. Energies 16:427

Farm Progress (2015) Digging into the science of cellulosic ethanol. Farm Progress. Retrieved from https://www.farmprogress.com/farm-business/digging-into-the-science-of-cellulosic-ethanol

Fredens J, Wang K, de la Torre D, Funke LF, Robertson WE, Christova Y et al (2019) Total synthesis of Escherichia coli with a recoded genome. Nature 569(7757):514–518. https://doi.org/10.1038/s41586-019-1192-5

Gonzalez-Contreras M et al (2021) Synthesis, design, and evaluation of intensified lignocellulosic biorefineries—case study: Ethanol production. Chem Eng Process Process Intensif 159:108220

GranBio (2014) About 2G ethanol. Retrieved from https://www.granbio.com.br/en/net-zero-solutions/etanol-2g/about-ethanol

Han M-J, Lee SY, Koh S-T, Noh S-G, Han WH (2010) Biotechnological applications of microbial proteomes. J Biotechnol 145(4):341–349

Hemalatha P, Abda EM, Shah S et al (2023) Multi-faceted CRISPR-Cas9 strategy to reduce plant-based food loss and waste for sustainable bio-economy—a review. J Environ Manag 332: 117382

Horinouchi T, Maeda T, Furusawa C (2018) Understanding and engineering alcohol-tolerant bacteria using OMICS technology. World J Microbiol Biotechnol 34:157. https://doi.org/10.1007/s11274-018-2542-4

Hughes TR et al (2000) Functional discovery via a compendium of expression profiles. Cell 102(1): 109–126. https://doi.org/10.1016/S0092-8674(00)00015-5

Jacob S, Rajeswari G, Rai A, Shweta S (2024) Paradigm of integrative OMICS of microbial technology towards biorefinery prospects. Biocatalysis Agric Biotechnol 58:103226. https://doi.org/10.1016/j.bcab.2024.103226

Javed MR, Noman M, Shahid M et al (2019) Current situation of biofuel production and its enhancement by CRISPR/Cas9-mediated genome engineering of microbial cells. Microbiol Res 219:1–11

Jiang Y et al (2020) Designing synthetic microbial consortia for biofuel production. Trends Biotechnol 38(8):828–831

Jilani SB et al (2020) Deletion of *pgi* gene in *E. coli* increases tolerance to furfural and 5-hydroxymethyl furfural in media containing glucose–xylose mixture. Microb Cell Factories 19:153

Joshi A et al (2021) Process consolidation approaches for cellulosic ethanol production. In: Ray RC (ed) Sustainable biofuels. Academic/Elsevier BV, London, pp 43–72

Joshi A et al (2022) Recent advances in metabolic engineering of microorganisms for advancing lignocellulose-derived biofuels. Bioengineered 13:8135–8163

Jung J et al (2016) Precise precursor rebalancing for isoprenoids production by fine control of *gapA* expression in *Escherichia coli*. Metab Eng 38:401–408

Karp SG et al (2022) New technologies for bioethanol production: patents and innovation. In: Biofuel and biorefinery technologies, vol 12. Springer, Cham, pp 488–515

Kim J et al (2008) Metabolic engineering of *Escherichia coli* for the production of ethanol. Biotechnol J 3(2):157–163

Kim J, Lee SY, Park SJ (2022) CRISPR-based multiplexed genome engineering for pathway optimization in industrial microbes. Trends Biotechnol 40(1):91–105. https://doi.org/10.1016/j.tibtech.2021.06.010

Kuyper M et al (2005) Metabolic engineering of a xylose-isomerase-expressing Saccharomyces cerevisiae strain for rapid anaerobic xylose fermentation. FEMS Yeast Res 5(4–5):399–409. https://doi.org/10.1016/j.femsyr.2004.11.003

LanazTech (2025) Biorefining: sustainable carbon recycling for fuels and chemicals. LanzaTech. Retrieved from https://lanzatech.com/biorefining/

Lee J et al (2012) Genomics and metabolic engineering of microorganisms for bioethanol production. Biotechnol Bioeng 109(6):1091–1103. https://doi.org/10.1002/bit.24361

Lee SM, Jellison T, Alper HS (2014) Systematic and evolutionary engineering of a xylose isomerase-based pathway in *Saccharomyces cerevisiae*. Biotechnol Biofuels 7:122

Li H et al (2016) Engineering a wild-type diploid *Saccharomyces cerevisiae* strain for second-generation bioethanol production. Bioresour Bioprocess 3:51

Li X, Zhou Z, Feng Y (2023) Precision genome editing in bacteria: the rise of CRISPR and beyond. Annu Rev Genet 57:345–365. https://doi.org/10.1146/annurev-genet-120922-115431

Liang F, Englund E, Lindberg P, Lindblad P (2018) Engineered cyanobacteria with enhanced growth show increased ethanol production and higher biofuel to biomass ratio. Metab Eng 46:51–59

Lopez-Hidalgo AM, Balderas Hernandez VE, De Leon-Rodriguez A (2021) Scale-up of hydrogen and ethanol co-production by an engineered *Escherichia coli*. Fuel 300:121002. https://doi.org/10.1016/j.fuel.2021.121002

Lu H et al (2022a) Bioengineered microbial platforms for biomass-derived biofuel production—a review. Chemosphere 288:132528

Lu L, Shen X, Sun X et al (2022b) CRISPR-based metabolic engineering in non-model microorganisms. Curr Opin Biotechnol 75:102698

Majidian P et al (2018) Metabolic engineering of microorganisms for biofuel production. Renew Sust Energ Rev 82:3863–3885

Mauger S, Raghunathan A, Palsson BO (2021) Systems biology of microbial metabolism: multi-omics and computational modeling perspectives. Curr Opin Syst Biol 25:67–76

Naghshbandi MP et al (2019) Progress toward improving ethanol production through decreased glycerol generation in *Saccharomyces cerevisiae* by metabolic and genetic engineering approaches. Renew Sust Energ Rev 115:109353

National Renewable Energy Laboratory (2015) Advancements in grid technology and renewable energy integration. NREL News Features. Retrieved from https://www.nrel.gov/grid/news/features/2015/16468

Nielsen J, Keasling JD (2016) Engineering cellular metabolism. Cell 164(6):1185–1197. https://doi.org/10.1016/j.cell.2016.02.004

Nielsen J et al (2016) Engineering cellular metabolism. Cell 164(6):1185–1197

Noman M et al (2021) Lignocellulosic biomass and microbial genome engineering for sustainable ethanol production: an overview. In: Arshad M (ed) Sustainable ethanol and climate change. Springer, Cham

Othoum G, Prigent S, Derouiche A et al (2019) Comparative genomics study reveals Red Sea *Bacillus* with characteristics associated with potential microbial cell factories (MCFs). Sci Rep 9:19254. https://doi.org/10.1038/s41598-019-55726-2

POET-DSM Project LIBERTY (2014) U.S. Department of Energy. Retrieved from https://www.energy.gov/eere/bioenergy/poet-dsm-project-liberty

Rai V, Karthikaichamy A, Das D, Noronha S, Wangikar PP, Srivastava S (2016) Multi-omics frontiers in algal research: techniques and progress to explore biofuels in the postgenomics world. OMICS 20(7):387–399. https://doi.org/10.1089/omi.2016.0065. Epub 2016 Jun 17. PMID: 27315140

Raízen (2025) Raízen official website. Retrieved from https://www.raizen.com.br/en/our-business/ethanol

Rajkumar AS et al (2016) Adaptation of industrial *Saccharomyces cerevisiae* for efficient fermentation of xylose and mixed sugars. Biotechnol Biofuels 9:1–12

Rodriguez GM et al (2014) Toward aldehyde and alkane production by removing aldehyde reductase activity in *Escherichia coli*. Metab Eng 25:227–237

Romaní A, Pereira F, Johansson B, Domingues L (2015) Metabolic engineering of *Saccharomyces cerevisiae* ethanol strains PE-2 and CAT-1 for efficient lignocellulosic fermentation. Bioresour Technol 179:150–158

Roy S, Manna S, Chowdhury S, Choudhury L (2023) Improvement of large-scale production of lignocellulosic bioethanol through synthetic biology approaches: a comprehensive review. World J Biol Pharm Health Sci 14(3):316–331. https://doi.org/10.30574/wjbphs.2023.14.3.0275

Sarwan J, Nair MV, Sharma S, Uddin N, Bose KJC (2024) Omics technology approaches for the generation of biofuels. In: Biomass energy for sustainable development, 1st edn. CRC Press, p 27. https://doi.org/10.1201/9781003406501

Shaw AJ et al (2008) Metabolic engineering of a thermophilic bacterium to produce ethanol at high yield. Proc Natl Acad Sci 105:13769–13774

Tanaka K, Bamba T, Kondo A, Hasunuma T (2024) Metabolomics-based development of bioproduction processes toward industrial-scale production. Curr Opin Biotechnol 85:103057. https://doi.org/10.1016/j.copbio.2024.103057

Tsegaye KN, Alemnew M, Berhane N (2024) *Saccharomyces cerevisiae* for lignocellulosic ethanol production: a look at key attributes and genome shuffling. Front Bioeng Biotechnol 12:1466644. https://doi.org/10.3389/fbioe.2024.1466644

Velmurugan R, Incharoensakdi A (2020) Co-cultivation of two engineered strains of *Synechocystis sp.* PCC 6803 results in improved bioethanol production. Renew Energy 146:1124–1133

Wu Y et al (2025) Robust *Saccharomyces cerevisiae* by rational metabolic engineering for effective ethanol production from undetoxified steam-exploded corn stover hydrolysate. Bioresour Technol 431:132605

Xia J et al (2019) Engineering *Zymomonas mobilis* for robust cellulosic ethanol production. Trends Biotechnol 37:960–972

Xu J-R et al (2022a) OMICs-based strategies to explore stress tolerance mechanisms of Saccharomyces cerevisiae for efficient fuel ethanol production. Front Energy Res 10:Article 884582. https://doi.org/10.3389/fenrg.2022.884582
Xu W, Brynjolfsson S, Fu W (2022b) Use of genetically modified algae: consumption safety challenges, current legislations, and socio-economic and environmental concerns. In: Shekh A, Dasgupta S (eds) Microalgae for sustainable products. Royal Society of Chemistry, pp 178–190
Yang Q et al (2020) Development and characterization of acidic-pH-tolerant mutants of *Zymomonas mobilis*. Biotechnol Biofuels 13:1–17
Zhang M et al (1995) Metabolic engineering of a pentose metabolism pathway in ethanologenic *Zymomonas mobilis*. Science 267:240–243
Zhang W, Li F, Nie L (2010) Integrating multiple 'omics' analysis for microbial biology: application and methodologies. Microbiology 156(2):287–301. https://doi.org/10.1099/mic.0.034793-0
Zhang Y, Gao J, Huang Y, Wang J (2018) Recent developments in single-cell RNA-seq of microorganisms. Biophys J 115(2):173–180
Zhao XQ, Bai FW (2009) Mechanisms of yeast stress tolerance and its manipulation for efficient ethanol production. J Biotechnol 144(1):23–30. https://doi.org/10.1016/j.jbiotec.2009.05.001
Zhao N et al (2014) Flocculating *Zymomonas mobilis* is a promising host to be engineered for fuel ethanol production from lignocellulosic biomass. Biotechnol J 9(3):362–371
Zhou YJ et al (2016) Production of fatty acid-derived oleochemicals and biofuels by synthetic yeast cell factories. Nat Commun 7:11709

Chapter 8
Design of Experiments in Lignocellulosic Bioethanol Production

Kofi Akodwaa-Boadi, Poonam Verma, Mohit Chaudhary, and Akshaya Kumar Verma

Abstract Global industrialisation and population growth have fuelled a significant rise in the demand for bioethanol, a sustainable and eco-friendly biofuel. Bioethanol production supports circular economy and zero-waste goals. Utilising ethanol as an alternative fuel reduces reliance on petrochemicals while bolstering energy security and global environmental sustainability. Commercial bioethanol production has traditionally relied on food-based agricultural crops such as sugarcane and sugar beet, often considered as resources primarily intended for human and animal consumption. Second-generation (2G) ethanol production, which uses lignocellulosic or agro-industrial biomass, involves a detailed three-step process: pretreatment, saccharification, and fermentation. Enzymatic saccharification is particularly critical for efficiently extracting fermentable sugars from biomass. The saccharification of biomass is significantly impacted by factors such as pretreatment methods, enzyme dosage, pH, temperature, substrate and surfactant concentrations, and reaction time. Optimising these variables is vital for maximising bioethanol production and minimising energy consumption. Design of Experiment (DoE) tools, combined with response surface methodology (RSM), offer an effective strategy for optimisation, providing a systematic means to understand the effect of several factors on

K. Akodwaa-Boadi
Akenten Appiah-Menka University of Skills Training and Entrepreneurial Development, Accra, Ghana

TCC-CIMET, Kwame Nkrumah University of Science and Technology, Accra, Ghana

P. Verma
Centre for Biotechnology, School of Pharmaceutical Sciences, Siksha 'O' Anusandhan (Deemed to be University), Bhubaneswar, Odisha, India

M. Chaudhary
Gogri Hub for Membranes Research, Indian Institute of Technology Bombay, Mumbai, Maharashtra, India

A. K. Verma (✉)
Department of Civil Engineering, ITER, Siksha 'O' Anusandhan (Deemed to be University), Bhubaneswar, Odisha, India
e-mail: akshayakumarverma@soa.ac.in

S. Prasad et al. (eds.), *Sustainable Bioethanol Production from Lignocellulosic Biomass*, Clean Energy Production Technologies,
https://doi.org/10.1007/978-981-95-9690-4_8

saccharification and enabling efficient process optimisation. DoE and RSM facilitate achieving optimisation goals with fewer experiments, reduced time, and lower costs. This chapter summarises the successful application of DoE and RSM tools for optimising biomass saccharification. It also provides an overview of the primary experimental design tools including the application of artificial intelligence (AI) and machine learning (ML), highlighting their significance in improving bioethanol production efficiency.

Keywords Design of experiment · Response surface methodology · Bioethanol · Lignocellulosic biomass

8.1 Introduction

Product development, like that in biorefineries, hinges on rigorous scientific methodologies and experimentation (de Souza et al. 2025). Unlike mechanistic approaches, scientists have successfully applied experimentation to investigate and clarify complex phenomena that are not fully understood, or develop new biorefinery products (de Souza et al. 2025; Dutta et al. 2025; Eberz et al. 2025; Khounani et al. 2025). In principle, conventional experimentations may yield desired outcomes or otherwise, contingent upon the cogent permutation of various input parameters and experimental conditions. Design of experiment (DoE) refers to a methodical approach for identifying the relationships between factors that influence a process and how input variables affect the final product (Politis et al. 2017). Researchers conduct experiments as systematic investigations comprising a series of trials where intentional modifications are applied to the independent variables. This is to understand and observe the reasons for various product formations as a response. DoE enables a structured analysis of the interplay between these factors, systematically classifying and quantifying cause-and-effect relations. Experimental designs are crucial for scientists to identify the key input variables that drive observed changes in response variables, facilitating the development of robust mathematical models for informed decision-making. Design of experiment methodologies allow for the systematic exploration of conditions to optimise both product yield and quality (Silva et al. 2023). This enables the identification of optimal operating parameters in complex systems. Consequently, researchers aim to establish a rigorous process that minimises the impact of external variables, and DoE facilitates this objective effectively, underscoring their critical role. DoE has many benefits over others, including but not limited to facilitating product design and development, process development, and continuous improvement (Czitrom 1999). Moreover, DoE is vital in the development of processes as it aids in recognising significant variables that impact results. Through meticulously designed experiments, researchers can identify areas needing enhancement and optimise their processes for increased efficiency and effectiveness. This leads to more consistent results and waste reduction. Using experimental design can reduce development time, cost, performance, and reliability of products (Cristea et al. 2018). Experimentation involves several factors, and the experimenters' objective is to investigate the impact of selected factors on the output, which is the response.

In modern experiments, scientists endeavour to understand how several factors influence a response (Serrano et al. 2025). As such, the factorial experimental design is practised more often. The factorial design of experiment enables all factors to be studied together, instead of one at a time. In a factorial design, scientists can investigate the individual effects of each factor on a response and determine if the factors interact and the extent of interaction to create the observed response. In bioethanol production, the simultaneous interaction of pretreatment heat and fermentation can be studied. In this case, the factorial experiment being referenced is a 2^2 factorial design, which indicates that two factors, each at two different levels, are being examined. Therefore, if there are n factors, with each factor having two levels, the factorial design will demand a total of 2^n runs.

8.2 Statistical Experimental Design

Statistical experimental design (SED) involves designing an experiment to ensure that suitable data is gathered and evaluated using statistical techniques, leading to a reliable and unbiased conclusion. SED has gone through about four eras. The first, which is known as the agricultural era, was spearheaded by Sir Ronald A. Fisher in the 1920s and 1930s (Bodmer et al. 2021; Parolini 2015). Fisher, a statistician and data analyst at an agricultural research station in England, identified numerous deficiencies in the data he encountered, stemming from the experimental design employed in various studies (Box 1980). Through his collaborations with scientists and researchers across diverse disciplines, he formulated three foundational principles of experimental design: randomisation, replication, and blocking. These principles are essential for minimising bias, controlling variability, and enhancing the reliability of experimental results (Hall 2007; Speed 1992). Fisher interacted with scientists and propounded the statistical thinking and principles into the design of experiments, factorial design, and analysis of variance.

The second era, known as the industrial era, was significantly advanced by Box and Wilson in 1952 through their development of response surface methodology (RSM) (Box 1980). This period marked the initiation of optimal DoE, which laid the groundwork for systematic experimentation in industrial applications. The third era, characterised by a focus on quality improvement, was championed by Genichi Taguchi. He promoted the application of DoE in robust parameter design, which emphasised the importance of minimising variability in product and process performance. His methodologies, along with contributions from notable figures like Wu and Kackar, facilitated the widespread adoption of DoE across various sectors, including automotive, aerospace, electronics, and semiconductors (Hisam et al. 2024; Ntemi et al. 2022). Taguchi's principles and evolving practices in DoE sparked renewed interest in experimental design during the fourth era, despite criticism. The integration of statistics in higher education and the development of computational software for experiment design and analysis have been crucial. Every experimental design consists of two phases: designing the experiment and statistical data analysis, based on four key principles: randomisation, replication, factorial design, and blocking.

8.2.1 Randomisation

Randomisation is a technique frequently utilised by experimenters to minimise the influence of experimental bias. This process allows for independent and randomly distributed errors, as required by statistical methods (Cousineau et al. 2023). Proper randomisation in experimental design is crucial for mitigating the influence of extraneous variables (Antony 2014). For instance, in the production of bioethanol from lignocellulosic biomass, if heavier biomass samples are exclusively subjected to acid hydrolysis while lighter samples undergo enzyme hydrolysis, the results will likely exhibit systematic bias. This biased distribution can lead to ambiguous outcomes about the efficacy of hydrolysis methods. Randomly assigning biomass weights to either treatment group ensures more reliable and valid comparisons. Randomisation enables the researcher to assign different weights of biomass to a type of hydrolysis, thereby eliminating bias. In modern times, computer programmes enable researchers to randomise experimental trials completely. However, some experiments may be challenging to randomise based on the nature of the variables completely. Modern software incorporates features to take care of such unique instances where complete randomisation is restricted.

8.2.2 Replication

Replication refers to conducting independent repeated trials for each unique combination of factors. Replication differs from repeated measurement. While repeated measurement entails duplicating or triplicating the same measurement within the same experimental framework, replication involves conducting independent trials with varied combinations of factors. This distinction is crucial for enhancing the robustness of experimental findings and ensuring the generalisability of results. Replication may consist, for example, of treating a specific weight of biomass against both enzyme and acid hydrolysis. Thus, if six (6) specimen weights are experimented against both enzyme and acid hydrolysis, then there are six (6) replicates. Replication has several advantages. Firstly, it offers a way to assess experimental error, which is crucial for ensuring the validity of hypothesis tests and confidence intervals for estimators. In the absence of replication, there is no means to gauge the underlying variability or "noise" in the estimates of treatment effects (Casler 2015). Additionally, it enhances the precision in estimating factor and interaction effects. Finally, increased replication reduces experimental error, improving overall precision (Antony 2014).

8.2.3 *Blocking*

Blocking is a methodological design technique aimed at enhancing the accuracy of comparisons among key factors of interest (Cousineau et al. 2023; Hesser 2020). By mitigating the variability introduced by nuisance factors, blocking effectively controls for confounding effects, allowing for a clearer interpretation of the primary factors under investigation, and improving efficiency (Bisgaard 1994). This approach is essential in experimental design, as it ensures that extraneous influences do not skew the comparisons made. Nuisance factors may not be of interest to the experiment, but they may influence the experimental response in some way. In a bioethanol production experimental design, different batches of enzymes may be required (Costa 2019). The same enzyme purchased from different suppliers may have variabilities. Thus, every batch of enzyme used in the experiment may form a block. This will allow the experimenter to manage the variability of enzyme batches within the block. The experimental runs are then executed in blocks. Blocking enables the experimenter to eliminate the potential influence of a nuisance. In experimental research, various blocking techniques are utilised to account for variability and improve the accuracy of the results. Some of the key types of blocking designs include:

8.2.3.1 Randomised Complete Block Design (RCBD)

This design consists of categorising experimental units into similar groups, or blocks, according to a particular characteristic. Within every block, treatments are allocated randomly, guaranteeing that each treatment is evaluated in each block. This design is especially beneficial for managing variability among subjects, thereby improving the accuracy of the treatment effects (Casler 2018; Festing 2020).

8.2.3.2 Randomised Block Design (RBD)

Like the randomised complete block design (RCBD), this design arranges experimental units into groups based on a specific characteristic. However, it can be utilised even when not every treatment is administered in each block. The goal continues to be minimising the effects of variability within treatments by maintaining similar units together (Casler 2018; Festing 2020).

8.2.3.3 Latin Square Design

This design is intricate and enables researchers to account for two types of variability. In a Latin Square, rows and columns denote various blocking factors. Each treatment is shown precisely once in each row and column, aiding in the balancing of the impacts from these extra variables.

8.2.3.4 Graeco-Latin Square Design

This represents a more sophisticated iteration of the Latin Square Design that incorporates two blocking factors. It involves both rows and columns, similar to a traditional Latin Square, but also adds an extra layer of treatment combinations. Thus, the number of treatments, rows, and columns is equal (AlAita et al. 2024). This design is especially useful in studies where it is important to investigate interactions between different factors.

8.2.3.5 Balanced Incomplete Block Design (BIBD)

BIBD is beneficial when there are more treatments than can be accommodated in each block. In scenarios where the availability of homogeneous experimental units is limited, the completely randomised design (CRD) and randomised block design (RBD) may prove inadequate due to their requirement for a substantial number of experimental units to encompass all treatments under investigation. When it is impossible to secure sufficient experimental units to apply all treatments within a block, incomplete block designs become a viable alternative. In these designs, each block is assigned a subset of treatments, rather than the entirety, allowing for a more efficient comparison while accommodating the constraints of the experimental setting. Each treatment is represented in a fixed number of blocks, but not every treatment appears in each block (Toutenburg and Shalabh. 2009). This format preserves balance and facilitates the efficient assessment of treatments, even when not all treatments are included in every block.

Blocking designs serve as vital tools in experimental research, as they enhance the reliability of findings by controlling for confounding variables and facilitating a clearer interpretation of the effects of the treatments administered.

8.2.4 Factorial Design

Experiments frequently encompass the manipulation of one or more factors to assess their effects on a response variable. In such cases, factorial designs are a highly effective approach because they allow investigators to recognise several factors simultaneously (Box and Draper 1971). A classical factorial design entails a systematic investigation of every possible combination of the factors and their respective levels within each experimental replicate (Cheng 2016). This comprehensive method enables the identification of both main effects, which are the individual impacts of each factor, and interaction effects, which reveal how the factors influence one another. By employing a factorial design, researchers can have deeper understandings into the dynamics of the experiment and optimise their analysis. In the lignocellulose experiments, three factors were considered: pretreatment heat, fermentation, and hydrolysis, each with two levels. In a typical replicate, all three factors and level combinations are considered.

8.3 Software and Tools Used in Experimental Design

There are several great software options for DoE, each catering to different needs. Here are some of the top choices:

8.3.1 John's Macintosh Project

John's Macintosh Project, referred to as JMP, offers a robust statistical analysis suite equipped with dynamic visualisations and seamless integration with SAS. The design of experiment (DoE) platforms within JMP facilitate the design, evaluation, and analysis of experimental data. The Easy DoE feature assists users in navigating the complete DoE workflow (Statistical Discovery LLC 2022). Below is a brief summary of various platforms offered in the DoE menu, emphasising their unique features and roles in the design of experiments (Box and Draper 1971; Statistical Discovery LLC 2022):

- The Custom Design platform creates a design that aligns with your unique requirements. It generates designs that consider a range of elements, including those that are challenging to adjust (like split plot scenarios). Users have the choice to determine the number of runs.
- The Definitive Screening Design platform presents a novel screening design that keeps main effects separate from one another as well as from two-way interactions. Additionally, these designs enable quadratic term estimation.
- The Screening Design, Response Surface Design, Full Factorial Design, Mixture Design, and Taguchi Arrays platforms create conventional DoE. For example, the Taguchi Arrays platform generates designs utilised for signal-to-noise ratio analysis using Taguchi's inner and outer array methodology.

8.3.2 *Minitab*

A well-established software for statistical analysis and predictive modelling, Minitab supports factorial designs to analyse multiple factors simultaneously, response surface designs, mixture designs, and Taguchi designs (Alin 2010). Minitab streamlines the experiment setup, guiding users through selecting factors, levels, and interactions. The software offers advanced automated analysis features, generating sophisticated visualisations such as interaction plots and contour graphs. It integrates robust statistical validation tools, ensuring the reliability and reproducibility of results. Sectors and researchers in manufacturing, pharmaceuticals, and engineering have applied Minitab to enhance quality metrics, minimise costs, and improve operational efficiency (Alin 2010; Nhien et al. 2017). By utilising its DoE capabilities, organisations can make informed, data-driven decisions, optimise workflows, and significantly improve outcomes (Minitab 2024).

8.3.3 *Design Expert*

Design Expert is widely utilised for process optimisation and enhancing product development efficiency. Its capabilities include factor screening, which effectively identifies the most significant variables influencing outcomes. The software employs response surface methodology (RSM) to refine experimental designs and ascertain optimal operating conditions. Additionally, robust process optimisation features ensure that manufacturing and research processes maintain stability and operational efficiency. Design Expert also offers dynamic 2D and 3D visualisation tools that facilitate in-depth analysis of experimental results (StatEase 2025). Design Expert has been applied widely in research studies, as reported elsewhere (Chettupalli et al. 2025; Duangjit et al. 2015; Ezeh et al. 2025; Williams 1991).

8.3.4 *MODDE*

MODDE is extensively utilised in the pharmaceutical, food production, petrochemical, and manufacturing industries to improve operational efficiency and decrease experimental expenses (Enochson et al. 2012; Fabbri et al. 2021). It emphasises quality-by-design (QbD) principles, which are essential for product development and meeting regulatory standards in industries. By employing advanced statistical methods, MODDE assists in developing strong experimental designs and enhancing process comprehension, resulting in improved quality outcomes. The key features of MODDE include a guided workflow, which enables users to set up experiments and obtain built-in recommendations. MODDE can help researchers assess experimental uncertainties, design visualisation tools, and optimise tools to identify the best experimental conditions for process improvement. Its QbD philosophy ensures that experiments meet industry standards.

8.3.5 Quantum Boost

Quantum boost uses AI to minimise the number of experiments needed while maximising insights. The platform employs sophisticated AI algorithms (Chatterjee et al. 2023) to significantly enhance research velocity, achieving experimentation speeds that are 2–5 times greater than conventional design of experiment (DoE) approaches (Biamonte et al. 2017). It offers a robust framework for project development, enabling users to iteratively refine experiments while retaining previously acquired knowledge (Biamonte et al. 2017; Wang et al. 2021). Its user-friendly interface demystifies complex statistical analyses, catering to both novice and seasoned researchers. Quantum Boost has become a valuable tool in sectors such as manufacturing, pharmaceuticals, and materials science, facilitating efficient optimisation of processes for enhanced research outcomes.

8.4 Artificial Intelligence (AI) and Machine Learning (ML) in Bioethanol Production

Researchers are harnessing artificial intelligence (AI) and machine learning (ML) tools to formulate experiments aimed at generating robust predictive models. Methods such as random forest algorithms (RF) and artificial neural networks (ANN) have demonstrated significant efficacy in modelling and optimising key parameters within bioethanol production systems. These approaches enable a more nuanced understanding of the complex interactions involved in the production process, facilitating improved efficiency and yield. For example, research on bioethanol production from sugarcane bagasse, *S. cerevisiae*, in a simultaneous hydrolysis and fermentation (SHF) utilised machine learning techniques to model the effect of variables namely temperature, feedstock composition, and enzyme loading on yield. The SHF predictions generated by these models were closely aligned ($R^2 > 0.90$), with the outcomes observed in experimental trials (Fischer et al. 2017), giving credence to the application of AI and ML in bioethanol production as well as the experimental design. Other experimental designs have also harnessed the power of ML in modelling and predicting bioethanol production, even for second-generation feedstocks. For example, ANN and RF algorithms were applied in the modelling of integrated process involving hydrolysis and fermentation. The modelled results were in excellent agreement with experimental results in predicting the bioethanol concentration with a coefficient of determination (R^2) value of 0.961 (Smuga-Kogut et al. 2021). This underscores the relevance of employing artificial intelligence and machine learning in the optimisation of bioethanol production, as well as in the design of experimental methodologies.

In another study, response surface methodology (RSM) and ANN were utilised to predict components in the production hydrolysis stage of oligosaccharides from sugar beet pulp. The study found that a multilayer perceptron (MLP) was effective in modelling oligosaccharide production through enzymatic hydrolysis. Additionally, MLP outperformed RSM in maximising reducing sugars from water hyacinth biomass, achieving an average prediction error of 3.08 and an optimum of 0.95 (Vani et al. 2015). Experimental findings from the pretreatment and hydrolysis of sugarcane bagasse demonstrate that an artificial neural network (ANN) with multilayer perceptron (MLP) architecture provides accurate estimations of glucose concentrations. The results indicate that optimal conditions, characterised by low initial biomass and acid concentrations, elevated enzyme concentrations, and a 72-h enzymatic hydrolysis period, significantly improve glucose yields (Plazas Tovar et al. 2018). Other studies have also concluded that MLP and AI-based solutions are accurate, simple, and powerful tool for designing experiments and modelling a cost-effective bioethanol production from substrates (Sebayang et al. 2017). ML and AI were employed to perform an economic feasibility analysis for optimal bioethanol production. A multilayer perceptron (MLP) model was developed and trained with the Levenberg-Marquardt algorithm, successfully predicting important parameters such as bioethanol content, yeast cell concentrations, and reducing sugar levels derived from the intermediates and by-products of sugar beets in a yeast batch fermentation process. The findings underscore the effective predictive capabilities of artificial intelligence and machine learning in generating critical insights for experimental design. This suggests their significant potential to optimise decision-making frameworks in the domain of bioethanol production (Gueguim Kana et al. 2012).

8.5 Model Building

In most of the experiments, more than two variables are interrelated, and it is important to model and investigate their relationship. For example, in bioethanol production using lignocelluloses is related to the type of lignocellulosic biomass, moisture content, pretreatment method, temperature, pressure, pH, etc. The investigator may want to develop a model relating the yield to aforementioned parameter (s). In general, when a single response y depends on k independent variables such as $x_1, x_2, \ldots\ldots\ldots x_k$, their relationship is described by a mathematical construct known as a regression model. In some instances, the investigator knows the exact form of the true functional relationship between response and regressor variable $\{y = f(x_1, x_2, \ldots\ldots\ldots x_k)\}$, while in most of the cases the true relationship is unknown (Montgomery et al. 2006). Regression models are commonly used to analyse the data from randomly designed experimentations.

8.5.1 Linear Regression Models

To demonstrate the linear regression model, an empirical relation might be developed involving bioethanol yield to pH and temperature. A model that might define this relation is

$$y = \beta o + \beta_1 x_1 + \beta_2 x_2 + \varepsilon \tag{8.1}$$

where y denotes the bioethanol yield, x_1 denotes the temperature, and x_2 denotes the pH, indicating a multiple regression model involving two independent variables. Such independent variables are also known as predictor variables or regressors. The model describes a plane in the 2-D x_1, x_2 space, where βo indicates the intercept of the plane, while β_1 and β_2 are the partial regression coefficients, dealing with the change in y per unit change in x_1 and x_2 (Montgomery 2017).

Commonly, the response variable y may be related to k regressor variables and the model will look like Eq. (8.2).

$$y = \beta o + \beta_1 x_1 + \beta_2 x_2 + \dots\dots\dots + \beta_k x_k + \varepsilon \tag{8.2}$$

The parameters βj, $j = 0, 1, 2, \dots\dots\dots, k$ are termed as the regression coefficients and represent the probable variation in response y per unit change in x_j when all the other independent variables are held constant. More complex model with interaction term to the first-order model can be denoted as:

$$y = \beta o + \beta_1 x_1 + \beta_2 x_2 + \beta_{12} x_1 \, x_2 + \varepsilon \tag{8.3}$$

While a second-order response surface model with two variables can be represented as:

$$y = \beta o + \beta_1 x_1 + \beta_2 x_2 + \beta_{11} {x_1}^2 + \beta_{22} {x_2}^2 + \beta_{12} x_1 \, x_2 + \varepsilon \tag{8.4}$$

If, $x_3 = {x_1}^2$, $x_3 = {x_2}^2$, $x_5 = x_1 \, x_2$, $\beta_3 = \beta_{11}$, $\beta_4 = \beta_{22}$ and $\beta_5 = \beta_{12}$, Eq. (8.4) will become

$$y = \beta o + \beta_1 x_1 + \beta_2 x_2 + \beta_3 x_3 + \beta_4 x_4 + \beta_5 x_5 + \varepsilon \tag{8.5}$$

Equation (8.5) is a linear regression model. A regression model is considered linear if it is linear with respect to its parameters (the β coefficients).

8.5.2 Hypothesis Testing in Multiple Regression

In multiple linear regression, hypothesis tests concerning the model parameters are useful for evaluating the model's effectiveness. These testing procedures assume that the errors (ε_i) are normally distributed, independent of one another, have a mean of zero, and share a constant variance (σ^2).

The test for the significance of regression determines whether a linear relationship exists between the response variable y and a subset of regressor variables x1, x2, ..., x_k. The suitable hypotheses are

Null hypothesis

$$H_0 : \beta_1 = \beta_2 = \dots\dots\dots = \beta_k = 0 \tag{8.6}$$

Alternative hypothesis

$$H_1 : \beta_j \neq 0 \text{ for at least one } j \tag{8.7}$$

Rejection of $H0$ indicates that at least one variable contributes significantly to the model. Analysis of variance (ANOVA) is used to perform, which partitions the total sum of squares (SST) into two components: the sum of squares attributed to the model and the sum of squares associated with residuals (errors),

$$\mathrm{SS_T} = \mathrm{SS_R} + \mathrm{SS_E} \tag{8.8}$$

If the null hypothesis is true, then SSR/σ^2 follows a chi-square distribution ($\chi 2k$) with k degrees of freedom, where k is the number of regressor variables in the model. Similarly, SSE/σ^2 follows a chi-square distribution with n-k-1 degrees of freedom. Moreover, SSR and SSE are independent.

The test procedure for null hypothesis is to compute F_0, where

$$F_0 = \frac{SSR/k}{SSE/(n-k-1)} = \frac{MSR}{MSE} \tag{8.9}$$

Reject H_0 if F_0 exceeds $F_{\alpha,k,n\text{-}k\text{-}1}$. Alternatively, the p-value approach can also be used: reject H_0 if the p-value corresponding to F_0 is less than α (Table 8.1).

Table 8.1 ANOVA summary table for assessing regression significance of a multiple linear regression model (Montgomery 2017)

Source of variation	Sum of squares	Degree of freedom	Mean square	F_0	Coefficient of multiple determination (R^2)
Regression	*SSR*	*k*	*MSR*	*MSR/ MSE*	*SSR/SST or 1-(SSE/SST)*
Error or residual	*SSE*	*n-k-1*	*MSE*		
Total	*SST*	*n-1*			

The coefficient of determination, denoted as R^2, measures how well the regression model fits the data.

8.5.3 *Confidence Interval in Multiple Regression*

In multiple regression, the combined influence of several explanatory variables on a response variable is evaluated. We may be interested in estimating confidence intervals for individual predictors, assessing whether specific variables have a significant impact on the response, or determining if the predictors collectively contribute meaningfully to the model.

$$\begin{aligned}\text{Confidence interval (CI)} = {} & \text{Estimated regression coefficient (true value of } \beta j) \\ & \pm \text{Critical t} - \text{t} - \text{value } (t_c) \\ & \times \text{Standard error of regression coefficient } (S_{\beta j})\end{aligned} \quad (8.10)$$

The t-statistic has *n-k-1* degrees of freedom, where *k* is the number of independent variables in the model.

Suppose we construct a 95% two-sided confidence interval for βj; this means that, in the long run, 95% of such intervals from repeated random samples would contain the true value of βj. Alternatively, the 95% two-sided confidence interval for βj represents the range of values that would not be rejected at the 5% significance level in a two-sided hypothesis test (Algina et al. 2007). With a large sample size, these interpretations hold reliably.

$$95\%\text{confidence interval for } \beta j = \left[\widehat{\beta j} - 1.96SE\left(\widehat{\beta j}\right), \widehat{\beta j} + 1.96SE\left(\widehat{\beta j}\right)\right] \quad (8.11)$$

8.6 Optimisation Strategies/Tools

After the models are fitted, the response is adjusted within the experimental section to achieve a specific objective such as maximising, minimising, or reaching a desired response value. In polynomial response surface models based on classical design of experiments (DoE), this entails locating the maximum or minimum value of the fitted function. Typically, this requires calculus techniques, such as differentiating the fitted function with respect to each input variable to obtain partial derivatives that describe how the response changes with respect to each input variable. By setting these partial derivatives to zero, we locate critical points that could potentially be maximum or minimum values of the model. By solving these equations, we can determine the values of the input variable that lead to the occurrence of the extremum. The nature of these points is then confirmed using second derivative tests: a positive second derivative indicates a local minimum, a negative value signals a local maximum, and a zero may require further analysis (Lundstedt et al. 1998). The ideal combination of input variables is determined to achieve the desired objective. Modern software tools like Minitab, Design-Expert, MATLAB, R, and Python automate these steps using numerical optimisation techniques that efficiently locate the extremum without the need for analytical solutions (Janković et al. 2025).

8.6.1 Response Surface Methodology (RSM)

Response surface methodology (RSM) encompasses a set of statistical and mathematical tools designed to explore and model the connection between one or more input factors and response variables. Its main objective is to identify the optimal settings of the factors that lead to the best possible response (Bakshi et al. 2020). RSM commonly employs quadratic models to capture curvature in the response surface, providing valuable insights into complex systems. This methodology is extensively used in industrial applications to enhance the product quality and lower the production costs (Pereira et al. 2021). Some of the examples include Central Composite Design, Box-Behnken Design, etc.

8.6.1.1 Objectives

- Response optimisation: To determine the levels of factors that produce the most desirable response, such as achieving the highest yield or the lowest cost.
- Exploring variable relationships: To build models and assess how individual factor and their interactions influence the response.
- Detection of key points: To identify regions on the response surface where the response is at a maximum, minimum, or the centre location.
- Enhancing process: To refine and adjust experimental procedures for greater efficiency or better overall performance.

8.6.1.2 Advantages

- *Model nonlinear patterns:* RSM can model curvature and complex interactions in the response, which traditional factorial designs cannot.
- *Reduced experimental runs:* It achieves optimisation with fewer experiments compared to full factorial approaches.
- *Graphical insights:* Generates response surfaces and contour plots that help illustrate how factors influence the response.
- *Efficient for multivariable analysis*: Allows efficient study of several factors and their interactions at the same time.

8.6.1.3 Limitations

- *Complexity:* Applying RSM involves more complex statistical and mathematical concepts than simpler experimental designs.
- *Higher experimentation Cost:* Developing a quadratic model often requires additional experimental runs compared to basic designs.
- *Best for gradual response patterns:* RSM is most effective when response varies smoothly with changes in factors; it may not handle abrupt changes.
- *Relies on prior factor selection:* RSM is typically used after key factors have been identified through preliminary screening experiments.

8.6.2 Central Composite Design (CCD)

Central composite design (CCD) is a widely used design within response surface methodology (RSM) for developing quadratic models in optimisation studies. It augments a factorial (whole or fractional) design by incorporating centre points and axial points, enabling the detection of curvature in the response surface. Its versatility and efficiency make it a preferred choice in optimisation studies (Bhattacharya 2021). CCD is frequently applied in manufacturing and related industries to determine ideal operating conditions and explore how factors influence the response. Components of CCD include:

- Factorial Points: These are 2^k or fractional factorial points positioned at the corners of the experimental region.
- Centre Points: Located at the middle of the design space, these points help in assessing curvature and provide a reference for comparison.
- Axial (star) Points: These points are placed beyond the factorial region to enable the estimation of quadratic effects. The distance of these points from the centre influences whether the design is rotatable.

8.6.2.1 Structure of CCD for two factors

Factors: (i) Factor A: Temperature (°C); (ii) Factor B: pH

Levels and Points: Factor levels: Low (−1), Centre (0), High (+1); Axial Points: Positioned at ($\pm\alpha$) to estimate curvature

Objective: To optimise ethanol production yield by examining the combined effect of temperature (A) and pH (B)

Factor Ranges: Temperature (A): 20–60 °C; pH (B): 4–10

Table 8.2 Trial run and response table

Trial no.	Point type	Factor A (Temperature)	Factor B (pH)	Response (% yield)
1	Factorial (−1, −1)	Low (20 °C)	Low (4)	[Measure]
2	Factorial (−1, +1)	Low (20 °C)	High (10)	[Measure]
3	Factorial (+1, −1)	High (60 °C)	Low (4)	[Measure]
4	Factorial (+1, +1)	High (60 °C)	High (10)	[Measure]
5	Axial (−α, 0)	Below Low (α = 10 °C)	Centre (7)	[Measure]
6	Axial (+α, 0)	Beyond High (α = 70 °C)	Centre (7)	[Measure]
7	Axial (0, −α)	Centre (40 °C)	Below Low (α = 3)	[Measure]
8	Axial (0, +α)	Centre (40 °C)	Beyond High (α = 11)	[Measure]
9	Centre (0, 0)	Centre (40 °C)	Centre (7)	[Measure]

Table 8.2 outlines a central composite design (CCD) for optimising ethanol yield using factor A (temperature) and factor B (pH). It includes factorial, axial, and centre points to model linear, interaction, and quadratic effects (Verma et al. 2015). Following trials cover combinations at specified levels to assess their influence on the production process.

After performing experiments at the specified factor combinations and levels, the percentage of ethanol yield is recorded. Analysing this response data allows for the construction of a quadratic model that captures the influence of the temperature and pH on yield, helping to determine the optimal conditions that maximise production.

8.7 Factors Affecting the Lignocellulosic Bioethanol Production and DoE-Supported Data Analysis and Optimisation

Lignocellulosic biomass typically contains about 35–50% cellulose, 25–30% hemicellulose, and 20–25% lignin (Nigam et al. 2009). The conversion of this biomass into ethanol encompasses several steps: pretreatment, hydrolysis (either acidic or enzymatic), fermentation, and separation. One of the major challenges in converting lignocellulose into fermentable monosaccharides lies in the hydrolysis step, as the breakdown of cellulose is impeded by multiple physical, chemical, and structural factors (Adaganti et al. 2014). Due to these inherent structural complexities, pre-treatment plays a vital role in enhancing the availability of sugars during hydrolysis, facilitating more efficient saccharification (Broda et al. 2022).

An ideal pretreatment technique should be robust, economically viable, and environmentally sustainable. It should also minimise the synthesis of inhibitory byproducts while enabling the maximum utilisation of lignocellulosic components, leading to efficient bioethanol production and effective lignin management. Pretreatment techniques are mainly categorised as: physical, chemical, physico-chemical, and biological. However, a single pretreatment method is generally not effective, and one or more methods need to be combined to achieve optimal outcomes (Beluhan et al. 2023). Despite ongoing progress, identifying the most effective pretreatment approach remains a growing area of research.

8.7.1 Physical Pretreatment

Physical pretreatment methods are used to break down the structural complexity of lignocellulosic biomass making it more accessible for enzymatic hydrolysis. These methods rely on mechanical actions, temperature, pressure, irradiation, or electric fields to reduce the particle size, increase the surface area, and improve the porosity. Common techniques include grinding, milling, extrusion, freezing, sonication, microwave irradiation, and pulsed electric field (PEF) treatment. Mechanical processes such as ball milling and hammer milling are effective in reducing cellulose crystallinity and increasing sugar yields; however, they are energy-intensive. Extrusion combines heat, shear, and pressure to disrupt biomass structure efficiently, without producing toxic byproducts, but it may not always be economically feasible due to high energy demand (Perez-Rodríguez et al. 2018). Freeze pretreatment is a promising and environmentally friendly technique that enhances the enzymatic breakdown of biomass while posing relatively low energy demand. It improves the availability of cellulose without the need for harmful chemicals. Sonication, on the other hand, utilises high-frequency sound waves to disrupt the structural integrity of lignocellulosic materials, leading to improved sugar release along with the reduced hydrolysis duration (Bussemaker and Zhang 2013). For example, applying a slightly higher temperature, around 50 °C, along with introducing alkalinity can further improve the pretreatment process. Microwave pretreatment works by evenly heating the biomass, which helps modify lignin and carbohydrate structures, thereby increasing their vulnerability to enzymatic attack (Jedrzejczyk et al. 2019). Pulsed electric field (PEF) treatment involves exposing biomass to brief, high voltage pulses, which increases the permeability of cell walls and facilitates better extraction of fermentable sugars (Kovačić et al. 2021). Each of these methods has unique advantages and limitations, and their effectiveness can vary depending on the type of lignocellulosic feedstock used. Often, these techniques are integrated or tailored in combination with others to optimise the pretreatment process and improve bioethanol production efficiency.

8.7.2 Chemical Pretreatment

Chemical pretreatment of lignocellulosic biomass uses various chemicals such as bases, acids, gases, salts, oxidising agents, etc. to break down the complex structure and enhance enzymatic hydrolysis of polysaccharides (Behera et al. 2014). Acid pretreatment, typically using diluted mineral acids like HCl and H_2SO_4 under high temperature (100–290 °C) and moderate pressure, primarily targets hemicelluloses and polysaccharide-lignin bonds, ensuring the improved availability of cellulose (Sun and Cheng 2002). However, it produces toxic by-products and requires costly acid recovery. Alkaline pretreatment uses dilute solutions of NaOH, KOH, $Ca(OH)_2$, or ammonia to selectively remove lignin and some hemicelluloses while swelling cellulose fibres, improving porosity and enzymatic digestibility (Behera et al. 2014). Although effective, this method typically demands longer reaction times. Organic solvent pretreatment employs solvents such as acetone, ethanol, and phenols, sometimes with catalysts, to fractionate biomass into lignin, cellulose, and hemicellulose with high purity and allows solvent recycling (Rezania et al. 2020). Despite its promise, it is hindered by high energy consumption and solvent costs. Ionic liquids composed of large organic cations and small anions can efficiently dissolve lignin and disrupt cellulose crystallinity, but face challenges in cost, waste, and viscosity management. Deep eutectic solvents (DES) such as choline chloride-based DESs (lactic acid, urea, or glycerol) offer a greener alternative at moderate temperatures to selectively solubilise lignin with low toxicity, biodegradability, and recyclability (Chen et al. 2020).

Metal salt pretreatment enhances lignin removal and hemicellulose degradation, improving sugar yield without requiring expensive reactors (Das et al. 2023). Oxidative methods like hydrogen peroxide hydrolysis generate hydroxyl radicals to degrade lignin, but with limited efficiency. Ozonolysis, a green oxidation method, effectively delignifies biomass under ambient conditions with minimal toxic byproducts, but its high ozone consumption limits economic viability (Travaini et al. 2016).

8.7.3 Biological Pretreatment

Biological pretreatment of lignocellulosic biomass offers a cost-effective, eco-friendly, and efficient alternative to conventional methods. It uses enzyme extracts or ligninolytic capabilities of specific microorganisms to improve the digestibility of the substrate. However, the cost of enzymes remains a major limitation to widespread application. Most often, filamentous fungi are used. However, a suitable microbial consortium or enzyme mix is selected based on the biomass composition with optimal temperature, pH, and incubation time (Olajuyigbe et al. 2019). The requirement of a large sterile area for pilot-scale biological pretreatment might make this process expensive (Usman et al. 2020). Although this process is

Table 8.3 Mode of action, benefits, and limitations of various biological pretreatment candidates

Bi-candidates for pretreatment	Mechanism of action	Advantages	Limitations
Fungi	Removes lignin, partially hemicellulose	High efficiency, low energy requirement, simple setup, minimal inhibitors	Slow process, sugar loss, requires controlled growth
Bacteria	Removes lignin, partially hemicellulose	Fast growth, cost-effective, tolerant to extreme conditions, low inhibitors	Slow to moderate, lower efficiency and yield, sugar loss
Actinomycetes	Decompose complex organic polymers, including cellulose and lignin	Strong lignocellulolytic activity, tolerant to extreme operating conditions	Slower growth rate compared to bacteria, sensitive to environmental changes
Microbial consortia	Removes lignin, partially hemicellulose	Broad degradation, high efficiency, better accessibility, improved yield	Competitive microbial action, sugar consumption, and sensitive to environmental conditions
Enzymatic	Removes lignin and hemicellulose	Specific, low-energy, simple process, minimal inhibitors	Expensive enzymes, slow processing

Adopted from Beluhan et al. (2023)

environmentally friendly, its slow degradation rate makes it unsuitable for large-scale application compared to conventional methods. Mechanisms, advantages, and disadvantages of various biological pretreatment candidates are outlined in Table 8.3.

Fungal pretreatment effectively degrades lignin in lignocellulosic biomass using ligninolytic enzymes such as laccase, lignin peroxidase, and manganese peroxidase (Liu et al. 2013). White rot fungi like *Phanerochaete chrysosporium* and *Pleurotus* species are remarkably efficient, utilising oxidative mechanisms to break down complex lignin structures (Castoldi et al. 2014).

Bacteria such as Bacillus, Clostridium, and Streptomyces decompose lignocellulosic biomass using hydrolytic and ligninolytic enzymes (Sari et al. 2016). Aerobic bacteria rely on free enzymes, whereas anaerobes use complex structures, such as cellulosomes. Their adaptability and diverse enzymatic systems make them efficient in breaking down cellulose, hemicellulose, lignin, and xenobiotic compounds (Lee et al. 2019). Enzyme-producing microbes, either single strains or consortia, are being used to break down lignocellulosic biomass (Ferdes et al. 2020). Microbial consortia show superior stability and efficiency, mimicking natural degradation. Mixed communities, including bacteria like Pseudomonas and fungi like Penicillium, enhance hydrolysis, though selection criteria for consortium members remain unclear (Jimenez et al. 2013). Enzymatic pretreatment uses purified or crude enzymes from bacteria and fungi to degrade lignocellulosic biomass without microbial growth. Enzyme cocktails are more effective than a single enzyme. Cellulose

Table 8.4 DoE-supported data analyses of the findings

Stage	DoE approach used	Key findings	References
Pretreatment	Full factorial and response surface methodology (RSM)	Optimisation of dilute acid pretreatment conditions (temperature, acid concentration, time) for maximising sugar yield	Yildirim et al. (2021)
Enzymatic hydrolysis	Central composite design (CCD) and Box-Behnken design	Optimisation of enzyme concentration, pH, temperature, and reaction time for the optimum enzymatic saccharification and production of biohydrogen	Izzi et al. (2023) and Pandiyan et al. (2014)
Fermentation	Taguchi method and factorial designs	Identification of optimal fermentation parameters (pH, temperature, inoculum size) for production of bioethanol from giant reed hydrolysate	Madian et al. (2022)
Inhibitor reduction	Full factorial design	Investigates the influence of major inhibitors acetic acid, furfural, and 5-hydroxymetilfurfural on fermentation by *Pichia stipits* and development of kinetic models	Abud et al. (2017) and Schneiderman et al. (2015)

degradation requires synergistic action of endoglucanases, exoglucanases, and β-glucosidases, while hemicellulose breakdown requires diverse depolymerising and debranching enzymes (Koupaie et al. 2019).

Actinomycetes play a pivotal role in lignocellulosic biomass degradation by producing diverse enzymes such as cellulases, hemicellulases, and lignolytic enzymes. For example, *Thermobifida fusca* synthesises multiple cellulases and xylanases, while *Streptomyces* spp. produce laccases and peroxidases capable of lignin breakdown (Saini et al. 2015). These enzymes act synergistically to depolymerise complex plant polymers, making actinomycetes valuable for biorefinery applications, biofuel production, and biomass valorisation (Lynd et al. 2002). Various lignocellulosic biomasses, including palm oil residue, corncob, wheat straw, and sugar cane bagasse, were pretreated using methods like acid hydrolysis, alkali treatment, or steam explosion as reported by different researchers. Enzymatic hydrolysis followed by fermentation using microorganisms such as *K. marxianus* and *S. cerevisiae* in SSF, SHF, or CBP configurations yielded ethanol concentrations ranging from 4.05 to 65.43 g/L, achieving up to 83.7% conversion, indicating effective bioethanol production across diverse substrates and conditions (Sukhang et al. 2020; Ballesteros et al. 2004; Boonchuay et al. 2021). Table 8.4 summarises different DoE-supported data analyses and findings specifically applied to lignocellulosic bioethanol production. The DoE approach enables the systematic study of factors and their interactions to optimise various stages of the process, such as pretreatment, enzymatic hydrolysis, and fermentation.

8.8 Optimisation Challenges and Bottlenecks

Batch-to-batch variability in lignocellulosic biomass composition presents a major challenge, arising not only from natural heterogeneity but also from the specific fractionation method employed, which can significantly alter the chemical composition of even the same biomass source (Krasznai et al. 2018). Such inconsistencies affect the reproducibility and reliability of experimental results. Furthermore, the selection of microbial strains influences process performance, introduces additional variability, and reduces the predictive power of design of experiment (DoE) models. Although techniques like response surface methodology (RSM) are effective in minimising the number of experimental trials, the inherently multifactorial nature of lignocellulosic bioethanol production still necessitates a substantial number of experiments to derive statistically robust conclusions (Mood et al. 2013). Another key bottleneck is the prolonged fermentation time, typically ranging from 48 to 72 h, which increases operational duration and associated costs (Balat et al. 2008). Moreover, the applicability of optimised lab-scale experimental parameters during scale-up remains a significant challenge. Scaling up to the pilot or industrial scale introduces additional challenges such as mass transfer limitations, inefficient heat transfer, non-uniform mixing, and increased stress to the microbial community (Garcia-Ochoa and Gomez 2009). These issues often result in reduced yields and increased inhibition effects at larger scales. Therefore, robust process control strategies and scale-up validation are essential to ensure the reproducibility and efficiency of bioethanol production beyond the laboratory environment.

8.9 Conclusions and Future Prospects

Lignocellulosic biomass from agricultural, industrial, and natural waste sources offers a sustainable, non-food-competing, and cost-effective option for bioethanol production. Unfortunately, lignocellulosic bioethanol production encounters several challenges, particularly during the process of pretreatment, hydrolysis, and fermentation, which significantly impact process efficiency and overall techno-economic viability. Design of experiment (DoE) offers a structured approach to optimise these stages by identifying the most suitable pretreatment methods based on feedstock type, thereby maximising fermentable sugar yield. It also facilitates the optimisation of hydrolysis parameters such as enzyme dosage, temperature, and pH to enhance sugar release from cellulose and hemicellulose. Additionally, DoE enables the integration of process steps to boost ethanol concentration and overall yield while lowering the production costs, improving competitiveness with fossil fuel. Advancements should focus on establishing integrated biorefineries for diversified product output, developing more efficient pretreatment methods, and exploring unconventional biomass sources like algae to support sustainability. Incorporating membrane-based technologies for ethanol separation and purification can further reduce energy demands and expenses. Ultimately, scaling up from laboratory to industrial applications requires feedstock availability, process scalability, and cost efficiency.

Acknowledgement The authors wish to acknowledge their esteemed institution/university for extending invaluable support that greatly contributed to the successful preparation of this book chapter, and extend their gratitude to the editors for the kind invitation to contribute to this volume.

References

Abud AKDS et al (2017) Effect of inhibitors on ethanol production by *Pichia stipitis* in a complex culture media. Acta Scientiarum Technol 39(2):223–229. https://doi.org/10.4025/actascitechnol.v39i2.29719

Adaganti SY et al (2014) Factors affecting bioethanol production from lignocellulosic biomass (*Calliandra calothyrsus*). Waste Biomass Valor 5:963–971. https://doi.org/10.1007/s12649-014-9305-8

AlAita A et al (2024) Analysis of Graeco-Latin square designs in the presence of uncertain data. J Big Data 11(1):109. https://doi.org/10.1186/s40537-024-00970-1

Algina J et al (2007) Confidence intervals for an effect size measure in multiple linear regression. Educ Psychol Meas 67(2):207–218. https://doi.org/10.1177/0013164406292030

Alin A (2010) Minitab. WIREs Comput Stat 2(6):723–727. https://doi.org/10.1002/wics.113

Antony J (2014) Fundamentals of design of experiments. In: Design of experiments for engineers and scientists. Elsevier, pp 7–17. https://doi.org/10.1016/B978-0-08-099417-8.00002-X

Bakshi A et al (2020) Electrocoagulation for removal of phosphate from aqueous solution: statistical modeling and techno-economic study. J Clean Prod 246:118988. https://doi.org/10.1016/j.jclepro.2019.118988

Balat M et al (2008) Progress in bioethanol processing. Prog Energy Combust Sci 34(5):551–573. https://doi.org/10.1016/j.pecs.2007.11.001

Ballesteros M et al (2004) Ethanol from lignocellulosic materials by a simulataneous saccharification and fermentation process (SFS) with Kluyveromyces marxianus CECT 10875. Process Biochem 39:1843–1848

Behera S et al (2014) Importance of chemical pretreatment for bioconversion of lignocellulosic biomass. Renew Sust Energ Rev 36:91–106

Beluhan S et al (2023) The production of bioethanol from lignocellulosic biomass: pretreatment methods, fermentation, and downstream processing. Energies 16:7003. https://doi.org/10.3390/en16197003

Bhattacharya S (2021) Central composite design for response surface methodology and its application in pharmacy. IntechOpen. https://doi.org/10.5772/intechopen.95835

Biamonte J et al (2017) Quantum machine learning. Nature 549(7671):195–202. https://doi.org/10.1038/nature23474

Bisgaard S (1994) Blocking generators for small 2k-p designs. J Qual Technol 26(4):288–296. https://doi.org/10.1080/00224065.1994.11979540

Bodmer W et al (2021) The outstanding scientist, R.A. Fisher: his views on eugenics and race. Heredity 126(4):565–576. https://doi.org/10.1038/s41437-020-00394-6

Boonchuay P et al (2021) Bioethanol production from cellulose-rich corncob residue by the thermotolerant Saccharomyces cerevisiae TC-5. J Fungi 7:547

Box JF (1980) R.A. Fisher and the design of experiments, 1922–1926. Am Stat 34(1):1–7. https://doi.org/10.1080/00031305.1980.10482701

Box MJ, Draper NR (1971) Factorial designs, the $|X'X|$ criterion, and some related matters. Technometrics 13(4):731–742. https://doi.org/10.1080/00401706.1971.10488845

Broda M et al (2022) Bioethanol production from lignocellulosic biomass-challenges and solutions. Molecules 27(24):8717. https://doi.org/10.3390/molecules27248717

Bussemaker MJ, Zhang D (2013) Effect of ultrasound on lignocellulosic biomass as a pretreatment for biorefinery and biofuel applications. Ind Eng Chem Res 52:3563–3580

Casler MD (2015) Fundamentals of experimental design: guidelines for designing successful experiments. Agron J 107(2):692–705. https://doi.org/10.2134/agronj2013.0114

Casler MD (2018) Blocking principles for biological experiments. In: Barry G, Yeater KM (eds) Applied statistics in agricultural, biological, and environmental sciences, pp 53–72. https://doi.org/10.2134/appliedstatistics.2015.0074.c3

Castoldi R et al (2014) Biological pretreatment of *Eucalyptus grandis* sawdust with white-rot fungi: study of degradation patterns and saccharification kinetics. Chem Eng J 258:240–246

Chatterjee S et al (2023) Quantum boosting using domain-partitioning hypotheses. Quant Mach Intell 5(2):33. https://doi.org/10.1007/s42484-023-00122-3

Chen Z et al (2020) Insights into structural changes of lignin toward tailored properties during deep eutectic solvent pretreatment. ACS Sustain Chem Eng 8:9783–9793

Cheng CS (2016) Theory of factorial design single-and multi-stratum experiments monographs on statistics and applied probability. In: Bunea F, Isham V, Keiding N, Louis T, Smith RL, Tong H (eds) Monographs on statistics and applied probability. CRC Press/Taylor & Francis Group, pp 1–47

Chettupalli AK et al (2025) Ritonavir loaded solid lipid nanoparticles for oral drug delivery and bioavailability enhancement. Discov Appl Sci 7(1):58. https://doi.org/10.1007/s42452-024-06322-1

Costa N (2019) Design of experiments—overcome hindrances and bad practices. TQM J 31(5): 772–789. https://doi.org/10.1108/TQM-02-2019-0035

Cousineau M et al (2023) Estimating causal effects with optimization-based methods: a review and empirical comparison. Eur J Oper Res 304(2):367–380. https://doi.org/10.1016/j.ejor.2022.01.046

Cristea G et al (2018) A study about how to use experimental design to increase the efficiency in a research project. IOP Conf Ser Mater Sci Eng 444:072007. https://doi.org/10.1088/1757-899X/444/7/072007

Czitrom V (1999) One-factor-at-a-time versus designed experiments. Am Stat 53(2):126–131. https://doi.org/10.1080/00031305.1999.10474445

Das N et al (2023) A comprehensive review of characterization, pretreatment and its applications on different lignocellulosic biomass for bioethanol production. Biomass Convers Biorefinery 13: 1503–1527. https://doi.org/10.1007/s13399-021-01294-3

de Souza EC et al (2025) One-pot production of Carboxymethyl Holocelluloses from mango and pineapple wastes optimized via design of experiments. Carbohydr Polym Tech Appl 10:100822. https://doi.org/10.1016/j.carpta.2025.100822

Duangjit S et al (2015) Application of design expert for the investigation of capsaicin-loaded microemulsions for transdermal delivery. Pharm Dev Technol:1–8. https://doi.org/10.3109/10837450.2015.1048552

Dutta S et al (2025) Microalgal biorefineries in sustainable biofuel production and other high-value products. New Biotechnol 87:39–59. https://doi.org/10.1016/j.nbt.2025.02.007

Eberz J et al (2025) Mini-batch settling cell for investigation of liquid-liquid phase separation. Chem Eng Sci 301:120751. https://doi.org/10.1016/j.ces.2024.120751

Enochson L et al (2012) Optimization of a chondrogenic medium through the use of factorial design of experiments. BioResearch 1(6):306–313. https://doi.org/10.1089/biores.2012.0277

Ezeh EM et al (2025) Process evaluation and analysis of variance of rice husk gasification using Aspen Plus and design expert software. Chem Afr 8(3):1085–1100. https://doi.org/10.1007/s42250-025-01189-0

Fabbri F et al (2021) Biocatalyzed synthesis of flavor esters and polyesters: a Design of Experiments (DoE) approach. Int J Mol Sci 22(16):8493. https://doi.org/10.3390/ijms22168493

Ferdeş M et al (2020) Microorganisms and enzymes used in the biological pretreatment of the substrate to enhance biogas production: a Review. Sustainability 12:7205
Festing MFW (2020) The "completely randomised" and the "randomised block" are the only experimental designs suitable for widespread use in pre-clinical research. Sci Rep 10(1): 17577. https://doi.org/10.1038/s41598-020-74538-3
Fischer J et al (2017) Machine learning techniques applied to lignocellulosic ethanol in simultaneous hydrolysis and fermentation. Braz J Chem Eng 34:53–63
Garcia-Ochoa F, Gomez E (2009) Bioreactor scale-up and oxygen transfer rate in microbial processes: an overview. Biotechnol Adv 27(2):153–176. https://doi.org/10.1016/j.biotechadv.2008.10.006
Gueguim Kana EB et al (2012) Modeling and optimization of biogas production on saw dust and other co-substrates using Artificial Neural network and Genetic Algorithm. Renew Energy 46: 276–281
Hall NS (2007) R. A. Fisher and his advocacy of randomization. J Hist Biol 40(2):295–325. https://doi.org/10.1007/s10739-006-9119-z
Hesser H (2020) Estimating causal effects of internet interventions in the context of nonadherence. Internet Interv 21:100346. https://doi.org/10.1016/j.invent.2020.100346
Hisam MW et al (2024) The versatility of the Taguchi method: optimizing experiments across diverse disciplines. J Stat Theory Appl 23(4):365–389. https://doi.org/10.1007/s44199-024-00093-9
Izzi AZ et al (2023) Thermophilic biohydrogen production from optimized enzymatic pretreatment of palm oil mill effluent via Box–Behnken design. J Eng Res 2(2):100054. https://doi.org/10.1016/j.jer.2023.100054
Janković A et al (2025) Optimization through classical design of experiments (DOE): an investigation on the performance of different factorial designs for multiobjective optimization of complex systems. J Build Eng:111931. https://doi.org/10.1016/j.jobe.2025.111931
Jedrzejczyk M et al (2019) Physical and chemical pretreatment of lignocellulosic biomass. In: Second and third generation of feedstocks: the evolution of biofuels. Elsevier, Amsterdam, pp 143–196
Jimenez DJ et al (2013) Novel multispecies microbial consortia involved in lignocellulose and 5-hydroxymethylfurfural bioconversion. Appl Microbiol Biotechnol 98(6):2789–2803
Khounani Z et al (2025) Lignin first-lignin modification outperforms lignin-first strategies: environmental and economic insights in second-generation ethanol biorefineries. J Environ Chem Eng 13(3):117092. https://doi.org/10.1016/j.jece.2025.117092
Koupaie EH et al (2019) Enzymatic pretreatment of lignocellulosic biomass for enhanced biomethane production—a review. J Environ Manag 233:774–784
Kovačić Đ et al (2021) Pulsed electric field: an emerging pretreatment technology in biogas production. Waste Manag 120:467–483. https://doi.org/10.1016/j.wasman.2020.10.009
Krasznai DJ et al (2018) Compositional analysis of lignocellulosic biomass: conventional methodologies and future outlook. Crit Rev Biotechnol 38(2):199–217. https://doi.org/10.1080/07388551.2017.1331336
Lee S et al (2019) Bacterial valorization of lignin: strains, enzymes, conversion pathways, biosensors, and perspectives. Front Bioeng Biotechnol 7:209
Liu J et al (2013) Fungal pretreatment of switchgrass for improved saccharification and simultaneous enzyme production. Bioresour Technol 135:39–45
Lundstedt T et al (1998) Experimental design and optimization. Chemom Intell Lab Syst 42(1–2): 3–40. https://doi.org/10.1016/S0169-7439(98)00065-3
Lynd LR et al (2002) Microbial cellulose utilization: fundamentals and biotechnology. Microbiol Mol Biol Rev 66(3):506–577
Madian HR et al (2022) Statistical optimization of bioethanol production from giant reed hydrolysate by *Candida tropicalis* using Taguchi design. J Biotechnol 360:71–78. https://doi.org/10.1016/j.jbiotec.2022.10.007
Minitab (2024) Unlock the value of your data with Minitab Statistical Software

Montgomery DC (2017) Design and analysis of experiments, 9th edn. Wiley
Montgomery DC et al (2006) Introduction to linear regression analysis, 4th edn. Wiley, Hoboken
Mood SH et al (2013) Lignocellulosic biomass to bioethanol, a comprehensive review with a focus on pretreatment. Renew Sust Energ Rev 27:77–93. https://doi.org/10.1016/j.rser.2013.06.033
Nhien LC et al (2017) Novel heat–integrated and intensified biorefinery process for cellulosic ethanol production from lignocellulosic biomass. Energy Convers Manag 141:367–377. https://doi.org/10.1016/j.enconman.2016.09.077
Nigam PS et al (2009) Pre-treatment of agro industrial residues. In: Nigam PS, Pandey A (eds) Biotechnology for agro-industrial residues utilization, 1st edn. Springer, Dordrecht, pp 13–33
Ntemi M et al (2022) Infrastructure monitoring and quality diagnosis in CNC machining: a review. CIRP J Manuf Sci Technol 38:631–649. https://doi.org/10.1016/j.cirpj.2022.06.001
Olajuyigbe FM et al (2019) Characterization of free and immobilized laccase from Cyberlindnera fabianii and application in degradation of bisphenol A. Int J Biol Macromol 125:856–864
Pandiyan K et al (2014) Optimization of enzymatic saccharification of alkali pretreated Parthenium sp. using response surface methodology. Enzyme Res:764898. https://doi.org/10.1155/2014/764898
Parolini G (2015) The emergence of modern statistics in agricultural science: analysis of variance, experimental design and the reshaping of research at Rothamsted Experimental Station, 1919–1933. J Hist Biol 48(2):301–335. https://doi.org/10.1007/s10739-014-9394-z
Pereira LS et al (2021) Using response surface methodology (RSM) to optimize 2G bioethanol production: a review. Biomass Bioenergy 151:Article 106166. https://doi.org/10.1016/j.biombioe.2021.106166
Perez-Rodríguez N et al (2018) Faster methane production after sequential extrusion and enzymatic hydrolysis of vine trimming shoots. Environ Chem Lett 16:295–299
Plazas Tovar L et al (2018) Prediction of overall glucose yield in hydrolysis of pretreated sugarcane bagasse using a single artificial neural network: good insight for process development. J Chem Technol Biotechnol 93:1031–1043
Politis SN et al (2017) Design of experiments (DoE) in pharmaceutical development. Drug Dev Ind Pharm 43(6):889–901. https://doi.org/10.1080/03639045.2017.1291672
Rezania S et al (2020) Different pretreatment technologies of lignocellulosic biomass for bioethanol production: an overview. Energy 199:117457
Saini A et al (2015) Actinomycetes: a source of lignocellulolytic enzymes. Enzyme Res 2015: 279381. https://doi.org/10.1155/2015/279381
Sari SLA et al (2016) Cellulolytic and hemicellulolytic bacteria from the gut of Oryctes rhinoceros larvae. Biodiversitas J Biol Divers 17:78–83
Schneiderman SJ et al (2015) Quantifying second generation ethanol inhibition: design of experiments approach and kinetic model development. Bioresour Technol 179:219–226. https://doi.org/10.1016/j.biortech.2014.11.087
Sebayang AH et al (2017) Optimization of bioethanol production from sorghum grains using artificial neural networks integrated with ant colony. Ind Crop Prod 97:146–155
Serrano ACCL et al (2025) The use of Design of Experiments (DoE) approaches for the development of Self-Emulsifying Drug Delivery Systems (SEDDS). Appl Nano 6(1):4. https://doi.org/10.3390/applnano6010004
Silva H et al (2023) Design of experiments: an overview and future paths, pp 328–341. https://doi.org/10.1007/978-3-031-45021-1_25
Smuga-Kogut M et al (2021) Use of machine learning methods for predicting amount of bioethanol obtained from lignocellulosic biomass with the use of ionic liquids for pretreatment. Energies (Basel) 14:243
Speed TP (1992) Introduction to Fisher (1926). In: The arrangement of field experiments, pp 71–81. https://doi.org/10.1007/978-1-4612-4380-9_7
StatEase (2025) Design of Experiments (DOE) made easy. https://Www.Statease.Com/Software/Design-Expert/
Statistical Discovery LLC, J (2022) Using JMP. http://support.sas.com/thirdpartylicenses

Sukhang S et al (2020) Bioethanol production from oil palm empty fruit bunch with SSF and SHF processes using *Kluyveromyces marxianus* yeast. Cellulose 27:301–314

Sun Y, Cheng J (2002) Hydrolysis of lignocellulosic materials for ethanol production: a review. Bioresour Technol 83:1–11

Toutenburg H, Shalabh. (2009) Incomplete block designs. In: Statistical analysis of designed experiments, Springer texts in statistics, 3rd edn. Springer, pp 181–244. https://doi.org/10.1007/978-1-4419-1148-3_6

Travaini R et al (2016) Ozonolysis: an advantageous pretreatment for lignocellulosic biomass revisited. Bioresour Technol 199:2–12

Usman M et al (2020) Characterization and biogas production potentials of aqueous phase produced from hydrothermal carbonization of biomass–major components and their binary mixtures. Chem Eng J 388:124201

Vani S et al (2015) Prediction of sugar yields during hydrolysis of lignocellulosic biomass using artificial neural network modeling. Bioresour Technol 188:128–135

Verma AK et al (2015) Effects of physicochemical pretreatment on the performance of an upflow anaerobic sludge blanket (UASB) reactor treating textile wastewater: application of full factorial central composite design. Can J Chem Eng 93(5):808–818. https://doi.org/10.1002/cjce.22168

Wang X et al (2021) Quantum speedup in adaptive boosting of binary classification. Sci China-Phys Mech Astron 64(2):220311. https://doi.org/10.1007/s11433-020-1638-5

Williams CL (1991) Design expert: an expert system application to clinical investigations. Expert Syst Appl 2(4):361–371. https://doi.org/10.1016/0957-4174(91)90042-D

Yildirim O et al (2021) Statistical optimization of dilute acid pretreatment of lignocellulosic biomass by response surface methodology to obtain fermentable sugars for bioethanol production. Int J Energy Res 45(15):8882–8899. https://doi.org/10.1002/er.6423

Chapter 9
From Fields to Fuel: Ethanol Production from Rice Residues for Sustainability: A Case Study

Reena Singh and Purvi Thangaraj

Abstract The rice-wheat cropping system encompasses about 10.6 million hectares (mha) in Punjab and Haryana, and plays a vital role in boosting per capita food grain availability in irrigated regions. Traditionally, farmers typically collected wheat and rice residues from their fields to use as cattle fodder, bedding for livestock, roofing material, or fuel. With the rise of mechanised harvesting, however, many rice growers now opt to burn the leftover residues directly in the fields, as these materials hinder tillage and sowing for the subsequent crop. This practice of residue burning is harmful, as it accelerates the depletion of soil organic matter and nutrients, heightens greenhouse gas (GHG) emissions, worsens air pollution, and diminishes soil microbial activity. In addition, large-scale burning of rice residue has also added health hazards, as evidenced by a reported increase in respiratory problems and itching of the eyes in the local population. By adopting ethanol blending practices, India is advancing towards a sustainable energy future. As the world's third-largest energy consumer, India has long relied on oil imports to meet its energy demands, resulting in energy security challenges and significant foreign currency outflows. Ethanol blending presents a promising solution, enabling India to reduce its dependency on imported oil while addressing environmental issues. Crop residues from rice could be an essential resource base for bioethanol production. This approach aligns seamlessly with circular economic strategies and zero-waste initiatives. This chapter sheds light on the opportunities and challenges of bioethanol production from rice residues in Punjab and Haryana, considering various factors, including biomass availability, production capacities, environmental constraints, economic factors, social and public policy influences, and supply chain factors.

Keywords Punjab · Haryana · Rice residues · Greenhouse Gas emissions · Crop residue burning · Ethanol blending program · Energy security · Sustainability

---

R. Singh (✉) · P. Thangaraj
Indian Council for Research on International Economic Relations (ICRIER), New Delhi, Delhi, India
e-mail: rsingh@icrier.res.in

S. Prasad et al. (eds.), *Sustainable Bioethanol Production from Lignocellulosic Biomass*, Clean Energy Production Technologies,
https://doi.org/10.1007/978-981-95-9690-4_9

9.1 Introduction

The diminishing global petroleum supply and the increasing GHG emissions have necessitated the demand for alternative energy sources. Biofuels are the fuels derived directly or indirectly from organic material such as agricultural residues, animal waste, organic fractions of municipal solid waste, industrial waste, and effluents (Prasad et al. 2007). Bioethanol is regarded as a highly promising renewable energy source, particularly for transportation, as it does not contribute additional carbon dioxide (CO_2) to the atmosphere. However, its combustion releases volatile organic compounds, nitrogen oxides (NOx), and carbon monoxide (CO), albeit in low concentrations. Sugarcane-derived materials serve as the primary feedstocks for bioethanol production in India, with C-heavy molasses, B-heavy molasses, and sugarcane juice or syrup being the most commonly utilised (Prasad et al. 2021). To diversify, the country has introduced separate pricing for maize-based ethanol, maize, damaged food grains, surplus rice, and agricultural residues like wheat and rice residue.

Agricultural crop residues refer to the materials remaining in fields after harvest, primarily including stalks, stubble, leaves, and seedpods. India possesses a vast and geographically varied biomass resource base, generating approximately 754.5 million tonnes (MT) of crop biomass annually (MNRE 2025). This huge volume is derived from a range of major crops, with wheat and rice accounting for nearly half of the total, producing 192.3 MT and 184.5 MT per year, respectively. Other significant contributors include maize (69.7 MT), sugarcane (17.9 MT), groundnut (17.7 MT), rapeseed and mustard (13.9 MT), pearl millet (*bajra*) (23.5 MT), and pigeon pea (*arhar/tur*) (7.2 MT). The distribution of this biomass is far from uniform across the country. Leading states in total crop biomass production include Uttar Pradesh (124.7 MT), Madhya Pradesh (70.2 MT), Rajasthan (59.5 MT), Punjab (53.0 MT), Maharashtra (52.5 MT), Tamil Nadu (52.1 MT), Gujarat (50.2 MT), West Bengal (47.5 MT), Andhra Pradesh (40 MT), and Haryana (36.2 MT). These states collectively form the backbone of India's agricultural residue landscape.

After factoring in customary uses of crop residues like fodder, fuel, and thatching, India has a surplus of approximately 228.5 MT of crop biomass each year (MNRE 2025). About one-third of this surplus comes from rice (41.7 MT) and wheat (33.4 MT), followed by maize (15.2 MT) and sugarcane (6.4 MT) (Fig. 9.1). Cotton (46.2 MT) and potato (17.4 MT) also contribute substantially, while pulses, oilseeds, and various other crops collectively account for an additional 39.2 MT of surplus biomass. Punjab emerges as the leading state in surplus biomass production, generating 22.3 MT, of which an overwhelming 84% is attributed to rice (18.8 MT), which is 52.2% of the country's total surplus rice biomass (Fig. 9.2). Other major states with high surplus biomass include Uttar Pradesh (21.6 MT), Madhya Pradesh (19.9 MT), Gujarat (21.7 MT), Maharashtra (21.5 MT), West Bengal (16.3 MT), Tamil Nadu (12.2 MT), and Haryana (10.9 MT).

The high surplus of rice residue in Punjab, driven by mechanised harvesting and intensive rice-wheat cropping, has made this state hotspot for stubble burning,

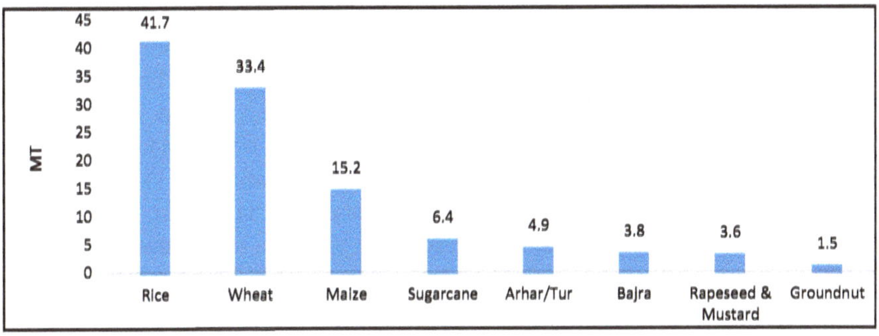

Fig. 9.1 Crop-wise surplus biomass residue. (Source: MNRE 2025)

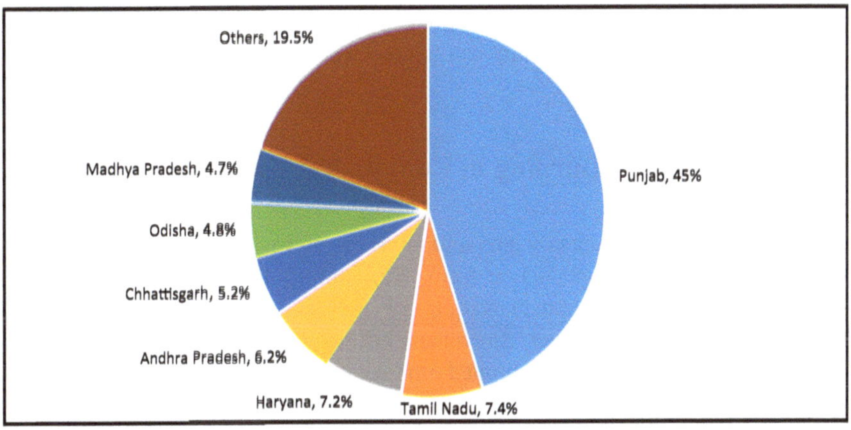

Fig. 9.2 State-wise share of surplus rice biomass residue. (Source: MNRE 2025)

contributing to severe air pollution episodes in North India each year. However, this surplus represents a significant opportunity for bioenergy and bioethanol production, aligning with India's policy push for second-generation (2G) biofuels. This chapter assesses the potential of ethanol production from rice residues in Punjab and Haryana as a crop residue management strategy in the north-western region of the country.

9.2 Agriculture Production and Biomass Availability in Punjab and Haryana

In Punjab, wheat and rice are the predominant crops, together constituting 80% of the state's gross cropped area (GCA) of 8.3 mha, with wheat occupying 3.5 mha and rice 3.2 mha (UPAg 2025). Similarly, wheat and rice account for 58% of the GCA (6.6 mha) in Haryana, with 2.3 mha under wheat and 1.5 mha under rice cultivation.

Despite representing only 13% of India's total wheat and rice cultivation area, Punjab and Haryana together produce 20% of the country's rice and wheat output, amounting to 49 MT out of the national total of 251 MT (Table 9.1). This remarkable productivity can be attributed to the Green Revolution, which began in Punjab and subsequently in Haryana, and was characterised by the widespread adoption of modern agricultural technologies such as high-yielding varieties (HYVs) of wheat and rice, expanded irrigation infrastructure, increased mechanisation, and intensive use of fertilisers and pesticides. As a result, the yields in these states far surpass national averages: Punjab's rice and wheat yields are 4516 kg/ha and 5045 kg/ha, respectively, compared to national averages of 2882 kg/ha and 3559 kg/ha, while Haryana achieves rice and wheat yields of 3822 kg/ha and 4723 kg/ha, respectively. This high productivity translates into substantial biomass generation of wheat and rice. At the same time, other crops such as maize, sugarcane, pigeon pea (*arhar/tur*), pearl millet (*bajra*), rapeseed and mustard, and groundnut occupy smaller cultivation areas and contribute proportionally less to the national biomass surplus (Table 9.1).

9.3 Rice Residue Burning in Punjab and Haryana

During Kharif, paddy nursery preparation starts from 20th May to 20th June, followed by paddy transplantation from 20th June to 10th July. Paddy varieties take 93–117 days after transplantation for maturity and are harvested during mid-October to mid-November. Since the wheat sowing starts in the first fortnight of November (Fig. 9.3), there is limited time window for disposal and farmers often burn residues—a practice that has grown due to labour scarcity, high removal costs, and mechanised harvesting.

Punjab experiences numerous fire incidents during paddy harvesting (Fig. 9.4). In 2016 alone, 102,379 fire events were reported between September and November. Since 2020, however, the numbers have shown a downward trend, with 10,909 cases recorded during the same period in 2023. During 2020, 25.44 MT of rice residue was produced, of which 12.84 MT was burnt, which means nearly 50.47% of the rice residue was burnt. In terms of area, out of 2.97 mha rice area, 1.65 mha (56% of the total rice cultivation area) was burnt in 2020.

The combustion of crop residues is a major source of GHG emissions, as well as chemically and radiatively active trace gases and aerosols such as methane (CH_4), nitrous oxide (N_2O), CO, NO_x, and various hydrocarbons. Beyond these gases, burning residue produces a substantial amount of particulate matter, which is composed of diverse organic and inorganic compounds (Table 9.2). Biomass smoke contains numerous pollutants that are either confirmed or suspected carcinogens, associated with respiratory disorders and lung diseases (NPMCR 2014).

Table 9.1 Crop production and biomass availability in Punjab and Haryana, 2023–2024

	States	Rice	Wheat	Maize	Sugarcane	Pigeon pea (*Arhar/Tur*)	Pearl millet (*Bajra*)	Rapeseed and Mustard	Groundnut
Area (000 ha)	Haryana	1563	2370	5	88	1	528	741	5
	Punjab	3179	3516	95	90	1	1	43	2
	India	47,828	31,833	11,241	5740	4131	7375	9183	4707
Production (000 tonnes)	Haryana	5976	11,191	18	7374	2	1161	1345	5
	Punjab	14,356	17,737	362	7510	1	1	69	3
	India	137,825	113,292	37,665	453,158	3417	10,716	13,259	10,180
Yield (Kg/ha)	Haryana	3822	4723	3685	83,582	1180	2200	1816	1066
	Punjab	4516	5045	3827	83,254	1127	785	1611	1740
	India	2882	3559	3351	78,953	827	1453	1444	2163
Available biomass (000 tonnes)	Haryana	5986	20,024	58	372	27	2025	1518	8
	Punjab	18,922	30,926	1013	366	12	0	81	6
	India	184,539	192,339	69,739	17,887	7234	23,481	13,880	17,710
Surplus biomass residue (000 tonnes)	Haryana	3003	2002	12	93	19	383	759	2
	Punjab	18,786	3093	203	92	8	0	41	1
	India	41,727	33,373	15,207	6379	4899	3842	3642	1543

Source: UPAg (2025) and MNRE (2025)

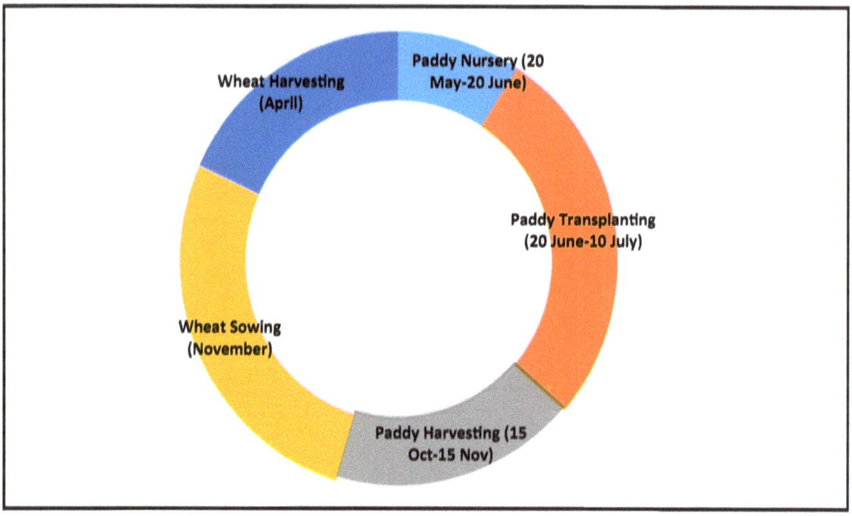

Fig. 9.3 Predominant crop cycle of Punjab and Haryana. (Source: PAU 2024)

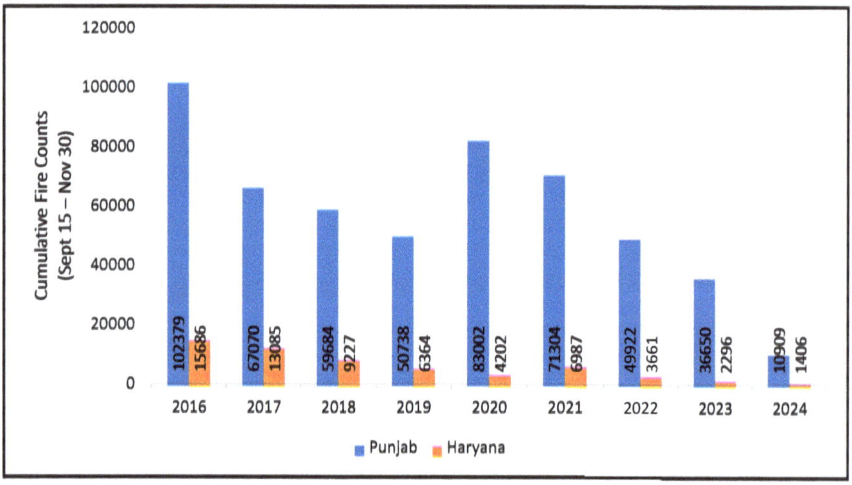

Fig. 9.4 Cumulative fire counts in Punjab during September–November (2016–2023). (Source: IARI 2023)

Table 9.2 Emission of various pollutants from the burning of rice residues in India

Pollutants	Emission estimates (Gg/year)
PM 2.5	418
PM 10	458.29
Sulphur-di-oxide (SO_2)	9.07
Carbon dioxide (CO_2)	59,275.74
Carbon monoxide (CO)	4683.64
Nitrous oxide (N_2O)	24.17
Nitrous oxides (NOx)	114.82
Ammonia (NH_3)	206.48
Volatile organic compounds (VOC)	352.53
Elemental carbon (EC)	25.68
Organic carbon (OC)	150.58
Polycyclic aromatic hydrocarbons (PAH)	0.026

Source: Dutta et al. (2022)

9.4 Potential of Rice Residues for Ethanol Production

Rice residues hold considerable promise for ethanol production in Punjab and Haryana, given the large volumes of surplus biomass generated in these states each year, existing policy frameworks, and alignment with national ethanol blending targets.

District-level discrepancies between the area of *kharif* rice and surplus biomass (Tables 9.3 and 9.4) suggest that factors such as alternative residue usage, local policy, and farmer practices result in significant geographical variation in the biomass available for alternative uses. The geographic concentration of surplus, especially in select districts of Punjab and Haryana, indicates that interventions focusing on collection logistics and biorefinery siting should prioritise these high-surplus zones to maximise efficiency and sustainability in residue-based ethanol production.

This district-level analysis from Punjab and Haryana is done with a clear focus on key drivers for bioethanol production: area under rice cultivation (UPAg 2025), surplus rice biomass available (TIFAC-IARI 2018), and the frequency of fire counts (IARI 2023). In Punjab, Sangrur, Ludhiana, Patiala, Bathinda, Firozpur, and Moga are the most prominent districts (Fig. 9.5), with a total of 11 MT dry biomass generating 9.7 MT of surplus biomass and bioethanol potential of 2668 million litres. Other districts such as Jalandhar, Amritsar, Faridkot, Gurdaspur, Muktsar Sahib, and Tarn Taran also contribute to surplus biomass (3.8 MT) and have bioethanol potential of 1053 million litres. In contrast, some districts (e.g., Pathankot, Sas Nagar, Rupnagar) contribute modest amounts to surplus biomass, aligned with their smaller paddy acreage. In Haryana, key districts include Fatehabad, Sirsa, Kurukshetra, Kaithal, Ambala, Jind, Panipat, and Yamunanagar. These districts have sizable rice cultivation areas, surplus biomass availability (1.6 MT), and corresponding bioethanol potential (439 million litres). The

Table 9.3 District-level analysis for biorefinery potential in Punjab

Punjab	Rank	*Kharif* Rice Area - (000 ha)	Surplus Biomass available (000 tonnes)	Fire Count
Sangrur	High potential	239	2296	9389
Ludhiana		259	2125	5817
Patiala		234	1584	5426
Bathinda		190	858	4481
Firozpur		187	1427	6289
Moga		183	1445	6515
Muktsar	Medium potential	183	465	4598
Tarn Taran		182	278	4117
Amritsar		181	793	2174
Gurdaspur		173	583	1396
Faridkot		116	754	3953
Jalandhar	Low potential	176	971	2548
Mansa		126	159	3217
Kapurthala		119	646	1798
Barnala		115	495	4326
Fazilka	No priority	108	420	2389
Fatehgarh Sahib		87	413	1724
Hoshiarpur		81	470	331
Nawashahr		63	275	355
Sas Nagar		42	113	148
Rupnagar		41	138	307
Pathankot		30	77	6

Source: UPAg (2025), TIFAC-IARI (2018), and IARI (2023)

classification of districts based on the criteria not only captures the agricultural scale of each district but also points to critical hotspots for both residue management challenges and biofuel opportunities. Therefore, targeting these regions for residue collection, value chain development, and setting of 2G ethanol biorefineries would simultaneously advance environmental and energy policy goals. The massive bioethanol potential documented points to enormous opportunities for decentralised, residue-based bioenergy industries that could reduce GHG emissions, address local air quality problems, and support regional rural economies.

Punjab's paddy residue management strategy is rapidly evolving, as evidenced by the most recent data for 2023 and 2024 (Table 9.5). Out of approximately 15.86 MT (2023) to an expected 19.52 MT (2024) of crop residue generated, the largest

Table 9.4 District-level analysis for biorefinery potential in Haryana

Haryana	Rank	*Kharif* Rice Area - (000 ha)	Surplus Biomass available (000 tonnes)	Fire counts
Fatehabad	High potential	140	361	1479
Sirsa		113	341	551
Kurukshetra		130	214	538
Kaithal		179	197	1157
Ambala		105	159	308
Jind		174	157	919
Panipat		80	39	254
Yamunanagar	Medium potential	87	135	147
Hisar		98	59	245
Karnal		181	32	955
Jhajjar	Low potential	53	34	7
Sonipat		120	22	219
Faridabad	No priority	16	26	3
Bhiwani		26	20	12
Rohtak		79	15	78
Rewari		2	12	-
Gurugram		5	1	-
Mahendragarh		0	0	-
Mewat		12	0	-
Palwal		41	0	115
Panchkula		13	3	-

Source: UPAg (2025), TIFAC-IARI (2018), and IARI (2023)

proportion is managed in situ. In 2023, about 11.5 MT were managed in situ, rising to 12.7 MT in 2024 (GoP 2024).

Ex situ uses encompass industrial boilers, biomass power plants, compressed biogas facilities, bioethanol plants, thermal power generation, brick kilns, and fodder production. It accounts for a growing share, rising from 3.66 MT in 2023 to an estimated 5.96 MT in 2024. Despite bioethanol plants' use increasing from 0.1 to 0.2 MT, it remains a small but emergent segment of total residue utilisation. This growth trajectory underscores the vast untapped potential in diverting larger shares of surplus biomass towards residue-based ethanol, simultaneously reducing air pollution and mitigating stubble burning.

A robust paddy straw management value chain is essential to bridge the gap between district-level surplus availability and the practical deployment of bioethanol solutions (Fig. 9.6). By integrating efficient storage, collection, and aggregation infrastructure, this value chain directly supports the mobilisation of crop residues

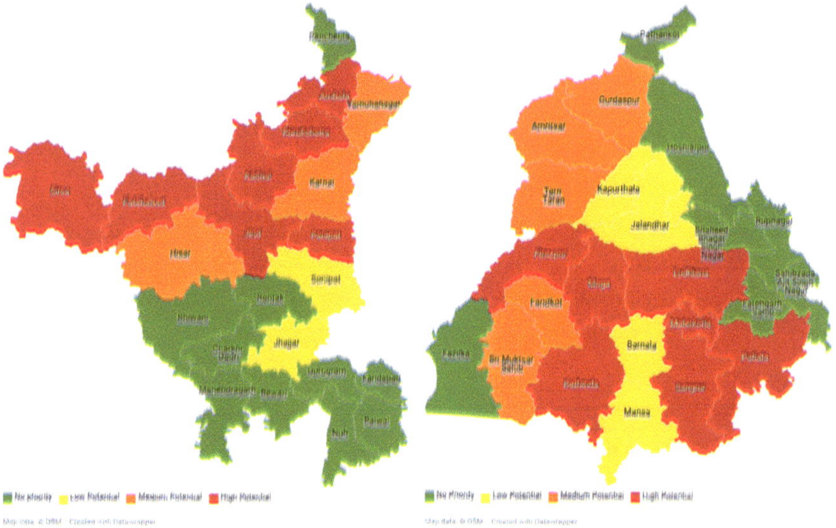

Fig. 9.5 Potential districts mapped in Punjab and Haryana. (Source: Authors' creation using UPAg 2025; TIFAC-IARI 2018; IARI 2023)

Table 9.5 Paddy straw utilisation in Punjab

S. No.	Strategy	2023 (Actual), MT	2024 (Estimated), MT
1	In Situ Management	11.5	12.7
2	Ex Situ Management	3.66	5.96
2.1	Industrial Boilers	2.1	3.17
2.2	Biomass Power Plant	0.96	1.17
2.3	Compressed Bio-gas Plant	0.3	0.54
2.4	Bio-Ethanol Plant	0.1	0.2
2.5	Thermal Power Plant	0.2	0.77
2.6	Brick Kilns	0	0.18
3	Fodder	0.7	0.86
	Total	15.86	19.52

Source: GoP (2024)

from high-surplus districts, making consistent feedstock available for bioenergy industries. Establishing such systems will not only enable the commercial viability of residue-based biorefineries but also help ensure year-round supply, reduce burning, and advance effective, sustainable residue management in both Punjab and Haryana (Box 9.1).

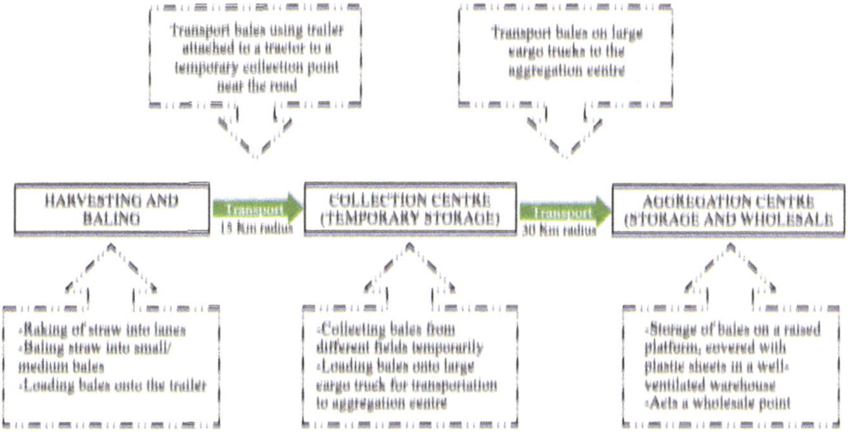

Fig. 9.6 Model value chain for rice straw. (Source: FAO 2022)

Box 9.1 Haryana Supply Chain for Residue Management

A farmer receives an incentive of ₹1000 for not burning crop residue under the crop diversification scheme, once the farmer submits the stamped weighbridge receipt issued by the residue collection centre as proof of the quantity of residue delivered. If the baler collects the residue from farmers' fields (generally from non-basmati fields), the farmer pays about ₹500 to the baler for collecting and clearing the residue from the fields. Balers obtain these machines at a subsidised rate: the actual cost of a baler is around ₹16 lakhs, on which eligible applicants receive a 50–60% subsidy from the government, with beneficiaries selected through a lottery system when applicants exceed the number of machines available. Other operating costs for the baler include fuel, labour, repair, and maintenance. The baler then aggregates, transports, and sells the residue to stubble collection centres, which further process, store, and organise for resale.

For basmati residues, farmers generally undertake manual harvesting, with labour costs around ₹8000 per acre. Farmers then aggregate, transport, and sell the residue directly to stubble collection centres. A typical collection centre covers a radius of about 50 km, operates approximately 12 collection points, and handles roughly 5000–6000 quintals per day of residue per day. Purchase prices at the collection centre are in the range of ₹123–200 per quintal for non-basmati straw (usually combine-harvested and mixed with soil) and about ₹350 per quintal for manually harvested basmati straw. Non-basmati residue is mainly used as fuel for power plants, while basmati straw is primarily sold as fodder and as a raw material for paper and pulp mills; depending on destination and quality, the residue can fetch around ₹430 per quintal at final sale, with rates varying across uses and demand conditions.

Source: Visit to weight bridge (Residue Collection Centre), Kaithal, Haryana, during October 2023.

The economic considerations for implementing this value chain are substantial but achievable. According to the United Nations Food and Agriculture Organisation's technical assessment in Punjab, the average collection cost, including raking, baling, on-field transport, subsequent loading and offloading, and temporary and long-term storage, ranges from ₹733/tonne when targeting the collection of 30% of the available rice straw to ₹1571/tonne for 5% collection target. Feedstock costs (collection, handling, and storage) typically account for 20–40% of total production costs, underlining the critical importance of efficient supply chains. Annualised costs also include labour, fuel/energy for machinery, and maintenance at aggregation centres. In the case of 2G ethanol facilities, capital investment is a significant requirement: establishing a medium-scale plant with a 5 million litre annual capacity demands about ₹264 crores, while a large-scale facility producing 100 million litres annually requires nearly ₹1592 crores (FAO 2022). Government controls ethanol prices in India, which differ based on the feedstock source. For example, average farm-gate prices for paddy straw in Punjab range from ₹1000 to ₹1200 per tonne, with peaks up to ₹2500 per tonne (FAO 2022). Farmer remuneration is a function of localised demand, logistical costs, and the existence of functional value chains connecting fields to processing plants.

Economic feasibility analyses reveal that producing 2G ethanol remains financially unviable in several districts, driven largely by high capital investments and the costs of necessary pre-treatment chemicals, which constitute the largest share of production expenses. The breakeven production capacity is estimated at approximately 93 million litres annually (FAO 2022). Costs per litre vary significantly based on plant scale and operational inputs like labour, energy, and materials. At current market prices,[1] only a few districts, including Sangrur, Ludhiana, Patiala, and Moga, show potential profitability due to sufficient biomass availability and scale. However, a 50% increase in ethanol pricing could render production economically feasible for most districts. The development of 2G ethanol facilities entails substantial capital expenditure, with investment needs exceeding ₹250 crores for a 5 million litre unit and surpassing ₹1000 crores for a 100 million litre plant. The current pricing structure restricts economic viability, emphasising the need for policy interventions to reduce capital expenses or increase ethanol prices to support large-scale industrial deployment. Ultimately, concerted investments in both the supply chain and refinery infrastructure, alongside policy incentives and financing mechanisms, are critical to unlocking Punjab and Haryana's vast potential for sustainable, residue-based ethanol production.

[1] ₹62.65 per litre for ethanol produced.

9.5 Environmental Laws, Policy Frameworks to Curb Air Pollution, and Stubble Burning

Under the 1981 Air (Prevention and Control of Pollution) Act, crop residue burning is a punishable offence. The Supreme Court established Environment Pollution (Prevention and Control) Authority (EPCA) in 1998 to combat NCR air pollution, and in 2014 the National Green Tribunal (NGT) launched National Policy for Management of Crop Residue (NPMCR) to address residue burning in Delhi's neighbouring states. NPMCR promotes low-cost machinery, sustainable residue management, and satellite monitoring of fires, while Central Pollution Control Board (CPCB) and National Remote Sensing Agency (NRSA) can penalise farmers for burning.

Since 2014, India has enacted regulations to curb crop burning and improve air quality. Figure 9.7 shows central and state policies targeting north-western states to reduce pollution in Delhi NCR. Launched in 2013–2014, the Department of Agriculture and Farmers Welfare's (DA&FW) Crop Diversification Programme, a sub-scheme under *Rashtriya Krishi Vikas Yojana* (RKVY) encourages farmers in Haryana, Punjab, and Western Uttar Pradesh to shift from paddy to crops like pulses, cotton, nutri cereals, and oilseeds.

The Indian government further facilitates crop diversification at the state level through initiatives such as the National Food Security Mission (NFSM), which promotes pulses, coarse cereals, nutri cereals, cotton, and oilseeds, and the Mission for Integrated Development of Horticulture (MIDH) for horticultural crops. In addition, states may customise programs under the RKVY to meet specific regional needs, subject to approval by the State Level Sanctioning Committee (SLSC), chaired by the Chief Secretary.

To strengthen sustainable residue management practices, the government launched the "Promotion of Agricultural Mechanization for In-Situ Management of Crop Residue" initiative across Punjab, Haryana, Uttar Pradesh, and the National

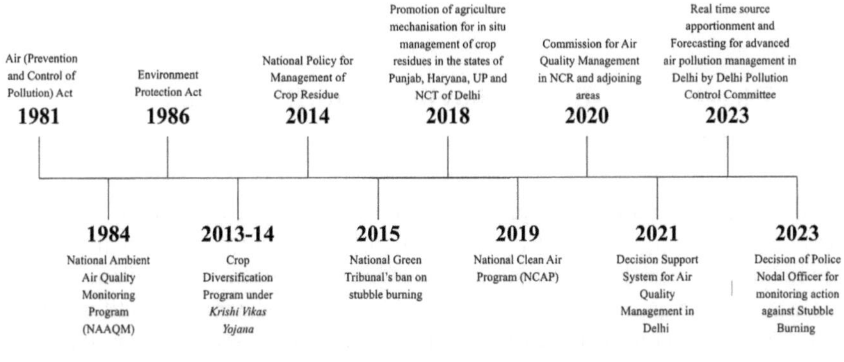

Fig. 9.7 Key policy interventions to curb air pollution and stubble burning in India. (Source: Authors' compilation)

Capital Territory of Delhi. This scheme provides subsidies for crop residue management machinery to support environmentally friendly disposal and processing of crop residues. Individual farmers receive a 50% subsidy on these machines, while Custom Hiring Centres (CHCs) are offered an 80% subsidy to ensure that small and marginal farmers, who cannot afford expensive equipment, have access to necessary machinery. Since the scheme's inception in 2018–2019, a total of 1,00,882 machines have been subsidised, including 31,510 machines distributed through 6794 CHCs and 69,372 to individual farmers, covering the period till 2024–2025. To promote awareness and adoption, the scheme allocates funds for Information, Education, and Communication activities. For 2024–2025, a funding allocation of ₹268.18 crore has been earmarked by the Government of India for implementing crop residue management strategies. During the current year itself, farmers have purchased 9844 machinery units under this subsidy program. The scheme also emphasises capacity building, establishment of supply chains for crop residue collection and utilisation, and deployment of bio-decomposers to reduce residue burning and its environmental impact. This initiative aids in sustainable agriculture by mitigating air pollution and preserving soil health across participating states.

Furthermore, financial aid is extended to State Governments' Krishi Vigyan Kendras (KVKs), Indian Council of Agricultural Research (ICAR) institutions, Central Government agencies, and Public Sector Undertakings (PSUs) for awareness programs on residue management. ICAR's Pusa Decomposer, available as liquid or capsules, speeds up paddy residue breakdown for better management.

For promoting paddy straw usage in distilleries/bioethanol and refineries/sugar mills, the Government of Punjab has mandated that all new distilleries, as well as existing distilleries installing new boilers, must use paddy straw as fuel. Additionally, under the Punjab Industrial and Business Development Policy (PBIP), the State Government is offering State Goods and Services Tax (SGST) reimbursement incentives for boilers using paddy straw in industries such as sugar mills, paper mills, and for steam-generating boilers with a capacity of over 25 tonnes per hour. The government is also considering a proposal to provide capital incentives for the installation of new paddy straw-based boilers in industrial units (GoP 2024).

9.6 Rice-Residue-Based Biorefinery for Bioethanol Production

"Biorefining is the sustainable processing of biomass into a spectrum of marketable products and energy" (IEA Bioenergy 2009). As society grapples with growing environmental, social, and economic challenges, their effects will continue to shape the way we live and work in the future. Advanced biorefineries that utilise diverse lignocellulosic, non-food feedstocks are widely recognised as environmentally responsible solutions for converting biomass into sustainable fuels, chemical intermediates, and essential products for society while fostering job creation and business growth. Shifting away from reliance on fossil fuels, this transition ensures that agriculture remains a key player in food security and supplying sustainable, renewable raw materials for various industries. This shift will be central to the future of a more sustainable economy.

The Ethanol Blended Petrol (EBP) Programme, launched in 2003, set out a mandate in 2006 to Oil Marketing Companies (OMCs) to sell 5% EBP across 20 states and 4 Union Territories. Despite this early initiation, the EBP programme could not take off owing to high taxation (18%), procurement challenges, limited availability of feedstock, exclusion of grain-to-ethanol conversion, dissatisfactory pricing, and infrastructural bottlenecks (MoPNG 2021). Later, in December 2014, with the EBP Programme stimulus, the government introduced administered ethanol pricing and broadened the range of permitted feedstocks beyond sugarcane molasses to include B-heavy molasses, sugarcane juice, surplus and damaged food grains, and, more recently, agricultural residues supported by efficient procurement processes. The programme was also extended across India (Andaman & Nicobar and Lakshadweep Islands), with a massive reduction in taxation to 5%.

India's 2018 biofuel policy advanced the 20% ethanol blending target to Ethanol Supply Year (ESY) 2025–2026. OMCs attained 10% blending in June 2022, ahead of schedule, with levels rising to 18.6% by April 2025. Supplies grew from 38 crore litres in 2013–2014 to 707.4 crore litres in 2023–2024 (MoPNG 2025a). Achieving E20 will need 1016 crore litres, or 1350 crore litres including other uses (NITI Aayog 2021). Since 2014–2015, the EBP Programme has paid farmers ₹1.04 lakh crore, saved ₹1.20 lakh crore in foreign exchange, replaced 200 lakh tonnes of crude oil, and cut CO_2 emissions by 626 lakh tonnes (MoPNG 2025b).

EBP is primarily produced from both sugarcane-based feedstocks, such as sugarcane juice, sugar syrup, sugar, and by-products like B-heavy and C-heavy molasses, and grain-based feedstocks, including maize, damaged food grains (DFG), and surplus rice (Fig. 9.8). In the last 5 years, from ESY 2019–2020 to ESY 2023–2024, sugarcane-based feedstock accounted for 75% of the ethanol produced and supplied

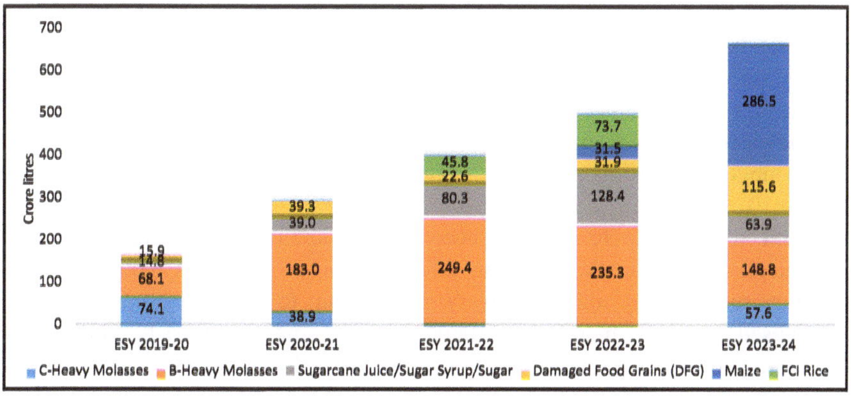

Fig. 9.8 The quantity of ethanol produced and supplied to OMCs. (Source: MoCAFPD 2025)

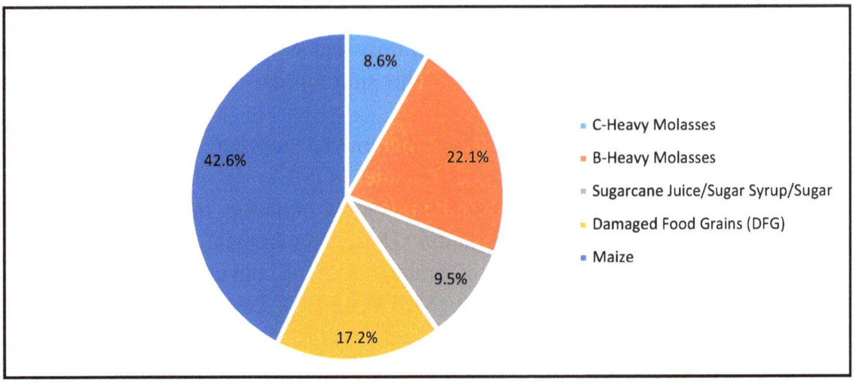

Fig. 9.9 Feedstock-wise share of ethanol produced and supplied to OMCs in ESY 2023–2024. (Source: MoCAFPD 2025)

to OMCs, leaving the remaining 25% with grain-based feedstock. Sugarcane-based feedstock was dominated by B-heavy molasses, followed by sugarcane juice/sugar syrup/sugar and C-heavy molasses. At the same time, DFG and FCI rice were the major contributors to grain-based feedstock. However, in ESY 2023–2024, 60% was supplied from grain-based feedstock (Fig. 9.9). Maize accounted for 42.6% of the total ethanol produced and supplied to OMCs, followed by B-heavy molasses (22.1%), DFG (17.2%), Sugarcane Juice/Sugar Syrup/Sugar (9.5%), and C-heavy molasses (8.6%).

India's broiler industry, growing 8–10% annually, will need 1–3 MT more maize, given its 60–65% share in feed with a Feed Conversion Ratio (FCR) of 1.65 (Gulati and Juneja 2023). Other uses of maize are livestock feed and starch. The diversion of maize towards biofuel would create challenges for poultry feed and other uses. Another concern is that this could encourage the cultivation of spring maize, a crop with high water requirements, which distilleries may then source for processing. In recent years, spring maize cultivation has been encouraged in Punjab from February to March. Many farmers cultivate early maturing wheat varieties, which are harvested during February-March. Soon after the wheat harvest, they grow spring maize, which requires 11–17 irrigations, and then cultivate paddy which requires 20–25 irrigations. This will add additional pressure on the already declining groundwater levels. During 2000–2022, a decline of −11.94 meters below ground level (mbgl) and −10.89 mbgl was recorded in Haryana and Punjab, respectively (Fig. 9.10).

From surplus rice residues, lignocellulosic biorefinery is an alternative for bioethanol production (Singh et al. 2024). Inspired by the oil industry, where refining succeeds by utilising every component to create commodities and specialised products, biorefining should adopt a similar approach. This means leveraging every fraction of biomass, optimising the valorisation of cellulose, hemicellulose, and lignins to ensure maximum utility and sustainability.

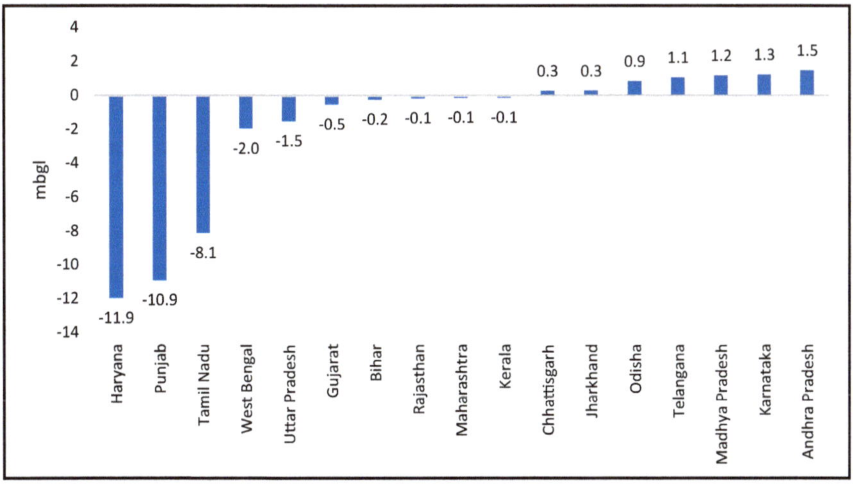

Fig. 9.10 Average decline in water level in major states during 2000–2022. (Source: Thangaraj and Gulati 2024)

Rice residues are a lignocellulosic material composed primarily of cellulose (24%), hemicelluloses (27.8%), and lignin (13.5%), along with a notable amount of ash (17%), predominantly silica (Rosado et al. 2021). Given its abundance, affordability, and high carbohydrate and lignin content, rice residue is acknowledged as an important feedstock for biofuel production, biochemicals, and biobased materials within lignocellulosic biorefineries (Singh et al. 2024).

9.7 Challenges for Scaling Up the Rice Residue-Based Biorefinery

Ensuring a reliable and sustainable feedstock supply is fundamental to the development of industrial bio-based products. For biomass to be considered sustainable, it must remain economically feasible, consistently available over time, and have minimal environmental impact (Singh et al. 2024). In addition, the extraction of biomass should align with global environmental principles in agriculture and forestry. This requires careful attention to soil health (e.g., maintaining or enhancing stable soil carbon levels, ensuring a steady return of fresh carbon to the soil, limiting nutrient loss, and preventing compaction and erosion) and water quality (such as reducing nitrogen-based inputs and herbicide use). Sustainability also involves lowering water demand, safeguarding biodiversity, including both common and rare species and habitats, minimising landscape alterations, cutting GHG emissions, and adapting to climate change through the most resource- and energy-efficient applications of biomass.

To evaluate whether the case study regions of Punjab and Haryana can support year-round bio refineries, the availability of surplus crop residues was analysed, with particular focus on rice residues and their competing uses. In these states, wheat residues are mainly utilised as animal feed and in the paper pulp industry, while rice residues are typically burned in fields—a practice not considered a competing use in this study. After accounting for residues allocated to alternative purposes, Punjab was found to have 22.3 MT of crop residues annually, comprising 84% rice, 14% wheat, and 2% from other crops. Haryana, on the other hand, has 10.9 MT of residues per year, distributed as 28% rice, 18% wheat, 7% rapeseed and mustard, 4% pearlmillet (*bajra*), and 44% from other crops.

However, the sustainable supply of feedstock availability for biorefinery will depend on many factors:

- Whether future biomass production will match or fall below current levels.
- No new land can be allocated for crop cultivation.
- Changes in residue availability from expanded utilisation in emerging applications such as packaging and biomass-based power generation.
- With the mechanisation of crop residue management, e.g., happy seeder, smart seeder, super seeder, etc., it is anticipated that the farmers will also be encouraged to incorporate the residues in the field, so the feedstock availability can be decreased.
- Emerging farming practices (e.g., organic farming, crop rotation diversification, and conservation techniques) often enhance environmental sustainability, but they can also diminish overall yields and the amount of residues produced.
- Externalities, such as climate change, water availability, etc., that impact residue generation.
- Potential challenges in securing long-term agreements with farmers, considering social limitations.
- Unforeseen challenges in logistics.
- Adoption of new policy might steer decisions towards the most efficient utilisation of biomass resources and energy.

Establishing a rice residue-based biorefinery in any region requires the development of a cost-efficient supply chain that encompasses not only biomass production and harvesting but also its collection, storage, and pre-processing, ensuring a consistent year-round feedstock supply. Achieving this goal necessitates consideration of multiple factors—such as geographical conditions, seasonal availability of biomass, logistics of transport and storage, and drying requirements—all of which play a crucial role in identifying optimal location for future bio refineries in Punjab and Haryana.

Integrating biorefineries within existing crop residue logistics systems (for instance, situating a facility near a paper mill) offers clear economic advantages. The primary component of crop residue costs arises from processes enabling production and delivery to the field boundary, encompassing harvesting expenses as well as the inherent value of residues when considered as potential soil organic matter. Logistics costs, in turn, are associated with activities required to move residues from farms to the biorefinery gate, namely storage and transportation, and are determined through optimisation of feedstock supply systems by residue type. Storage costs are represented as surcharges paid to farmers for land allocated to residue storage, while transportation costs, proportional to delivery distance and freight volume, indirectly capture land-use patterns, local biomass availability, and infrastructure quality. Drawing on insights from case studies, two transport scenarios emerge: one relying exclusively on road networks for local supply, and another incorporating rail transport to source biomass from other states. Furthermore, while residues in the first scenario are limited to semi-baling, the second scenario envisions the adoption of advanced baling technologies.

9.8 Conclusion

Agriculture in Punjab and Haryana is characterised by intensive practices, particularly the rice–wheat cropping system, which places considerable strain on the environment. A major source of this stress is the burning of rice residues in fields, a practice that contributes to water depletion, pollution, and biodiversity loss. Locally, residue burning diminishes soil health, depletes soil carbon and water reserves, and generates substantial gaseous emissions, including GHGs such as methane and nitrous oxides. Establishing biorefineries in the region is therefore expected to yield significant environmental benefits, improving both air quality and soil conditions.

When rice straw is burned, most of its carbon is released as CO_2 (around 70%), with smaller proportions emitted as CO (7%) and CH_4 (0.66%). Additionally, approximately 2.09% of the nitrogen content is discharged as N_2O. The combustion of crop residues also produces large amounts of particulate matter containing diverse organic and inorganic compounds. Many of these pollutants, present at high concentrations in biomass smoke, are either confirmed or suspected carcinogens, posing serious risks to respiratory health (NPMCR 2014). Importantly, the establishment of biorefineries in this northwestern region would reinforce existing legislation against residue burning and support the Ethanol Blended Petrol (EBP) program, which aims to achieve 20% ethanol blending. By assigning economic value to rice residues, biorefineries would also discourage farmers from burning them, offering both environmental and economic incentives.

Development of affordable and efficient technologies for converting rice residues into biorefinery feedstock should be a central priority, accompanied by research focused on generating high-value co-products. Rigorous economic evaluations are necessary to determine the practicality of producing bioethanol and related outputs at a commercial scale using surplus residues. Equally important are in-depth investigations into the environmental implications of ethanol and by-product generation—particularly through life cycle assessment approaches—to ensure that sustainability goals are achieved in the context of bioenergy and biofuel-oriented biorefineries.

References

Dutta A et al (2022) A state of the art review in crop residue burning in India: previous knowledge, present circumstances and future strategies. Environ Challenges 8:100581

FAO (2022) Establishing residue supply chains to reduce open burning. The case of rice straw and renewable energy in Punjab, India. Environment and Natural Resources Management Working Paper No. 95. Rome. https://doi.org/10.4060/cb9570en

GoP (2024) Action plan for control of crop residue burning in the State of Punjab (2024–25). Government of Punjab

Gulati A, Juneja R (2023) Poultry revolution in India: lessons for smallholder production systems. In Centre for Development Research, University of Bonn (Ed.), ZEF working paper series-225

IARI (2023) Consortium for research on agroecosystem monitoring and modeling from space. Division of Agricultural Physics at Indian Council for Agricultural Research—Indian Agricultural Research Institute. Retrieved from: https://creams.iari.res.in

IEA Bioenergy (2009) Biorefineries: adding value to the sustainable utilisation of biomass. International Energy Agency Bioenergy. T42:2009:01. Retrieved from https://www.ieabioenergy.com/wp-content/uploads/2013/10/Task-42-Booklet.pdf

MNRE (2025) Biomass resource atlas of India. National Institute of Bio-Energy. Ministry of New and Renewable Energy. Government of India. Retrieved from https://nibe.res.in/english/biomass-atlas.php

MoCAFPD (2025) Ethanol production from sugar byproducts. Lok Sabha Unstarred question No. 3044. Department of Food and Public Distribution. Ministry of Consumer Affairs, Food and Public Distribution. Government of India. Retrieved from https://sansad.in/getFile/loksabhaquestions/annex/184/AU3044_cVI3Ca.pdf?source=pqals

MoPNG (2021) Ethanol growth story: fulfilling Aatmanirbhar Bharat dream. Ministry of Petroleum & Natural Gas. Government of India

MoPNG (2025a) India's oil & gas ready reckoner. Oil & Gas Industry Information at a Glance. Petroleum Planning & Analysis Cell. Ministry of Petroleum & Natural Gas. Government of India

MoPNG (2025b) Impact of ethanol blending on food security and the environment. Rajya Sabha Unstarred question No. 2689. Ministry of Petroleum & Natural Gas. Government of India. Retrieved from: https://sansad.in/getFile/annex/267/AU2689_2woJ8x.pdf?source=pqars

National Policy for Management of Crop Residues (2014) Department of Agriculture & Cooperation (Natural Resource Management Division). Ministry of Agriculture. Government of India

NITI Aayog (2021) Roadmap for ethanol blending in India 2020–25. Ministry of Petroleum and Natural Gas. Government of India

PAU (2024) Package of practices for crops of Punjab. Punjab Agricultural University

Prasad S, Singh A, Joshi HC (2007) Ethanol as an alternative fuel from agricultural, industrial and urban residues. Resour Conserv Recycl 50(1):1–39

Prasad S, Gupta VC, Bhatt R, Dhanya MS (2021) Ethanol production from sugarcane: an overview. In: Input use efficiency for food and environmental security, pp 665–686

Rosado MJ et al (2021) Structural characteristics of the guaiacyl-rich lignins from rice (Oryza sativa L.) husks and straw. Front Plant Sci 12:640475

Singh S et al (2024) Exploring rice straw's potential from a sustainable biorefinery standpoint: towards valorisation and diverse product production. Process Saf Environ Prot 184:314–331

Thangaraj P, Gulati A (2024) India's food-water-energy conundrum. Agrifood Trends Analyt Bull 3(4). Re-aligning Agri-food Policies for Protecting Soil, Water, Air & Biodiversity (SWAB). Indian Council for Research on International Economic Relations (ICRIER)

TIFAC-IARI (2018) Estimation of surplus crop residues in India for biofuel production. Technology Information, Forecasting & Assessment Council-Indian Agricultural Research Institute. New Delhi

UPAg (2025) Unified Portal for Agricultural Statistics. Department of Agriculture & Farmers Welfare. Ministry of Agriculture and Farmers' Welfare. Government of India. Retrieved from: https://upag.gov.in

Chapter 10
Potent Inhibitors in Lignocellulosic Ethanol Production and Their Removal Techniques

Neelesh Prashant, Mohd. Zafar, Vikas Chandra Gupta, and Avinash Singh

Abstract Lignocellulosic biomass is a renewable feedstock for second-generation bioethanol production. However, pretreatment processes produce various inhibitory compounds, primarily weak organic acids, furan derivatives, and phenolics, resulting from the breakdown of cellulose, hemicellulose, and lignin during thermo-chemical pretreatment. The profiles of these compounds depend on factors such as feedstock type, catalyst, temperature, and residence time. These potent inhibitors can significantly impede both enzymatic hydrolysis and yeast fermentation, even at low concentrations, thereby reducing efficiency and ethanol yield. Consequently, these inhibitors pose a significant challenge to the commercialization of cellulosic ethanol. Developing effective detoxification methods is crucial for enhancing the economic viability of lignocellulosic biorefineries. A combination of physical, chemical, and biological approaches, including engineered tolerant strains, is now commonly used worldwide as detoxification strategies to mitigate these effects, balancing efficiency, cost, and proper sugar preservation.

Keywords Lignocelluloses · Lignin · Biopolymers · Depolymerization · Bioethanol · Biorefineries · Circular economy

N. Prashant (✉)
Molecular Systematics Lab, Department of Botany, Central University of Punjab, Bathinda, Punjab, India

M. Zafar
Department of Biotechnology, College of Applied Sciences and Pharmacy, University of Technology and Applied Sciences, Sur, Oman
e-mail: Mohammed.zafar@utas.edu.om

V. C. Gupta
Department of Biotechnology, IILM University, Greater Noida, Uttar Pradesh, India

A. Singh
Department of Biotechnology, Meerut Institute of Engineering and Technology (MIET), Meerut, Uttar Pradesh, India

S. Prasad et al. (eds.), *Sustainable Bioethanol Production from Lignocellulosic Biomass*, Clean Energy Production Technologies,
https://doi.org/10.1007/978-981-95-9690-4_10

10.1 Introduction

Lignocellulosic biomass (LCB) conversion to ethanol is a vital second-generation biofuel pathway. As a feedstock, LCB contains about 40% cellulose, 20–30% hemicellulose, and 15–30% lignin, requiring pretreatment to break down the dense cellulose-hemicellulose-lignin matrix before enzymatic hydrolysis and fermentation can occur effectively (Saucedo-Gutierrez et al. 2024; Li et al. 2021). Although a pretreatment process is necessary to access cellulose and hemicellulose, it also produces inhibitors like furfural, 5-hydroxymethylfurfural (HMF), acetic acid, formic acid, and phenolic compounds, which can significantly hinder yeast performance and reduce ethanol yields by up to 60% (Parawira and Tekere 2011; Kim 2018). This underscores the urgent need for broad-spectrum detoxification methods to address cross-pathway inhibition (Guo et al. 2022), which hinders microbial growth and fermentation (Lee et al. 2015; Li et al. 2017).

The inhibition problem is a significant obstacle to large-scale bioethanol production. Kim (2018) points out that these inhibitors raise overall production costs because additional enzyme inputs are required to maintain hydrolysis rates, while also reducing sugar recovery and ethanol yields. Understanding the chemical reactions that form these inhibitors and their mechanisms is vital for developing effective detoxification methods or strain-engineering strategies. Standard yeast strains, including *Saccharomyces cerevisiae*, can tolerate ethanol but remain sensitive to these inhibitors (Kim 2018). Therefore, detoxification is necessary, although current physical, chemical, and biological methods are often expensive and complex (Pilap et al. 2022). Notably, furfural is enzymatically reduced rather than metabolically incorporated in yeast and bacterial cells. Modig et al. (2002) found that NADPH-dependent oxidoreductases help convert furfural into furfuryl alcohol, a less toxic substance. This reduction process competes with NADPH-dependent biosynthetic reactions, imposing a metabolic burden.

Notably, *Saccharomycodes ludwigii* APRE2 has demonstrated tolerance to various stresses, including high temperatures and common fermentation inhibitors. Using such stress-tolerant yeast strains could improve ethanol production from lignocellulosic biomass and decrease reliance on extensive detoxification processes (Tian et al. 2009). Incorporating detoxification into biorefinery operations can lower costs by enabling the co-production of lignin and hemicellulose fractions (Pan et al. 2005). This approach enables the simultaneous production of biofuels and bioproducts, thereby maximizing biomass utilization and enhancing economic viability (Pan et al. 2005). Advances in metabolic engineering, such as the development of *Saccharomyces cerevisiae* strains with improved xylose fermentation and inhibitor tolerance, have further increased ethanol yields from lignocellulosic feedstocks. Additionally, the literature underscores the importance of pH, especially its interaction with acetic acid, in optimizing the cofermentation of glucose and xylose to ethanol using engineered *Saccharomyces cerevisiae* strains (Casey et al. 2010). Proper pH management is therefore crucial for designing effective detoxification processes and boosting yeast performance (Huang et al. 2011).

This chapter highlights the management of potent inhibitors in lignocellulosic biomass conversion and assesses the techno-economic feasibility of bioethanol production. It emphasizes recent detoxification techniques, the use of stress-tolerant microbes, and co-product integration to improve fermentation performance and yields. Ongoing research is essential to address the challenges posed by these inhibitors.

10.2 Major Categories of Inhibitors

Potent inhibitors such as furans, weak acids, and phenolics pose major challenges to lignocellulosic ethanol production. Furan compounds, including furfural (from pentose sugars) and 5-hydroxymethylfurfural (HMF, from hexose sugars), are produced through acid-catalyzed dehydration reactions during pretreatment.

10.2.1 Furan Derivatives

Furfural and 5-hydroxymethylfurfural (HMF) are the most common furan derivatives produced during acid-catalyzed pretreatment of lignocellulosic biomass. Furfural results from the dehydration of pentose sugars (mainly xylose), while HMF comes from hexose sugars (glucose and mannose). Studies have shown that furfural is more toxic than HMF to fermenting microorganisms. Wang et al. (2020) reported that the inhibitory effect of furfural was more evident than that of HMF during ethanol fermentation of acid-pretreated corn stover. However, HMF stays in the fermentation medium about four times longer than furfural due to lower microbial conversion rates (Lee et al. 2013; Park et al. 2017; Guo et al. 2022).

10.2.2 Weak Acids

Acetic acid is the main weak-acid inhibitor in lignocellulosic hydrolysates, released mainly from the acetyl groups of hemicellulose during pretreatment. Other weak acids include formic acid and levulinic acid, which are produced through the further degradation of furan compounds. Saucedo-Gutierrez et al. (2024) demonstrated that acetic acid at 0.3% (v/v) completely inhibited the growth of *Scheffersomyces stipitis*, while furfural mainly affected mitochondrial function without significantly reducing biomass. Li et al. (2021) found that acetic acid and furfural have synergistic inhibitory effects when present together in hydrolysates, significantly reducing cell growth rate, sugar consumption, and ethanol yield (Li et al. 2016; Wang et al. 2020; Saucedo-Gutierrez et al. 2024).

10.2.3 Phenolic Compounds

Lignin-derived phenolic compounds form a diverse group of inhibitors, including vanillin, syringaldehyde, 4-hydroxybenzaldehyde, ferulic acid, and p-coumaric acid. These aromatic compounds are released during lignin degradation and significantly inhibit both enzymatic hydrolysis and microbial fermentation. Wang et al. (2020) identified the inhibitory strength hierarchy as: furfural > acetic acid > ferulic acid > p-coumaric acid > HMF. Phenolic compounds are more inhibitory than soluble sugars, furan derivatives, and organic acids because they can cause enzyme precipitation and irreversible inhibition (Taherzadeh 1999; Lee et al. 2013).

Figure 10.1 illustrates the pathways for the formation of major inhibitors during lignocellulosic pretreatment. It should show the chemical degradation routes, including: (a) dehydration of xylose to furfural, followed by further degradation to formic acid; (b) dehydration of glucose to HMF and its subsequent conversion to levulinic acid and formic acid; and (c) lignin fragmentation, which results in the production of phenolic compounds such as vanillin, syringaldehyde, 4-hydroxybenzaldehyde, ferulic acid, and p-coumaric acid (Li et al. 2021). The process of converting lignocellulosic biomass into ethanol involves multiple stages, with pretreatment being essential for enhancing the biodegradability of cellulose and hemicellulose.

As shown in Table 10.1, the formation of inhibitory by-products is closely associated with pretreatment severity, which depends on factors like temperature, pH, residence time, and catalyst concentration. Although pretreatment offers benefits, it often produces inhibitory by-products that can hinder both microbial fermentation and enzymatic hydrolysis.

Kim (2018) noted that phenolic concentrations increase proportionally with pretreatment severity. For instance, liquid hot-water pretreatment of maple at 200 °C yielded phenolic concentrations ranging from 1.3 g/L at lower severity to 5.9 g/L in hardwood at 195 °C, while steam explosion of wheat straw generated

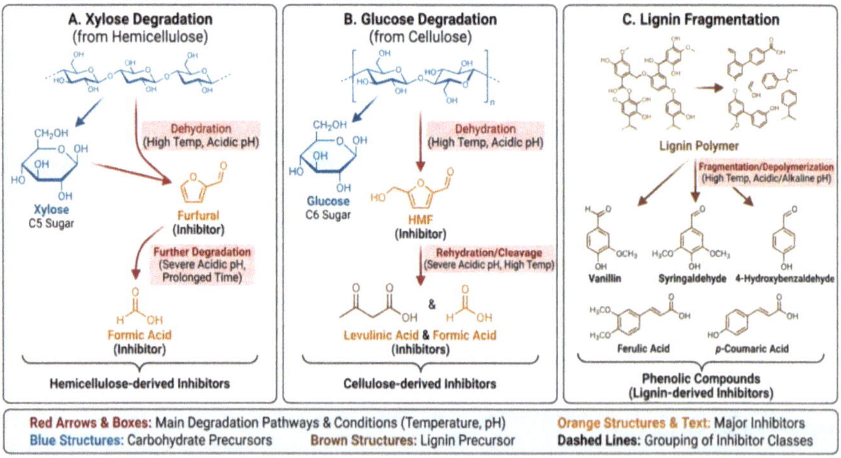

Fig. 10.1 Formation pathways of major inhibitors during lignocellulosic pretreatment

Table 10.1 Major factors affecting inhibitor formation in lignocellulosic biomass treatment

Inhibitor category	Specific inhibitor	Primary source	Promoting factors and conditions	Relative toxicity	References
Furan derivatives	Furfural	Degradation of pentose sugars (xylose, arabinose)	High temperatures (130–210 °C), acidic conditions (pH 2.5–3.5), severe pretreatment, and extended residence times	Most potent among furans	Palmqvist and Hahn-Hägerdal (2000), Kammoun et al. (2023), Kim (2018), Wang et al. (2020), and Lee et al. (2013)
Furan derivatives	5-Hydroxymethyl furfural (5-HMF)	Hexose sugars (glucose, mannose) degradation	High temperature (>180 °C), strong acid, prolonged heating, and severe acid pretreatment (0.2–2.5% w/w acid)	Less toxic than furfural, longer persistence	Jönsson et al. (2016), Kammoun et al. (2023), Brodeur et al. (2011), Taherzadeh (1999), and Park et al. (2017)
Weak organic acids	Acetic acid	Deacetylation of hemicellulose acetyl groups	Steam explosion (160–260 °C), autohydrolysis, high temperature and pressure, acid-catalyzed pretreatment	Severe growth inhibition at ≥0.1%	Kim (2018), Guo et al. (2022), Saucedo-Gutiérrez et al. (2024), and Wang et al. (2020)
Weak organic acids	Formic acid	Degradation of furfural and 5-HMF; sugar decomposition	Severe pretreatment: high temperature, long reaction time, acidic hydrolysis	Moderate inhibition	Jönsson et al. (2016) and Piotrowski et al. (2019)
Weak organic acids	Levulinic acid	Degradation of 5-HMF under acidic conditions	High acid, prolonged heat, and rehydration of HMF	Lower toxicity than acetic acid	Jönsson et al. (2016), Park et al. (2017), and Siqueira and Reginatto (2015)
Phenolic compounds	4-Hydroxybenzoic acid	Lignin degradation and fragmentation	High pretreatment severity, oxidative conditions, pH variations, and elevated temperature	Moderate toxicity	Jönsson et al. (2016), Kim (2018), Li et al. (2016), and Taherzadeh (1999)
Phenolic compounds	Vanillin	Degradation of guaiacyl (G) lignin units	Moderate to severe pretreatment; oxidative lignin breakdown; acidic or alkaline processing	High enzyme and cell toxicity	Palmqvist and Hahn-Hägerdal (2000), Jönsson et al. (2016), Taherzadeh (1999), and Li et al. (2016)
Phenolic compounds	Syringaldehyde	Degradation of syringyl (S) lignin units	High-temperature pretreatment; oxidative conditions; acid-catalyzed lignin degradation	High toxicity	Jönsson et al. (2016), Kim (2018), and Qin et al. (2018)

0.16–2.14 g/L at 190–210 °C. The severity factor (log R_0), defined as $R_0 = t \times \exp((T - 100)/\omega)$, where T is temperature, t is residence time, and ω is activation energy, is directly linked to inhibitor formation (Kim et al. 2012). Implementing practical mitigation strategies can reduce this formation. Lowering the temperature from 200 °C to 180 °C and increasing the residence time from 20 to 30 min can significantly decrease furfural yield while maintaining efficient processing throughput, giving engineers actionable temperature-time trade-offs (Faheem et al. 2020).

Depolymerization of lignin releases phenolic monomers by breaking ether linkages, especially β-O-4, β-β, and β-5 bonds, during pre-treatment, where temperatures range from 160 to 260 °C (USDoE 2025). This process involves oxidative and reductive reactions that produce monomeric and oligomeric phenolic compounds (Hafezisefat et al. 2025). During pre-treatment, hemicellulose undergoes autohydrolysis, in which acetyl groups (–O–CO–CH_3) attached to the xylan backbone are cleaved by hydronium ions (H_3O^+) generated from water dissociation, releasing free acetic acid (Faheem et al. 2020).

Prolonged high-temperature pre-treatment causes HMF to degrade excessively, forming levulinic acid, which then fragments into formic acid. This occurs through aldol condensation and retro-aldol cleavage of the HMF structure (Zhou et al. 2013; Xu et al. 2013). Importantly, each inhibitor can serve as a platform chemical for further exploration. Additionally, converting furfural and HMF into valuable fuels via downstream catalytic processes can transform these inhibitors into potential feedstocks, thereby creating opportunities for process innovation and sustainable solutions (Zhu et al. 2021), aligning with DOE's vision for lignin-first biorefineries.

10.3 Mechanisms of Inhibitor Toxicity

The degradation products formed during pretreatment include weak acids, furan derivatives, and phenolic compounds, which collectively impair enzymatic activity (Jørgensen et al. 2007) during the enzymatic hydrolysis by hindering cellulase efficiency through multiple mechanisms. Mechanism of inhibitor toxicity, for instance, in yeast cells, are illustrated in Fig. 10.2. This schematic diagram shows: (a) weak acid penetration through cell membranes and subsequent intracellular acidification; (b) furan aldehyde-induced oxidative stress and depletion of cofactors (NADH/NADPH); (c) interactions of phenolic compounds with cell membranes and enzymes; (d) mitochondrial dysfunction and ATP depletion; (e) DNA and protein damage pathways (Wang et al. 2020; Parawira and Tekere 2011).

First, interactions at the catalytic site, especially those involved in the enzymatic breakdown of cellulose, are crucial. Product inhibition is a primary mechanism affecting enzymatic hydrolysis, mainly due to the accumulation of cellobiose and glucose released during cellulose degradation. The enzymatic hydrolysis of cellulose occurs through a series of reactions catalyzed by endo-1,4-beta-D-glucanase (EC 3.2.1.4), exo-1,4-beta-D-glucanase, and beta-glucosidase (EC 3.2.1.21), with

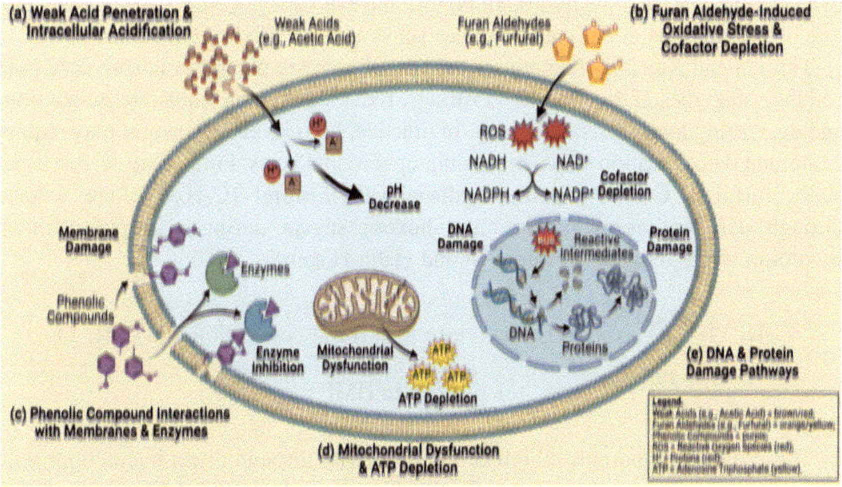

Fig. 10.2 Mechanisms of inhibitor toxicity, for example, in yeast cells

cellobiose acting as a competitive inhibitor that reduces overall conversion efficiency (Andrić et al. 2010). The hydrolysis reaction can be represented as:

$$(C_6H_{10}O_5)_n + nH_2O \overset{\text{cellulase}}{\rightarrow} nC_{12}H_{22}O_{11} \overset{\beta-\text{glucosidase}}{\rightarrow} 2nC_6H_{12}O_6$$

Where cellobiose ($C_{12}H_{22}O_{11}$) accumulates, it inhibits both endoglucanase and cellobiohydrolase activities (Corazza et al. 2005). For example, the accumulation of cellobiose can reduce overall yield by approximately 20%, thereby affecting cost efficiency in biorefinery operations. Likewise, Kristensen et al. (2009) showed that glucose concentrations of 50 g/L significantly slow initial hydrolysis rates and reduce final glucan conversion, with cellobiose having a stronger inhibitory effect than glucose due to its feedback inhibition on cellobiohydrolases. In the industry, this can lead to higher operational costs due to reduced throughput. Despite these issues, new mitigation strategies are being explored. Advances in detoxification and enzyme engineering, such as adaptive laboratory evolution and laccase-based treatments, are promising research areas that could help overcome these inhibitory effects and enhance biorefinery efficiency.

Phenolic compounds released from lignin during pretreatment also strongly inhibit through various pathways, including irreversible enzyme denaturation and competitive adsorption onto cellulose surfaces (Qin et al. 2016). According to Qin et al., vanillin, a phenolic aldehyde, markedly reduces cellulase activity during enzymatic hydrolysis, potentially increasing enzyme requirements and production costs in industrial biorefinery applications. The mechanism behind this inhibition involves nonproductive binding of cellulases to phenolic compounds through hydrophobic interactions, hydrogen bonds, and π-π stacking, which alters the structure of the catalytic domain and reduces substrate accessibility (Li et al. 2024).

Tejirian and Xu (2010) demonstrated that oligomeric phenolics adsorb onto both cellulase enzymes and cellulose substrates, reducing enzyme-substrate interactions and decreasing hydrolysis efficiency. In practice, high levels of furfural may require additional detoxification steps, increasing operational costs. Furan derivatives, especially furfural ($C_5H_4O_2$) and 5-hydroxymethylfurfural ($C_6H_6O_3$), are formed through dehydration of pentose and hexose sugars during acid pretreatment, according to the reactions (Palmqvist and Hahn-Hägerdal 2000):

$$C_5H_{10}O_5 \overset{-3H_2O}{\rightarrow} C_5H_4O_2 \text{ for xylose to furfural conversion and } C_6H_{12}O_6 \overset{-3H_2O}{\rightarrow} C_6H_6O_3, \text{ for glucose to HMF formation.}$$

These furan compounds inhibit cellulase activity through direct interactions with enzyme active sites and conformational changes, with HMF showing somewhat lower toxicity than furfural at similar concentrations (Qin et al. 2018). Inhibition mechanisms during microbial fermentation of lignocellulosic hydrolysates result from complex interactions between fermentative microorganisms and toxic compounds produced during biomass pretreatment.

10.3.1 Cellular Damage and Metabolic Disruption

Furan aldehydes cause multiple harmful effects, including damage to cellular membranes, DNA, RNA, and proteins through oxidative stress. Furfural and HMF reduce intracellular ATP levels and disturb redox balance by consuming NADH and NADPH cofactors needed for their reduction. Ma and Liu (2010) found that genes encoding NADPH-dependent alcohol dehydrogenases (ADH6, ADH7) and aldo-keto reductases (ARI1, GRE2, OYE3) showed the highest increase in expression during adaptation to HMF stress in *Saccharomyces cerevisiae* (Parawira and Tekere 2011; Wang et al. 2020; Chavan and Gaikwad 2021). Among these inhibitors, furfural is the most potent, with concentrations of 3–6 g/L causing 34–86% reductions in ethanol production in *Saccharomyces cerevisiae* strains during the fermentation of hydrolysates from bagasse, corn cobs, and spruce (Vanmarcke et al. 2021). The inhibitory effect of furfural involves generating reactive oxygen species (ROS) within yeast cells, disrupting glycolytic enzyme activities such as alcohol dehydrogenase and pyruvate decarboxylase, and depleting cellular redox power through the ongoing reduction of furfural to furfuryl alcohol, as follows (Modig et al. 2002):

$$C_5H_4O_2 + 2H \overset{ADH}{\rightarrow} C_5H_6O_2.$$

The absence of a hydroxymethyl group in furfural makes it more hydrophobic than HMF, potentially leading to preferential membrane permeation. Because of this enhanced permeability, furfural rapidly accumulates within cells, leading to oxidative stress characterized by the formation of ROS. While HMF inhibits fermentation through similar mechanisms, it is less toxic; for instance, 6 g/L HMF reduces ethanol yields by 33%, whereas equivalent furfural concentrations result in 77–86% reductions (Vanmarcke et al. 2021). The conversion pathway for HMF detoxification proceeds through alcohol dehydrogenase-catalyzed reduction to 2,5-bis-hydroxymethylfuran according to (Almeida et al. 2007)

$$C_6H_6O_3 + 2H \overset{ADH}{\rightarrow} C_6H_8O_3,$$

which drains NADH pools and disrupts cellular redox balance.

According to a study, alcohol dehydrogenases in *Scheffersomyces stipitis* are involved in the reduction of HMF and furfural to less toxic compounds during detoxification (Ma et al. 2013). Therefore, the reduction of HMF to 2,5-bis-hydroxymethylfuran imposes a substantial energetic penalty, highlighting the importance of engineering strains with balanced cofactor usage to improve fermentation efficiency.

10.3.2 Membrane Permeability and pH Homeostasis

Weak acids like acetic acid exist in both dissociated and undissociated forms, with the lipophilic undissociated form freely diffusing across cell membranes. Once inside the neutral-pH cytoplasm, these acids dissociate, releasing protons that cause intracellular acidification, disrupt the membrane potential, and inhibit enzymatic activities. Caspeta and Nielsen (2015) demonstrated that this disruption triggers complex cellular stress responses involving multiple transcription factors, including MSN2/4, YAP1, HSF1, and RLM1 (Qin et al. 2018; Schroyen et al. 2017). Weak acids inhibit fermentation through both specific and nonspecific mechanisms, with inhibition influenced by external pH and acid dissociation constants. Acetic acid, the most common inhibitor in lignocellulose hydrolysates at concentrations of 4–11 g/L, crosses cell membranes in its undissociated form and dissociates in the neutral cytoplasm, as shown below.

$$CH_3COOH \rightleftharpoons CH_3COO^- + H^+$$

Acetic acid causes intracellular acidification and disrupts transmembrane pH gradients (Pampulha and Loureiro-Dias 1989). This acidification blocks key glycolytic enzymes, such as phosphofructokinase, and increases cellular energy demand, which must actively export protons via plasma membrane H^+-ATPase to maintain pH homeostasis (Russell 1992). According to research published by Xiao et al.

(2025) in Microbial Cell Factories, formic acid is more inhibitory to ethanol production during glucose fermentation than acetic acid. Formic acid has even stronger inhibitory effects than acetic acid at equal molar concentrations because of its lower pKa (3.75 vs 4.76), resulting in greater undissociated acid that readily crosses membranes and causes more severe intracellular acidification (Casal et al. 1996). Vanmarcke et al. (2021) showed that 6 g/L of formic acid decreased ethanol yields by 42–58% across different hydrolysates, with xylose fermentation being more sensitive than glucose fermentation. Levulinic acid is formed through acid-catalyzed breakdown of HMF via:

$$C_6H_6O_3 + H_2O \xrightarrow{H^+} C_5H_8O_3 + HCOOH,$$

Levulinic acid exhibited minimal inhibition at industrially relevant concentrations (<3 g/L), suggesting that it is a minor contributor to overall fermentation toxicity (Almeida et al. 2009).

10.3.3 Enzyme Inhibition

Phenolic compounds can both reversibly and irreversibly inhibit cellulases during enzymatic hydrolysis. These phenolic compounds, produced from partial lignin depolymerization—such as vanillin ($C_8H_8O_3$), syringaldehyde ($C_9H_{10}O_4$), 4-hydroxybenzaldehyde ($C_7H_6O_2$), and p-coumaric acid ($C_9H_8O_3$)—interfere with fermentation by damaging cell membrane integrity, blocking essential metabolic enzymes, and causing oxidative stress (Ando et al. 1986). The presence of hydroxyl, carbonyl, and methoxy groups on aromatic rings significantly influences the strength of inhibition. Cellulases are more susceptible to phenolic inhibition than β-glucosidases, with vanillin reducing enzyme activity by 13–45% over 72 h of hydrolysis (Taherzadeh 1999). Kim et al. (2011) demonstrated that phenolic aldehydes are more inhibitory than their corresponding phenolic acids; for example, 1 g/L of vanillin decreased the specific growth rate of engineered xylose-fermenting *S. cerevisiae* by 75%, while 4-hydroxybenzoic acid at the same level had little effect.

Palmqvist and Hahn-Hägerdal (2000) found that the inhibitory effects of hydroxymethoxybenzaldehydes on enzymes depend heavily on the position of their substituents, which influence their interaction with and inhibition of enzymes during fermentation. Phenolic amides formed during ammonia fiber expansion (AFEX) pretreatment exhibit distinct inhibition profiles: they exert more potent effects on enzymatic hydrolysis but lower toxicity during fermentation than phenolic acids, likely due to differences in hydrophobicity and membrane permeability (Li et al. 2020). Multiple inhibitors can interact synergistically, significantly increasing toxicity beyond their individual effects; for instance, combinations of weak acids and furan aldehydes at sub-inhibitory levels cause severe fermentation inhibition, exceeding predictions based on single inhibitors (Almeida et al. 2007).

10.4 Detoxification and Removal Techniques

Current research has shown that various physical, chemical, and biological techniques are used to effectively remove and detoxify inhibitors such as furfural, HMF, acetic acid, phenolic aldehydes, and phenolic acids, thereby improving sugar recovery, enhancing microbial viability, and improving ethanol yield (Zhang et al. 2018; Qin et al. 2018; Fernández-Sandoval et al. 2024; Haq et al. 2024a, b) and are discussed in the following subheading. A comparison of removal percentages of different inhibitor classes by various detoxification methods, such as activated carbon, overliming, sequential overliming, enzymatic (laccase), microbial biodetoxification, and water washing studies, is shown in Table 10.2.

10.4.1 Physical Methods

10.4.1.1 Activated Carbon Adsorption

Activated carbon (AC) effectively removes inhibitors, such as phenolic compounds, from pretreated hydrolyzates via physical adsorption. For instance, Zhang et al. (2018) reported that a 5% (w/v) AC treatment eliminated 77.9% of furan derivatives and 98.6% of aromatic monomers from poplar prehydrolyzates. Fernández-Sandoval et al. (2024) found that AC application completely removed cinnamic acid, dihydroxybenzoic acid, and coumaric acid, along with an 85.74% reduction in vanillic acid. Due to its high surface area and porous structure, AC selectively adsorbs hydrophobic inhibitors while minimizing sugar loss. Research by Zhang et al. (2012) also indicated that both commercial and polymer-derived ACs reduced furfural concentrations by more than 80% in model solutions, thereby increasing ethanol yields from 38% to 84% of the theoretical maximum by reducing microbial inhibition. These treatments can also lower production costs, reducing the per-liter cost of ethanol in large-scale applications.

The oxygen-containing functional groups on the carbon surface enhance selectivity for inhibitors over sugars, and regeneration via ethanol stripping in fixed-bed columns enables multiple cycles without loss of performance (Zhang et al. 2012). Compared to overliming, AC more effectively removes certain phenolics from lignocellulosic hydrolysates, with process optimization achieving up to 97% reduction in inhibitors (Hong et al. 2021). These results suggest that AC's adsorption efficiency and selectivity make it a more effective choice for phenolic removal than overliming. Recent studies also show that waste-derived activated carbon maintains high adsorption efficiency for vanillin and furfural in rice straw hydrolysates (Hou et al. 2024).

Table 10.2 Comparison of detoxification methods for lignocellulosic hydrolysates

Method	Target inhibitors	Removal efficiency	Advantages	Disadvantages	References
Activated carbon (5% w/v)	Phenolic compound, furans	98.6% aromatics, 77.9% furans	High efficiency for phenolics, physical process	Costly, sugar loss, and disposal issues	Zhang et al. (2018) and Fernández-Sandoval et al. (2024)
Overliming (pH 10)	Dialdehydes, diketones	High removal of specific furans	Low cost, effective for dialdehydes	Less effective for phenolic acids, forms precipitates	Zhang et al. (2018) and Fernández-Sandoval et al. (2024)
Sequential overliming + AC	Multiple inhibitor classes	96.8% phenolic acids, high furan removal	Synergistic effect, broad spectrum	Multi-step process, higher cost	Zhang et al. (2018) and Fernández-Sandoval et al. (2024)
Laccase enzymes	Phenolic compounds	Effective phenolic oxidation	Mild conditions, targeted, low waste	Enzyme cost, process optimization needed	Zhang et al. (2025) and Jönsson et al. (2013)
Microbial (*K. huakuii*)	Phenolic aldehydes, weak acids	Transforms to less toxic acids	Converts to eco-friendly, benign products	Time-consuming, requires optimization	Qin et al. (2018)
Water washing	All soluble inhibitors	Very high removal (>90%)	Most effective removal	Large wastewater, biomass/sugar loss	Qin et al. (2018) and Haq et al. (2024a, b)
Genetic engineering (PDR1/YAP1/RPN4)	Furfural, HMF, acids, phenolics	Enhanced tolerance, shorter lag	No extra processing, permanent solution	Strain development time, regulatory concerns	Zhang et al. (2025) and Liu et al. (2021a, b)
In-situ SSF	Furans, aldehydes	Continuous biological reduction	Integrated process, reduced inhibitor accumulation	Process optimization complexity	Caspeta and Nielsen (2015) and Jilani and Olson (2023)

10.4.1.2 Water Washing

Water washing is a simple and commonly used physical method for detoxifying lignocellulosic biomass hydrolysates by removing inhibitory compounds formed during acidic thermochemical pretreatment (Jönsson et al. 2013; Haq et al. 2024a, b).

The process involves separating the pretreated solid biomass from the liquid phase and then rinsing it with fresh water to wash away soluble inhibitors from the cellulosic fibers (Frederick 2013). Because inhibitory byproducts are highly soluble, even small amounts of wash water can effectively remove water-soluble inhibitors (Frederick 2013). For example, washing dilute-acid-pretreated high-density poplar with water at 1.5 times the biomass volume increased glucose yield to 3–3.5 grams per gram of dry biomass, compared with 0.7 grams per gram without washing—a fivefold improvement (Frederick 2013). This method removes fermentation inhibitors while mostly preserving cellulose, typically recovering at least 75% of the original cellulose in the solid fraction (Chen et al. 2022). However, water washing has drawbacks, including high freshwater consumption, large volumes of wastewater containing diluted inhibitors and dissolved sugars, and incomplete removal of strongly adsorbed phenolic compounds (Ujor and Okonkwo 2022).

The technique also results in significant sugar losses due to the dissolution and washout of hemicellulose- derived sugars, which can reduce ethanol yields by 10–15% depending on the washing intensity (Jönsson et al. 2013). Despite these limitations, water washing remains appealing due to its simplicity, lack of chemical additives, and compatibility with standard solid-liquid separation equipment, such as belt filters, screw presses, and centrifuges (Broda et al. 2022). Its efficiency can be further enhanced by combining it with other detoxification methods, such as overliming or biological strategies, to maximize inhibitor removal while minimizing resource use (Jönsson et al. 2013). Recent advances in liquid hot water pretreatment demonstrate that carefully controlling washing parameters can increase lignocellulosic biomass conversion by 40–65% relative to untreated biomass, underscoring the continued importance of water-based separation in modern biorefineries (Chen et al. 2022).

10.4.1.3 Ion Exchange and Membrane Filtration

Ion exchange and membrane filtration methods selectively remove charged inhibitor species but can be expensive and require regeneration steps (Haq et al. 2024a, b). Ion exchange is widely regarded as an effective technique for removing inhibitory compounds prior to fermentation. Larsson et al. (2003) reported that removal efficiencies for furan aldehydes indicate that furfural can be removed at rates ranging from 53% to 99%. In contrast, HMF removal varies from 37% to 100%, depending on the resin and process conditions (Larsson et al. 2003). Sequential application of different resins has enabled comprehensive detoxification; for instance, one method removed 82.1% of furfural, 66.5% of HMF, and substantially reduced other toxic compounds through systematic resin use (Marton et al. 2003). More recently, ultrasound-assisted regeneration of ion-exchange resins has been explored to enhance efficiency and sustainability in biorefinery detoxification, with a focus on improving resin reusability in continuous processes (Han et al. 2025).

Membrane filtration technologies offer complementary or alternative approaches to detoxification by physically separating inhibitory compounds and concentrating fermentable sugars. Ultrafiltration (UF) and nanofiltration (NF) have been employed in two-stage processes to treat lignocellulosic hydrolysates. For instance, PES10 UF membranes operating at 25.6 $L/m^2{\cdot}h$ achieved 73.9% recovery of cellulase protein and permitted complete glucose transmission (Qi et al. 2012). Subsequent nanofiltration using DL270 membranes at 13.3 $L/m^2{\cdot}h$ increased glucose concentration from 30 to 52 g/L, thereby enhancing fermentation efficiency (Qi et al. 2012). Supported liquid membrane (SLM) systems employing hybrid polyethersulfone-graphene membranes removed approximately 95% of acetic acid from 10 g/L solutions and demonstrated stability for over 116 h of continuous operation (Abdulhameed et al. 2022) for the removal of volatile organic acids, membrane contactors with reactive stripping solutions achieved up to 86% acetic acid removal, with temperature identified as the most influential factor for mass transfer efficiency (Mohammadi et al. 2022).

10.4.2 Chemical Methods

10.4.2.1 Overliming

Overliming detoxification is a widely used method for removing inhibitory compounds from lignocellulosic biomass hydrolysates to enhance bioethanol production. Its popularity comes from its high efficiency, simple operation, and low cost (Chong et al. 2025). The process involves raising the pH to 9–11 using calcium hydroxide, $Ca(OH)_2$, followed by neutralization. It has been observed that overliming removes approximately 75.6% of furan derivatives and 68.1% of aromatic monomers from poplar prehydrolysates (Zhang et al. 2018). Specifically, furfural content was reduced by 78%, from 3360.9 to 736.8 mg/L, and hydroxymethylfurfural decreased by 65%, from 625.5 to 216.8 mg/L. For aromatic compounds, overliming removed vanillin, syringaldehyde, and acetosyringone by 65%, 67%, and 53%, respectively (Zhang et al. 2018). Zhang et al. (2018) found that overliming was especially effective at removing dialdehydes and diketones, significantly reducing levels of 2,5-furandicarboxaldehyde, 5-ethylfuran-2-carbaldehyde, and 2,5-hexanedione compared to activated carbon treatment. However, overliming removed fewer phenolic acids than activated carbon, indicating the need for sequential treatment methods (Zhang et al. 2018; Fernández-Sandoval et al. 2024).

Although overliming alone is among the most effective detoxification methods, its efficiency can be significantly improved when combined with additional strategies (Jonsson et al. 2013). Research shows that sequential treatment with overliming followed by activated carbon removes 90.7% of furans and 99.5% of aromatic monomers, compared to 75.6% and 68.1% removal by overliming alone (Zhang et al. 2018). The fermentability of overliming-treated hydrolysates varies widely with the microorganism used, as different organisms tolerate residual inhibitors to varying degrees. Jiang et al. (2024) compared overliming with biological detoxification methods, noting that while overliming is effective, biodetoxification using engineered microorganisms like *Rhodococcus aetherivorans* N1-S can degrade 80.5% of furfural, 50.7% of HMF, and 71.5% of phenolic compounds from corn cob hydrolysates. The study demonstrated that detoxified hydrolysates could increase product yields by 6.5-fold relative to non-detoxified feedstocks. Despite its advantages, overliming has notable limitations, including moderate glucose retention and potential sugar loss at high pH levels. However, the limitations and challenges of overliming detoxification remain areas needing further research and the development of improved methods (Chong et al. 2025).

10.4.2.2 Organic Solvent Extraction

Organic solvent extraction is a primary method for removing inhibitory compounds generated during lignocellulosic biomass pretreatment. Ethyl acetate, which aligns with Principle 5 of green chemistry (safer solvents), is widely used to extract inhibitors from lignocellulosic hydrolysates because of its selectivity for phenolic compounds and furans while minimizing sugar loss (Zha et al. 2014). Ethyl acetate extraction achieves 96–100% removal efficiencies of phenolic acids, such as p-hydroxybenzoic acid and cinnamic acid, within 30 min under optimal conditions (Cañadas et al. 2021). Comparative studies show that chloroform exhibits higher extraction efficiency for furans, alcohols, and phenolics, with values exceeding 94% for guaiacol and 87% for 3-ethylphenol. In contrast, ethyl acetate is especially effective for phenolic compounds, reaching 100% for phenol and 90% for 3-ethylphenol (Ren et al. 2017). Advanced solvent screening using COSMO-RS predictive modeling has identified cyclohexanone, isophorone, and methyl isobutyl ketone (MIBK) as promising green solvents for furan extraction.

Cyclohexanone has a partition coefficient of 3.66 for HMF, while MIBK achieves 7.80 for furfural over four extraction cycles, with performance declines of less than 18% (Chan et al. 2024). Residual MIBK and cyclohexanone at concentrations below 500 ppm cause minimal inhibition of yeast fermentation, highlighting the importance of solvent partitioning to optimize microbial yields. Sequential extraction with multiple organic solvents allows fractionation of different inhibitor classes. For example, chloroform extracts 62.2% of original furans, ketones, alcohols, and phenolics, followed by ethyl acetate, which extracts 62% of acetic acid. This method supports targeted removal and potential valorization of separated inhibitor fractions (Ren et al. 2017). Choosing suitable organic solvents requires balancing extraction efficiency, inhibitor selectivity over sugars, solvent toxicity, recovery and reuse, regulatory compliance, and compatibility with downstream fermentation organisms to prevent added inhibitory effects from residual solvents (Palmqvist and Hahn-Hägerdal 2000).

10.4.3 *Biological Methods*

Biological methods offer a cost-effective approach to removing inhibitors from lignocellulosic hydrolysates and improving bioethanol fermentation (Nasir et al. 2022). Unlike chemical methods, biological detoxification avoids sugar loss and additional processing costs (Ujor and Okonkwo 2022).

10.4.3.1 Enzymatic Detoxification

Enzymatic detoxification using oxidative enzymes is an effective biological method for removing phenolic inhibitors from hydrolysates (Díaz et al. 2023). Ligninolytic enzymes, especially laccases and peroxidases, oxidize phenolic compounds into less toxic derivatives. Laccase-mediated detoxification has successfully reduced inhibitor concentrations in hydrolysates, while versatile peroxidase proved sensitive to high phenolic levels. Enzymes offer mild reaction conditions and target-specific inhibitors, without producing secondary waste streams (Haq et al. 2024a, b; Zhang et al. 2025; Jönsson et al. 2013). Laccases, which are multicopper oxidases from white-rot fungi such as Trametes versicolor, oxidize phenolic compounds to phenoxy radicals that polymerize into larger, less toxic molecules (Larsson et al. 2001). Expressing a laccase gene from *T. versicolor* in *S. cerevisiae* enhanced detoxification of spruce hydrolysate and increased ethanol production, although residual furanic aldehydes still inhibited growth (Larsson et al. 2001). Peroxidases, another class of oxidative enzymes, also detoxify phenolic inhibitors through similar mechanisms (Wang et al. 2020). Both laccases and peroxidases have been studied for their ability to improve hydrolysate fermentability without consuming fermentable sugars (Jönsson et al. 2013).

Recent studies indicate that low-dose acidic fungal laccases are more effective than neutral- and alkaline-variant laccases in detoxifying inhibitors. Fang et al. (2015) identified a novel laccase, Glac15, from *Ganoderma lucidum* 77002, which exhibited strong detoxification capabilities. In a comprehensive review, Haq et al. (2024a, b) reported that microbial laccases reduced phenolic compounds in pre-hydrolysates by 88–92%. Laccases from *Coltricia perennis* achieved a 76% reduction in phenols, while those from *Pycnoporus cinnabarinus* and *Coltricia rigida* reached reductions of 92–95%. The detoxification process involves the oxidation of phenolic compounds to phenoxy radicals, which subsequently polymerize into less inhibitory oligomers. Liu et al. (2021a, b) compared three fungal laccases with different pH optima for corncob detoxification and found that the acidic laccase rLacA (pH 4.0) was the most effective, reducing total phenol concentration to 0.18 g/L and increasing ethanol production to 8.02 g/L. This result surpassed the performance of both the neutral (rLcc9, pH 6.5) and alkaline (PIE5, pH 8.5) laccases. Additionally, the acidic laccase achieved the highest volumetric ethanol productivity at 1.34 g/L/h.

10.4.3.2 Microbial Detoxification

Microbial detoxification through in situ bioconversion during fermentation is gaining attention because fermenting microbes can convert inhibitors into less harmful forms. Recent studies identified *Kurthia huakuii* LAM0618^{T} as an effective biodetoxification agent that converts phenolic aldehydes, such as syringaldehyde, hydroxybenzaldehyde, and vanillin, into less toxic carboxylic acids, including syringic acid, hydroxybenzoic acid, and vanillic acid (Jönsson et al. 2013). *Issatchenkia occidentalis* reduced 66.67% of syringaldehyde, 73.33% of furfural, 62% of furfural and 85% of HMF after 24 h (Haq et al. 2024a, b). *Saccharomyces cerevisiae,* a widely used ethanologenic yeast, reduces furfural to furfuryl alcohol under anaerobic conditions via NAD(P)H-dependent aldehyde reductases (Nilsson et al. 2005). HMF is similarly reduced to 2,5-bis(hydroxymethyl)furan through NADPH-dependent pathways, although this process diverts electrons from ethanol production, thereby reducing fermentation efficiency (Gorsich et al. 2006). While this method reduces toxicity, furfuryl alcohol remains toxic at concentrations above 2 g/L (Saini and Dagar 2023) in solventogenic *Clostridium* species (Zhang and Ezeji 2014).

Fungal biodetoxification involves specific fungi degrading or transforming inhibitors before fermentation. *Amorphotheca resinae* ZN1 effectively degrades lignocellulose-derived inhibitors under both aerobic and anaerobic conditions, with oxygen availability dictating the pathway (Koopman et al. 2010). Aerobically, *A. resinae* ZN1 converts furfural and HMF to their respective alcohols (Ran et al. 2014), then further to furoic acid and HMF acid via aldehyde oxidase. Anaerobically, degradation stops at the alcohol stage (Koopman et al. 2010). The bacterium *Cupriavidus basilensis* also shows effective two-stage detoxification, using various inhibitors without significantly consuming fermentable sugars (Wierckx et al. 2010). Simultaneous Saccharification and Fermentation (SSF) processes enable in situ detoxification by quickly converting released sugars into ethanol, while fermenting microorganisms simultaneously reduce inhibitor concentrations. This approach maintains lower sugar and inhibitor levels throughout the process, reducing feedback inhibition and cellular stress (Caspeta and Nielsen 2015; Jilani and Olson 2023).

10.4.4 Metabolic Engineering Approaches

Metabolic engineering has been widely used to improve the detoxification ability of fermenting microorganisms (Wang et al. 2024). *Paenibacillus polymyxa* DSM 365, a soil-dwelling nitrogen-fixing bacterium, can utilize HMF as its sole carbon source and produce 2,3-butanediol, making it a strong candidate for engineering strains that both mineralize inhibitors and generate target chemicals (Okonkwo et al. 2021). Zou et al. (2021) engineered *Pseudomonas putida* KT2440 to be unable to utilize sugars but highly efficient at degrading lignocellulose-derived inhibitors.

When co-cultured with *Bacillus coagulans* for lactic acid production from corn stover hydrolysate, this system achieved complete removal of furanic aldehydes and 90% detoxification of lignin-derived monoaromatic inhibitors, yielding approximately 36 g/L of lactic acid. In contrast, *B. coagulans* alone could not ferment untreated hydrolysate (Zou et al. 2021). This synthetic consortium approach highlights the potential of engineering complementary metabolic functions across organisms to enable efficient detoxification and high-yield fermentation without sugar loss (Zhu et al. 2021).

Combining multiple detoxification mechanisms through metabolic engineering proves more effective than single-gene strategies (Liu et al. 2024). For example, expressing stress-responsive genes like ycfR in *E. coli,* which supports membrane repair and stress tolerance, increased resistance to acetic acid, furfural, and phenolic compounds (Chang et al. 2021). In *Clostridium tyrobutyricum*, co-expressing a short-chain dehydrogenase/reductase from *C. beijerinckii* NCIMB 8052 with the heat shock protein GroESL raised butyrate output by 28% from undetoxified corncob hydrolysate compared to wild-type strains, demonstrating the benefits of combining detoxification and product tolerance (Suo et al. 2019). In *S. cerevisiae*, overexpressing trehalose-6-phosphate synthase (tps1) and aldehyde reductase (ari1), along with inactivating neutral trehalase (nth1), increased trehalose levels, reduced membrane stress from inhibitors, and improved growth and ethanol production in the presence of furfural and HMF (Divate et al. 2017).

Zhang et al. (2025) showed that co-overexpression of transcription factors PDR1, YAP1, and RPN4 significantly enhanced *S. cerevisiae's* tolerance to furfural, HMF, acetic acid, formic acid, and phenolic compounds. The engineered strains demonstrated shorter lag phases and higher growth rates under inhibitor-stress conditions (Piotrowski et al. 2019). Ma and Liu (2010) reported that ADH7 expression increased by more than 30-fold during HMF adaptation, playing a key role in aldehyde biotransformation and detoxification. Other important genes, including ARI1, ADH6, GRE2, and OYE3, were co-regulated by YAP1 and contributed to inhibitor tolerance.

Upregulation of the PDR gene family members (TPO1, TPO4, RSB1, PDR5, PDR15, YOR1, SNQ2) enhances cellular adaptation by promoting the active efflux of inhibitors (Chavan and Gaikwad 2021). For example, the cellular stress response network in *S. cerevisiae* during inhibitor exposure is shown in Fig. 10.3. This network diagram illustrates: (a) membrane sensors detecting stress signals; (b) signaling cascades involving transcription factors (YAP1, MSN2/4, HSF1, PDR1/3, RPN4); (c) downstream responses including oxidoreductase genes (ADH6, ADH7, ARI1, GRE2, OYE3), PDR transporters (PDR5, PDR15, YOR1, SNQ2), and heat shock proteins; (d) metabolic adaptations enabling inhibitor detoxification and tolerance (Chavan and Gaikwad 2021; Qin et al. (2018).

The cellular stress response network in *Saccharomyces cerevisiae* during inhibitor exposure is a hierarchical system involving membrane sensors, transcription factor-driven signaling cascades, and downstream effectors. When exposed to lignocellulose-derived inhibitors such as furfural and hydroxymethylfurfural (HMF), membrane sensors including Mid2, Wsc1, and Hkr1 detect stress and

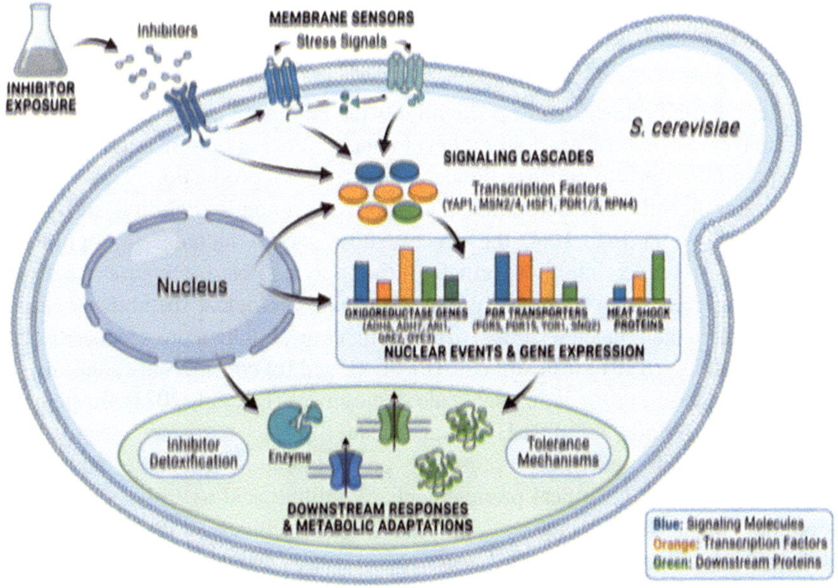

Fig. 10.3 Cellular stress response network in *S. cerevisiae* during inhibitor exposure

activate signaling pathways (Zhang et al. 2012). These pathways trigger transcriptional responses mediated by key transcription factors: YAP1 (oxidative stress tolerance), MSN2/4 (general stress response), HSF1 (heat shock response), PDR1/3 (drug resistance), and RPN4 (proteasome biogenesis). YAP1 activates oxidoreductase genes, such as ADH6, ADH7, ARI1, GRE2, and OYE3, which support detoxification of inhibitors (Zhang et al. 2018). PDR1/3 regulate ABC transporters, including PDR5, PDR15, YOR1, and SNQ2, to export toxic compounds and confer multidrug resistance (Decottignies et al. 2008). RPN4 regulates proteasome- and ubiquitin-related gene expression, including SHP1, to promote the degradation of damaged proteins (Xie and Varshavsky 2001). HSF1 induces heat shock proteins such as HSP26 and SSA4 to maintain proteostasis under stress (Brandman et al. 2021). This integrated network enables metabolic adaptation by reducing aldehyde inhibitors via enzymatic reduction, promoting active efflux of xenobiotics, enhancing protein quality control, and facilitating chaperone-mediated protein refolding (Nguyen et al. 2025).

10.5 Integrated Biorefinery Strategy for Inhibitor Management

The integrated biorefinery strategy highlights the co-design of pretreatment, inhibitor management, robust yeast strain development, process optimization, fermentation, and utility loops to enhance lignocellulosic ethanol production. This approach aims to significantly enhance efficiency, improve economic viability, and reduce risk. As shown in Fig. 10.4, the process flow should include: (a) optimizing pretreatment to reduce inhibitor formation; (b) applying post-pretreatment detoxification techniques, such as physical, chemical, and biological methods; (c) using inhibitor-tolerant engineered strains; (d) implementing simultaneous saccharification and fermentation (SSF) for in situ detoxification; and (e) considering economic and environmental factors for industrial-scale deployment (Li et al. 2021; Caspeta and Nielsen 2015; Lam et al. 2021; Haq et al. 2024a, b).

In integrated biorefineries, optimizing pretreatment is crucial to minimizing inhibitor formation. Chemical pretreatments like dilute acid hydrolysis effectively solubilize hemicellulose but can produce high levels of furan derivatives at 100–180 °C and 0.5–10% acid. Careful optimization balances sugar release and inhibitor production (Kumar et al. 2023). Alkaline pretreatments with sodium hydroxide, calcium hydroxide, or ammonia remove lignin and generate fewer inhibitors but require longer reaction times and neutralization, thereby increasing costs (Sindhu et al. 2023). Biological pretreatments with fungi or bacteria are environmentally friendly and yield minimal inhibitor formation, but they require 30–60 days and yield lower sugar concentrations, limiting industrial scalability (Kumar et al. 2023). New techniques such as ionic liquids and deep eutectic solvents enable selective lignin extraction with fewer inhibitors, but the high cost of solvent recovery hampers commercial implementation (Brodeur et al. 2011).

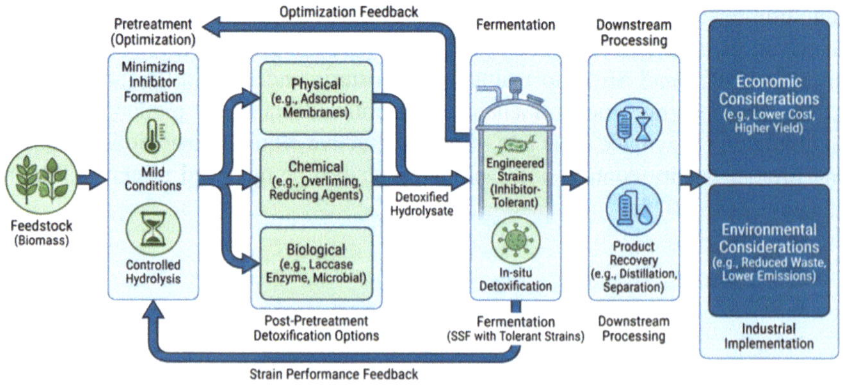

Fig. 10.4 Integrated biorefinery strategy for inhibitor management

If pretreatment optimization does not sufficiently suppress inhibitor formation, post-pretreatment detoxification is needed to prepare hydrolysates for microbial fermentation in integrated biorefineries. Comparing standalone and in situ biological detoxification can improve process integration, with in situ approaches potentially streamlining operations. Whole-cell biological detoxification using *Trichoderma reesei* or *Coniochaeta ligniaria* can simultaneously hydrolyze lignocellulose and detoxify inhibitors, achieving furfural removal rates of 3–8 g/L/h. However, this additional fermentation step raises capital costs by 15–25% compared to direct fermentation (Linares-López et al. 2022). Co-cultivation in these processes might reduce overall residence time, partly offsetting the increased capital costs.

Developing inhibitor-tolerant engineered strains is a transformative strategy that enables fermentation of toxic hydrolysates without extensive detoxification, thereby significantly improving the economics of integrated biorefineries. Metabolic engineering efforts have focused on overexpressing aldehyde reductases, such as GRE2 from *Saccharomyces cerevisiae*, which convert furfural and HMF into less-toxic alcohols. Engineered strains have demonstrated a 32% increase in ethanol production under inhibitory conditions compared to wild-type controls (Lam et al. 2021). Directed evolution of GRE2 via PCR mutagenesis and selection has produced variants with amino acid substitutions at positions 48 and 290, resulting in 39% faster detoxification rates and ethanol titers exceeding 100 g/L from highly toxic lignocellulosic hydrolysates (Lam et al. 2021). Additional tolerance has been attained by enhancing cellular stress responses, such as overexpressing the YAP1 transcription factor, which improves oxidative stress resistance and supports fermentation in the presence of up to 3 g/L phenolic compounds (Smith et al. 2014).

Simultaneous saccharification and fermentation (SSF) offers significant benefits for inhibitor control in integrated biorefineries by enabling in situ detoxification through several synergistic mechanisms. During SSF, the continuous consumption of sugars prevents the accumulation and degradation of furan inhibitors, thereby maintaining glucose and xylose levels below 20 g/L. In contrast, in separate hydrolysis and fermentation, levels can exceed 100 g/L (Miah et al. 2022). High biomass loading, typically 15–20% dry matter, allows yeast populations at an OD600 of 20 to fully convert furfural and HMF into their respective alcohols within 24 h, serving as a biological detox step alongside ethanol production (Lam et al. 2021). Temperature optimization in SSF involves balancing cellulase activity, which peaks at 45–50 °C, with yeast fermentation, which performs best at 30–35 °C. Using thermotolerant *Saccharomyces cerevisiae* strains at 40 °C facilitates faster enzymatic hydrolysis and continuous fermentation, leading to ethanol productivities of 1.8–2.3 g/L/h from pretreated biomass (Miah et al. 2022).

Economic analysis indicates that selecting among pretreatment optimization, detoxification, and the deployment of tolerant strains is crucial for determining the minimum ethanol selling price and overall profitability of integrated biorefineries. Techno-economic modeling indicates that reducing inhibitor effects through process intensification can cut production costs by $0.15–0.30 per gallon compared to traditional detoxification, with actual savings depending on feedstock and pretreatment severity (Humbird et al. 2021). Investing in tolerant engineered strains offers significant economic benefits, as modest development costs and zero marginal costs per unit can decrease expenses by \$0.08–0.12 per gallon by eliminating overliming and activated carbon treatment (Brodeur et al. 2011). Industrial-scale biological detox systems require capital investments of \$8–15 million for a 50-million-gallon-per-year plant, versus \$3–6 million for physical detoxification equipment. However, biological methods have lower operating costs at \$0.03–0.05 per gallon compared to \$0.06–0.10 for physical methods, mainly due to reduced energy use (Chen et al. 2022). Life cycle assessments show that removing detoxification steps can reduce greenhouse gas emissions by 8–12% by lowering chemical and energy consumption.

Environmental assessments indicate that integrated biorefineries optimized through mild pretreatment, SSF configuration, and inhibitor-tolerant strains can cut the global warming potential by 40–55% compared to petroleum fuels. When using inhibitor-tolerant strains in SSF, total emissions for cellulosic ethanol drop from 18–22 g CO_2-equivalent per MJ to 15–19 g CO_2-eq/MJ (Morales et al. 2021). Water use in biorefineries employing biological detoxification or tolerant strains is 15–25% lower than with physical-chemical methods, decreasing freshwater consumption from 6–8 L per liter of ethanol to 5–6 L. This is especially valuable in water-scarce agricultural areas (Morales et al. 2021). Additional advantages include better nutrient management via agricultural residues and support for rural bioeconomies (Cherubini et al. 2024). Incentives like the U.S. Renewable Fuel Standard and the EU Renewable Energy Directive provide bonuses of \$0.50–1.20 per gallon for cellulosic ethanol over corn ethanol, promoting inhibitor management investments (Johnson et al. 2023). Market analysis suggests that, with optimized inhibitor management, integrated biorefineries can be cost-competitive with petroleum gasoline when crude oil prices exceed \$55–\$65 per barrel. Applying combined tolerance strategies could lower the breakeven price to \$45–\$55 per barrel, a crucial factor amid rising global carbon prices (Anderson et al. 2024).

10.6 Conclusions and Future Perspectives

Effective management of inhibitors produced during the pretreatment of lignocellulosic materials presents a significant opportunity to enhance commercial bioethanol production. Integrated biorefinery strategies that combine pretreatment, targeted detoxification, metabolic engineering, and fermentation systematically address challenges posed by inhibitors such as furfural, 5-hydroxymethylfurfural (HMF), acetic acid, and phenolic compounds. Detoxification methods have evolved from traditional physical and chemical techniques to advanced biological and genetic engineering approaches. Biological detoxification using enzymes and microorganisms offers environmentally friendly solutions and works under mild conditions, making it a practical choice. Developing inhibitor-tolerant strains through metabolic engineering is a promising approach that could remove the need for separate detoxification steps. Introducing new enzymes and engineered microbial strains that tolerate or break down these inhibitors is crucial for increasing ethanol yields. To improve production conditions and ensure large-scale reliability, real-time monitoring and adaptive control systems should be incorporated into the process. Combining detoxification and fermentation into a single bioprocess can streamline operations, reduce complexity, and cut costs. Implementing these innovative methods within biorefinery frameworks can significantly enhance the scalability, cost-effectiveness, and sustainability of lignocellulosic ethanol production. Ongoing interdisciplinary research and collaboration with industry partners are vital for turning these advancements into commercially viable, environmentally sustainable bioethanol systems.

References

Abdulhameed AS, Jawad AH, Kashi E, Radzun KA, Ismail K (2022) Removal of acetic acid from aqueous solution using a supported liquid membrane with a hybrid PES-graphene membrane. J Environ Chem Eng 10(3):107644. https://doi.org/10.1016/j.jece.2022.107644

Almeida JRM, Modig T, Petersson A, Hähn-Hägerdal B, Lidén G, Gorwa-Grauslund MF (2007) Increased tolerance and conversion of inhibitors in lignocellulosic hydrolysates by Saccharomyces cerevisiae. J Chem Technol Biotechnol 82(4):340–349. https://doi.org/10.1002/jctb.1676

Almeida JRM, Bertilsson M, Gorwa-Grauslund MF, Gorsich S, Lidén G (2009) Metabolic effects of furaldehydes and impacts on biotechnological processes. Appl Microbiol Biotechnol 82(4):625–638. https://doi.org/10.1007/s00253-009-1875-1

Anderson ST, Chen R, Liu J (2024) Market competitiveness of advanced biofuels under carbon pricing scenarios. Energy Econ 128:107142. https://doi.org/10.1016/j.eneco.2024.107142

Ando S, Arai I, Kiyoto K, Hanai S (1986) Identification of aromatic monomers in steam-exploded poplar and their influences on ethanol fermentation by Saccharomyces cerevisiae. J Ferment Technol 64(6):567–570. https://doi.org/10.1016/0385-6380(86)90119-8

Andrić P, Meyer AS, Jensen PA, Dam-Johansen K (2010) Reactor design for minimizing product inhibition during enzymatic lignocellulose hydrolysis: I. Significance and mechanism of cellobiose and glucose inhibition on cellulolytic enzymes. Biotechnol Adv 28(3):308–324. https://doi.org/10.1016/j.biotechadv.2010.01.002

Brandman O, Stewart-Ornstein J, Wong D, Larson A, Williams CC, Li LS, Zhou S, King D, Shen PS, Weibezahn J, Goldberg AL (2021) Hsf1 activation by proteotoxic stress requires concurrent protein synthesis. Cell Rep 36(1):109335. https://doi.org/10.1016/j.celrep.2021.109335

Broda M, Yelle DJ, Serwańska K (2022) Bioethanol production from lignocellulosic biomass—challenges and solutions. Molecules 27(24):8717. https://doi.org/10.3390/molecules27248717

Brodeur G, Yau E, Badal K, Collier J, Ramachandran KB, Ramakrishnan S (2011) Chemical and physicochemical pretreatment of lignocellulosic biomass: a review. Enzyme Res 2011:Article ID 787532. https://doi.org/10.4061/2011/787532

Cañadas R, González-Miquel M, González EJ, Díaz I, Rodríguez M (2021) Bio-based solvents for phenolic acids extraction. Sep Purif Technol 275:119179. https://doi.org/10.1016/j.seppur.2021.119179

Casal M, Paiva S, Queirós O, Soares-Silva I (1996) Transport of carboxylic acids in yeasts. FEMS Microbiol Rev 19(4):271–283. https://doi.org/10.1016/S0168-6445(96)00022-8

Casey E, Sedlák M, Ho N, Mosier N (2010) Effect of acetic acid and pH on the cofermentation of glucose and xylose to ethanol by a genetically engineered strain of Saccharomyces cerevisiae. FEMS Yeast Res 10(4):385–393. https://doi.org/10.1111/j.1567-1364.2010.00623.x

Caspeta L, Nielsen J (2015) Modifying yeast tolerance to inhibitory conditions of ethanol production processes. Front Bioeng Biotechnol 3:184. https://doi.org/10.3389/fbioe.2015.00184

Chan YJ, Yap CL, Ang WL, Ngoh GC (2024) A COSMO-RS guided approach to greener solvent selection for the extraction of furans from aqueous media. ACS Sustain Chem Eng 12(8): 3218–3230. https://doi.org/10.1021/acssuschemeng.3c06609

Chang YH, Xu Y, Xu J, Lei P, Liu J (2021) Engineering Escherichia coli for improved tolerance to lignocellulosic inhibitors through overexpression of stress-responsive gene ycfR. Bioresour Technol 326:124770. https://doi.org/10.1016/j.biortech.2021.124770

Chavan S, Gaikwad A (2021) Extraction and recovery of lignin derived phenolic inhibitors to enhance enzymatic glucose production. Biomass Bioenergy 144:105897. https://doi.org/10.1016/j.biombioe.2020.105897

Chen Y, Wang L, Zhang M (2022) Techno-economic analysis of biological versus physical detoxification in cellulosic ethanol production. Bioresour Technol 351:126943. https://doi.org/10.1016/j.biortech.2022.126943

Cherubini F, Peters GP, Berntsen T, Strømman AH, Hertwich E (2024) Multidimensional life-cycle assessment of biorefineries across generations. Appl Energy 346:121355. https://doi.org/10.1016/j.apenergy.2024.121355

Chong KF, Lu Y, Han Y, Shen Y, Thangalazhy-Gopakumar S, Shi S, Han L (2025) A review on the over-liming detoxification of lignocellulosic biomass prehydrolysate for bioethanol production. Appl Biochem Biotechnol 197(6):3581–3613. https://doi.org/10.1007/s12010-025-05212-5

Corazza FC, Calsavara LPV, Moraes FF, Zanin GM, Neitzel I (2005) Determination of inhibition in the enzymatic hydrolysis of cellobiose. Braz J Chem Eng 22(4):501–508. https://doi.org/10.1590/S0104-66322005000400012

Decottignies A, Grant A, Nichols J, Wet HD, McIntosh D, Goffeau A (2008) Compensatory activation of the multidrug transporters Pdr5p, Snq2p, and Yor1p by Pdr1p in Saccharomyces cerevisiae. J Biol Chem 283(9):5740–5749. https://doi.org/10.1074/jbc.M708019200

Díaz T, Martín C, Hernández M, Caminal G (2023) Minor phenolic compounds in detoxified lignocellulosic hydrolysates are the determining factor in cell growth and metabolic activity of *Escherichia coli*. Appl Environ Microbiol 89(1):e01956–e01922. https://doi.org/10.1128/AEM.01956-22

Divate RD, Keller NP, Cavka A (2017) Combining engineered acetyl-xylan esterase and a trehalose synthesis pathway in Saccharomyces cerevisiae for improved bioconversion of lignocellulosic materials. Biotechnol Biofuels 10:176. https://doi.org/10.1186/s13068-017-0862-9

Faheem M, Shimasaki T, Tachibana R, Sonomoto K (2020) Acid hydrolysis of lignocellulosic biomass: sugars and furfurals formation. Catalysts 10(4):437. https://doi.org/10.3390/catal10040437

Fang Z, Li T, Wang Q, Zhang X, Yu Y, Zhang X (2015) A novel laccase (Glac15) from Ganoderma lucidum 77002: purification, characterization, and its application in the detoxification of bioethanol production. Bioresour Technol 197:318–324. https://doi.org/10.1016/j.biortech.2015.08.112

Fernández-Sandoval MT, García A, Teymennet-Ramírez KV, Arenas-Olivares DY, Martínez-Morales F, Trejo-Hernández MR (2024) Removal of phenolic inhibitors from lignocellulose hydrolysates using laccases for the production of fuels and chemicals. Biotechnol Prog 40(1): e3406. https://doi.org/10.1002/btpr.3406

Frederick N (2013) Minimizing wash water usage after acid hydrolysis pretreatment: a life cycle optimization approach. Undergraduate Honors Thesis, University of Arkansas. ScholarWorks@UARK. https://scholarworks.uark.edu/baeguht/5

Gorsich SW, Dien BS, Nichols NN, Slininger PJ, Liu ZL, Skory CD (2006) Tolerance to furfural-induced stress is associated with pentose phosphate pathway genes ZWF1, GND1, RPE1, and TKL1 in Saccharomyces cerevisiae. Appl Microbiol Biotechnol 71(3):339–349. https://doi.org/10.1007/s00253-005-0142-3

Guo H, Zhao Y, Chen X, Shao Q, Qin W (2022) Inhibitor formation and detoxification during lignocellulose biorefinery: a review. Bioresour Technol 361:127666. https://doi.org/10.1016/j.biortech.2022.127666

Hafezisefat P, Rahic E, Brown RC (2025) Oxidative fractionation of biomass to produce phenolic monomers and processable carbohydrate pulp. React Chem Eng 10:2644–2656. https://doi.org/10.1039/D4RE00413B

Han J, Liu S, Zhou X, Zhang Y, Wang X (2025) Ultrasound-assisted regeneration of ion exchange resins for efficient detoxification of lignocellulosic hydrolysate in biorefinery. Renew Energy 220:119627. https://doi.org/10.1016/j.renene.2023.119627

Haq IU, Durrani R, Mukhtar H, Khan MR, Saleem M (2024a) Recent advances in microbial laccases and their applications: a review. J Clean Prod 432:138993. https://doi.org/10.1016/j.jclepro.2023.138993

Haq IU, Saleem A, Chaudhary R, Alessa AH, Nawaz A, Du C (2024b) Role of microbial laccases in valorization of lignocellulosic biomass to bioethanol. Front Bioeng Biotechnol 12:1441075. https://doi.org/10.3389/fbioe.2024.1441075

Hong J-W, Gam D-H, Kim J-H, Jeon S-J, Kim H-S, Kim J-W (2021) Process development for the detoxification of fermentation inhibitors from acid-pretreated microalgae hydrolysate. Molecules 26(9):2435. https://doi.org/10.3390/molecules26092435

Hou J et al (2024) Detoxification of fermentation inhibitors using a waste adsorbent derived from enzymatic hydrolyzed residue of rice straw. Ind Crop Prod. https://doi.org/10.1016/j.indcrop.2024.119250

Huang H, Guo X, Li D, Liu M, Wu J, Ren H (2011) Identification of crucial yeast inhibitors in bio-ethanol and improvement of fermentation at high pH and high total solids. Bioresour Technol 102(16):7486–7493. https://doi.org/10.1016/j.biortech.2011.05.023

Humbird D, Davis R, Tao L, Kinchin C, Hsu D, Aden A (2021) Process design and economics for biochemical conversion of lignocellulosic biomass to ethanol. National Renewable Energy Laboratory Technical Report, NREL/TP-5100-47764

Jiang W, Lei Z, Gao H, Jiang Y, Lin CSK, Zhang W, Xin F, Jiang M (2024) Biodetoxification of lignocellulose hydrolysate for direct use in succinic acid production. Biodesign Res 6:0044. https://doi.org/10.34133/bdr.0044

Jilani SB, Olson DG (2023) Mechanism of furfural toxicity and metabolic strategies to engineer tolerance in microbial strains. Microb Cell Factories 22(1):221. https://doi.org/10.1186/s12934-023-02116-z

Johnson TL, Keith DW, Reichert M (2023) Policy mechanisms for the deployment of advanced biofuels under renewable fuel standards. Nat Energy 8:445–456. https://doi.org/10.1038/s41560-023-01245-w

Jonsson LJ, Alriksson B, Nilvebrant NO (2013) Bioconversion of lignocellulose: inhibitors and detoxification. Biotechnol Biofuels 6(1):16. https://doi.org/10.1186/1754-6834-6-16

Jönsson LJ, Alriksson B, Nilvebrant N-O (2016) Pretreatment of lignocellulose: formation of inhibitory by-products and strategies for minimizing their effects. Bioresour Technol 199:103–112. https://doi.org/10.1016/j.biortech.2015.10.009

Jørgensen H, Vibe-Pedersen J, Larsen J, Felby C (2007) Liquefaction of lignocellulose at high-solids concentrations. Biotechnol Bioeng 96(5):862–870. https://doi.org/10.1002/bit.21261

Kammoun M, Aguedo M, Bettaieb T, Richel A (2023) The key role of pretreatment for the one-step and multi-step conversions of European lignocellulosic materials into furan compounds. Green Chem 25:5340–5364. https://doi.org/10.1039/D3GC00949A

Kim D (2018) Physico-chemical conversion of lignocellulose: inhibitor effects and detoxification strategies: a mini review. Molecules 23(2):309. https://doi.org/10.3390/molecules23020309

Kim Y, Ximenes E, Mosier NS, Ladisch MR (2011) Soluble inhibitors/deactivators of cellulase enzymes from lignocellulosic biomass. Enzym Microb Technol 48(4–5):408–415. https://doi.org/10.1016/j.enzmictec.2011.01.007

Kim Y, Ximenes E, Mosier NS, Ladisch MR (2012) Effect of pretreatment severity on accumulation of major degradation products from dilute acid pretreated corn stover and subsequent inhibition of enzymatic hydrolysis of cellulose. Bioresour Technol 114:329–336. https://doi.org/10.1016/j.biortech.2012.02.019

Koopman F, Wierckx N, de Winde JH, Ruijssenaars HJ (2010) Identification and characterization of the furfural and 5-(hydroxymethyl) furfural degradation pathways of Cupriavidus basilensis HMF14. Proc Natl Acad Sci 107(11):4919–4924. https://doi.org/10.1073/pnas.0913039107
Kristensen JB, Felby C, Jørgensen H (2009) Yield-determining factors in high-solids enzymatic hydrolysis of lignocellulose. Biotechnol Biofuels 2(1):11. https://doi.org/10.1186/1754-6834-2-11
Kumar D, Singh B, Korstad J (2023) Different pretreatment and kinetic models for bioethanol production from lignocellulosic biomass. Energy 273:127235. https://doi.org/10.1016/j.energy.2023.127235
Lam FH, Turanlı-Yıldız B, Liu D, Resch MG, Fink GR, Stephanopoulos G (2021) Engineered yeast tolerance enables efficient production from toxified lignocellulosic feedstocks. Sci Adv 7(26): eabf7613. https://doi.org/10.1126/sciadv.abf7613
Larsson S, Reimann A, Nilvebrant NO, Jönsson LJ (2001) Comparison of different methods for the detoxification of lignocellulose hydrolysates of spruce. Appl Biochem Biotechnol 77(1): 91–103. https://doi.org/10.1385/ABAB:77:1-3:91
Larsson S, Reimann A, Nilvebrant NO, Jönsson LJ (2003) Evaluation of ion exchange resins for removal of inhibitory compounds in dilute-acid hydrolysates from spruce. Appl Biochem Biotechnol 105–108:615–625. https://doi.org/10.1385/ABAB:105:1-3:615
Lee HJ et al (2013) Improvement of ethanol fermentation from lignocellulosic hydrolysates by the removal of inhibitors. J Ind Eng Chem 19(6):2010–2015. https://doi.org/10.1016/j.jiec.2013.03.016
Lee S, Lee J, Yang X, Kim S, Lee J, Yoo H et al (2015) Phenolic compounds: strong inhibitors derived from lignocellulosic hydrolysate for 2,3-butanediol production by Enterobacter aerogenes. Biotechnol J 10(12):1920–1928. https://doi.org/10.1002/biot.201500090
Li H et al (2016) Inhibition of lignin-derived phenolic compounds to cellulase. Biotechnol Biofuels 9:70. https://doi.org/10.1186/s13068-016-0488-5
Li Y, Gou Z, Zhang Y, Xia Z, Tang Y, Kida K (2017) Inhibitor tolerance of a recombinant flocculating industrial Saccharomyces cerevisiae strain during glucose and xylose co-fermentation. Braz J Microbiol 48(4):791–800. https://doi.org/10.1016/j.bjm.2016.11.011
Li M, Cao Q, Wang F, Tan L, Jia X, Zhang C, Zhang J, Huang L (2020) Understanding the structural characteristics of water-soluble phenolic compounds from four pretreatments of corn stover and their inhibitory effects on enzymatic hydrolysis and fermentation. Bioresour Technol 304:122975. https://doi.org/10.1016/j.biortech.2020.122975
Li B et al (2021) Improving acetic acid and furfural resistance of xylose-fermenting Saccharomyces cerevisiae. Appl Environ Microbiol. https://doi.org/10.1128/AEM.00294-21
Li Z, Chen Y, Luo X, Huang R, Qiu X (2024) Unveiling the inhibition of enzymatic hydrolysis of cellulose by lignin-derived phenolics: interfacial kinetics and molecular simulations. ACS Sustain Chem Eng 12(26):9735–9745. https://doi.org/10.1021/acssuschemeng.4c02447
Linares-López JC, Ballesteros I, Tourán J, Cara C, Castro E, Ballesteros M, Romero I (2022) Optimization of uncatalyzed steam explosion pretreatment of rapeseed straw for biofuel production. Bioresour Technol 190:97–105. https://doi.org/10.1016/j.biortech.2022.126845
Liu Y, Wang Q, Zhang J, Li Y, Zhao L (2021a) Comparative study of three fungal laccases with different pH optima in corncob detoxification for bioethanol production. Renew Energy 173: 932–941. https://doi.org/10.1016/j.renene.2021.04.060
Liu Z, Fels M, Dragone G, Mussatto SI (2021b) Effects of inhibitory compounds derived from lignocellulosic biomass on the growth of the wild-type and evolved oleaginous yeast Rhodosporidium toruloides. Ind Crop Prod 170:113799. https://doi.org/10.1016/j.indcrop.2021.113799
Liu Y, Liu Y, Liu Y (2024) Synthetic biology approaches to improve tolerance of inhibitors in lignocellulosic hydrolysates. Front Bioeng Biotechnol 12:112. https://doi.org/10.3389/fbioe.2024.112
Ma M, Liu ZL (2010) Comparative transcriptome profiling during the lag phase identifies YAP1, PDR1, PDR3, RPN4, and HSF1 as key regulatory genes in Saccharomyces cerevisiae's

genomic adaptation to the lignocellulose-derived inhibitor HMF. BMC Genomics 11:660. https://doi.org/10.1186/1471-2164-11-660

Ma M, Wang X, Zhang X, Zhao X (2013) Alcohol dehydrogenases from Scheffersomyces stipitis involved in the detoxification of aldehyde inhibitors derived from lignocellulosic biomass conversion. Appl Microbiol Biotechnol 97(18):8411–8425

Marton JM, Felipe MGA, Almeida e Silva JB, Pessoa A Jr (2003) Use of ion-exchange resins for removing toxic compounds from sugarcane bagasse hemicellulosic hydrolysate for xylitol production. Biotechnol Prog 19(5):1628–1633. https://doi.org/10.1021/bp034049c

Miah R, Tanvir RU, Hossain MB, Rahman MM, Ahmed S (2022) Development of high temperature simultaneous saccharification and fermentation (SSF) process for bioethanol production using thermotolerant *Saccharomyces cerevisiae*. Sci Rep 12:3579. https://doi.org/10.1038/s41598-022-07589-3

Modig T, Lidén G, Taherzadeh MJ (2002) Inhibition effects of furfural on alcohol dehydrogenase, aldehyde dehydrogenase and pyruvate dehydrogenase. Biochem J 363(3):769–776. https://doi.org/10.1042/bj3630769

Mohammadi R, Raisi A, Aroujalian A (2022) Removal of acetic acid from aqueous post-fermentation waste streams using membrane contactors. J Chem Technol Biotechnol 97(6): 1612–1621. https://doi.org/10.1002/jctb.7013

Morales M, Quintero J, Conejeros R, Aroca G (2021) Life cycle assessment of lignocellulosic bioethanol production: environmental impacts and energy balance. Renew Sustain Energy Rev 142:110837. https://doi.org/10.1016/j.rser.2021.110837

Nasir IM, Said M, Hamid SA (2022) Review of lignocellulosic biomass pretreatment using physical, thermal and chemical methods for higher yields in bioethanol production. Sustainability 17(1):287. https://doi.org/10.3390/su17010287

Nguyen M, Patel R, Kim S (2025) Chaperone overexpression boosts heterologous small molecule production in Saccharomyces cerevisiae. Microb Cell Factories 24:Article 2728. https://doi.org/10.1186/s12934-025-02728-7

Nilsson A, Gorwa-Grauslund MF, Hahn-Hägerdal B, Lidén G (2005) Cofactor dependence in furan reduction by Saccharomyces cerevisiae in fermentation of acid-hydrolyzed lignocellulose. Appl Environ Microbiol 71(12):7866–7871. https://doi.org/10.1128/AEM.71.12.7866-7871.2005

Okonkwo CC, Azam MM, Ezeji TC, Qureshi N (2021) Enhancing ethanol production from cellulosic sugars using Scheffersomyces (Pichia) stipitis. Bioprocess Biosyst Eng 44:1079–1090. https://doi.org/10.1007/s00449-020-02504-0

Palmqvist E, Hahn-Hägerdal B (2000) Fermentation of lignocellulosic hydrolysates. II: inhibitors and mechanisms of inhibition. Bioresour Technol 74(1):25–33. https://doi.org/10.1016/S0960-8524(99)00161-3

Pampulha ME, Loureiro-Dias MC (1989) Combined effect of acetic acid, pH and ethanol on intracellular pH of fermenting yeast. Appl Microbiol Biotechnol 31(5–6):547–550. https://doi.org/10.1007/BF00270793

Pan X, Arato C, Gilkes N, Gregg D, Mabee W, Pye K et al (2005) Biorefining of softwoods using ethanol organosolv pulping: preliminary evaluation of process streams for manufacture of fuel-grade ethanol and co-products. Biotechnol Bioeng 90(4):473–481. https://doi.org/10.1002/bit.20453

Parawira W, Tekere M (2011) Biotechnological strategies to overcome inhibitors in lignocellulose hydrolysates for ethanol production: review. Crit Rev Biotechnol 31(1):20–31. https://doi.org/10.3109/07388551.2010.510025

Park YC, Kim JS, Seong DH, Kim SW, Jin YS (2017) Evaluation of the main inhibitors from lignocellulose pretreatment on enzymatic hydrolysis and yeast fermentation. Biotechnol Biofuels 10:81. https://doi.org/10.1186/s13068-017-0768-4

Pilap W, Thanonkeo P, Klanrit P (2022) The potential of multistress-tolerant yeast, Saccharomycodes ludwigii, for second-generation bioethanol production. Sci Rep 12(1). https://doi.org/10.1038/s41598-022-26686-x

Piotrowski JS, Okekeogbu I, Wilken PM, Zylstra GJ, Slininger PJ (2019) Exploring the tolerance of marine yeast to inhibitory compounds for improved bioethanol production. Sustain Energy Fuels 3:1545–1561. https://doi.org/10.1039/C9SE00156A

Qi B, Luo J, Chen X, Hang X, Wan Y (2012) Application of ultrafiltration and nanofiltration for recycling cellulase and concentrating glucose from enzymatic hydrolyzate of steam exploded wheat straw. Bioresour Technol 104:466–472. https://doi.org/10.1016/j.biortech.2011.10.092

Qin L, Li WC, Zhu JQ, Liang JN, Li BZ, Yuan YJ (2016) Inhibition of lignin-derived phenolic compounds to cellulase. Biotechnol Biofuels 9:70. https://doi.org/10.1186/s13068-016-0488-5

Qin L, Li WC, Zhu JQ, Liang JN, Li BZ, Yuan YJ (2018) Biodetoxification of phenolic inhibitors from lignocellulose pretreatment by Kurthia huakuii LAM0618^T. Front Microbiol 9:2405. https://doi.org/10.3389/fmicb.2018.02405

Ran H, Zhang J, Gao Q, Lin Z, Bao J (2014) Analysis of biodegradation performance of furfural and 5-hydroxymethylfurfural by *Amorphotheca resinae* ZN1. Biotechnol Biofuels 7(1):51. https://doi.org/10.1186/s13068-014-0051-0

Ren S, Ye XP, Borole AP (2017) Separation of chemical groups from bio-oil water extract via sequential organic solvent extraction. J Anal Appl Pyrolysis 123:30–39. https://doi.org/10.1016/j.jaap.2016.11.011

Russell JB (1992) Another explanation for the toxicity of fermentation acids at low pH: anion accumulation versus uncoupling. J Appl Bacteriol 73(5):363–370. https://doi.org/10.1111/j.1365-2672.1992.tb04990.x

Saini J, Dagar GO (2023) Mechanism of furfural toxicity and metabolic strategies to engineer tolerance in microbial strains. Microb Cell Factories 22(1):223. https://doi.org/10.1186/s12934-023-02223-x

Saucedo-Gutiérrez JJ, Lira-Navarrete E, Díaz-Navarrete BA, Martínez JL (2024) Differential impacts of furfural and acetic acid on the bioenergetic and fermentation parameters of Scheffersomyces stipitis. J Biotechnol 381:55–64. https://doi.org/10.1016/j.jbiotec.2024.01.002

Schroyen M, Van Hulle SW, Holemans S, Vervaeren H, Raes K (2017) Laccase enzyme detoxifies hydrolysates and improves biogas production from hemp straw and miscanthus. Bioresour Technol 244:597–604. https://doi.org/10.1016/j.biortech.2017.08.010

Sindhu R, Binod P, Pandey A, Ankaram S, Duan Y, Awasthi MK (2023) Recent developments and emerging methodologies in the pretreatment of lignocellulosic biomass for bioethanol production. Fuel 342:127835. https://doi.org/10.1016/j.fuel.2023.127835

Siqueira MR, Reginatto V (2015) Inhibition of fermentative H2 production by hydrolysis byproducts of lignocellulosic substrates. Renew Energy 80:109–116. https://doi.org/10.1016/j.renene.2015.01.061

Smith J, Lam FH, Cho KM, Wen J, Sjaarda C, Du J, Turner TL, Jin YS (2014) Simultaneously improving xylose fermentation and tolerance to lignocellulosic inhibitors through evolutionary engineering of recombinant *Saccharomyces cerevisiae*. BMC Biotechnol 14:41. https://doi.org/10.1186/1472-6750-14-41

Suo Y, Luo S, Zhang Y, Liao Z (2019) Enhanced butyric acid production from lignocellulosic hydrolysates by engineered Clostridium tyrobutyricum overexpressing Class II heat shock protein GroESL. Bioresour Technol 284:328–335. https://doi.org/10.1016/j.biortech.2019.03.141

Taherzadeh MJ (1999) Ethanol from lignocellulose: physiological effects of inhibitors and fermentation strategies. Doctoral dissertation, Chalmers Tekniska Hogskola

Tejirian A, Xu F (2010) Inhibition of cellulase-catalyzed lignocellulosic hydrolysis by iron and oxidative metal ions and complexes thereof. Appl Environ Microbiol 76(23):7673–7682. https://doi.org/10.1128/AEM.01452-10

Tian S, Zhou G, Yan F, Yu Y, Yang X (2009) Yeast strains for ethanol production from lignocellulosic hydrolysates during in situ detoxification. Biotechnol Adv 27(5):656–660. https://doi.org/10.1016/j.biotechadv.2009.05.004

Ujor VC, Okonkwo CC (2022) Microbial detoxification of lignocellulosic biomass hydrolysates: biochemical and molecular aspects, challenges, exploits and future perspectives. Front Bioeng Biotechnol 10:1061667. https://doi.org/10.3389/fbioe.2022.1061667

USDoE (2025) U.S. Department of Energy, Office of Scientific and Technical Information. Lignin Valorization Reshapes Sustainable Biomass Refining. https://www.osti.gov/servlets/purl/2538169

Vanmarcke G, Demeke MM, Foulquié-Moreno MR, Thevelein JM (2021) Identification of the major fermentation inhibitors of recombinant 2G yeasts in diverse lignocellulose hydrolysates. Biotechnol Biofuels 14:92. https://doi.org/10.1186/s13068-021-01961-1

Wang LQ, Wang L, Zhang Y, Zhang Z, Xue Y (2020) Study on inhibitors from acid pretreatment of corn stalk on ethanol production. RSC Adv 10:26860–26868. https://doi.org/10.1039/D0RA05323A

Wang Y, Liu J, Ma Y, Liu Y, Liu L (2024) Metabolic engineering of microorganisms to produce pyruvate and derived compounds. Metab Eng 80:157–170. https://doi.org/10.1016/j.ymben.2023.10.006

Wierckx N, Koopman F, Ruijssenaars HJ, de Winde JH (2010) Microbial degradation of furanic compounds: biochemistry, genetics, and impact. Appl Microbiol Biotechnol 92(6):1095–1105. https://doi.org/10.1007/s00253-011-3632-5

Xiao XY, Li B, Xia ZY, Zhang Q, Xie CY, Tang YQ (2025) Regulatory mechanism of Haa1p and Hap4p in Saccharomyces cerevisiae to mixed acetic acid and formic acid when fermenting mixed glucose and xylose. Microb Cell Factories 24(1):156. https://doi.org/10.1186/s12934-024-02306-0

Xie Y, Varshavsky A (2001) RPN4 is a ligand, substrate, and transcriptional regulator of the 26S proteasome: a negative feedback circuit. Proc Natl Acad Sci 98(6):3056–3061. https://doi.org/10.1073/pnas.071022298

Xu H, Selig MJ, Green DJ, Fan ZL, Wyman CE (2013) Severity factor coefficients for subcritical liquid hot water pretreatment of hardwood chips. Bioresour Technol 142:1–7. https://doi.org/10.1016/j.biortech.2013.05.019

Zha Y, Muilwijk B, Coulier L, Punt PJ (2014) Identifying inhibitory compounds in lignocellulosic biomass hydrolysates using an exometabolomics approach. BMC Biotechnol 14:22. https://doi.org/10.1186/1472-6750-14-22

Zhang J, Ezeji TC (2014) Investigating effect of furfural-derived inhibitors on Clostridium beijerinckii growth and butanol production. J Anim Sci Biotechnol 5:15. https://doi.org/10.1186/2049-1891-5-15

Zhang Y, Liu J, Zhang Y, Liu Y, Liu Y, Liu Y (2012) Furfural and 5-hydroxymethylfurfural induce stress responses in Saccharomyces cerevisiae. Appl Environ Microbiol 78(1):1–9. https://doi.org/10.1128/AEM.06204-11

Zhang Y, Xia C, Lu M, Tu M (2018) Effect of overliming and activated carbon detoxification on inhibitors removal and butanol fermentation of poplar prehydrolysates. Biotechnol Biofuels 11(1):178. https://doi.org/10.1186/s13068-018-1182-0

Zhang Z et al (2025) Enhanced tolerance of Saccharomyces cerevisiae to industrial inhibitors through transcription factor engineering. Bioresour Technol 417:131039. https://doi.org/10.1016/j.biortech.2024.131039

Zhou S, Yang Q, Chen J, Wang Y, Li Y, Yuan T, Sun RC (2013) Formation of degradation compounds from lignocellulosic biomass in the biorefinery: sugar reaction mechanisms. Bioresour Technol 150:1–13. https://doi.org/10.1016/j.biortech.2013.08.019

Zhu X, Li X, Zhao X, Gu H (2021) The advanced performance of microbial consortium for simultaneous utilization of glucose and xylose to produce lactic acid directly from dilute sulfuric acid pretreated corn stover. Biotechnol Biofuels 14:85. https://doi.org/10.1186/s13068-021-02085-8

Zou W, Yu L, Zhou J, Stephanopoulos G (2021) Metabolic engineering of Pseudomonas putida for efficient lactic acid production from lignin-derived aromatics and sugars. Metab Eng 67:72–82. https://doi.org/10.1016/j.ymben.2021.06.001

Chapter 11
Advanced Artificial Intelligence-Based Process Modeling of Lignocellulosic Bioethanol Production

Mohd. Zafar, Vikas Chandra Gupta, and Shiv Prasad

Abstract Artificial intelligence (AI) techniques, based on machine learning (ML) and deep learning (DL), are integrated across various phases of the lignocellulosic bioethanol supply chain, including pretreatment, hydrolysis, and fermentation. These incorporations enhance the predictive ability of process accuracy, process optimization, and economic viability, thereby boosting the commercialization of sustainable 2G bioethanol production systems. This book chapter presents an extensive review of the global bioethanol market, with a particular focus on lignocellulosic biomass. The basic concept of AI and its various approaches, utilizing ML and DL algorithms, are discussed for the 2G bioethanol process intensification. Additionally, a bibliometric study systematically evaluated research publications on AI applications in lignocellulosic biomass over the past decade. In addition, challenges, opportunities, and future directions for integrating AI into 2G-bioethanol production are briefly discussed. Therefore, by combining advanced AI tools into LCB-based biorefinery systems, a continuous 2G bioethanol production system can be developed despite seasonal variability in feedstocks and contribute to a sustainable energy future.

Keywords Lignocellulosic biomass · 2G Bioethanol · Biorefinery · Artificial Intelligence · Bibliometric study

---

M. Zafar (✉)
Department of Biotechnology, College of Applied Sciences and Pharmacy, University of Technology and Applied Sciences, Sur, Oman
e-mail: Mohammed.zafar@utas.edu.om

V. C. Gupta
Department of Biotechnology, IILM University, Greater Noida, Uttar Pradesh, India

S. Prasad
Division of Environmental Sciences, ICAR-Indian Agriculture Research Institute, New Delhi, Delhi, India

S. Prasad et al. (eds.), *Sustainable Bioethanol Production from Lignocellulosic Biomass*, Clean Energy Production Technologies,
https://doi.org/10.1007/978-981-95-9690-4_11

11.1 Introduction

Lignocellulosic biomass (LCB) is one of the most abundant and non-food-based feedstocks for second-generation bioethanol, attracting global research interest as a way to meet rising ethanol demand. Commercial bioethanol production from LCB is a complex, challenging process that involves several steps, from LCB pretreatment to the energy-intensive downstream separation of bioethanol. To overcome challenges in the commercial deployment of LCB-based bioethanol, advanced AI-driven data techniques must be adopted throughout the process. AI mimics human intelligence using advanced biologically inspired algorithms and computer vision (Arias et al. 2023). Key components of AI include machine learning (ML), generative AI, language processing, and expert systems, all designed to process information, improve over time, and provide actionable insights in industrial contexts. It is widely used for modeling and simulating bioprocesses and bioenergy systems, as well as for optimizing complex biological systems. Advanced AI methods, including machine learning (ML), deep learning, and hybrid models, can facilitate comprehensive process optimization, real-time control, and predictive maintenance, thereby eliminating key bottlenecks.

Machine learning (ML) is an artificial intelligence-based modelling tool that offers a valuable solution for understanding the complex behavior of factors such as environmental conditions, chemicals, and enzymes used in the pretreatment of LCB and in the upstream and downstream processes of bioethanol (Coşgun et al. 2023; Broda et al. 2022). ML is an advanced, data-driven approach that helps map nonlinear relationships between input and output variables at different stages of 2G bioethanol production. Because of its self-learning capabilities, simplicity, robustness, and high-speed computation, ML can replace traditional models in biorefinery process studies (Azad et al. 2025; Niu et al. 2025). ML models can accurately predict the behavior of the 2G bioethanol production system under various conditions without needing explicit knowledge of the physical system (Aghbashlo et al. 2021).

Integrating AI into biorefinery processes can facilitate multi-objective optimization of various process, environmental, and social factors to promote market adoption of developed technologies. Therefore, an AI-based bioethanol production model utilizing LCB could be viewed as a sustainable, circular economy-driven "smart biorefinery process" that delivers quality products (bioethanol) in response to market needs. AI tools can analyze vast amounts of data quickly and identify process patterns under different scenarios, helping to boost the productivity of 2G-bioethanol production from complex LCB. AI can improve bioethanol production from LCB by optimizing bioresources and selecting the best pretreatment conditions, production stages, and downstream processes through precise monitoring of process parameters, reducing waste, and effectively managing the raw material supply chain.

The integration of AI into 2G-bioethanol production improves capacity, transportation, process logistics, and marketability (Arias et al. 2023). An AI and machine learning (ML) model for bioethanol production from LBS, including pretreatment to

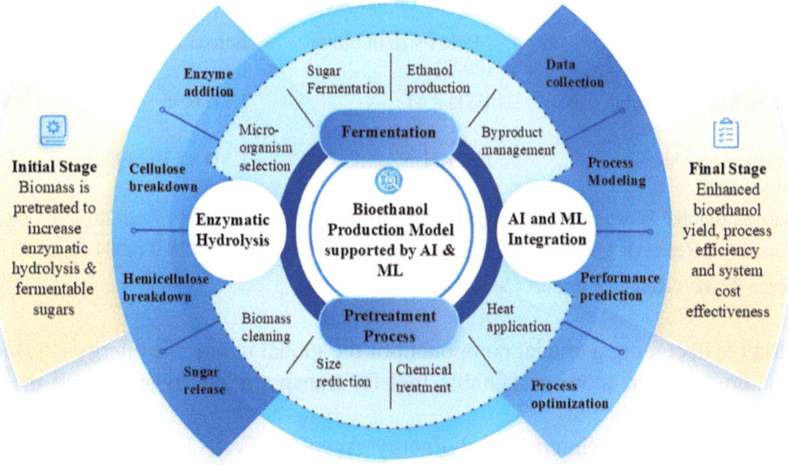

Fig. 11.1 AI and ML-based bioethanol production from lignocellulosic biomass model

prepare feedstock for pretreatment, enzymatic hydrolysis, fermentable sugar recovery, and efficient bioethanol production, enhances process efficiency with high purity and cost-effectiveness, as shown in Fig. 11.1.

This chapter presents an extensive review of (i) the global market outlook of LCB-based bioethanol production; (ii) basic concept of AI and their various approaches implemented for 2G bioethanol process intensification; (iii) bibliometric study on AI applications in lignocellulosic (2G) bioethanol production process; (iv) different 2G bioethanol process stages and implemented AI tools; (v) case studies of potential AI applications in 2G-bioethanol productions and (vi) industrial challenges and new directions in the field of 2G-bioethanol production handle by AI.

11.2 Global Market Outlook of LCB-Based Bioethanol Production

The demand for alternative transportation fuels, such as ethanol, has been recognized globally since the mid-1990s, driven by rising oil prices and growing environmental concerns. The global biofuels market size is expected to grow from USD 131.0 billion in 2023 to around USD 307.3 billion by 2033, with a CAGR of 8.9% (Table 11.1). Similarly, the global bioethanol market is expected to increase to USD 80.47 billion in 2024, up from USD 121.93 billion in 2023, with a CAGR of 5.33% (Ethanol Market Research 2025). The key drivers of the global biofuel market are government regulations and incentives aimed at reducing carbon emissions and promoting renewable energy (Ethanol Market Research 2025). The 2G-bioethanol

Table 11.1 Global bioethanol market based on lignocellulosic biomass

Report source	Market size in billion (2025)	Forecast in billion (2030)	CAGR
Future Market Insights (LBS Market)	USD 4.61	USD 9.76	7.8%
Research and Markets (Bioethanol)	USD 51.65	USD 82.39 (2029)	12.4%
Markets (Global Bioethanol)	USD 83.4 (2023)	USD 114.7 (2028)	6.6%
Precedence Research	USD 66.99	USD 178.9 (2034)	11.7%

Source: Market Research Report (2024)

market is becoming a crucial component of the global renewable energy landscape, providing cleaner, more sustainable fuel for transportation, aviation, and the fuel industry.

According to the International Energy Agency (IEA), 2G bioethanol plays a pivotal role in decarbonizing the fuel sector by providing a low-carbon, cleaner solution for both existing and emerging technologies. It is noteworthy that biofuel demand in 2022 reached a record high of 4.3 EJ (170 billion liters), underscoring the sector's resilience and the significance of biofuels in the global energy mix, particularly in transportation fuel. However, significant scaling of biofuel production to 10 EJ by 2030 is expected to achieve Net Zero Emission (NZE) by 2050 (Ethanol Market Research 2025). Given these scenarios, an annual growth rate of 11% in LCB-derived bioethanol use is required, increasing the share from 9% in 2021 to over 40% by 2030. The major biofuel-producing countries, such as the United States, India, and Brazil, play a crucial role in achieving this growth through their targeted policies and investments in the biofuel sector.

In India, the government is emphasizing the blending of biofuels with petrol to help the country reduce millions of tons of CO_2 emissions. Recently, India achieved its 20% ethanol-blended petrol (E20) objective 5 years ahead of its target (Deb 2025). Since the introduction of ethanol blending with petrol, India has reduced CO_2 emissions by 69.8 million tons and saved 1.36 trillion rupees in foreign exchange (Deb 2025). Similarly, Brazil planned to increase biodiesel blending to 15% by 2026 from 10% in 2022. The United States has committed $9.4 billion in investment to support biofuel industries through its Inflation Reduction Act (Market Research Report 2024). It is expected that biofuels will account for 9% of global energy demand, up from 3.5% in 2022, while bio-jet fuel aims to account for 10% of aviation fuel by 2030 (Market Research Report 2024).

It is noteworthy that liquid biofuels account for 46% of the biofuel market, primarily because they are used as transportation fuels. Among liquid biofuels, ethanol has the largest share of 106 billion liters, followed by biodiesel at 45.1 billion liters. It is noteworthy that bioethanol leads the biofuel market with a 41.3% share in 2023, due to its widespread applications in gasoline, primarily in the United States and Brazil, the global leaders in bioethanol production.

11.3 Bibliometric Study on AI Applications in the 2G-Bioethanol Production Process

A bibliometric study is a comprehensive examination of research work, including the mining of research data (literature), processing, and mapping approaches to explore both qualitative and quantitative research information in a specific field of investigation. In this study, a bibliometric investigation is conducted to describe the recent trends and advancements in AI applications for lignocellulosic bioethanol production as shown in Fig. 11.2. The Scopus database was selected for its comprehensive coverage of literature across various disciplines, including environmental, agricultural, and biological sciences, as well as energy, chemical engineering, materials sciences, chemistry, economics, econometrics, and finance. In this study, a search was conducted through the Scopus database in August 2025, focusing on articles published between 2015 and 2025.

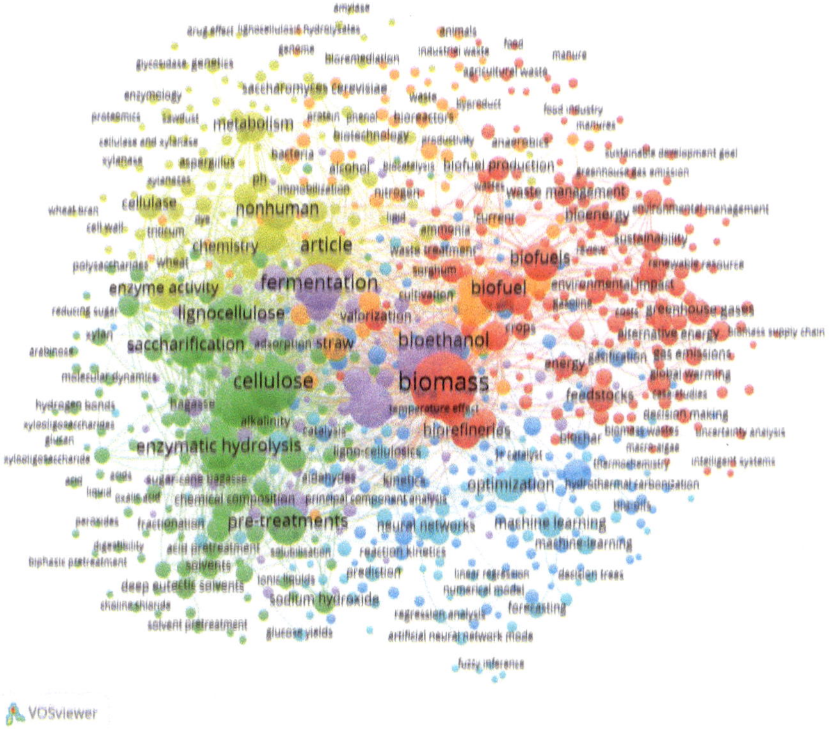

Fig. 11.2 Co-occurrence network visualization map of the key terms appeared during search of advanced research on AI applications in lignocellulosic (2G) bioethanol production process using Scopus database [("Artificial Intelligence" OR "AI" OR "Machine learning" OR "Deep learning") AND ("lignocellulosic Biomass") AND ("bioethanol production " OR "2G bioethanol production")]

It was observed that the number of published articles applying AI approaches in LCB based bioethanol production was 20 times greater in 2025 than in 2015 observed query string ("Artificial Intelligence" OR "AI" OR "Machine learning" OR "Deep learning") AND ("lignocellulosic Biomass") AND ("bioethanol production " OR "2G bioethanol production ")—the search criteria excluded conference papers, review papers, and book chapters. A total of 1029 articles were identified as relevant to recent applications of AI approaches in LCB-based bioethanol production. In order to investigate the bibliometric trends, these relevant articles were downloaded in CSV format and examined using VOSviewer software developed by Nees Jan van Eck and Ludo Waltman (Leiden University) in the year 2010 (https://www.vosviewer.com/).

LCB is an abundant, readily available non-food feedstock for 2G bioethanol production; however, its commercial deployment is challenging due to its complex structure. Advanced AI tools can enable comprehensive process optimization, real-time control, and predictive maintenance, helping overcome key bottlenecks associated with LCB's complex structure. In recent years, advanced AI tools such as machine learning (ML), artificial neural networks (ANN), adaptive neuro-fuzzy inference systems (ANFIS), and hybrid algorithms have offered a powerful approach to model, predict, and optimize each process stage, ultimately increasing ethanol yields and process efficiency (Goncü et al. 2025; Yang et al. 2025; Lee et al. 2025).

As shown in Fig. 11.2, the network map in this study identifies six clusters related to AI applications in lignocellulosic (2G) bioethanol production process. Cluster 1 (represented by 194 keywords, colored red) illustrates biomass feedstocks and their bioconversion and pretreatment for sustainable bioenergy production. Cluster 2 (including 150 keywords), symbolized by green color, represents chemical treatment and enzymatic hydrolysis aspects of lignocellulosic materials. Cluster 3 (comprising 132 keywords) is represented by blue and focuses on pyrolysis, biochar, and chemical treatment of biomass waste. Cluster 4 (including 134 keywords), symbolized by a yellow color, deals with bioethanol production, biorefinery, lignocellulosic degradation, and metabolic pathway analysis. Cluster 5 (representing 97 keywords), symbolized by purple color, includes bioprocess, bioethanol, cellulosic ethanol, lignocellulosic biomass, process optimization, and process simulation. Cluster 6 (including 77 keywords), symbolized by turquoise color, includes optimization, algorithm, artificial neural network, correlation analysis, fuzzy neural network, genetic algorithm, linear programming, multi-objective optimization, neuro-fuzzy interference, numerical modelling, random forest, and support vector machine. Final cluster 7 (including 48 keywords), symbolized by orange color, includes anaerobic digestion, bioenergy, rice straw, and biogas production.

Adeleke et al. (2026) examined the effects of chemical pretreatment on the combustion properties of a biofuel derived from Amaranthus hybridus. They used various data analysis methods, including correlation analysis, feature importance analysis (FIA), principal component analysis (PCA) for dimensionality reduction, and neuro-fuzzy modeling to explore the relationships among different biofuel properties (Adeleke et al. 2026). In this study, a grid-partitioned (GP)-clustered Adaptive Neuro-Fuzzy Inference System (ANFIS) combined with Particle Swarm

Optimization (PSO) was developed to predict the calorific value of solid biofuel. Xie and Fan (2025) conducted a comparative assessment of the enzymatic bioconversion efficiency of LCB using multiple-input machine learning architectures, including compositional signatures. They identified support vector regression (SVR) as the best model overall. A software tool was developed to predict glucose and xylose yields after various pretreatments accurately. This research improved the understanding of how pretreatment variables influence process efficiency and enzymatic hydrolysis yields from LCB. Selwal et al. (2025) presented AI-based optimization of integrated thermochemical and biochemical LCB conversion technologies, offering a scalable, cost-effective, and environmentally friendly energy solution through a sustainable, optimized LCB conversion system.

Madadi et al. (2025) optimized biphasic pretreatment conditions for LCB fractionation using machine learning models. They focused on how biphasic pretreatment influences the release of different components and the delignification process. In this study, a maximum cellulose release of 10.7%, 98.9% hemicellulose degradation, and 91.2% delignification were achieved through gradient boosted regression (GBR)-based optimization. A notable reduction in experimental time and costs was achieved through efficient biofuel production enabled by the developed machine learning model and optimized techniques.

11.4 The Various 2G Bioethanol Process Stages and the Implemented AI Tools

The application of AI and ML in bioethanol production, particularly in feedstock characterization, pretreatment, and hydrolysis, represents a significant advancement toward improving process efficiency and cost-effectiveness. It is noteworthy that characterizing feedstock based on LCB is critical and complex for bioethanol production. The various pretreatment and saccharification steps of LCB, including chemical and enzymatic hydrolysis, directly affect the variable composition and bioethanol titers. Recently developed AI and machine learning (ML) tools have revolutionized conventional, slow, and traditional pretreatment and saccharification methods, enabling fast, predictive, and high-throughput analysis. This technology has been successfully utilized for the characterization and optimization of pretreatment/saccharification of LCB feedstock (Xie and Fan 2025). The core challenges related to LCB feedstock characterization such as chemical composition (e.g., concentration of cellulose, hemicellulose, lignin, ash and extractives), physical properties of biomass (e.g., particle size, morphology, porosity, and crystallinity), maximization and prediction of sugar yields, and high throughput screening (e.g., identification of suitable non-food crops and its supply chain) can be overcome by recent AI and ML tools (Pancholi et al. 2023). The various Machine learning approaches, such as artificial neural network (ANN), Artificial Intelligence Decision-Making System (AIDMS), combined fuzzy logic with ANN (ANFIS), support vector machine (SVM), random forest (RF), and evolutionary algorithms (genetic algorithm, particle swarm optimization) coupled with ANN/ANFIS were used extensively for multi-objective optimization of pretreatment and hydrolysis process of different biomass feedstocks and their influence on saccharification and bioethanol yields (Xie and Fan 2025; Pancholi et al. 2023).

Near-infrared spectroscopy combined with machine learning has revolutionized real-time feedstock characterization. Gaussian Process Regression (GPR) predicts corn stover abundance with 95% accuracy, outperforming Partial Least Squares Regression (PLSR) at 92% (Cousins et al. 2022). Long et al. (2023) achieved accuracies of 0.8 or higher ($R^2 > 0.8$) in predicting ethanol yield and NaOH consumption during bioethanol processing using Random Forests, XGBoost, and ANNs, resulting in an 18% increase in yield. Fischer et al. (2017a, b) applied machine learning in the simultaneous saccharification and fermentation (SSF) of sugarcane bagasse, achieving R^2 values over 0.90 and producing 12.1 g/L ethanol. ANNs effectively optimized fermentation parameters, predicting a bioethanol yield of 36.45 g/L, close to the experimental value of 36.43 g/L, while Response Surface Methodology (RSM) estimated 35.43 g/L (Bahafi et al. 2024). Sebayang et al. (2017) improved bioethanol production from sorghum using an ANN, achieving $R^2 = 0.99$. Table 11.2 represents the applications of AI/ML algorithms for LCB feedstock characterization, pretreatment, hydrolysis, and bioethanol production.

Gaussian Process Regression (GPR) has shown great promise for predicting bioethanol yields from agricultural wastes, achieving an R^2 of 0.999 with minimal error metrics. The squared exponential GPR model demonstrated both low training time and fast prediction speed on 50 refined data points. When expanding the dataset to 1615 points using synthetic data from Gretel ACTGAN, artificial neural network (ANN) models also achieved near-zero MSE and high regression R values (Yeboah et al. 2024). Additionally, a recurrent neural network-based model predictive control (RNN-MPC) for continuous fermentation achieved an MSE of 7.14×10^{-5}, outperforming traditional PID control in settling time and disturbance robustness. The integration of machine learning with life cycle assessment (LCA) has produced synergistic benefits, increasing ethanol yield by 16–19% while decreasing greenhouse gas emissions by 8–14% for switchgrass and corn stover (Long and Liu 2023). XGBoost models have excelled in optimizing ethanol production from *Saccharomyces cerevisiae*, achieving R^2 values of 0.74 for ethanol production and 0.90 for glucose consumption, significantly reducing the workload by narrowing down valid strains from 216 to just 31. Comparative studies reveal that machine learning algorithms' strengths vary by context; Random Forest, XGBoost, and ANN models generally achieve R^2 values over 0.80 for predicting ethanol yields, while GPR excels with smaller datasets. C5.0 classification trees, ANNs, and Random Forests showed R^2 values above 0.90 in simultaneous saccharification and fermentation processes, indicating the effectiveness of machine learning in enhancing bioethanol production efficiency.

An artificial intelligence decision-making system (AIDMS) algorithm that combines a simple linear regression (SLR) and a weighted rank-order aggregate (WROA) function is used to optimize the LCB-based bioethanol process (Vinitha

Table 11.2 AI/ML algorithms used for feedstock characterization, pretreatment, hydrolysis, and bioethanol production

AI/ML approaches	Feedstock	Processes	Process variables (conditions)	Highlights	References
ML models—BPANN, MARS, DT, BN, GBM SVR	Wheat straw, corn straw, sugarcane bagasse, and poplar	Enzymatic bioconversion efficiency of LCB	CC, HC, LC, PH, Reaction temperature; RT, CRY, HRY, and LRY	SVR was superior ($R^2 = 0.90$) in performance to other models Solution pH was a primary factor affecting biomass composition Enzymatic bioconversion released cellulose and hemicellulose, which were determinants of glucose and xylose yields, respectively	Xie and Fan (2025)
ML models combined with orthogonal experimental design (OED)	Sugarcane bagasse	Triton-X 100 (NaOH-catalyzed) pretreatment and enzymatic hydrolysis	NaOH, pH, Triton-X, temperature, and time	Optimized conditions: Solid loading at 100 g/L; NaOH at 45 g/L; pH 13.8; Triton-X 100 at 200 mL; temperature 175 °C; and duration 45 min resulted in 88.5% and 81.8% recoveries of cellulose and hemicellulose, respectively, with 92.3% delignification	Azad et al. (2025)
MLP Random Forest (RF), Support Vector Machine (SVM), Multilayer Perceptron (MLP), Xtreme Gradient Boosting (XGBoost), and Decision Tree (DT)	Sugarcane bagasse	Delignification ability after deep eutectic solvents pretreatments	Delignification and hemicellulose removal	The highest reported enzymatic hydrolysis efficiency was 78.17% Lignin recovery using Lewis-acid-enhanced DESs showed greater antioxidant activity and UV protection	Zhao et al. (2025)
	Pine wood		Cellulose, hemicellulose, and lignin degradation	The cellulose model performed best, followed by	

(continued)

Table 11.2 (continued)

AI/ML approaches	Feedstock	Processes	Process variables (conditions)	Highlights	References
ANN, RF, and DT Support vector machines (SVM) models		Prediction of wood composition with spectral pretreatment		the hemicellulose and lignin models The polynomial kernel was the best-performing kernel, with R^2 values of 0.92, 0.77, and 0.74 for the cellulose, hemicellulose, and lignin models, respectively	Hernandez-Diaz et al. (2025)
ANN	Corn straw and wheat straw	Prediction of glucose release by the ACSE pretreatment method	Biomass: cellulose, hemicellulose, lignin, particle size. Acid impregnation: solid load, acid concentration, time, and moisture. Steam explosion: temperature, time, solid loading, reactor, and size. Enzymatic hydrolysis: cellulase and β-glucosidase dosages	Model prediction: (i). For corn stover—94% glucose release with 22% inhibitors (ii). For wheat straw—100% glucose with 13% inhibitors	Huang et al. (2024)
ANN, Genetic algorithms, fuzzy logic, and random forest (RF)	LCB	Nanomaterials-based enhanced ethanol production	Dolomite, temperature, methyl-functionalized cobalt ferrite–silica, calcium alginate	Optimized conditions: Dolomite, methyl-functionalized cobalt ferrite–silica, and calcium alginate yielded 106%, 126.9%, and 100% at 121 °C, 37 °C, and 37 °C, respectively ANFIS and NARX models successfully predicted bio-ethanol yield	Bosu et al. (2022)
XGBoost and Random Forest	Pistachio shells	Ultrasonic, hot water, and enzymatic pretreatments	Ultrasonic energy (W), time (min), Temperature (°C), Pressure (atm), Enzyme	36% of lignin removal is achieved by ultrasonic treatment, compared to 27% by	Goncü et al. (2025)

			concentration (FPU), and Substrate Concentration (%)	hot-water pretreatment Under enzymatic pretreatment, ultrasonic and hot water pretreatments yielded 7.34 g/L and 5.98 g/L of reducing sugar, respectively	
ANN, RF, and XG Boost genetic algorithms (GA), particle swarm optimization (PSO), and simulated annealing (SA)	Grass and corn stover	Feedstock characterization, pretreatment, hydrolysis, and fermentation	Ethanol yield and NaOH consumption	The highest ethanol yield of 0.47 g/g (18% increase) was reported About a 19% increase in ethanol yield using corn stover as the substrate is reported by the LCA (GREET1) model, along with a significant 14% reduction in GHG emissions	Long and Liu (2023)
ML-based- Gradient Boosted Regression (GBR) model	Lignocellulosic biomass (LCB)	Biphasic pretreatment	Solid loading and temperature for cellulose degradation, hemicellulose removal, and delignification	The GBR-based optimization technique reports approximately 10.7%, 98.9%, and 91.2% for cellulose degradation, hemicellulose removal, and delignification, respectively	Madadi et al. (2025)
Partial least squares regression (PLSR) and artificial neural networks (ANN)	Lignocellulosic biomass	Multi-feedstock composition	Cellulose, hemicellulose, and lignin composition	Developed ANN models using ASD-predicted cellulose, hemicellulose, and lignin composition, achieving R^2 values of 99.21%, 99.27%, and 99.23%, respectively	Pushpa et al. (2024)

Abbreviations: *BPANN* backpropagation artificial neural networks, *MARS* multivariate adaptive regression splines, *DT* decision tree, *BN* Bayesian networks, *GBM* gradient boosting machine, *SVR* support vector regression, *CC* cellulose content, *HC* hemicellulose content, *LC* lignin content, *PH* solution pH, *Temp* reaction temperature, *RT* retention time, *CRY* cellulose recovery yield, *HRY* hemicellulose removal yield, *LRY* lignin removal yield, *ACSE* acid-catalyzed steam explosion, *ASD* augmented spectroscopic dataset

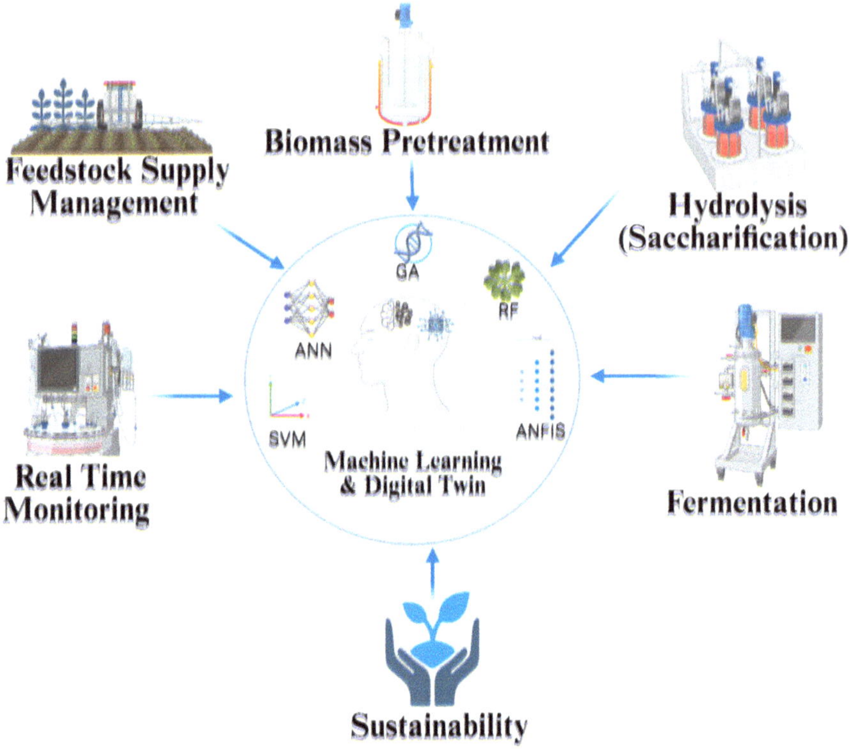

Fig. 11.3 Application of AI-augmented tools for feedstock characterization, pretreatment, hydrolysis, real-time monitoring, fermentation, and sustainability of 2G bioethanol production. (Created in BioRender. Zafar (2025) https://BioRender.com/xxezek4)

et al. 2025). Experimental data from various LCB feedstocks were used to train the AIDMS algorithm, enabling the exploration of the complex relationship between biomass characteristics and process conditions. This study identified several parameters—cellulose (%), temperature (°C), acid concentration (%), lignin (%), and hemicellulose (%)—as crucial for glucose yield. Additionally, the ranked parameters, including cellulose (%), fermentation temperature (°C), fermentation time (h), lignin (%), and hemicellulose (%), were explored for bioethanol production. The application of AI-augmented tools for feedstock characterization, pretreatment, hydrolysis, real-time monitoring, fermentation, and sustainability in 2G bioethanol production is illustrated in Fig. 11.3.

The significant challenges in LCB-based biorefineries include biomass availability, technical maturity, scale-up challenges, and a lack of investment (Singh et al. 2022). The issues related to biomass availability year-round can be addressed by selecting an appropriate LCB, managing its transportation, logistics, and supply chain, and controlling costs. Various AI/ML algorithms, such as ANN, GA, SVM, and RF, address aspects of LCB and identify suitable process variables across different scenarios to ensure LCB availability throughout the year for continuous bioethanol production. Similarly, challenges related to technological maturity include a lack of suitable pretreatment technology, limited reproducibility, and process complexity and scale-up.

Recent AI advancements, including ML, artificial neural networks (ANN), adaptive neuro-fuzzy inference systems (ANFIS), and hybrid algorithms, offer powerful tools to model, predict, and optimize each process stage, ultimately increasing ethanol yields and process efficiency (Selwal et al. 2025; Madadi et al. 2025; Azad et al. 2025; Göncü et al. 2025; Vinitha et al. 2025). David et al. (2025) optimized pretreatment conditions to maximize cellulose release and reduce sugar from peanut shells using an ML approach that efficiently optimized the alkali pretreatment process.

LCB logistics, supply chain, and costs are significant challenges in LCB-based biorefineries, which can be overcome through an AI-based prototype design that considers the size of the product (bioethanol), the materials used (LCB), the manufacturing process, packaging, and transportation system. The ANN tool can efficiently evaluate these process variables using a random function implemented in suitable software (e.g., MATLAB). Another suitable AI algorithm, ant colony optimization (ACO), can be used to achieve an optimized solution for these tasks, followed by probabilistic techniques combined with LCA to minimize environmental impact (Arias et al. 2023).

ANN can capture the nonlinear relationship between operating variables, such as temperature, pH, and enzyme load, and outputs, including saccharification rate and ethanol concentration, during the LCB-based biorefinery process (Pradhan et al. 2022). Pradhan et al. (2022) have reported the high predictive capability of ANN and ANFIS models for various LCB-based biorefinery processes, including pretreatment (microwave-assisted, organosolv-, and ultrasound-assisted processes), thermal processes (pyrolysis and gasification), enzymatic hydrolysis, and fermentation processes. Process conditions of LCB-based biorefinery were classified using support vector machine (SVM) and random forests (RF) (Fischer et al. 2017a, b).

Deep learning approaches, such as CNNs and LSTMs, can be used for time-series monitoring and fault detection in the bioethanol production process (Owusu and Marfo 2023). Additionally, various evaluation AI algorithms, such as genetic algorithms (GA) and particle swarm optimization (PSO), are combined with ANN/ANFIS for the multi-objective optimization of bioethanol yield from LCB (Pradhan et al. 2022). Xie and Fan (2025) conducted a comparative evaluation of six machine learning models to predict enzymatic bioconversion efficiency of LCB and identified SVR as a superior performer. This study provided novel insights into the critical factors and their synergistic relationships that affect the pretreatment and enzymatic hydrolysis of LCB.

AI-based modeling of the fermentation process aims to maximize bioethanol yield from the sugars released during pretreatment processes. ANN and PSO frameworks have been reported for optimizing industrial fermentation variables, resulting in a 10% increase in bioethanol yield (Fischer et al. 2017a, b). Deep learning enables real-time prediction of ethanol concentration based on hydrolysate volatile profiles.

A model predictive control (MPC) system integrated with recurrent neural networks (RNNs) that allowed real-time adjustment of temperature and pH is reported (Fischer et al. 2017a, b). This control strategy resulted in a 10% increase in ethanol titer, accompanied by a 20% reduction in by-product formation.

Similarly, the downstream process (e.g., a distillation column) for bioethanol production using LCB can be optimized by predicting column tray efficiency with AI-based neural network models. A process simulator can be fine-tuned via Bayesian optimization to minimize the energy input in DSP. SVM-based membrane separation modeling is used to predict fouling behavior and schedule cleaning cycles in DSP. Additionally, AI algorithms (e.g., artificial neural networks, ANNs) are used to optimize dehydration columns (e.g., monoethylene glycol processes), and hybrid genetic algorithms combined with machine learning are employed for energy-efficient distillation (Arias et al. 2023).

11.5 Potential of Artificial Intelligence (AI) in 2G-Bioethanol Production

Commercial applications of artificial intelligence (AI) in lignocellulosic (2G) bioethanol production are popular because AI can optimize complex biorefinery processes, improve yields, and reduce costs. Potentials of artificial intelligence (AI) in 2G-bioethanol production, particularly process optimization (during monitoring of biomass pretreatment and fermentation) and cost reduction (predictive maintenance and supply chain management are shown in Fig. 11.4.

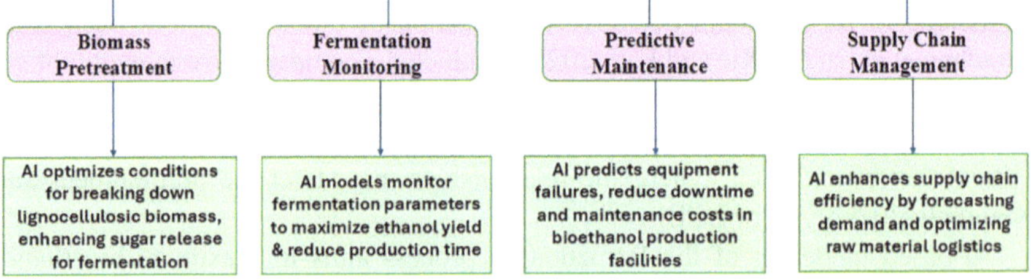

Fig. 11.4 Potentials of AI in process optimization and cost reduction for 2G bioethanol

AI models based on machine learning (ML) techniques (e.g., ANNs and RF algorithms) optimized enzymatic hydrolysis and simultaneous saccharification and fermentation (SSF) processes. These algorithms facilitated the selection of optimal pretreatment conditions, thereby enhancing process efficiency. Companies are adopting such AI-driven approaches to improve process scalability and economic viability, making 2G bioethanol more competitive with fossil fuels (Owusu and Marfo 2023). Brazil is a global leader in lignocellulosic bioethanol production, particularly from sugarcane, and has successfully integrated AI into 2G bioethanol production from sugarcane bagasse.

A hybrid ANN-GA modeling approach and ASPEN Plus-based simulation were used to assess the techno-economic feasibility of bioethanol plants and optimize pretreatment conditions and fermentation efficiency (Meena et al. 2021). These AI-based technologies supported Brazil's bioethanol industries by improving biomass feedstock conversion efficiency and reducing operational costs. Brazilian bioethanol producers, such as Raízen, leverage AI to enhance supply chain management and process optimization, thereby improving Brazil's bioeconomy (Meena et al. 2021). An AI framework model, including ANN and PSO algorithms, was used to optimize industrial bioethanol production in the state of São Paulo (Brazil) (Pereira et al. 2020). These findings demonstrated a 10% increase in bioethanol concentrations. This study enhances knowledge of industrial fermentation processes and can serve as a tool to guide increased biofuel production in states with nonlinear process behavior.

Jabbarzadeh and Shamsi (2025) studied the complexity of integrating multiple feedstocks, such as wheat straw, corn stover, and microalgae, for 2G bioethanol production to address challenges related to supply chain design and resilience (Jabbarzadeh and Shamsi 2025). A hybrid optimization approach, including data envelopment analysis (DEA), ANNs, and robust stochastic-probabilistic programming, is used to design and plan a resilient and sustainable bioethanol supply chain. This AI-driven approach is relevant to firms such as POET or DuPont and has demonstrated its ability as a decision-support tool for large-scale LCB biorefineries. Fermentation of lignocellulosic hydrolysates is typically inhibited by various compounds (e.g., acetate) and requires precise control to minimize inhibition and maximize ethanol yield. In a study on bioethanol production from corn cob hydrolysate, an ANNs algorithm is used to model the interaction of different ions (e.g., Na, Ca, Mg) on fermentation efficiency (Owusu and Marfo 2023).

In another study, an integrated biorefinery related to poly(3-hydroxybutyrate-co-3-hydroxyvalerate) production, along with bioethanol, was optimized using an ANNs algorithm. These AI models enable biorefineries to fine-tune fermentation parameters, thereby reducing the impact of inhibitors and enhancing yields in integrated processes. A popular industry (e.g., Novozymes) that supplies enzymes for bioethanol production can integrate these models to provide tailored solutions for clients, thereby enhancing process reliability and profitability (Cheng et al. 2023). Real-time monitoring of the bioethanol production process is crucial for maintaining stable and efficient operations, given the variability in biomass composition. Deep learning (DL) and near-infrared reflectance spectroscopy (NIRS)-based ML models (e.g., ANNs) integrated for real-time monitoring of lignocellulosic fermentation process to provide a better insight into metabolic states, enabling dynamic process adjustments (Rosales-Calderon and Arantes 2019). LCB-based bioethanol producers (e.g., Sekab, a Nordic company) can integrate AI-based technologies to enhance process control in their biorefineries, thereby improving consistency of LCB-based products and reducing environmental impact. A pilot bioethanol plant using rice straw as feedstock employed Aspen HYSYS with an embedded ANN algorithm for dynamic feedstock modeling. The simulated model can be effectively applied to SSF processes operating with different biomass feedstocks (Le et al. 2020). Therefore, AI technologies can transform the 2G bioethanol production process and the logistics supply chain to enhance the availability of feedstocks. In the real world, several global companies, such as Raízen and Sekab, have demonstrated the potential of artificial intelligence to improve bioethanol efficiency, reduce costs, and enhance sustainability (Rosales-Calderon and Arantes 2019).

A comparison of the economic impacts of AI-driven and traditional lignocellulosic bioethanol production methods is presented in Table 11.3.

11.6 Challenges and New Directions in AI Applications in 2G Bioethanol Production

The integration of AI into 2G bioethanol production encountered several challenges due to the inherent complexity of the process and limited data. The composition of lignocellulosic biomass is highly variable, leading to inconsistencies in datasets. Most studies rely on small datasets with limited descriptors, such as temperature, pH, and time, which limits the model's generalization ability. The prediction ability of multi-step processes, such as pretreatment and hydrolysis, is challenging due to noise in experimental/literature-led data sources. Similarly, diversity in production routes (e.g., thermochemical and fermentation) and configurations (e.g., SSF, SHF, etc.) fragmented research objectives and hindered the accumulation of large, homogeneous datasets.

The use of AI in 2G bioethanol production from lignocellulosic biomass is a quickly developing field with great potential for process improvement. However, it faces significant technical and economic challenges that need innovative solutions. Table 11.4 highlights key challenges and new directions for AI in 2G bioethanol production based on recent research.

Advanced AI modeling tools integrating ANNs, RF, and GAs are prone to overfitting when handling small datasets. Therefore, choosing an appropriate algorithm and adequate validation is challenging and often leads to unreliable results. In the current climate change scenario, yield prediction (for both feedstocks and products) using AI tools is challenging because datasets are unbalanced, high-dimensional, and noisy, and conditions vary. It is necessary to model the process using image-based ML approaches that require high-quality data and substantial

Table 11.3 A comparison of the economic impacts of AI-driven and traditional lignocellulosic bioethanol production methods

Economic parameter	Traditional methods	AI-driven processes	Economic impact/ improvement	Citations
Ethanol yield optimization	Baseline yield 0.41 g/g (typical)	Up to 0.48 g/g with ML optimization	18% yield increase; 63% improvement in specific fermentation conditions	Long and Liu (2023), Khamwachirapithak et al. (2023a, b), and Owusu and Marfo (2023)
Production cost structure	Total costs: Production 43%, Import 25%, Transport 17%, Infrastructure 15%, Inventory 0.43%	Optimized through an AI-driven supply chain and process control	Cost reduction through operational efficiency and reduced waste	Gumte and Mitra (2019)
Fermentation efficiency	Standard fermentation cycle duration	10% increment in industrial-scale production with MLP-PSO optimization ($R^2 = 0.91$)	Shorter fermentation cycles with equal or better conversion rates	Owusu and Marfo (2023) and Imubit (2025)
Energy consumption	High energy use in distillation (33% of natural gas budget)	Double-digit fuel reduction in distillation through AI process control	Significant reduction in steam and energy requirements; lower carbon intensity	Imubit (2025)
Process optimization accuracy	Manual or RSM-based optimization	ANN/ML models with $R^2 > 0.96$ for bioethanol prediction; RMSE < 0.05	Enhanced prediction accuracy reduces experimental iterations and costs	Owusu and Marfo (2023) and Wang et al. (2023)
NaOH/Chemical consumption	Standard chemical usage rates	ML prediction accuracy > 0.8 for NaOH optimization	Reduced chemical costs through precise dosage optimization	Long and Liu (2023)
Feedstock-to-ethanol efficiency	Current technology: 35% efficiency (HHV); Overall with cogeneration: 60%	AI-optimized conditions improve to 48% ethanol efficiency; 68% overall	13% improvement in ethanol efficiency; 8% improvement in overall process	Klein-Marcuschamer et al. (2011)
Capital investment requirements	€2.1k/kW (at 400 MW HHV input)	Future AI-optimized plants: €900/ kW (at 2 GW HHV)	57% reduction in specific capital investment through scale and optimization	Klein-Marcuschamer et al. (2011)

(continued)

Table 11.3 (continued)

Economic parameter	Traditional methods	AI-driven processes	Economic impact/ improvement	Citations
Experimental & Development costs	216 combinatorial strains tested for optimization	ML-guided workflow reduced experimental load to 31 strains (86% reduction)	Significant cost savings in the R&D phase through reduced experimental requirements	Khamwachirapithak et al. (2023a, b)
Supply chain optimization	Conventional supply chain management	Integrated ML and optimization for sustainable bioethanol supply chain	Enhanced feedstock logistics and inventory management efficiency	Momenitabar et al. (2023)

training resources. AI tools without robust datasets often struggle to mitigate environmental impacts during acid hydrolysis, particularly regarding acid recovery and the risk of sugar degradation. Overall, AI applications for biorefinery systems with limited datasets are challenging due to their nonlinear and dynamic nature, as well as the challenges of scaling from the lab to industry.

In order to overcome these challenges, extensive research work is required to develop innovative strategies that incorporate advanced deep and machine learning approaches in 2G bioethanol production. Advanced computational tools, including density functional theory (DFT), can generate synthetic data related to complex LCB hydrolysis and fermentation, supplementing data limitations during AI-based model training and testing. A digital twin is a virtual representation of a physical system that can interact with the real world through data exchange, thereby enabling the redefinition of complex biomass conversion processes (Selwal et al. 2025). Digital mapping is a modeling approach that includes Digital Twins (DT) for understanding complex biorefinery processes. Integrated machine learning and DT technologies lead to a paradigm shift in LCB conversion, predictive analytics, and operational flexibility through real-time optimization and scenario building (Selwal et al. 2025). A developed DT model can successfully optimize biorefinery processing using continuous process data to improve efficiency and minimize adverse environmental impacts (Gamero et al. 2024).

Advanced machine learning-based models that have been successfully implemented in 1 G bioethanol production can be extended to 2G processes, which incorporate additional features for handling inhibitor challenges in lignocellulosic feedstocks. Deep learning, combined with explainable AI approaches, can be leveraged to enhance the reliability and understanding of cause-and-effect relationships by integrating mathematical models for improved feedstock yield prediction under climatic variability. Incorporating future innovative technologies, such as

Table 11.4 Key challenges and new directions in AI applications in 2G bioethanol production

S. N.	Challenge	Description	New direction/solution	Citation
1.	Data scarcity and quality	Limited experimental datasets and sparse, corrupted data hinder ML model predictions	Synthetic data generation with VAE and GAN enhanced datasets, improving soft sensor predictions by 34% and reducing variability by 82%	Kwon et al. (2024)
2.	Biomass feedstock complexity	High variability in lignocellulosic composition complicates the assessment of efficiency due to limited data. Feedstock choice influences process efficiency and sustainability	Using AI with RF and ANN for rapid biomass characterization and bioethanol yield prediction by integrating genomic data	Tušek et al. (2024) and Vani et al. (2015)
3.	Pretreatment optimization challenges	Complex interactions among acid concentration, temperature, and reaction time; formation of fermentation inhibitors; high energy demands and cost issues	Hybrid AI models combining ENN with metaheuristic optimization (GA, PSO) improve sugar yield prediction and the accuracy of biomass-to-bioethanol conversion	Ahmad et al. (2023), Makhatov et al. (2019), and Jain and Kumar (2024)
4.	Process modeling complexity	The nonlinear bioethanol production stages (pretreatment, hydrolysis, fermentation, recovery) are complex and challenging to manage due to poorly understood kinetics and inhibition mechanisms	Artificial neural networks (ANNs) offer high tolerance to errors and flexibility for nonlinear processes. Random Forest algorithms attain $R^2 > 0.90$ for SHF modeling with an average deviation of $<8.1\%$	Vani et al. (2015)
5.	Real-time monitoring limitations	Difficulty predicting real-time process quality, like ethanol concentration; delayed response to deviations; inadequate sensors for continuous monitoring	Deep learning soft sensors predict fermentation quality in real time; IoT sensors and digital twin enable continuous data collection and analysis; feedforward neural networks (FNN) monitor processes dynamically	Kwon et al. (2024) and Le et al. (2024)
6.	Model interpretability and explainability	Deep learning's black-box nature limits transparency in decision-making, parameter importance, and guidance for optimization	Explainable AI methods like SHAP analysis, Pearson heatmaps, partial dependence plots, and feature importance to identify key process parameters	Ramalingam et al. (2025)

(continued)

Table 11.4 (continued)

S. N.	Challenge	Description	New direction/solution	Citation
7.	High production costs	2G bioethanol has higher production costs than first-gen biofuels, raising economic concerns; infrastructure deficits hinder biomass handling	AI-driven optimization using machine learning and life cycle assessment increases yield by 18%, reduces emissions by 14%, and utilizes predictive models for cost-efficient processes	Long et al. (2023) and Kwon et al. (2024)
8.	Enzyme loading optimization	Vast amounts of enzymes are needed for hydrolysis; optimizing enzyme levels, temperature, and substrate loading is challenging; high enzyme costs affect economic viability	Machine learning optimization with Random Forest and ANN predicted optimal enzyme loading and conditions: 35 °C, 36 h SHF time, 99.8% enzymatic load, yielding 12.1 g/L ethanol	Vani et al. (2015)
9.	Fermentation process control	Complex interactions among pH, temperature, inoculum size, and fermentation time hinder ethanol production; inhibitor buildup affects microbes	Hybrid AI models combining MLP with PSO; R^2 of 0.91 for accurate concentration predictions; neural networks for control systems	Owusu and Marfo (2023)
10.	Scale-up and industrial implementation	Laboratory AI models often struggle to scale for industrial use, facing challenges in sustainable polymer extraction from biomass and integrating with existing infrastructure	Transfer learning improves generalization across scales and feedstocks, creates adjustable AI models for different engines, regions, and fuels, and employs ensemble learning for scalability	Chaudhary and Kumar (2025) and Awogbemi et al. (2023)
11.	Data integration and standardization	Inconsistent data formats, data silos, and missing values lead to inaccurate predictions	Cloud and edge computing for data integration; Big Data Analytics platforms for diverse data sources; CPS for seamless data flow between physical processes and digital models	Ma et al. (2022) and Le et al. (2024)
12.	Multi-objective optimization	Need to simultaneously optimize ethanol yield, minimize greenhouse gas emissions, reduce NaOH consumption, and maintain economic feasibility	Advanced metaheuristic algorithms such as GA, PSO, and Simulated Annealing optimize feedstock traits and process conditions, achieving an 18% yield improvement	Long et al. (2023)

near-infrared (NIR) sensing or UAV imagery, into ML frameworks can provide better insight into the real-time optimization of pretreatment, fermentation, and distillation processes. The ongoing research on 2G bioethanol production emphasizes the use of advanced AI approaches for holistic process modeling, biomass feedstock analysis, and the evaluation of production routes, scenarios, and expenditures.

The convergence of AI with Industry 4.0 technologies marks a transformative shift for 2G bioethanol production (BioBiz 2024). Digital twin technology, combined with IoT sensors, allows real-time monitoring, simulation, and optimization of the entire production value chain from feedstock cultivation through fermentation and recovery (Le et al. 2024). Smart sensors installed at key production stages enable continuous data collection, supporting adaptive process control, and predictive maintenance (Le et al. 2024). Integrating machine learning with life cycle assessment (LCA) tools provides a comprehensive evaluation of environmental impacts (Long and Liu 2023). Random Forest, Extreme Gradient Boosting (XGBoost), and artificial neural networks achieve prediction accuracy over 0.8 for ethanol yield and chemical consumption, promoting sustainable process design with reduced greenhouse gas emissions (Long and Liu 2023).

Beyond traditional neural networks, advanced boosted machine learning algorithms such as CatBoost, XGBoost, AdaBoost, and Gradient Boosting Machines (GBM) show superior predictive accuracy ($R^2 = 0.955$) in optimizing biodiesel and bioethanol production (Ramalingam et al. 2025). These ensemble methods dynamically adjust model weights based on prediction errors, ensuring continuous improvement and better generalization (Ramalingam et al. 2025). AI tools can enable biorefinery systems to become more autonomous, leading to the development of a "bio-intelligence manufacturing system" (Wegener et al. 2024). A self-learning expert system integrated with advanced sensor technology will be used for process modeling, optimization, and scenario analysis—considering seasonal feedstock availability—of a lignocellulosic biomass-based biorefinery. Additionally, incorporating satellite-derived spatial data into AI models helps stabilize the feedstock supply chain and minimize environmental impacts (Velidandi et al. 2023). These innovations can improve the efficiency and viability of second-generation bioethanol production, supporting sustainable energy solutions. However, realizing the full potential of these technologies for the circular economy requires multidisciplinary collaboration and a supportive policy environment (Selwal et al. 2025; Yadav et al. 2023).

11.7 Conclusion

Advanced AI tools and techniques are continually addressing the challenges of bioethanol production from lignocellulosic biomass. The ongoing integration of AI tools with advanced industrial automation systems, digital twins, supply-chain modeling, and life-cycle assessment is considered a vital force in achieving commercial viability and sustainability. Continuous research toward hybrid and explainable AI, along with collaboration among academic institutions, industry leaders, and policymakers, will play a significant role in scaling up and deploying sustainable models of 2G bioethanol production. Thus, lignocellulosic bioethanol production could achieve higher yields, be less expensive, and be closer to commercial success by leveraging cutting-edge AI models and optimization techniques.

Acknowledgment The facilities and infrastructural support provided by the University of Technology and Applied Sciences-Sur, Oman is acknowledged. One of the authors also acknowledged the facilities and support provided by the Director, ICAR-IARI, and the AICRP, IARI-Center, New Delhi, India.

References

Adeleke O, Bamisaye A, Ige AR, Adegoke KA, Jen T-C (2026) Enhancing the energy yield of Amaranthus hybridus-derived biofuel using alkaline pretreatment: experimental and data-driven investigation. Fuel 405:136491. https://doi.org/10.1016/j.fuel.2025.136491

Aghbashlo M, Peng W, Tabatabaei M, Kalogirou SA, Soltanian S, Hosseinzadeh Bandbafha H, Mahian O, Lam SS (2021) Machine learning technology in biodiesel research: a review. Prog Energy Combust Sci 85:100904. https://doi.org/10.1016/j.pecs.2021.100904

Ahmad N, Aslam S, Hussain N, Bilal M, Iqbal HM (2023) Transforming lignin biomass to value: interplay between ligninolytic enzymes and lignocellulose depolymerization. BioEnergy Res 16(3):1246–1263. https://doi.org/10.1007/s12155-023-10507-6

Arias A, Feijoo G, Moreira MT (2023) How could Artificial Intelligence be used to increase the potential of biorefineries in the near future? A review. Environ Technol Innov 32:103277. https://doi.org/10.1016/j.eti.2023.103277

Awogbemi O, Olayemi AD, Adeyemo A, Oladimeji D (2023) Application of machine learning technologies in biodiesel production: a comprehensive review. Front Energy Res. Advance online publication. https://doi.org/10.3389/fenrg.2023.1122638

Azad S, Madadi M, Rahman A, Sun C, Sun F (2025) Machine learning-driven optimization of pretreatment and enzymatic hydrolysis of sugarcane bagasse: analytical insights for industrial scale-up. Fuel 390:134682. https://doi.org/10.1016/j.fuel.2025.134682

Bahafi A, Muddapur UM, Hombalimath VS, Shaikh IA, Manjunath S, Patil LR, Iqubal SS (2024) Optimizing fermentation parameters for bioethanol production from Areca nut leaves using artificial neural networks and response surface methodologies. Asian J Pharm 18(4):1230

Bosu S, Pooja RP, Rajasimman M (2022) Role of nanomaterials in enhanced ethanol production through biological methods—review on operating factors and machine learning applications. Fuel 320:123905. https://doi.org/10.1016/j.fuel.2022.123905

Broda M, Yelle DJ, Serwańska K (2022) Bioethanol production from lignocellulosic biomass: challenges and solutions. Molecules 27(24):8717. https://doi.org/10.3390/molecules27248717

Chaudhary V, Kumar D (2025) Artificial intelligence in biofuel production. In: Optimizing biofuel production with artificial intelligence, pp 33–51. https://doi.org/10.1007/978-3-031-23560-3_3

Cheng Y, Bi X, Xu Y, Liu Y, Li J, Du G, Lv X, Liu L (2023) Artificial intelligence technologies in bioprocess: opportunities and challenges. Bioresour Technol 369:128451. https://doi.org/10.1016/j.biortech.2022.128451

Coşgun A, Günay ME, Ramazan Y (2023) A critical review of machine learning for lignocellulosic ethanol production via fermentation route. Biofuel Res J 10(2):1859–1875. https://doi.org/10.18331/BRJ2023.10.2.5

Cousins DS, Otto WG, Rony AH, Pedersen KP, Aston JE, Hodge DB (2022) Near-infrared spectroscopy can predict anatomical abundance in corn stover. Front Energy Res 10:836690. https://doi.org/10.3389/fenrg.2022.836690

David AJ, Kannam AK, Sam S, Narasimhan MK, Krishnamurthi T (2025) Optimization of delignification and organic acid synthesis from *Arachis hypogaea* L. shell: a comparison of machine learning approaches and experimental techniques. Biomass Convers Biorefinary 15: 20515–20526. https://doi.org/10.1007/s13399-025-06586-6

Deb A (2025) India's biofuel drive is saving billions but also sparking worries. BBC News, Delhi. https://www.bbc.com/news/articles/cvgv873y02eo. Accessed 9 Sept 2025

Ethanol Market Research (2025) Bioethanol market size, share, growth and global industry analysis by type & application, regional insights and forecast to 2024–2032. Fortune Business Insights Pvt. Ltd. https://www.marketresearch.com/Fortune-Business-Insights-Pvt-Ltd-v4286/Bioethanol-Size-Share-Growth-Global-39210960/#description. Accessed 9 Sept 2025

Fischer J, Lopes VS, Cardoso SL, Filho UC, Cardoso VL (2017a) Machine learning techniques applied to lignocellulosic ethanol in simultaneous hydrolysis and fermentation. Braz J Chem Eng 34(1):53–63. https://doi.org/10.1590/0104-6632.20170341s20150475

Fischer J, Santos Moreira LR, Cardoso TF, Costa PS, Guillermo Rojas Pino L, Santos Bernardes O (2017b) Machine learning techniques applied to lignocellulosic ethanol production. Braz J Chem Eng 34(1):31–43. https://www.scielo.br/j/bjce/machine-learning-lignocellulosic

Gamero E, Shoshi A, Full J, Sauer A, Miehe R (2024) Data management in biorefineries: conceptual thoughts on lean digital twinning. Proc CIRP 125:48–53. https://doi.org/10.1016/j.procir.2024.08.009

Göncü B, Sarıışık G, Hoşgün EZ (2025) Valorization of lignocellulosic feedstock in bioethanol production via separate hydrolysis and fermentation: an integrated machine learning approach. Int J Hydrog Energy 142:1039–1053. https://doi.org/10.1016/j.ijhydene.2025.04.340

Gumte K, Mitra K (2019) Bio-Supply Chain Network Design to tackle ethanol deficiency in India: a mathematical framework. J Clean Prod 234:208–224. https://doi.org/10.1016/j.jclepro.2019.06.160

Hernandez-Diaz JA, Musah M, Bello F, Via B, Peresin MS (2025) Loblolly pine downed timber: Box-Behnken design and first derivative pretreatment for predictive modeling with near-infrared spectroscopy and machine learning (NIRS-ML). Holzforschung 79(2–3):102–115. https://doi.org/10.1515/hf-2024-0014

Huang X-Y, Zhang X, He Y, Yao J-W, Xing L, Bai F-W, Dong J-J, Liu C-G (2024) Leveraging machine learning for acid catalyzed steam explosion pretreatment: towards supporting fermentation by the trade-off between glucose and inhibitors. J Clean Prod 448:141530. https://doi.org/10.1016/j.jclepro.2024.141530

Imubit (2025) Top ways to increase ROI in ethanol production with AI. Industrial AI Platform. Retrieved from https://imubit.com/article/increase-roi-ethanol-production/. Accessed 30 Dec 2025

Jabbarzadeh A, Shamsi M (2025) Designing a resilient and sustainable multi-feedstock bioethanol supply chain: integration of mathematical modeling and machine learning. Appl Energy 377: 123794. https://doi.org/10.1016/j.apenergy.2024.123794

Jain S, Kumar S (2024) Advances and challenges in pretreatment technologies for bioethanol production: a comprehensive review. Sustain Chem Clim Action 5:100053. https://doi.org/10.1016/j.sccaj.2024.100053

Jurinjak Tušek A, Petrus A, Weichselbraun A, Mundani R-P, Müller S, Barkow I, Bucić-Kojić A, Planinić M, Tišma M (2024) Systematic review of machine-learning techniques to support development of lignocellulose biorefineries. Chem Biochem Eng Q 38(3):241–263. https://doi.org/10.15255/CABEQ.2023.1930

Khamwachirapithak P, Chantharakasem S, Tiwari S, Charoen R (2023a) Optimizing ethanol production in Saccharomyces cerevisiae using a machine learning approach for promoter fine-tuning. ACS Synth Biol 12(10):2897–2908. https://doi.org/10.1016/j.apbior.2023.100045

Khamwachirapithak P, Sae-Tang K, Mhuantong W, Tanapongpipat S, Zhao XQ, Liu CG, Wei D-Q, Champreda V, Runguphan W (2023b) Optimizing ethanol production in *Saccharomyces cerevisiae* at ambient and elevated temperatures through machine learning-guided combinatorial promoter modifications. ACS Synth Biol 12(10):2897–2908. https://doi.org/10.1021/acssynbio.3c00014

Klein-Marcuschamer D, Simmons BA, Blanch HW (2011) Techno-economic analysis of a lignocellulosic ethanol biorefinery with ionic liquid pretreatment. Biofuels Bioprod Biorefin 5(5): 562–569. https://doi.org/10.1002/bbb.303

Kwon HJ, Shiu JH, Yamakawa CK, Rivera EC (2024) Enhancing fermentation process monitoring through data-driven modeling and synthetic time series generation. Bioengineering 11(8):803. https://doi.org/10.3390/bioengineering11080803

Le PK, Le TDT, Nguyen QD, Tran VT, Mai PT (2020) Process simulation of the pilot-scale bioethanol production from rice straw by Aspen Hysys. IOP Conf Ser Mater Sci Eng 778: 012095. https://doi.org/10.1088/1757-899X/778/1/012095

Le CA, Le CH, Nguyen VD, Zlatov N, Le TH, Nguyen TA, Chu AM (2024) Digital twins for real-time monitoring and operation of the coffee value chain and supply chain. Int J Mechatron Appl Mech 17:114–124. https://doi.org/10.2991/jmam.k.230912.001

Lee Y, Ko J, Kwon O, Park H, Lee H, Jeong S, Ha B, Hwangbo S, Han J (2025) Biological-chemical conversion process design and machine learning-related life cycle assessment: bio-lubricant production in a real case study of South Korea. J Environ Manag 379:124877. https://doi.org/10.1016/j.jenvman.2025.124877

Long F, Liu H (2023) An integration of machine learning models and life cycle assessment for lignocellulosic bioethanol platforms. Energy Convers Manag 292:117379. https://doi.org/10.1016/j.enconman.2023.117379

Long F, Wang L, Cai W, Lesnik KL, Liu H (2023) An integration of machine learning models and life cycle assessment for sustainable bioethanol production. Bioresour Technol 381:129137. https://doi.org/10.1016/j.biortech.2023.129137

Ma S, Ding W, Liu Y, Ren S, Yang H (2022) Digital twin and significant data-driven sustainable smart manufacturing based on information management systems for energy-intensive industries. Appl Energy 326:119986. https://doi.org/10.1016/j.jclepro.2021.128262

Madadi M, Kargaran E, Al Azad S, Saleknezhad M, Zhang E, Sun F (2025) Machine learning-driven optimization of biphasic pretreatment conditions for enhanced lignocellulosic biomass fractionation. Energy 326:136241. https://doi.org/10.1016/j.energy.2025.136241

Makhatov Z, Kedelbayev B, Lieberzeit P, Dzhakashyeva M, Elemanova Z, Abildayeva R, Akhmedova Z (2019) Biosynthesis of cellulase with Trichoderma viride and Aspergillus awamori micromycetes in co-cultivation. EurAsian J BioSciences 13(2)

Market Research Report (2024) Global biofuels market by type (Biodiesel, Ethanol, Bioethanol, Propanol, Butanol, Methanol, Biogas), by form (Solid, Liquid, and Gaseous), by feedstock type (Palm Oil, Jatropha, Sugar Crop, Coarse Grain, and Others), by region and companies—industry segment outlook, market assessment, competition scenario, trends, and forecast 2024–2033. https://market.us/report/biofuels-market/#overview. Accessed 9 Sept 2025

Meena M, Shubham S, Paritosh K, Pareek N, Vivekanand V (2021) Production of biofuels from biomass: predicting the energy employing artificial intelligence modelling. Bioresour Technol 340:125642. https://doi.org/10.1016/j.biortech.2021.125642

Momenitabar M, Ebrahimi ZD, Abdollahi A, Helmi W, Bengtson K, Ghasemi P (2023) An integrated machine learning and quantitative optimization method for designing sustainable bioethanol supply chain networks. Decision Analyt J 7:100236. https://doi.org/10.1016/j.dajour.2023.100236

Niu Y, Tye YY, Hemashin T, Leh CP (2025) Feasibility of use limited data to establish a relationship between chemical composition and the enzymatic glucose yield using machine learning. Biomass Bioenergy 200:107956. https://doi.org/10.1016/j.biombioe.2025.107956

Owusu WA, Marfo SA (2023) Artificial intelligence application in bioethanol production. Int J Energy Res 2023(1):7844835. https://doi.org/10.1002/er.7844835

Pancholi MJ, Khristi AM, Athira KM, Bagchi D (2023) Comparative analysis of lignocellulose agricultural waste and pretreatment conditions with FTIR and machine learning modeling. Bioenergy Res 16:123–137. https://doi.org/10.1007/s12155-022-10444-y

Pereira RD, Badino AC, Cruz AJG (2020) Framework based on artificial intelligence to increase industrial bioethanol production. Energy Fuel 34(4):4670–4677. https://doi.org/10.1021/acs.energyfuels.0c00033

Pradhan D, Jaiswal S, Jaiswal AK (2022) Artificial neural networks in valorization process modeling of lignocellulosic biomass. Biofuels Bioprod Biorefin 16:1849–1868. https://doi.org/10.1002/bbb.2417

Pushpa SR, Awoyale AA, Lokhat D, Sukumaran RK, Savithri S (2024) Infrared-based machine learning models for the rapid quantification of lignocellulosic multi-feedstock composition. Bioresour Technol Rep 25:101747. https://doi.org/10.1016/j.biteb.2023.101747

Ramalingam K, Abdullah MZ, Elumalai PV, Yong X, Prasad KS, Chan CK, Prabhakar S, Yuqi B (2025) An evaluation of maximizing production and usage of biofuel by machine learning and experimental approach. Sci Rep 15(1):33265. https://doi.org/10.1038/s41598-025-05689-y

Rosales-Calderon O, Arantes V (2019) A review on commercial-scale high-value products that can be produced alongside cellulosic ethanol. Biotechnol Biofuels 12:240. https://doi.org/10.1186/s13068-019-1529-1

Sebayang AH, Masjuki HH, Ong HC, Dharma S, Silitonga AS, Mahlia TMI, Aditiya HB (2017) Optimization of bioethanol production from sorghum grains using artificial neural networks integrated with ant colony optimization. Ind Crop Prod 97:146–155. https://doi.org/10.1016/j.indcrop.2016.11.064

Selwal N, Sultana H, Rahayu F, Hariyono B, Riajaya PD, Kadarwati FT, Herwati A, Latifah E, Indriani FC, Saeri M, Dar MA, Chopra C, Abass KS, Wani AK (2025) Emerging technologies in biomass conversion: bioengineering and nanocatalysts to AI-driven process optimization. Biomass Bioenergy 200:108054. https://doi.org/10.1016/j.biombioe.2025.108054

Singh N, Singhania RR, Nigam PS, Dong C-D, Patel AK, Puri M (2022) Global status of lignocellulosic biorefinery: challenges and perspectives. Bioresour Technol 344:126415. https://doi.org/10.1016/j.biortech.2021.126415

Vani S, Sukumaran R, Savithri S (2015) Prediction of sugar yields during hydrolysis of lignocellulosic biomass using artificial neural network modelling. Bioresour Technol 188:128–135. https://doi.org/10.1016/j.biortech.2015.01.015

Velidandi A, Gandam PK, Chinta ML, Konakanchi S, Bhavanam AR, Baadhe RR, Sharma M, Gaffey J, Nguyen QD, Gupta VK (2023) State-of-the-art and future directions of machine learning for biomass characterization and for sustainable biorefinery. J Energy Chem 81:42–63. https://doi.org/10.1016/j.jechem.2023.02.020

Vinitha N, Vasudevan J, Gopinath KP, Arun J, Madhu S, Naveen S (2025) Enhancing the dilute-acid hydrolysis process using a machine learning approach: investigating the influence of different biomass feedstocks on glucose and ethanol yields. Biomass Convers Biorefinery 15:9159–9171. https://doi.org/10.1007/s13399-024-05714-y

Wang C, Zhang X, Zhao G, Chen Y (2023) Mechanisms, methods and applications of machine learning in bio-alcohol production and utilization: a review. Chemosphere 342:140191. https://doi.org/10.1016/j.chemosphere.2023.140191

Wegener K, Spierings A, Weiss L, Knüttel D (2024) Bio-intelligent machine tool: vision and steps towards realisation. Proc CIRP 125:16–23. https://doi.org/10.1016/j.procir.2024.08.004

Xie T, Fan M (2025) Machine learning models for predicting enzymatic hydrolysis yields of lignocellulosic biomass after various pretreatments. Ind Crop Prod 235:121644. https://doi.org/10.1016/j.indcrop.2025.121644

Yadav A, Sharma V, Tsai M-L, Chen C-W, Sun P-P, Nargotra P, Wang J-X, Dong C-D (2023) Development of lignocellulosic biorefineries for the sustainable production of biofuels: towards circular bioeconomy. Bioresour Technol 381:129145. https://doi.org/10.1016/j.biortech.2023.129145

Yang R, Zhao N, Li J, Liu M, Li X (2025) Boosting corn stover bioconversion to fatty acids with Gaussian Process optimization. Ind Crop Prod 227:120767. https://doi.org/10.1016/j.indcrop.2025.120767

Yeboah MK, Asiedu NY, Dogbe S, Addo A (2024) Performance of machine learning based-modelling approach in consolidated bioprocessing with microbial consortium for bioethanol production. Ind Biotechnol 20(2):77–97. https://doi.org/10.1089/ind.2023.0009

Zhao Q, Zhu M-J, Zhao L (2025) In-situ conversion of hemicellulose to furfural by Lewis acid-enhanced deep eutectic solvents to maintain stable pretreatment performance and trigger profitable biorefining processes. Int J Biol Macromol 311(2):143731. https://doi.org/10.1016/j.ijbiomac.2025.143731

Chapter 12
Industry 4.0 Based on Real-Time Process Monitoring and Control for Sustainable Bioethanol Production

Kush Kumar Nayak, Reecha Sahu, Mohd. Zafar, and Ravi Kant Singh

Abstract During sustainable bioethanol production, Industry 4.0 (I4.0) technologies can address process complexity, raw material variability, and energy efficiency challenges, enhancing yield and sustainability. Despite these interests, the adaptation of I4.0 technologies in the bioethanol manufacturing industry is scarce and often limited to first- and second-generation bioethanol production. This chapter incorporates the application of I4.0 principles in process monitoring and control for bioethanol production processes, with relevant literature information. Real-time monitoring is critical for the optimization of bioethanol production, especially during pretreatment, enzymatic hydrolysis, fermentation, and distillation. The IoT-enabled sensors are helpful in real-time data acquisition of key process parameters, viz. temperature, pH, sugar concentration, and ethanol yield. Besides, sensors integrated with IoT systems enable predictive maintenance and anomaly detection during bioethanol production. The various AI and ML-based data analytics models, such as artificial neural networks (ANNs) and long short-term memory (LSTM) networks, can analyze sensor data and develop predictive models. These models can efficiently address process accuracy and computational efficiency in bioprocess control, including nonlinearities in bioethanol production. Advanced process control systems based on model predictive control (MPC) and fuzzy logic systems on the MATLAB-Simulink platform can efficiently optimize bioethanol production by

K. K. Nayak
School of Studies in Biotechnology, Shaheed Mahendra Karma Vishwavidyalaya, Bastar, Jagdalpur, Chhattisgarh, India

R. Sahu
Biomedical Engineering and Bioinformatics, Chhattisgarh Swami Vivekanand Technical University, Durg, Chhattisgarh, India

M. Zafar
Department of Biotechnology, College of Applied Sciences and Pharmacy, University of Technology and Applied Sciences, Sur, Oman

R. K. Singh (✉)
Amity Institute of Biotechnology, Amity University Uttar Pradesh, Noida, UP, India
e-mail: rksingh1@amity.edu

S. Prasad et al. (eds.), *Sustainable Bioethanol Production from Lignocellulosic Biomass*, Clean Energy Production Technologies,
https://doi.org/10.1007/978-981-95-9690-4_12

adjusting process parameters in real-time. ML-based MPC solutions can offer faster computation than traditional ODE/PDE-based models, enabling real-time control in complex bioethanol production processes. Cyber-physical systems (CPS) integrate physical processes with digital control systems, creating a connected bioethanol production ecosystem for condition monitoring in smart factories. Digital twins can simulate bioethanol production processes, enabling virtual testing of control strategies. The application of cloud systems for processing large datasets from bioethanol plants enables efficient energy analysis and process optimization. Therefore, I4.0 technologies can efficiently enhance the sustainability of an advanced bioethanol production process (e.g., combining enzymatic hydrolysis with simultaneous saccharification and fermentation (SSF)) by reducing energy consumption and waste management.

Keywords Industry 4.0 · Internet of Things (IoT) · Artificial intelligence (AI) · Machine learning (ML) · Cyber-physical systems (CPS) · Digital twins (DTs) · Bioethanol manufacturing

12.1 Introduction

In the current sustainable development scenario, the bioprocess industries are transforming for quality assurance and shifting from traditional "Quality by Testing (QbT)" to "Quality by Design (QbD)". QbD is a systematic approach toward sustainable bioprocess development that integrates Artificial Intelligence (AI)-based advanced process design and control guided by a regulatory framework. Industry 4.0 represents the contemporary automation and data exchange trend in manufacturing technologies. It embodies the integration of cyber-physical systems (CPS), the Internet of Things (IoT), cloud computing, and cognitive computing. This digital transformation facilitates the creation of "smart factories" where machinery and equipment are capable of improving processes through self-optimization and self-configuration, and even autonomous decision-making (Muhammad and Abdullah 2023).

The integration of machine learning, Internet of Things, and digital twins (DTs) enables real-time bioprocess monitoring and data analysis, modeling, and process optimization under QbD implementation. DTs are known for bidirectional data communication and enable real-time adjustments in bioprocess systems; however, they possess challenges in system integration and data security. These challenges can be handled by advancements in AI, virtual reality/augmented reality, and communication technologies. Incorporation of human expertise, I4.0 emerged as I5.0, which can emphasize human intervention in bioprocess systems through the integration of machine performance with human critical thinking. Notably, I4.0 focuses on cyber-physical systems; however, I5.0 emphasizes the integration of collaborative robots (cobots).

In the context of bioethanol production, Industry 4.0 technologies enable real-time process monitoring and adaptive control strategies that are crucial for optimizing the complex biochemical and thermochemical processes involved. The implementation of sensors, automated feedback loops, and machine learning algorithms allows for precise control of parameters such as temperature, pH, substrate concentration, and fermentation time. These advancements result in higher yields, reduced operational costs, enhanced safety, and improved environmental performance (Kagermann et al. 2013). Bioethanol production has evolved through four distinct generations, each improving upon the sustainability and efficiency of the previous generation. The first generation (1G) utilizes food crops like sugarcane, corn, and wheat for ethanol production via fermentation. While well established and commercially viable, it raises concerns about food security and land use. The second generation (2G) employs lignocellulosic biomass such as agricultural residues, forest waste, and non-food energy crops. This generation mitigates the food-versus-fuel issue but involves complex pretreatment and enzymatic hydrolysis processes to release fermentable sugars. The third generation (3G) explores the use of algal biomass due to its high growth rate, minimal land requirement, and ability to utilize waste streams. Algal bioethanol remains in the experimental stage, facing challenges related to cultivation and harvesting techniques. However, the fourth generation (4G) involves genetically engineered feedstock and integrates carbon capture and storage (CCS) technologies. It aims to produce carbon-neutral or even carbon-negative bioethanol, offering the highest sustainability potential (Gupta and Verma 2015; Kumar et al. 2020). Industry 4.0 technologies are especially relevant for managing the complexities of 2G to 4G bioethanol production processes. Real-time monitoring tools can address the variability in feedstock composition and process conditions, ensuring consistent quality and output.

Cleaner production (CP) refers to continuously applying integrated preventive environmental strategies to processes, products, and services. In bioethanol production, CP strategies include optimizing energy and water use, minimizing waste and emissions, and substituting hazardous materials with safer alternatives. Through Industry 4.0, cleaner production can be significantly enhanced. For example, predictive analytics can anticipate equipment failures, reducing downtime and waste. IoT-enabled devices can track resource usage in real-time, identifying inefficiencies, and enabling swift corrective actions. This approach not only reduces the environmental impact but also improves economic performance (UNEP 2011; Ma et al. 2018). Cleaner production (CP) is a proactive, integrated approach to preventing pollution and minimizing waste generation at the source. It emphasizes process and product innovation to reduce environmental impacts while maintaining economic performance. In bioethanol production, CP involves: (i) using renewable and low-impact raw materials, (ii) enhancing energy efficiency through process integration, (iii) recycling water and chemical reagents, (iv) reducing by-product formation, and (v) employing non-toxic catalysts and reagents. Industry 4.0 enhances CP by leveraging real-time monitoring to identify inefficiencies and environmental hotspots. For example, smart meters and flow sensors track water and energy usage, while digital twins simulate scenarios to test CP interventions. Advanced analytics can predict emissions trends, enabling proactive interventions to maintain environmental compliance (UNEP 2011; Ma et al. 2018).

Resource recovery is essential for the economic and environmental sustainability of bioethanol production. It involves extracting and reusing valuable by-products such as lignin, glycerol, and biogas. Advanced separation and purification technologies, when integrated with Industry 4.0, can enhance the efficiency of recovery processes. For instance, real-time data analytics can be used to monitor the quality and quantity of side streams, guiding optimal recovery strategies. Lignin can be repurposed for energy generation or as a raw material for value-added products. At the same time, CO_2 from fermentation can be captured and utilized in other industries, such as beverage carbonation or algae cultivation (Chen et al. 2021). Resource recovery transforms by-products of bioethanol production into valuable co-products, contributing to economic viability and environmental sustainability. Industry 4.0 technologies like process sensors, automated sorting systems, and AI-driven optimization models enable precise control over resource recovery operations. Real-time data analysis ensures consistency in by-product quality and yield, supporting circular business models and maximizing the utility of every input (Chen et al. 2021).

The concept of a circular economy aims to redefine growth by decoupling economic activity from the consumption of finite resources. It emphasizes reducing waste and pollution, keeping products and materials in use, and regenerating natural systems. In the bioethanol industry, adopting a circular economy model means closing the loop of resource use. This includes recycling process water, converting waste into energy, and integrating bioethanol plants into biorefineries. Industry 4.0 facilitates this transformation by enabling traceability, lifecycle analysis, and real-time optimization of circular strategies. Digital twins, for instance, can simulate various circular economy scenarios, helping stakeholders make informed decisions about waste valorization, energy integration, and product design. By aligning bioethanol production with circular economy principles, we can significantly enhance its sustainability profile (Geissdoerfer et al. 2017). The circular economy framework reshapes bioethanol production to minimize waste, extend resource lifecycles, and regenerate ecosystems. This model includes: (i) designing for disassembly and recyclability, (ii) recycling process water and chemicals, (iii) valorizing all waste streams, and (iv) co-locating with other industries for symbiotic resource exchange. For instance, spent grains from fermentation can feed livestock or serve as substrate for composting, while waste heat can be recovered for other processes. Industry 4.0 enables this shift by deploying tools for lifecycle analysis, traceability systems for waste and emissions tracking, and predictive models for material flow optimization. Simulation platforms and AI can model various reuse and recycling scenarios, supporting evidence-based decision-making for circular integration (Geissdoerfer et al. 2017).

12.2 Industry 4.0 Technologies in Bioethanol Production

The deployment of Industry 4.0 technologies in bioethanol production begins with the digital modelling and design of core processes such as pretreatment, hydrolysis, fermentation, and distillation. Advanced simulation tools and digital twins allow engineers to virtually test and optimize system configurations under varying operational conditions before physical implementation. Process integration is further enhanced by embedding smart sensors and actuators throughout the production line, enabling continuous data acquisition and automated control. Artificial intelligence and machine learning models analyze operational data in real-time to forecast performance, optimize process parameters, and detect anomalies, thus reducing variability and maximizing efficiency. Cloud platforms and edge computing facilitate distributed data processing, supporting robust decision-making across networked biorefineries.

Core processes in bioethanol production, like pretreatment, hydrolysis, fermentation, and distillation, are interconnected and sensitive to numerous variables. Digital modeling tools like Aspen Plus, COMSOL Multiphysics, and digital twins enable virtual replication and simulation of these processes, reducing trial-and-error in pilot and commercial scales. Smart manufacturing integrates these models with: (i) cyber-physical systems for continuous data capture, (ii) edge computing for localized analytics, (iii) AI algorithms for self-learning process optimization, and (iv) cloud-based platforms for centralized data sharing and control. This integration fosters adaptive process control, predictive fault detection, and remote diagnostics, allowing operators to maintain optimal conditions even in dynamic environments. Such closed-loop systems enhance efficiency, reduce energy consumption, and support sustainability goals. Nanotechnology offers innovative solutions to enhance bioethanol production by improving catalyst performance, enhancing enzyme stability, and enabling novel separation techniques. Nanomaterials, such as magnetic nanoparticles and nanostructured catalysts, can be used for efficient biomass conversion and targeted delivery of hydrolytic enzymes. Additionally, nanosensors can provide high-resolution monitoring of fermentation dynamics, enabling precision control. Nanotechnology enhances multiple aspects of bioethanol production: (i) Nanocatalysts improve reaction kinetics in hydrolysis and fermentation. (ii) Nano-encapsulation prolongs enzyme stability and reusability. (iii) Nanosensors provide ultra-sensitive monitoring of fermentation parameters. These technologies reduce enzyme dosages and processing time, thus lowering costs and improving conversion efficiency.

Genetically modified (GM) feedstock represents another critical frontier. GM crops engineered for higher cellulose content, reduced lignin, or increased resistance to pests can significantly improve feedstock availability and processing efficiency. Furthermore, genetically engineered microorganisms tailored to withstand harsh industrial conditions or metabolize a broader range of sugars contribute to higher ethanol yields and process robustness. Genetically modified feedstock and microbes enable: (i) increased cellulose-to-lignin ratio in crops, (ii) enhanced sugar conversion pathways in microbes, and (iii) tolerance to fermentation inhibitors and extreme pH/temperature. CRISPR/Cas9 and synthetic biology enable precision genome editing, tailoring organisms for specific biomass types and fermentation pathways. Industry 4.0 tools support this with genomic data integration, real-time performance tracking, and AI-assisted strain selection. Together, the integration of nanotechnology and genetically engineered organisms, empowered by Industry 4.0 capabilities, lays the foundation for a transformative shift in bioethanol production—one that is more sustainable, productive, and adaptive to global energy needs.

12.3 IoT-Based Real-Time Monitoring System of Bioethanol Production

The Internet of Things (IoT) serves as a backbone for the real-time monitoring of bioethanol production. IoT-enabled systems comprise interconnected sensors, data acquisition devices, communication networks, and cloud-based analytics platforms. These systems continuously collect critical process data such as temperature, pressure, pH, dissolved oxygen, biomass concentration, and ethanol yield. Using wireless sensor networks (WSNs), the IoT system allows for seamless data flow from various production units to centralized control systems.

Cloud computing services then analyze this data using advanced algorithms to provide actionable insights. Real-time dashboards enable operators to monitor the process remotely, identify bottlenecks, and take proactive measures. Moreover, IoT platforms can integrate predictive maintenance protocols by analyzing equipment vibration, thermal imaging, and acoustic signals to foresee malfunctions. This minimizes downtime and improves operational reliability. Overall, IoT facilitates enhanced visibility, traceability, and responsiveness in bioethanol production, aligning with sustainability and efficiency goals. A schematic representation of the IoT-based real-time monitoring system layout for bioethanol production is presented in Fig. 12.1. IoT-based monitoring systems offer transformative benefits for bioethanol production (Fig. 12.2). By embedding sensors in equipment and pipelines, critical parameters such as temperature, pH, dissolved oxygen, and ethanol concentration have continuously tracked and transmitted to central control units. Components of an IoT system include: (i) sensor nodes for data acquisition, (ii) gateways for data aggregation and transmission, (iii) cloud platforms for storage and processing, and (iv) user interfaces for real-time visualization and control. Data collected is analyzed using AI and machine learning models to detect anomalies, predict failures, and suggest corrective measures. The IoT network also supports: (i) remote diagnostics and control, (ii) predictive maintenance, (iii) energy optimization, and (iv) process synchronization across multiple plants.

These capabilities improve operational efficiency and align with sustainability goals by reducing waste, enhancing energy use, and ensuring safer production environments.

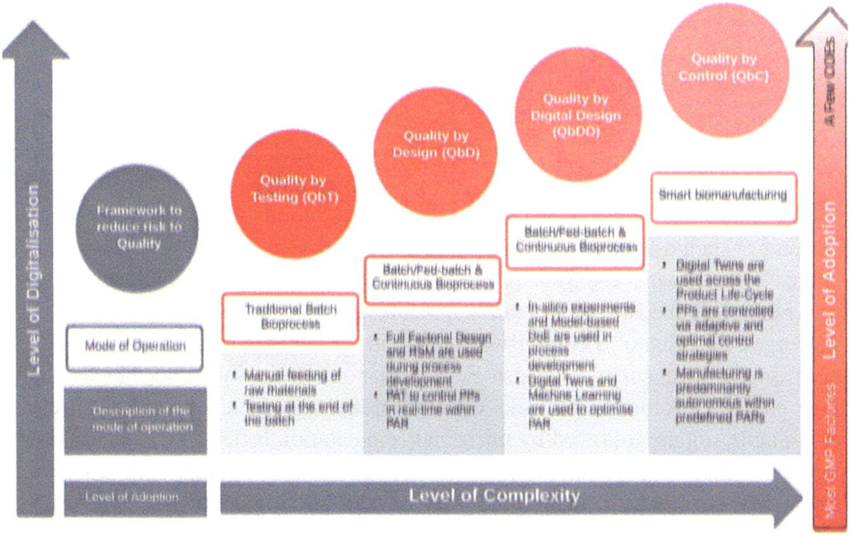

Fig. 12.1 Schematic representation of IoT-based real-time monitoring system layout for bioethanol production

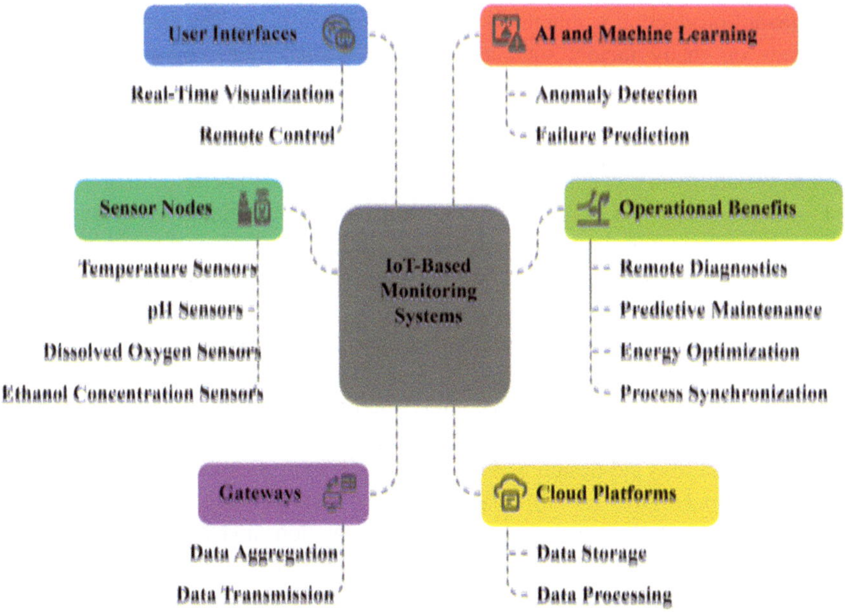

Fig. 12.2 Schematic representation of different components used in the IoT-based monitoring systems for bioethanol production

12.4 Data Analytics on Bioethanol Production Using AI and ML Approaches

Global energy demand continues to rise, driving reliance on finite fossil resources and contributing to climate change. Sustainable biofuels have attracted significant interest as renewable alternatives (Owusu and Marfo 2023). Bioethanol is a widely used biofuel derived from biomass (plants, agricultural wastes, or algae). It has advantages such as a higher-octane rating and lower vehicle emissions compared to pure gasoline (Owusu and Marfo 2023). Unlike fossil fuels, bioethanol has produced from renewable feedstocks and can help address environmental and energy security challenges. Bioethanol is currently blended with gasoline in many countries and can serve as a platform chemical feedstock. Bioethanol is typically produced via a multi-stage process. Biomass feedstocks undergo pretreatment and hydrolysis to release fermentable sugars, followed by microbial fermentation (often by yeast) to convert sugars into ethanol and carbon dioxide. First-generation bioethanol uses food crops (sugarcane, corn, wheat) that are easy to ferment but compete with food supplies. Second-generation bioethanol uses lignocellulosic residues (grass, corn stover, wood), which are more sustainable and abundant, but require more complex and costly pretreatment and enzymatic hydrolysis. Third-generation bioethanol comes from algal biomass, which can yield high productivity without using arable land, though cultivation costs are still high. The choice of feedstock and processing route greatly influences yields and costs.

Ethanol production is inherently complex and nonlinear. The fermentation yield depends on many factors such as feedstock composition, sugar concentration, temperature, pH, microbial strain, inoculum, and residence time. Pretreatment and hydrolysis conditions (acid concentration, enzyme loading) also strongly affect sugar release. In practice, achieving high and stable ethanol yields is challenging due to variability in raw materials and process conditions. Experimental optimization of all parameters is labor-intensive and time-consuming. Modelling these bioprocesses using traditional mechanistic equations (e.g., Monod kinetics) is difficult due to un-modeled dynamics (substrate inhibition, mixed substrates) and large-scale variability. For example, Li et al. (2023) noted that classical kinetic models poorly predict yeast growth when initial glucose varies, and that kinetic parameters may change during fermentation (X. Li et al. 2023). These limitations motivate the use of AI and ML, which can learn complex input-output relationships directly from data without requiring detailed first-principles models.

AI/ML can thus aid design and control by providing flexible predictive models. As Owusu and Marfo (2023) observe, "ANN flexibility and high accuracy" are beneficial in modelling the pretreatment, fermentation, and conversion stages of bioethanol production. Similarly, recent studies have applied various ML algorithms from classical techniques to deep learning to capture fermentation dynamics and optimize yields.

12.4.1 Machine Learning Techniques in Bioethanol Research

Various ML methods have been used to model bioethanol production. A common approach is to treat fermentation predictions as regression problems, using algorithms like ANNs, support vector machines, random forests, and gradient-boosting trees (Liu et al. 2025). For example, a recent work trained multiple models (Gradient Boosting, Random Forest, XGBoost, Gaussian Process Regression, SVR, and a multi-layer perceptron) to predict ethanol yield from gas fermentation inputs (Ma et al. 2024). After tuning, the support vector regression model achieved the best test data accuracy ($R^2 \approx 0.84$). This underscores that different models may excel on different datasets. Another study combined data-driven and kinetic models: Xinzhe Li et al. (2023) built an XGBoost model to predict biomass concentration and a neural network to update kinetic parameters, achieving low RMSE in predicting biomass, ethanol, and glucose time profiles. Feed-forward ANNs or multi-layer perceptrons are widely used to model nonlinear relationships between process inputs (e.g., sugar concentration, temperature) and outputs (e.g., ethanol yield). ANNs consist of layers of interconnected neurons that learn weights via back propagation. They can approximate complex biochemical processes without explicit equations (Sewsynker-Sukai et al. 2016). For instance, ANN models have been used to predict ethanol yield based on fermentation conditions in both lab and industrial settings. However, ANNs require careful tuning (number of layers, neurons, and regularization) and sufficient data to avoid overfitting. Liu et al. (2025) noted that while their multi-layer perceptron had rich capacity, it risked overfitting on their dataset. Support vector regression (SVR) extends support vector machines to regression by fitting a function within an ε-insensitive tube around the data. SVR can handle nonlinearities via kernel functions. It often works well on smaller datasets. In bioethanol applications, SVR has been used to map fermentation conditions to yield (Balabin and Smirnov 2011). For example, Liu et al. found SVR to outperform other regressors, achieving $R^2 \approx 0.84$. SVR offers robust performance on noisy data and provides a good baseline model when data is limited. Gradient-Boosting Trees (XGBoost) is a powerful ensemble of decision trees that fits residuals in sequence. It is known for handling heterogeneous and missing data well and for its scalability. (Mitchell and Frank 2017). In bioethanol production, XGBoost has been applied to predict fermentation dynamics (Momenitabar et al. 2023). An XGBoost model was developed to predict biomass concentration using different fermentation parameters (Li et al. 2023). By providing accurate intermediate predictions, the XGBoost component aided a hybrid model to forecast ethanol and glucose concentrations in a better way (Zhang et al. 2024). Long short-term memory (LSTM) networks are recurrent neural networks designed for sequence data, capturing time dependencies via memory cells. In fermentation, LSTMs can model the entire time-course of a batch process by learning from sequential inputs (e.g., pH and sugar profiles) to predict future outputs (Sousa et al. 2021).

12.4.2 Datasets for Bioethanol Modeling

The quality and quantity of data critically influence ML model performance. Real-world datasets come from laboratory or industrial fermentations, typically comprising sensor or analytical measurements over time. Data features may include time-varying profiles of sugar concentrations, ethanol concentration, cell biomass, pH, temperature, agitation, nutrient levels, and other inputs. Spectroscopic sensors (e.g., near-infrared) are increasingly used to provide rapid estimates of substrates or products; in one study, a PLS model converted spectral data to glucose concentration in real time (Lopez et al. 2020). Industrial datasets can be large (continuous sensor logs), but often remain proprietary. Simulated or synthetic data can supplement limited experiments. Mechanistic models (e.g., mass balances and kinetics) can generate in-silico fermentation data under varying conditions. More recently, deep generative models have been applied. Kwon et al. (2024) employed a variational autoencoder (VAE) to create synthetic fermentation time series, which augmented the training set and significantly improved model generalization. Their results showed ~30% increased robustness, when models were trained on both real and synthetic data. Data augmentation in this way helps overcome the challenges of small datasets. In practice, datasets are often of modest size (tens to hundreds of samples), which makes robust cross-validation and careful train/test splitting essential.

12.4.3 Methodology: Model Development and Evaluation

In applying AI/ML to bioethanol production, a typical methodology involves: (1) data collection and pre-processing, (2) feature selection/engineering, (3) training ML models, (4) hyperparameter tuning, (5) evaluating performance on unseen data, and (6) interpreting the models (Fig. 12.3). Common practices include normalizing inputs, handling missing values, and using k-fold cross-validation or train-test splits to assess generalization. Hyperparameters (e.g., tree depth in XGBoost, number of neurons in an ANN) are often tuned via grid search or Bayesian optimization.

Data Collection and Preprocessing Effective AI/ML integration in bioethanol production follows a structured pipeline from raw data to actionable insights.

The first stage is Data Collection and Preprocessing, where diverse datasets are gathered and cleaned. Typical data include biomass and feedstock properties (e.g., composition of cellulose, hemicellulose, lignin), microbial/strain information, and fermentation process parameters (pH, temperature, agitation, nutrient concentrations, inoculum type, pretreatment conditions, etc.). These data may come from experimental runs, literature surveys, sensor logs, or public databases. For example, Vinitha et al. (2022) compiled 250 *datasets* from published studies and experiments covering biomass characteristics and enzymatic hydrolysis conditions (Vinitha et al. 2022), while Smuga-Kogut et al. (2021) used feedstock details (ionic liquids,

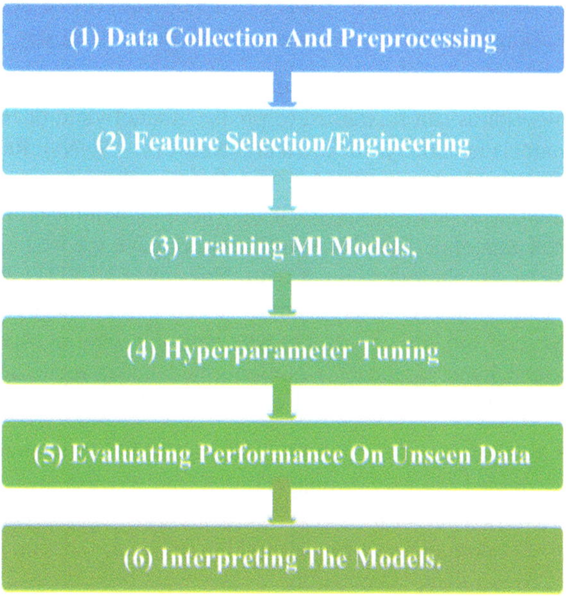

Fig. 12.3 A typical methodology involves bioethanol production using an AI/ML approach

enzymes, yeast type, pretreatment time/temperature) as inputs. After collection, preprocessing steps harmonize the data: missing values are imputed or filtered, outliers are detected and handled, and continuous variables are scaled (e.g., using Z-score standardization or min-max normalization) to stabilize model training. Categorical inputs (e.g., yeast strain, ionic liquid type) are encoded (one-hot or label encoding). Correlation analysis (e.g., Pearson's r) can detect redundant variables. Tools like Python (pandas, NumPy), R (tidyverse, data.table), or MATLAB can be used for cleaning and exploratory analysis.

Feature Selection and Engineering After preprocessing, feature selection and engineering are crucial to reduce dimensionality and enhance model performance. Common techniques include correlation filtering (removing highly collinear features) and principal component analysis (PCA) for unsupervised dimensionality reduction. For instance, PCA was used on high-dimensional Raman-spectral data from lignocellulosic fermentations, where the first principal components captured 89% of variance and correlated well with ethanol concentrations (Ewanick et al. 2013). Supervised methods like recursive feature elimination (RFE) and LASSO regression can also rank features by importance. Domain knowledge often guides engineering of new features: e.g., combining cellulose/hemicellulose ratios or calculating yield per unit biomass. Feature extraction pipelines (e.g., image-based morphological descriptors as in Itto-Nakama et al. 2021) may be applied before selection. In practice, one might: (i) Compute a correlation matrix and drop nearly duplicate variables. (ii) Apply PCA on spectroscopic or high-dimensional inputs to capture key variance. (iii) Use feature importance from a preliminary Random Forest to shortlist predictors. Tools like Scikit-learn (PCA, SelectKBest, RFE) or R's caret package (variable importance, rfe) facilitate these steps.

Training Machine Learning Models With a curated feature set, various ML algorithms can be trained to predict ethanol yield or fermentation metrics. Typical models include Artificial Neural Networks (ANNs), Random Forests (RF), Support Vector Machines (SVMs), and Gradient Boosting (e.g., XGBoost). ANNs (e.g., multilayer perceptrons) are widely used for nonlinear process data; for example, Itto-Nakama et al. found a neural network gave the best ethanol-forecast R^2 (>0.9) on yeast morphological inputs (Itto-Nakama et al. 2021). Random forests and boosting methods handle heterogeneous features well and give built-in importance scores. Several models on yeast-engineering data, noting that an XGBoost model (with tuned hyperparameters) outperformed linear regression, decision trees, RF, and SVM (Khamwachirapithak et al. 2023). A typical model configuration might be:

- ANN: e.g., 2–3 hidden layers with ReLU activations, learning rate 0.001–0.01, trained via back propagation (Keras, TensorFlow, or MATLAB's Neural Network Toolbox).
- Random Forest: e.g., 100–500 trees, max depth tuned, using Scikit-learn or Weka's implementation.
- SVM: e.g., Gaussian (RBF) kernel with box constraint C and kernel width γ optimized.
- XGBoost: e.g., 100–1000 trees, learning rate (eta) ~0.01–0.3, max_depth tuned. Configurations depend on data size and complexity. Tools: Python's Scikit-learn (RF, SVM), xgboost library, Keras/TensorFlow (ANNs/CNNs), R packages (randomForest, e1071, caret), or GUI tools like Weka. Each model is trained on a training set (e.g., 2/3 of the data) and its predictions compared to known outputs (e.g., ethanol concentration (Smuga-Kogut et al. 2021)).

Hyperparameter Tuning Each model has hyperparameters (e.g., number of trees, learning rate, layer sizes) that must be tuned for optimal performance. Common strategies are:

- Grid Search: Exhaustively search a parameter grid (e.g., via Scikit-learn's Grid-Search-CV).
- Random Search: Sample hyperparameters randomly (often more efficient).
- Bayesian Optimization: Model-based search (e.g., Tree-structured Parzen Estimator in hyperopt or scikit-optimize's Bayes-Search-CV). For example, "intensive hyperparameter tuning" of XGBoost to achieve top performance over other models. Tuning is usually done with cross-validation (see next section) to prevent overfitting. Automated tools like Auto-sklearn or open-source libraries (Optuna, Hyperopt, Scikit-Optimize) can streamline this process.

Evaluating Model Performance Model quality is assessed on hold-out or cross-validated data using regression metrics. Common metrics for bioethanol yield prediction include:

- Coefficient of Determination (R^2): fraction of variance explained.
- Root Mean Square Error (RMSE): the square root of the average squared error.
- Mean Absolute Error (MAE): average absolute deviation. For example, Vinitha et al. reported a regression model (OD-MS algorithm) with $R^2 \approx 0.976$ on test data, and Smuga-Kogut et al. saw RF $R^2 \approx 0.95$. Cross-validation (e.g., 5- or 10-fold CV) is used to estimate generalization error. The data are partitioned into *k* subsets, models are trained on *k–1* and tested on the remainder, rotating through all folds. This yields averaged metrics (mean RMSE, R^2) and variability. Tools: Scikit-learn's cross_val_score or R's trainControl (caret) automate CV. It is also helpful to plot predictions vs. actual yields to inspect the fit visually. In classification tasks (e.g., predicting optimal vs. suboptimal fermentation regimes), metrics like accuracy, precision/recall, and confusion matrices would be used instead, but ethanol yield is typically a regression problem.

Model Interpretation Model interpretability is important for insight and trust, as well as understanding which inputs most influence the outputs. Such analyses highlight process drivers and can guide experimental focus. Key interpretability techniques include:

- Feature Importance Scores: Many models (RF, XGBoost) provide the global importance of each input (e.g., the mean decrease in impurity or gain). This reveals which factors (cellulose content, pH, etc.) most influence ethanol yield.
- SHAP Values: "SHapley Additive exPlanations" quantify each feature's contribution to individual predictions (Lundberg and Lee 2017). Methods like SHAP (SHapley Additive exPlanations) assign each feature a contribution to a particular prediction, based on game-theoretic Shapley values. SHAP provides global importance (which features generally matter) and local explanations for individual predictions. In bioethanol studies, SHAP has been applied to identify key factors: Liu et al. used SHAP to show that initial CO and acetic acid concentrations positively correlated with ethanol yield. At the same time, H_2 and CO_2 had complex impacts. For instance, SHAP can show that high cellulose content pushes the predicted yield higher, while high lignin content pushes it lower.
- LIME: Locally Interpretable Model-agnostic Explanations approximate the model locally around a prediction by a simple, interpretable surrogate, highlighting feature effects in that neighborhood.
- Partial Dependence Plots: Show the marginal effect of a feature by averaging out others. For example, if an RF model predicts high ethanol yield, SHAP/LIME can identify which inputs (e.g., optimal pH, temperature) led to that. Lundberg and Lee (2017) emphasize that methods like SHAP assign each feature an "importance value for a particular prediction". In practice, one can generate a SHAP summary plot (Python shap library) or use LIME's explain_instance to probe individual cases. Interpretable modeling helps engineers validate that the model's

logic agrees with chemical understanding (e.g., confirming that agitation speed is important) and can suggest new experimentation (e.g., testing a predicted optimum).

Tools and Software (Per Stage) Data handling often uses Python (pandas, NumPy) or R (tidyverse). Feature selection/engineering can use Scikit-learn (PCA, feature_selection) or MATLAB toolboxes. Model training typically uses Scikit-learn (RF, SVM), XGBoost/LightGBM libraries, or deep learning frameworks (TensorFlow/Keras, MATLAB Neural Network Toolbox). Hyperparameter tuning can use Scikit-learn's GridSearchCV/RandomizedSearchCV, or specialized libraries (Hyperopt, Optuna). Evaluation metrics are available in Scikit-learn (metrics. r2_score, mean_squared_error) and R (caret). Interpretation tools include the Python SHAP and LIME packages, or R packages like DALEX or iml. Weka (Java) and other GUI tools (RapidMiner, KNIME) offer integrated environments for smaller projects.

Model Training Each ML model is trained to minimize prediction error on training data. For regression tasks (predicting ethanol concentration, yield, etc.), loss functions such as mean squared error (MSE) or mean absolute error (MAE) are used. The chosen algorithm learns parameters/weights (e.g., tree splits or network weights) to fit the training data. For instance, in Li et al. (2023), the XGBoost model was trained on measured biomass data to predict cell growth, while the ANN was trained to learn optimal kinetic parameters. After training, both components were combined in a hybrid mechanism to forecast ethanol and glucose profiles. Liu et al. similarly trained SVR and other regressors on experimental fermentation results, optimizing kernel parameters for best test accuracy.

Performance Metrics Models are evaluated using metrics such as root mean squared error (RMSE), coefficient of determination R^2, and mean absolute error (MAE). RMSE measures the typical magnitude of prediction error, while R^2 indicates the fraction of output variance explained by the model ($R^2 = 1$ is a perfect fit). Lower RMSE/MAE and higher R^2 denote better performance. (Chai and Draxler 2014). For example, Li et al. reported RMSE values of 0.33 g/L (biomass), 1.93 g/L (ethanol), and 3.05 g/L (glucose) for their hybrid model (Li et al. 2023). Liu et al. achieved $R^2 \approx 0.84$ on the test set with their SVR model. Generally, reported models often achieve R^2 values in the range 0.7–0.9 for well-fitted tasks, reflecting reasonably high predictive ability.

12.4.4 Real-Time Monitoring and Digital Twins

Modern biofuel facilities increasingly utilize process data for real-time control. A soft sensor is an ML model that estimates hard-to-measure variables (e.g., ethanol concentration) from easy-to-measure signals (e.g., pH, turbidity). Soft sensors can be implemented via ANNs, PLS, or hybrids to monitor fermentation continuously. For instance, developed a hybrid soft sensor for lignocellulosic ethanol production: a PLS model estimated glucose from spectroscopy in real time, and a kinetic model (fitted recursively) predicted future concentrations of glucose, xylose, and ethanol. By continuously updating predictions with live data, this approach embodies a digital twin, i.e., a virtual replica of the process that mirrors its current state and forecasts its evolution (Spyrou et al. 2023). The authors noted that this digital-twin soft sensor enabled a better understanding of inhibitory effects from biomass variability and could adapt scheduling accordingly (Lopez et al. 2020).

Digital twins combine first-principles (mechanistic) models with data-driven components. They can integrate sensor networks, IoT devices, and ML algorithms to provide real-time feedback and optimization (Okolie 2024). As Owusu and Marfo (2023) emphasize, AI-driven models are "necessary in the design, handling, and optimization" of bioethanol systems. In practice, a digital twin might use ML models (e.g., ANNs or LSTMs) to predict short-term trends and mechanistic models for long-term behavior, allowing operators to run scenario analyses and control loops. For example, if a decline in sugar concentration is detected earlier than expected, the twin could suggest adjusting feed rates or temperature to maximize yield. As data availability and sensor technology improve, digital twins will likely play a growing role in scaling up bioethanol production and achieving real-time optimization.

12.4.5 Challenges and Future Directions in Model Development

Data Limitations A significant challenge is the scarcity and variability of data. Fermentation experiments are costly and time-consuming, so many studies have only tens of samples (Kwon et al. 2024) observed that fermentation data acquisition "is often constrained by limitations in availability and resources". Small datasets can lead to overfitting and poor generalization. Synthetic data augmentation (e.g., via VAEs) is one promising solution, as is collaboration to build shared databases.

Model Robustness Biological processes are noisy and subject to disturbances (contamination, sensor drift). ML models must be robust to such variability. Complex models like deep neural networks can be sensitive to changes outside the training domain. Interpretability remains a concern, as "black-box" models must still be trusted by engineers. Tools like SHAP mitigate this by revealing which features drive predictions, but more work is needed on uncertainty quantification and safe AI in bioprocessing.

Integration and Scalability Translating lab-scale models to industrial scale is nontrivial. Scale-up introduces new physics (mass/heat transfer) and variability. Future research should explore transfer learning or multi-scale modelling, where ML models incorporate known engineering principles. Hybrid approaches—combining mechanistic equations with ML corrections—are promising.

Real-Time Implementation While many studies report offline predictions, deploying ML in real time (online monitoring and control) poses engineering challenges. Models must be updated with streaming data and operate with low latency. Auto ML and lightweight model architectures could facilitate real-time use.

Emerging Trends Future work may involve reinforcement learning for adaptive control of fermentation processes, where agents learn optimal feeding or temperature strategies to maximize ethanol output. Integrating ML with economic and life-cycle analysis could optimize yield and overall sustainability. Genomics and metabolic engineering advances may also feed richer data (e.g., yeast gene expression) into predictive models.

Bioethanol production is a key component of renewable energy systems, but its complex bioprocesses present optimization challenges beyond traditional modeling. Recent research demonstrates that AI and ML techniques (ANNs, SVR, XGBoost, LSTMs, etc.) can effectively model fermentation and pretreatment processes, enabling more accurate prediction and control. By leveraging both experimental and synthetic data, these methods achieve high predictive performance (e.g., low RMSE, high R^2) while tools like SHAP enhance the interpretability of models. Real-time monitoring using ML-based soft sensors and digital twins further extends capabilities for adaptive control. However, challenges remain in data availability, model robustness, and scaling to industrial settings. Future work should focus on integrating data-driven models with mechanistic understanding, expanding datasets, and advancing explainable AI techniques. With continued progress, AI-driven optimization promises to make bioethanol production more efficient, sustainable, and economically viable.

12.5 Circular Economy Meets Industry 4.0—Big Data Drives Industrial Symbiosis

Circular economy (CE) models aim to transform linear "take–make–dispose" processes into resource-efficient, closed-loop systems. A key CE strategy is industrial symbiosis (IS), where the by-products or waste of one process become inputs for another (Castellet-Viciano et al. 2022). Industry 4.0 technologies, incorporating big data analytics, Internet of Things (IoT), artificial intelligence (AI), and blockchain, can accelerate CE implementation by enabling real-time monitoring, predictive optimization, and transparent tracking of raw materials (Demestichas and Daskalakis 2020). These principles and technologies intersect in bioethanol production, which utilizes different biomass (e.g., corn, sugarcane, or lignocellulosic waste) to produce a renewable fuel. Bioethanol facilities generate co-products like distillers' grains (animal feed) and capturable CO_2, and consume large amounts of energy and water; thus, they present opportunities for circular flows and data-driven efficiency.

The traditional linear economy, characterized by "take–make–dispose" practices, is increasingly seen as unsustainable due to resource depletion and environmental impact. Circular economy (CE) models propose that products and materials be kept at high value for as long as possible, with waste minimized and outputs reused or recycled. CE is often summarized by "R" frameworks (3R, 4R, 6R, etc.), emphasizing reduce, reuse, recycle, and recover strategies. A concrete manifestation of CE in industry is industrial symbiosis (IS), where multiple firms collaboratively manage resources: by-product streams, shared utilities, or services. In IS networks, one company's waste becomes another's raw material, sometimes coordinated by a third-party platform. Classic examples include the Kalundborg (Denmark) eco-industrial park, where waste heat, gypsum, and other flows are mutually exchanged, drastically reducing collective resource use.

"Industry 4.0" (I4.0)—also called the Fourth Industrial Revolution—is the recent trend of integrating cyber-physical systems, IoT, AI, and cloud computing into manufacturing. It transforms factories into smart, interconnected systems capable of real-time data collection and intelligent control. For example, cyber-physical systems allow continuous sensor monitoring and automatic adjustment of processes; big data analytics uncover patterns for efficiency; AI algorithms optimize operations; and blockchain can securely record transactions for traceability. In practice, I4.0 is about *"integrating physical production and operations with intelligent digital technologies, such as big data and machine learning, to establish a comprehensive and interconnected [enterprise] environment"*. These digital enablers can reduce waste and resource use by enabling predictive maintenance, flexible production scheduling, and new business models (e.g., product-as-a-service)—making them natural allies of CE goals.

Bioethanol production refers to converting biomass into ethanol fuel, which is inherently linked to agriculture and energy systems. It contributes to renewable energy goals (displacing some fossil fuels) but also faces sustainability challenges (land use, water use, emissions). However, bioethanol plants generate valuable co-products (e.g., animal feed, biomethane, carbon dioxide) and can utilize waste biomass (e.g., straw, bagasse, fruit waste). These features fit well into circular approaches: using bagasse for combined heat & power (cogeneration) or feeding spent yeast and unfermented solids to livestock (Dennis et al. 2024). By embedding Industry 4.0 tools, bioethanol facilities can digitize these circular flows: sensors and data analytics can optimize fermentation and energy use; digital platforms can match waste outputs with local users; and blockchain ledgers can certify sustainability attributes of ethanol (e.g., origin of feedstock). CE theory, I4.0 technology, and bioethanol case studies to illustrate how *big data-driven industrial symbiosis* can improve the sustainability of ethanol production.

In the bioenergy sector, industrial symbiosis and circular economy overlap with the bioeconomy concept, using biological resources sustainably for fuels, chemicals, and materials. An integrated biorefinery illustrates this: for instance, a sugarcane mill might generate ethanol, electricity from bagasse, animal feed from remaining pulp, and even capture CO_2 for algae production. Each output adds value and closes loops. The Pomacle-Bazancourt (France) biorefinery is a classic case: it integrates multiple actors (farmers, co-generators, feed producers) so that sugar beet processing yields ethanol, animal feed, and biogas, with heat and CO_2 used by adjacent greenhouses (Allais et al. 2021). This kind of multi-stream valorization exemplifies circularity: biomass is used fully, with minimal waste. However, realizing that IS requires information, companies must know which waste streams exist and who could use them. This is where digital tools play a critical role. Digital *platforms* and databases can catalog by-products and match them with potential users across industries. For example, a product lifecycle management (PLM) system adapted for CE can list industrial outputs; an online marketplace can list available heat or sludge; and an analytics dashboard can highlight opportunities. Demestichas et al. note that such digital platforms facilitate the exchange of by-products between businesses, as well as the synergies in the context of industrial symbiosis. In summary, CE/IS provides the "why" (sustainability goals), and Industry 4.0 provides the "how" (the enabling technologies).

12.5.1 Industry 4.0: Big Data, IoT, AI, and Blockchain

Industry 4.0 encompasses a range of digital technologies that transform manufacturing and supply chains. Key components include:

Big Data Analytics Modern plants generate massive amounts of data (sensor readings, process logs, maintenance records). Big data tools store and process these large, fast, and varied datasets ("4Vs": Volume, Variety, Velocity, Veracity). Advanced analytics and machine learning can uncover hidden patterns, such as predicting yield variations from feedstock characteristics. In bioethanol, big data can optimize fermentation curves, predict equipment failures, and optimize energy use (Jabbour et al. 2019).

Internet of Things (IoT) and IIoT IoT refers to the network of interconnected devices. In industrial contexts (IIoT), sensors and smart instruments monitor temperature, pressure, flow, and other variables in real-time. For example, an ethanol plant might have wireless sensors on motors, conveyors, and tanks, all reporting to a central system. This connectivity enables continuous monitoring and control. Demestichas et al. observe that IoT and blockchain *"can play an integral role in embracing CE concepts"* by enabling real-time data collection and sharingmdpi.com. Indeed, *"Industrial Internet of Things (IIoT) encompasses various disciplines including energy production, manufacturing, agriculture, ..."* (Anttila 2019), highlighting its broad reach.

Artificial Intelligence (AI) and Machine Learning (ML) AI algorithms can analyze and process data to optimize operations. Based on feedstock quality, ML models might predict the best yeast strain or fermentation conditions in bioethanol plants. AI can also enable predictive maintenance: by analyzing vibration or thermal data, it can warn operators of impending pump failures, reducing downtime and waste. AI systems can even autonomously adjust process set points to maximize yield or minimize energy use.

Blockchain and Smart Contracts Blockchain is a distributed ledger technology that records transactions in a tamper-proof way. In supply chains, blockchain can track the origin and journey of materials. For bioethanol, this means every barrel could carry a digital certificate of its carbon footprint and sustainable sourcing. Databases built on blockchain have been prototyped for sustainable biofuels compliance (Roundtable on Sustainable Biomaterials (RSB) and Bioledger 2021). Blockchain also enables *smart contracts*: self-executing code that automatically enforces rules (e.g., releasing payment when ethanol quality meets a contract, or triggering carbon credit issuance when a CO_2 capture ton is verified). In CE, blockchain can ensure transparency and trust, which is crucial when verifying that waste-based feedstocks or emission reductions are real (Demestichas et al. 2020).

Digital Twins and Simulation A digital twin is a virtual replica of a physical process or plant. It combines real-time data with simulation models to test scenarios virtually. Bioethanol plants can use digital twins to evaluate changes (e.g., a new cogeneration unit) or to optimize energy flows. Simulation tools can also help design circular processes (e.g., optimizing heat exchanger networks) before investing in equipment.

These technologies overlap heavily with CE objectives. For example, big data and IoT support the "reduce" R by optimizing resource use and preventing losses. AI can help "design out" waste through process control. Data platforms act as digital IS enablers: a company could post its surplus steam or nutrient-rich effluent and find a neighbor who needs it. Industry 4.0 provides the digital infrastructure to make an industrial symbiosis dynamic and measurable.

12.5.2 *Circular Bioethanol: Feedstocks, Processes, and By-products*

Bioethanol is produced by fermenting biomass sugars into ethanol. Common feedstocks include starch (corn, wheat), sugar (sugarcane, beet), and cellulosic residues (corn stover, straw, bagasse, wood chips). The production process (see Fig. 12.1) typically involves pretreatment, saccharification, fermentation, and distillation. Each step has inputs (water, enzymes, energy) and outputs (ethanol, unfermented solids, heat, CO_2). From a circular perspective, the key is to valorize all outputs.

For example, corn ethanol plants produce distillers' grains (grain solids after fermentation) and corn oil. These have long been sold as animal feed, significantly contributing to plant economics. Indeed, Dennis et al. report that revenues from feed-use co-products rose from ~12% of ethanol plant revenue in 2005 to over 22% by 2022. During the COVID-19 downturn, co-products even temporarily supplied nearly 40% of revenues. One bushel of corn (≈25 kg) yields roughly 2.9 gallons (≈11 L) of ethanol and 15.2 pounds of distillers' grains. This illustrates how a significant fraction of the biomass carbon and value is diverted into valuable by-products rather than waste (Dennis et al. 2024). Innovations have further increased value: high-protein yeast from fermentation, de-oiled DDGS, and refined corn oil (for biodiesel) are now common. These co-products help stabilize bioethanol's profitability and embody circular principles by converting potential waste into feed and chemicals.

Similarly, sugarcane ethanol (dominant in Brazil) is inherently circular: after extracting juice from cane, the leftover fiber (bagasse) is burned to produce process steam and electricity, often exceeding the plant's own needs. Surplus power is exported to the grid, as seen in Brazilian mills (Li et al. 2020). Some plants capture fermentation CO_2 for sale (e.g., carbonating drinks or growing algae). Cellulosic ethanol plants (like those using corn stover or energy crops) generate even more lignin-rich residues; these can be burned or gasified for energy, or processed into bio-based chemicals. GranBio in Brazil runs a second-generation (2G) facility using sugarcane straw and bagasse (Basso et al. 2013). GranBio's plant is designed as a mini refinery: it produces ethanol *and* uses advanced automation to integrate operations, embodying a circular and digital approach. Figure 12.4 illustrates a generalized circular bioethanol production system, showing how crop residues, co-products, and waste streams can loop back. For instance, unfermented solids can be anaerobically digested into biogas (used for heat), while greenhouse-grade CO_2 can be captured. Industry 4.0 tools would layer on top of this: sensors monitor fermentation, data analytics optimize steam usage, and blockchain tracks biomass origin.

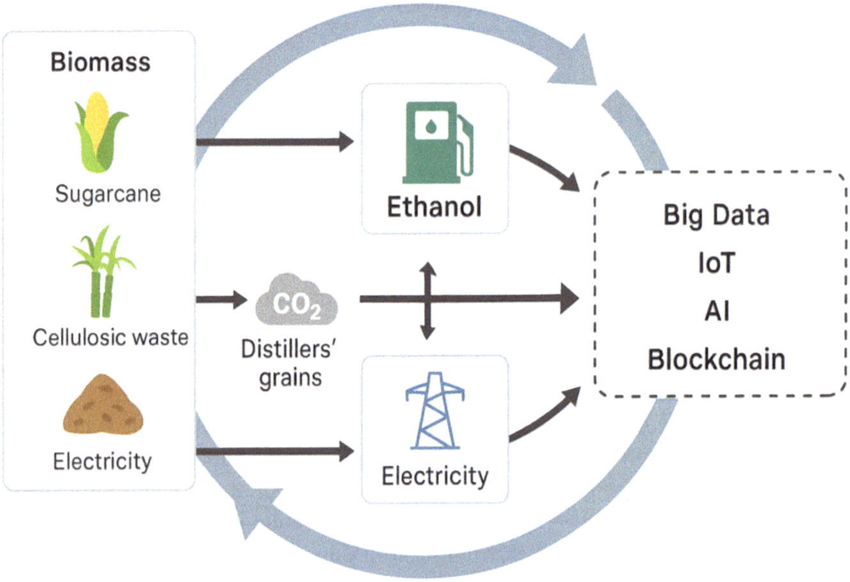

Fig. 12.4 Circular flows in bioethanol production: biomass feedstocks are converted to ethanol; co-products (animal feed, biogas, electricity, CO_2) are recovered or reused; and digital technologies (big data, IoT, blockchain) connect actors in the industrial symbiosis network

12.5.2.1 Case Study: GranBio Cellulosic Ethanol Plant (Brazil)

GranBio (Gran Investimentos S.A.) operates a pioneering second-generation (2G) bioethanol plant in Alagoas, Brazil. Commissioned in 2014 (Bioflex 1 unit), it uses sugarcane leaves and bagasse (usually waste) as feedstock, thus not competing with food crops. The plant's design exemplifies circularity: it converts non-food biomass into fuel and uses integrated cogeneration to power its operations. GranBio also embodies Industry 4.0 principles in automation and data integration. An Automation World case study describes how GranBio implemented a field device management and asset tracking solution using FDT (Field Device Tool) technology.

This system unifies data from thousands of sensors: the plant's control architecture includes 60,000 I/O points across distributed control and safety systems. By integrating all these smart devices through standardized interfaces, operators can monitor valve positions, pressures, and diagnostics on one platform. In practical terms, this meant GranBio could perform preventive maintenance and rapid troubleshooting. In other words, less downtime and fewer errors during startup ("Bioflex 1 plant commissioning was fast") were achieved thanks to the data-driven approach. Moreover, GranBio's management aimed for an "intelligent strategy" in its expansion. They planned 12 new plants with full digital integration of production data, ensuring optimal operation of assets. By treating the plant as a cyber-physical system, GranBio can optimize fermentation and energy flows.

For instance, data from temperature and pH sensors in bioreactors feed into algorithms that control enzyme addition and fermentation time, maximizing ethanol yield from the straw. Unfortunately, the case study does not disclose quantitative results (e.g., yield % improvement), but the qualitative gains (reduced maintenance labor, rapid commissioning) are clear. Importantly, GranBio's example shows how IoT-enabled automation and big data analytics can amplify circular practices. By having fine-grained operational data, the plant ensures minimal waste (e.g., precise energy usage) and supports its circular use of biomass.

12.5.2.2 Case Study: Western Plains Energy—Data-Driven Optimization (USA)

Western Plains Energy (WPE) is a corn-ethanol plant in Kansas, USA. WPE partnered with a data analytics company (dataPARC) to build an enterprise data infrastructure. Before this, plant data were siloed across control systems and spreadsheets. The new system integrated all process data (tank levels, flows, quality measurements) into a unified historian and dashboard. WPE's general manager stated that the integration "allows operators and engineers to spend less time tracking down data and more time troubleshooting". Quantitatively, the case report highlights "Saved 5–10 hours/week" per engineer from reduced data wrangling. In industry terms, this is a significant efficiency gain: if each of 5 staff saves 10 h, that is 50 engineer-hours weekly freed. More importantly, with real-time visibility, the plant improved process control. For example, continuous quality data from the distillation column and fermentation tanks are now monitored, so if the ethanol concentration begins to drift, adjustments to feed rate or yeast dosing can be made quickly. Though WPE's case is anecdotal, it suggests that data democratization—making all plant data visible and analyzable—is a foundational step toward circular outcomes. With access to data, operators can identify inefficiencies that previously went unnoticed.

For instance, spotting slowly drifting energy usage patterns could indicate heat exchanger fouling. While WPE's report did not give hard energy-saving numbers, other datapoints are instructive: Energy Transfer (a major ethanol company) reported *"$10M in annualized savings"* after implementing a similar enterprise data platform across 100 sites. Some of those savings were from process optimization and reduced downtime, which translates directly into lower fuel and raw material use per gallon of ethanol. The WPE example underscores how Industry 4.0 (big data collection) creates a single source of truth that, once harnessed, can drive energy efficiency—a key circular metric.

12.5.3 Blockchain for Biofuels Supply Chains

Beyond individual plants, digital tools are being applied at the *supply chain* level to enforce circularity. A notable initiative is the Blockchain Database for Sustainable Biofuels, a 2020 pilot by the Roundtable on Sustainable Biomaterials (RSB) and startup Bioledger. This prototype allowed participating companies to record transactions (feedstock origin, verification of sustainability criteria) on a blockchain. Although not directly related to plant operations, it is relevant for circular bioethanol because it tackles *traceability and fraud prevention*. By using blockchain, the project demonstrated how a secure digital ledger can replace paper-based chain-of-custody documents. For example, bilaterally agreed sustainability data (e.g., a promise that the ethanol was made from certified waste oils or eligible crop residues) were written immutably. Survey results from the pilot showed stakeholders recognized the value of data transparency—a component of the "monitoring and accountability" needed for a truly circular supply chain.

Crucially, this project highlighted practical challenges: feeding all required compliance data into a blockchain (e.g., EU Renewable Energy Directive fields) is complex and costly. The RSB report noted that, while blockchain can "mitigate fraud, restore trust, and build transparency", building such a system at scale for biofuels remains expensive. Nonetheless, this case shows how digital ledger technology can enforce circular economy principles (by ensuring that "waste-to-fuel" claims are genuine) and how it complements big data (ensuring data integrity). In the future, a blockchain could also automate incentives: for instance, by issuing digital credits when a plant captures and sequesters CO_2 or adheres to zero-discharge rules.

12.5.4 Synergies of I4.0 and Circularity in Bioethanol

Real-time resource monitoring enables optimization. Sensors and IoT devices in the plant feed a constant data stream. In the GranBio case, the management of 60,000 I/O points means that virtually every process variable is visible. When combined with big data analytics, this allows operators to fine-tune the process. For example, if the moisture content of incoming bagasse fluctuates, the control system can adjust the boiler firing rate automatically, rather than relying on manual intervention. Such responsive control reduces waste (e.g., over-burning or incomplete combustion) and improves yield. Similarly, in WPE, continuous data availability cuts 5–10 labor hours/week and likely reduces off-spec products (though not quantified) by enabling quicker corrective actions. Overall, studies show that big data analytics in manufacturing can improve energy efficiency by 10–30%, which could significantly lower the carbon footprint if applied to ethanol distillation.

Predictive maintenance reduces downtime and scrap: IoT devices (vibration, acoustic, thermal sensors) predict equipment failures before they occur. In GranBio, the asset management system could alert staff to a valve needing recalibration. Preventing a pump failure avoids a costly shutdown and product loss. This extends to ancillary equipment like conveyors and filters. Fewer emergency repairs mean the plant operates more continuously, using inputs (water, feedstock) more efficiently (less wasted feedstock in restart and clean-up). In circular terms, this means higher utilization of resources (every kilogram of corn or cane yields more ethanol) and lower waste (less spoiled mash).

Digital platforms facilitate industrial symbiosis: The collected data can reveal otherwise-hidden opportunities. For instance, by analyzing heat flow, a plant might have excess steam at a certain time of day. A digital noticeboard (or marketplace) could then connect this surplus with another facility that needs steam. While we have not seen a published example of this in ethanol specifically, the principle is established: digital industrial symbiosis platforms have been developed in Europe to map waste streams across industries (Halstenberg et al. 2017). One can imagine a future where an ethanol plant's database is integrated with a chemical plant's system, so that lignin-rich stillage is automatically offered for use in bioplastics. The *conceptual model* emerging is a data-driven loop: sensors → data integration → analytics → actionable insight → symbiotic exchange.

Blockchain and certification promote circular business models: The RSB/Bioledger case shows that transparency is valued by both industry and regulators. A certified circular bioethanol might command a market premium or meet regulatory quotas. As EU regulations tighten (e.g., by 2030 requiring a 14.5% cut in transport fuel emissions and specific sub-targets for advanced biofuels), companies will need robust tracking of emissions and feedstocks. Blockchain-enabled traceability thus becomes part of the circular system architecture. Economic benefits align with sustainability outcomes. The cases reveal that digital investments, though sometimes costly, pay off. GranBio's upfront cost of an advanced asset management system was offset by lower maintenance and startup costs.

WPE's data integration tool cost is confidential, but the stated savings in labor and implied process efficiency likely provided a return on investment within a few years. Energy Transfer's reported \$10M savings from data integration (across multiple plants) quantifies the scale: presumably, this includes fuel and raw material savings, which directly reduce carbon emissions. Moreover, Dennis et al. noted that co-products now comprise over 22% of ethanol plant revenues. Digitally optimizing co-product recovery (e.g., improving feed drying systems to sell more DDGS) can further boost that share. In summary, the synergy between big data and circularity is that both improve the resource-efficiency metric (ethanol per ton of biomass) and add economic value (higher sale of co-products, lower utility costs).

Integrating circular economy principles with Industry 4.0 technologies offers a promising route to more sustainable bioethanol production. Ethanol plants can significantly reduce their environmental footprint while improving profitability by repurposing waste streams, sharing resources, and optimizing processes in real time. Theoretical frameworks (CE, IS) combined with technological advances (big data, IoT, AI, blockchain) create a synergy: digital intelligence turns linear factories into regenerative ecosystems.

Our review indicates that advanced bioethanol facilities are beginning to realize this potential. For example, GranBio's 2G plant in Brazil uses sophisticated automation and data management to maximize yields from agricultural residues. US examples show that even commodity corn-ethanol plants can cut costs and energy usage by embracing analytics. Blockchain prototypes illustrate how supply chains can ensure that biofuels truly meet sustainability criteria. However, fully closing the loop requires continued investment in sensors, connectivity, skills, and supportive policies that value circular outcomes (e.g., carbon offsets, renewable fuel standards). Policy makers and industry leaders should note that the data infrastructure of ethanol plants can be just as important as the physical infrastructure. Just as pipelines and boilers define classic ethanol refineries, in the circular Industry 4.0 era, cloud platforms and AI engines will define the next-generation facilities.

Encouraging open standards and data sharing agreements (possibly through industry consortia or public databases) will be key to building broad industrial symbiosis networks beyond single plants. In conclusion, big data really can drive industrial symbiosis for circular bioethanol. When a refinery knows its material and energy flows down to the minute, and when data flows to partners across the value chain, every drop of ethanol can be produced more cleanly, and every residue can find a new use. The examples and analysis here demonstrate that this vision is feasible and already underway in parts. Future research should continue to quantify the environmental and economic benefits of digitized circular systems and explore how emerging technologies (e.g., 5G connectivity, edge AI) might further close gaps. With the twin pressures of climate change and resource scarcity, the marriage of CE and I4.0 in biofuels is not just an academic curiosity but a practical imperative.

12.6 Recent Advancements and Challenges in Industry 4.0 Bioethanol Production Processes

Industry 4.0 integrates IoT sensors, robotics, AI, and other advanced tools into manufacturing (including bioprocessing) to enable real-time monitoring and automation (Clauser et al. 2022).

In bioethanol plants, smart sensors and data analytics can optimize fermentation, distillation, and supply chains, potentially creating "smart" biorefineries that are more cost-effective and flexible. For example, Clauser et al. (2022) describe how cyber-physical systems and big data analytics aim to increase production efficiency and quality in the biorefinery context. Nevertheless, fully achieving these benefits requires overcoming challenges in four key dimensions as discussed below.

Cost and Scalability

Industry 4.0 applications have delivered concrete cost savings in bioethanol operations. For instance, detailed simulation (a digital-twin approach) of an ethanol plant's separation unit identified a simple retrofit—adding a vacuum distillation column—that enabled ~70% of output to be upgraded to higher-value solvent-grade ethanol (Bisgaard et al. 2017). Likewise, a European ethanol producer used plant-wide data and an optimization algorithm to tweak its heat-exchanger CIP procedures, boosting energy efficiency by 6%. This saved roughly €400 per day in steam costs (for a 0.1–0.5 million L/day plant), a payback achieved with only modest investment. In another case, replacing conventional heat exchangers in a distillation column with modern, wider-gap plate exchangers eliminated frequent clogging: maintenance stops dropped by 80% and uptime soared, translating into significantly higher yield and throughput. These examples show that digital analysis and equipment upgrades—hallmarks of Industry 4.0—can sharply lower operating costs and improve productivity in real ethanol facilities.

At the same time, the scalability of Industry 4.0 solutions depends on upfront investment and infrastructure. Clauser et al. (2022) note that digital integration (IoT, AI, cloud, etc.) can reduce production, logistics, and management costs by up to 30%. However, it requires capital for sensors, IT hardware, and skilled personnel. Many biorefineries remain at an "Industry 3.0" level because upgrading to smart sensors and connectivity is expensive: typical plants require reliable control systems and cyber-infrastructure, and the initial costs (plus training) can be a barrier. Thus, while Industry 4.0 tools promise significant cost reductions and more straightforward scale-up of optimized processes, these must be balanced against the significant investment and change management required to deploy them effectively in bioethanol production.

Data Security

The connectivity of Industry 4.0 inherently raises data security concerns. As manufacturers embrace data-driven processes, "manufacturing is becoming more vulnerable to digital threats", making cybersecurity a critical challenge (Toussaint et al.

2024). For bioethanol plants, which often collect and share large volumes of process and quality data, cyber-attacks or tampering could disrupt operations or compromise product integrity. Toussaint et al. (2024) emphasize that data manipulation is a particularly dangerous threat in Industry 4.0, requiring robust frameworks to protect industrial data flows. In practice, this means implementing secure networks, encryption, and anomaly detection—for example, blockchains have been proposed to ensure traceable, tamper-proof records of biomass feedstocks and fuel batches.

Biofuel producers must therefore weigh the benefits of cloud-based analytics and remote monitoring against vulnerability to attacks. As Gargalo et al. (2020) noted, the biochemical industry often lags in digital adoption partly due to regulatory compliance and data integrity concerns. Any Industry 4.0 retrofit must include cybersecurity safeguards: unauthorized access to SCADA controls or IoT devices could lead to costly downtime or safety incidents. In summary, securing the data and networks of a smart ethanol plant is a significant challenge that parallels those faced by other smart factories (Maddikunta et al. 2021).

Process Complexity

Bioprocesses like ethanol fermentation are inherently complex, which complicates Industry 4.0 implementation. Unlike discrete manufacturing, biomanufacturing involves living cells and biological variability, so each step (pretreatment, enzymatic hydrolysis, fermentation, distillation) can be sensitive to subtle changes. One industry survey observes that "biomanufacturing amplifies the complexity and risks associated with adopting Industry 4.0 technologies". Strict regulatory controls also apply (e.g., contamination control, food-safety standards for ethanol used in beverages), so every IoT integration or process modification must comply with tight production guidelines. In practice, even monitoring a few key variables often requires elaborate sensors and models. For example, soft sensors (inferential models) are used to estimate hard-to-measure parameters in real time, but developing and validating them adds complexity.

This complexity means many bioethanol plants remain at earlier automation levels. Gargalo et al. (2020) report that many biochemical firms still operate in "Industry 3.0" mode because upgrading is nontrivial: the intrinsic complexity of biological systems, stringent regulations, and concerns about scalability all slow digital adoption. In other words, unlike simpler chemical processes, ethanol production cannot always be "plugged in" to IoT without careful re-engineering. Hybrid modeling (combining mechanistic and data-driven models) is one approach under study to manage this complexity. Overall, the heterogeneity and tight control requirements of bioethanol processes impose significant challenges on Industry 4.0 projects, requiring more careful implementation and validation than in many other sectors (Caccavale et al. 2024).

Energy Efficiency

Energy optimization is a clear opportunity for Industry 4.0 in ethanol plants. Advanced monitoring and predictive control can reduce waste heat and improve the use of utilities. In practice, Industry 4.0 auditing tools have already realized considerable energy savings: as noted above, reconfiguring a beer-preheater at one plant via data analytics cut steam usage by up to 6%. That kind of improvement—modest changes yielding hundreds of euros per day saved—scales across hundreds of plants, implying a substantial global impact on CO_2 emissions. More generally, Industry 4.0 methods (real-time analytics, digital twins, AI) target heat and power integration: sensor networks can identify heat recovery opportunities, and smart algorithms can fine-tune fermentation and distillation to minimize energy per liter of ethanol.

Indeed, simulations of future biorefineries suggest significant efficiency gains are possible. Clauser et al. (2022) highlight that smart manufacturing (sensors + AI) can increase operational flexibility and resource efficiency across the production chain. For example, dynamic control of feedstock preprocessing and recycle loops—enabled by real-time data—can save steam and cooling water. Case studies also show improved uptime (hence more product per unit energy) after digital interventions. In short, by continuously monitoring energy flows and using predictive analytics, Industry 4.0 approaches make ethanol plants more energy efficient. However, achieving maximum efficiency gains again requires investment in instrumentation and control systems, so these projects must balance payback periods against long-term sustainability goals.

Industry 4.0 offers powerful tools to enhance bioethanol production—lowering costs, improving yields, and cutting energy use. Real-world examples (from process simulations to plant retrofits) already demonstrate these benefits. At the same time, the unique complexities of biological processes and the need for secure, robust data systems mean that rolling out these technologies is challenging. Future work must continue to address cybersecurity, regulatory, and modelling issues if smart biorefineries are to reach their full potential.

12.7 Challenges and Future Prospects

Challenges Despite the transformative potential of Industry 4.0 in sustainable bioethanol production, several challenges hinder its widespread implementation:

High Capital Investment—Integration of advanced sensors, process automation systems, and AI/ML infrastructure demands significant upfront financial resources, which can be a barrier for small and medium-scale bioethanol plants.

Data Integration and Standardization—Bioethanol plants often involve heterogeneous data sources (biochemical, thermochemical, environmental), and ensuring compatibility between IoT devices, legacy systems, and cloud platforms remains a challenge.

Scalability of AI/ML Models—ML models trained on laboratory or pilot-scale data may not generalize effectively to industrial-scale production due to variability in feedstock quality and fluctuating process conditions.

Cybersecurity Concerns—As production systems become more digitized and network-linked, they are increasingly vulnerable to cyberattacks, data breaches, and system disruptions.

Sustainability of Genetically Modified Feedstocks—While GM crops and engineered microorganisms improve yield, they also raise ecological, regulatory, and socio-economic concerns regarding biosafety, biodiversity, and farmer adoption.

Skilled Workforce Requirement—Effective implementation of Industry 4.0 technologies in bioprocess industries requires professionals with specialized interdisciplinary skills covering biotechnology, data analytics, automation, and cyber-physical systems.

Future Prospects Looking ahead, Industry 4.0-driven bioethanol production offers promising opportunities to enhance sustainability and circular economy principles:

Smart Biorefineries—Fully integrated digital biorefineries can achieve dynamic process optimization using real-time monitoring, advanced simulations, and adaptive control strategies.

Digital Twins—Developing digital twins of fermentation and distillation units will enable predictive maintenance, rapid scenario analysis, and resilient process control.

Blockchain for Biomass Traceability—Blockchain-based platforms can ensure feedstock traceability, authenticity, and carbon footprint transparency along the ethanol supply chain.

Hybrid Modelling Approaches—Combining mechanistic biochemical models with AI-driven data analytics will provide more interpretable, accurate, and scalable predictive frameworks for ethanol yield.

Circular Economy Integration—Expansion of co-product valorization strategies (bio-based chemicals, high-protein yeast, refined oils, and biogas) can reinforce the transition to "zero-waste" biorefineries.

Global Collaboration and Policy Support—Cross-institutional, multinational collaborations supported by enabling policies will accelerate the adoption of Industry 4.0 in ethanol production, particularly in regions emphasizing renewable energy independence and green growth.

12.8 Conclusion

Industry 4.0 played a significant role in the transition of integrated biomanufacturing of bioethanol production into a smart digital biorefinery equipped with advanced AI technologies and digital twins. Industry 4.0 has become a reality that can promise higher-order productivity of different generations of bioethanol production under the advanced industrial revolution. Bioethanol production depends on different feedstocks that require a central hub interconnected through advanced technologies. Thus, integrating these digital and physical systems ensured increased efficiency, productivity, and flexibility in bioethanol production. As a result of bioethanol 4.0 production, advanced technologies such as machine learning, big data analytics, the Internet of Things (IoT), cloud computing, and digital twins can be significantly utilized in the bioethanol industry.

However, in the future, the emergence of Industry 5.0 can be prioritized in industrial bioethanol production through the collaboration of human expertise with advanced machinery and technology. It ensures the merging of machine accuracy with human critical thinking. It is focused on large-scale customization of diverse biomass/feed conversion valorization for bioethanol production for selective product delivery to consumers. Thus, a new paradigm can foster a deep-level human-machine collaboration, thereby increasing the performance of the bioethanol production system.

Acknowledgment The authors sincerely acknowledge the constant support and encouragement provided by the Institutional heads of Shaheed Mahendra Karma Vishwavidyalaya, Bastar; Chhattisgarh Swami Vivekanand Technical University, Bhilai; University of Technology and Applied Sciences—Sur, Oman, and Amity University Uttar Pradesh, Noida. Their vision, infrastructure, and collaborative spirit were instrumental in facilitating this interdisciplinary research work and ensuring its successful completion.

References

Allais F, Lescieux-Katir H, Chauvet J (2021) The continuous evolution of the Bazancourt–Pomacle site is rooted in the commitment and vision of pioneering farmers. When reality shapes the biorefinery concept. EFB Bioecon J 1:100007. https://doi.org/10.1016/j.bioeco.2021.100007

Anttila J (2019) An intelligent circular economy upstream monitoring & optimization system based on industrial Internet of Things. Master's thesis, University of Turku, Turku, Finland

Balabin RM, Smirnov SV (2011) Variable selection in near-infrared spectroscopy: benchmarking of feature selection methods on biodiesel data. Anal Chim Acta 692(1–2):63–72. https://doi.org/10.1016/j.aca.2011.03.006

Basso T, Basso T, Gallo C, Basso L (2013) Towards the production of second generation ethanol from sugarcane bagasse in Brazil. In: InTech eBooks. https://doi.org/10.5772/54179

Bioethanol Producer Employs Advanced Automation Technology to Optimize Integration, Operation and Lifecycle Management of Critical Assets-automationworld.com

Bisgaard T, Mauricio-Iglesias M, Huusom JK, Gernaey KV, Dohrup J, Petersen MA, Abildskov J (2017) Adding value to bioethanol through a purification process revamp. Ind Eng Chem Res 56(19):5692–5704. https://doi.org/10.1021/acs.iecr.7b00442

Caccavale F, Gargalo CL, Gernaey KV, Krühne U (2024) Towards education 4.0: the role of large language models as virtual tutors in chemical engineering. Educ Chem Eng 49:1–11. https://doi.org/10.1016/j.ece.2024.07.002

Castellet-Viciano L, Hernández-Chover V, Bellver-Domingo Á, Hernández-Sancho F (2022) Industrial Symbiosis: a mechanism to guarantee the implementation of circular economy practices. Sustainability 14(23):15872. https://doi.org/10.3390/su142315872

Chai T, Draxler RR (2014) Root mean square error (RMSE) or mean absolute error (MAE)? Res Gate. https://doi.org/10.5194/gmdd-7-1525-2014

Chen H, Liu Z, Zhang Y (2021) Waste valorization and resource recovery in biorefineries: a review. Renew Energy 172:1095–1107

Clauser NM, Felissia FE, Area MC, Vallejos ME (2022) Integrating the new age of bioeconomy and Industry 4.0 into biorefinery process design. Bioresources 17(3):5510–5531. https://doi.org/10.15376/biores.17.3.clauser

Demestichas K, Daskalakis E (2020) Information and communication technology solutions for the circular economy. Sustainability 12(18):7272. https://doi.org/10.3390/su12187272

Demestichas K, Peppes N, Alexakis T, Adamopoulou E (2020) Blockchain in agriculture traceability systems: a review. Appl Sci 10(12):4113. https://doi.org/10.3390/app10124113

Dennis E, Gertner D, Erickson G (2024) Economic research on ethanol feed-use coproducts: a review, synthesis, and path forward. Animals 14(11):1551. https://doi.org/10.3390/ani14111551

Ewanick SM, Thompson WJ, Marquardt BJ, Bura R (2013) Real-time understanding of lignocellulosic bioethanol fermentation by Raman spectroscopy. Biotechnol Biofuels 6(1). https://doi.org/10.1186/1754-6834-6-28

Gargalo CL, De Las Heras SC, Jones MN, Udugama I, Mansouri SS, Krühne U, Gernaey KV (2020) Towards the development of digital twins for the bio-manufacturing industry. In: Advances in biochemical engineering, biotechnology, pp 1–34. https://doi.org/10.1007/10_2020_142

Geissdoerfer M, Savaget P, Bocken NMP, Hultink EJ (2017) The circular economy—a new sustainability paradigm? J Clean Prod 143:757–768

Gupta A, Verma JP (2015) Sustainable bio-ethanol production from agro-residues: a review. Renew Sust Energ Rev 41:550–567

Halstenberg FA, Lindow K, Stark R (2017) Utilization of product lifecycle data from PLM systems in platforms for industrial symbiosis. Proc Manuf 8:369–376. https://doi.org/10.1016/j.promfg.2017.02.047

Itto-Nakama K, Watanabe S, Kondo N, Ohnuki S, Kikuchi R, Nakamura T, Ogasawara W, Kasahara K, Ohya Y (2021) AI-based forecasting of ethanol fermentation using yeast morphological data. Biosci Biotechnol Biochem 86(1):125–134. https://doi.org/10.1093/bbb/zbab188

Jabbour C, Jabbour A, Sarkis J, Filho M (2019) Unlocking the circular economy through new business models based on large-scale data: an integrative framework and research agenda. Technol Forecast Soc Chang 144:546–552

Kagermann H, Wahlster W, Helbig J (2013) Recommendations for implementing the strategic initiative INDUSTRIE 4.0. Final report of the Industrie 4.0 Working Group

Khamwachirapithak P, Sae-Tang K, Mhuantong W, Tanapongpipat S, Zhao X, Liu C, Wei D, Champreda V, Runguphan W (2023) Optimizing ethanol production in saccharomyces cerevisiae at ambient and elevated temperatures through machine learning-guided combinatorial promoter modifications. ACS Synth Biol 12(10):2897–2908. https://doi.org/10.1021/acssynbio.3c00199

Kumar A et al (2020) Advancements in bioethanol production from lignocellulosic biomass: a review. Bioresour Technol 316:123928

Kwon H, Shiu J, Rivera EC, Yamakaya C (2024) Soft sensor for ethanol fermentation monitoring through data-driven modeling and synthetic data generation. Proceedings 104(1):30. https://doi.org/10.3390/proceedings2024104030

Li X, Jin X, Lu S, Li Z, Wang Y, Cao J (2020) Carbon-efficient production scheduling of a bioethanol plant considering diversified feedstock pelletization density: a case study. Processes 8(9):1189. https://doi.org/10.3390/pr8091189

Li X, Dong Y, Chang L, Chen L, Wang G, Zhuang Y, Yan X (2023) Dynamic hybrid modeling of fuel ethanol fermentation process by integrating biomass concentration XGBoost model and kinetic parameter artificial neural network model into mechanism model. Renew Energy 205:574–582. https://doi.org/10.1016/j.renene.2023.01.113

Liu Y, Feng C, Fang C, Ding W, Zhu F, Zhang G, An M (2025) Interpretable machine learning for optimizing carbon source design to enhance bioethanol yield in gas fermentation. SSRN. https://doi.org/10.2139/ssrn.5159136

Lopez PC, Udugama IA, Thomsen ST, Roslander C, Junicke H, Mauricio-Iglesias M, Gernaey KV (2020) Towards a digital twin: a hybrid data-driven and mechanistic digital shadow to forecast the evolution of lignocellulosic fermentation. Biofuels Bioprod Biorefin 14(5):1046–1060. https://doi.org/10.1002/bbb.2108

Lundberg SM, Lee S (2017) A unified approach to interpreting model predictions. arXiv (Cornell University). https://doi.org/10.48550/arxiv.1705.07874

Ma J, Liu Y, He W (2018) Cleaner production in the bioenergy industry. J Clean Prod 197:1681–1693

Ma H, Liu Y, Zhao J, Fei F, Gao M, Wang Q (2024) Explainable machine learning-driven predictive performance and process parameter optimization for caproic acid production. Bioresour Technol 410:131311. https://doi.org/10.1016/j.biortech.2024.131311

Maddikunta PKR, Pham QBP, Deepa N, Dev K, Gadekallu TR, Ruby R, Liyanage M (2021) Industry 5.0: a survey on enabling technologies and potential applications. J Ind Inf Integr 26: 100257. https://doi.org/10.1016/j.jii.2021.100257

Mitchell R, Frank E (2017) Accelerating the XGBoost algorithm using GPU computing. PeerJ Comput Sci 3:e127. https://doi.org/10.7717/peerj-cs.127

Momenitabar M, Ebrahimi ZD, Abdollahi A, Helmi W, Bengtson K, Ghasemi P (2023) An integrated machine learning and quantitative optimization method for designing sustainable bioethanol supply chain networks. Decision Analyt J 7:100236. https://doi.org/10.1016/j.dajour.2023.100236

Muhammad ZY, Abdullah A (2023) Industry-4.0-enabled digital transformation: prospects, instruments, challenges, and implications for business strategies. Sustainability 15(11):8553. https://doi.org/10.3390/su15118553

Okolie JA (2024) Introduction of machine learning and artificial intelligence in biofuel technology. Curr Opin Green Sustain Chem 47:100928. https://doi.org/10.1016/j.cogsc.2024.100928

Owusu WA, Marfo SA (2023) Artificial intelligence application in bioethanol production. Int J Energy Res:1–8. https://doi.org/10.1155/2023/7844835

Roundtable on Sustainable Biomaterials (RSB), Bioledger (2021) Blockchain database for sustainable biofuels: a case study [Report]. https://rsb.org/wp-content/uploads/2021/03/Blockchain-Database-for-Sustainable-Biofuels-A-Case-Study-March-2021.pdf#:~:text=Bioledger%20is%20further%20developing%20the,the%20lessons%20learnt%20through%20this

Sewsynker-Sukai Y, Faloye F, Kana EBG (2016) Artificial neural networks: an efficient tool for modelling and optimization of biofuel production (a mini review). Biotechnol Biotechnol Equip 31(2):221–235. https://doi.org/10.1080/13102818.2016.1269616

Smuga-Kogut M, Kogut T, Markiewicz R, Słowik A (2021) Use of machine learning methods for predicting the amount of bioethanol obtained from lignocellulosic biomass with the use of ionic liquids for pretreatment. Energies 14(1):243. https://doi.org/10.3390/en14010243

Sousa FMM, Fonseca RR, Da Silva FV (2021) Empirical modeling of ethanol production dynamics using long short-term memory recurrent neural networks. Bioresour Technol Rep 15:100724. https://doi.org/10.1016/j.biteb.2021.100724

Spyrou O, Hurst W, Verdouw C (2023) Virtual Reality-Based Digital Twins: a case study on pharmaceutical cannabis. Big Data Cogn Comput 7(2):95. https://doi.org/10.3390/bdcc7020095

Toussaint M, Krima S, Panetto H (2024) Industry 4.0 data security: a cybersecurity frameworks review. J Ind Inf Integr 39:100604. https://doi.org/10.1016/j.jii.2024.100604

UNEP (2011) Cleaner production: key elements. United Nations Environment Programme

Vinitha N, Vasudevan J, Gopinath KP (2022) Bioethanol production optimization through machine learning algorithm approach: biomass characteristics, saccharification, and fermentation conditions for enzymatic hydrolysis. Biomass Convers Biorefinery 13(8):7287–7299

Zhang P, Zhang T, Zhang J, Liu H, Chicaiza-Ortiz C, Lee JTE, He Y, Dai Y, Tong YW (2024) A machine learning assisted prediction of potential biochar and its applications in anaerobic digestion for valuable chemicals and energy recovery from organic waste. Carbon Neutrality 3(1). https://doi.org/10.1007/s43979-023-00078-0

Chapter 13
Lignin Valorization in Lignocellulosic Bioethanol Production Within Integrated Biorefinery and Circular Economy Systems

Shiv Prasad, Mohd. Zafar, and Vikas Chandra Gupta

Abstract The effective use of lignin biopolymers in integrated lignocellulosic bioethanol biorefineries offers a significant opportunity to improve sustainability and support a circular bioeconomy. Converting lignin, a biopolymer derived from lignocellulosic biomass, into high-value products at a commercial scale offers a promising alternative to chemically synthesized polymers and supports waste management. Fortunately, innovative research and technologies are emerging to address these challenges. Expanding the range of lignin-derived products and byproducts that can be produced at a large scale is key to unlocking the full potential of lignin-rich biomass waste as a renewable resource. Still, challenges remain due to lignin's complex and rigid structure, particularly its reactive, condensation-prone degraded fractions. There is an urgent need to adopt integrated strategies, including advanced depolymerization technologies, to improve production processes. Additionally, evaluating lignin's compatibility with various pretreatment methods—such as chemical, thermal, and biological techniques—is essential. Thorough techno-economic analyses are needed to maximize the use of lignin and support the circular economy, ultimately aiming for zero waste and aligning with global sustainability goals.

Keywords Lignocelluloses · Lignin · Biopolymers · Depolymerization · Bioethanol · Biorefineries · Circular Economy

S. Prasad (✉)
Division of Environmental Sciences, ICAR-Indian Agriculture Research Institute, New Delhi, Delhi, India

M. Zafar
Department of Biotechnology, College of Applied Sciences and Pharmacy, University of Technology and Applied Sciences, Sur, Oman
e-mail: Mohammed.zafar@utas.edu.om

V. C. Gupta
Department of Biotechnology, IILM University, Greater Noida, Uttar Pradesh, India

S. Prasad et al. (eds.), *Sustainable Bioethanol Production from Lignocellulosic Biomass*, Clean Energy Production Technologies,
https://doi.org/10.1007/978-981-95-9690-4_13

13.1 Introduction

Lignocellulosic biorefineries play a pivotal role in promoting energy sustainability and the bioeconomy, with growing emphasis on lignin utilization for both economic and environmental benefits (Ragauskas et al. 2014). Lignin, the second most abundant natural biopolymer after cellulose, constitutes approximately 15–30% of lignocellulosic biomass and provides structural integrity to plant cell walls (Tamo et al. 2025). Despite extensive research on carbohydrate conversion, lignin remains underutilized, often discarded or incinerated due to its structural complexity. Huang et al. (2019) describe lignin as a complex aromatic heteropolymer derived from three types of cinnamyl alcohol monomers, which form an amorphous, cross-linked three-dimensional network. Broda et al. (2022) note that the composition of lignin varies depending on the biomass source: hardwood lignin mainly contains guaiacyl and syringyl units, softwood lignin primarily consists of guaiacyl units, and grass lignin includes p-hydroxyphenyl, guaiacyl, and syringyl units.

The biorefinery industry and wood pulping processes produce more than 50 million tons of lignin annually, but only ~2% is recovered for utilization in applications (Global Lignin Markets Report 2021). Despite its abundance, lignin is often underutilized in bioethanol production; it is typically burned for low-value energy recovery, whereas carbohydrates are converted into fuels (Jiju et al. 2025). This practice represents a missed opportunity, as lignin's aromatic structure has considerable potential for producing high-value bio-based chemicals, materials, and fuels, including aromatics, phenolics, biofuels, and macromolecules (Zhang et al. 2024). The transition from fossil-based to bio-based resources has intensified research on lignin utilization within integrated biorefineries. Currently, most biorefineries prioritize carbohydrate conversion to biofuels and regard lignin as waste, with minimal efforts directed toward its conversion or separation (Wu et al. 2024). Emerging technologies, including catalytic conversion, ionic-liquid-based processes, and biocatalytic methods, have significant potential to transform lignin into valuable products (Li et al. 2018).

Effective lignin valorization within biorefineries is crucial for decreasing dependence on petrochemicals and fostering a circular bioeconomy. Barbera and Elena (2025) assessed the economic potential of lignin utilization, highlighting the significance of lignin-derived products. Nonetheless, substantial challenges persist in achieving economically viable lignin conversion. Huang et al. (2019) identified conversion efficiency, product pricing, production costs, and energy consumption as key determinants. Their analysis suggests that lignin valorization becomes economically feasible when products are priced at $1.30 per kilogram or higher, with an 80% yield (Beckham and Salvachua 2023). Achieving these benchmarks may be supported by carbon pricing or biofuel incentives, which can reduce emissions costs and encourage investment and innovation. The study also indicated that energy requirements often constrain optimal biorefinery processes, highlighting the necessity for technological advancements to improve efficiency and reduce costs. This chapter focuses on recent advances in lignin valorization, which are really exciting for improving lignocellulosic bioethanol production within integrated systems. These developments could create new opportunities for sustainable energy and more efficient processes, benefiting all stakeholders in biorefinery and circular-economy systems.

13.2 Lignin Structure and Composition

Lignin is a complex amorphous heteropolymer composed of phenylpropanoid monomers: guaiacyl (G), p-coumaryl (H), and syringyl (S) alcohol units. These monomers interconnect via ether and carbon-carbon bonds, forming a heterogeneous, three-dimensional aromatic network in plant cell walls, as shown in Fig. 13.1.

According to Vanholme et al. (2010), lignins from softwoods (e.g., gymnosperms) mainly consist of G-units, with minor amounts of H-units. In contrast, lignins from hardwoods (e.g., angiosperm dicots) contain a mixture of G- and S-units. Grass lignins are unique as they incorporate all three structural units (H, G, and S) in varying proportions. The composition of lignin varies significantly based on the biomass source: softwood lignin is primarily composed of guaiacyl units (80–90%), hardwood lignin features a combination of guaiacyl (25–50%) and syringyl units (50–70%), and grass lignin includes all three units, with guaiacyl (25–50%), syringyl (25–50%), and p-hydroxyphenyl units (10–25%) (Palonen 2004). This structural complexity presents both challenges and opportunities for lignin valorization strategies (Liu et al. 2021a, b, c).

The primary building blocks of lignin are three hydroxycinnamyl alcohols known as monolignols: p-coumaryl alcohol, coniferyl alcohol, and sinapyl alcohol (Vanholme et al. 2010). These monolignols are synthesized from phenylalanine through general phenylpropanoid and specific monolignol biosynthetic pathways. The degree of methoxylation on the aromatic ring varies among them: p-coumaryl alcohol has no methoxy groups, coniferyl alcohol has one methoxy group, and sinapyl alcohol has two methoxy groups (Feng et al. 2014).

The structural complexity of lignin arises from its heterogeneous aromatic polymer, which consists of phenylpropanoid units interconnected by diverse carbon-carbon and carbon-oxygen bonds (Ralph et al. 2004). This complexity poses significant challenges for efficient depolymerization and valorization (Zhao et al. 2025). Although lignin exhibits high antioxidant capacity and biodegradability, making it a promising feedstock for value-added bioproducts, its valorization is constrained by processing inefficiencies and scalability limitations (Smith et al. 2022a, b; Ruwoldt et al. 2023). Scalability remains the primary obstacle to industrial adoption. Addressing these challenges necessitates integrated approaches that combine technological innovation, interdisciplinary collaboration, industrial integration, and supportive policy measures.

Lignin polymerization occurs through the oxidative radicalization of phenolic monolignols, followed by combinatorial radical coupling. The most common linkage type in lignin is the β-O-4 (aryl ether) bond, which accounts for approximately

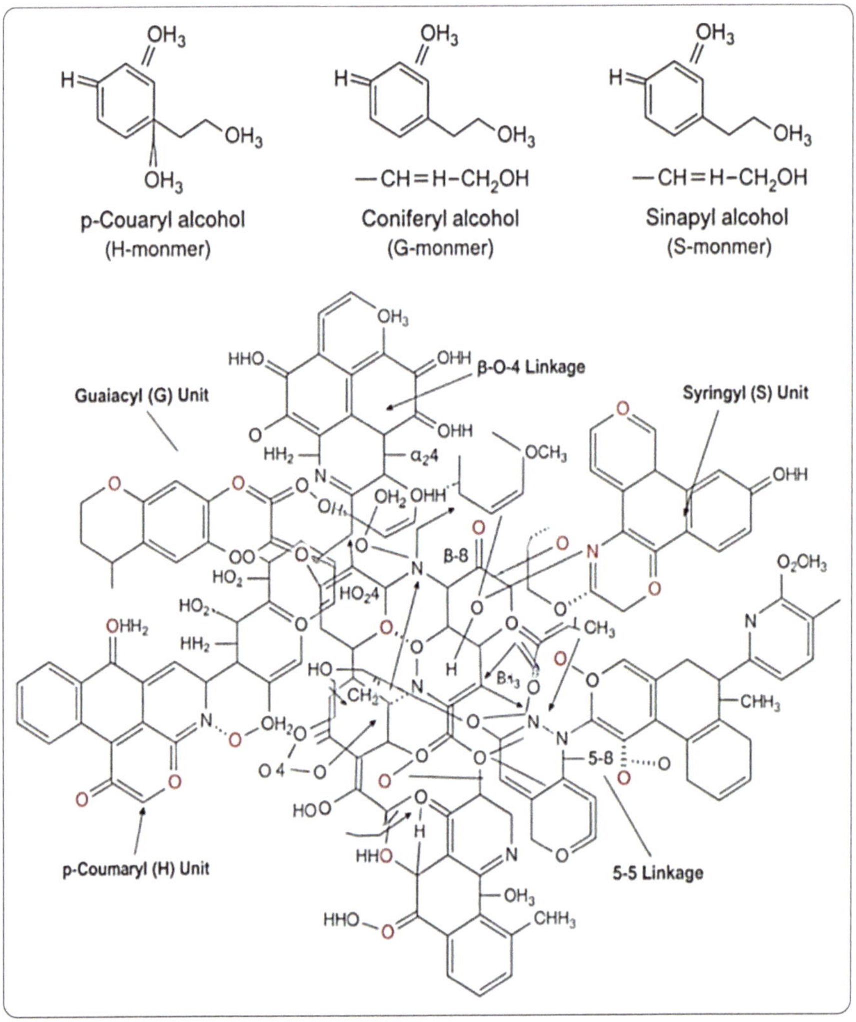

Fig. 13.1 Lignin phenylpropanoid monomers with a three-dimensional aromatic network

60% of all inter-unit connections. Smith et al. (2022a, b) observed that monolignol radicals tend to couple at their β positions, leading to β-β, β-O-4, and β-5 dimers as the primary linkage types. Less frequent linkages include 5-5, 4-O-5, and α-O-4 bonds. The distribution of these linkages varies by plant type; for instance, wheat straw lignin primarily contains β-O-4 linkages and fewer carbon-carbon bonds compared to wood lignin (Smith et al. 2022a, b). Lignin features key functional groups, including phenolic and aliphatic hydroxyls, methoxy, carbonyl, and carboxyl groups, which determine its chemical reactivity and physical properties (Table 13.1).

Table 13.1 Key functional groups and their properties description

Functional group	Properties description	References
Hydroxyl (-OH)	Phenolic (aromatic) and aliphatic compounds are key for reactivity, adhesion, and antioxidants	Ruwoldt et al. (2023) and Zhao et al. (2025)
Methoxy (-OCH_3)	Increases hydrophobicity, varying with monolignol (G/S units)	Ruwoldt et al. (2023) and Zhao et al. (2025)
Carbonyl (C=O)	Provides chemical reactivity in quinone forms post-oxidation	Liu et al. (2019a, b)
Carboxyl (-COOH)	Enhances pH responsiveness and solubility; minor but modifiable	Zhao et al. (2025) and Ruwoldt et al. (2023)

Lignin's molecular weight varies widely, affecting its physicochemical properties (Stewart et al. 2009). In poplar, linear lignin chains average 13–20 monomeric units. Molecular weight distribution is assessed by number-average (Mn), weight-average (Mw), and polydispersity index (PDI). Depending on source and extraction method, Mn typically ranges from 6000 to 10,000 g/mol (Stewart et al. 2009), making lignin a versatile macromolecule. In this range, for example, lignin serves as a reinforcing filler or compatibilizer in thermoplastics, enhancing mechanical strength and UV stability. Higher-molecular-weight fractions are especially suitable for fiber spinning and carbonization, providing a bio-based alternative to polyacrylonitrile (PAN) and enabling binder production.

13.3 Lignin from Bioethanol Biorefineries and Other Sources

In lignocellulosic biorefineries, removing lignin improves enzymatic hydrolysis of carbohydrate polymers and increases biofuel yields (Sun and Cheng 2002). After pretreatment and cellulose extraction, the residual material is primarily lignin, which is often incinerated for energy recovery. However, current processes produce lignin that is poorly suited for bioconversion due to inefficient separation, limited processability, and low product yields (Liu et al. 2021a, b, c). Harsh pretreatment conditions can further condense lignin, reducing its bioconversion efficiency (Zhang 2010). Despite these challenges, there is potential to convert lignin into higher-value products. Lignin from kraft pulping and sulfite processes includes kraft lignin, lignosulfonates, and organosolv lignin (Yoon and Bae 2015). Each type has distinct chemical properties and solubility, making them suitable for different industrial applications (Liu et al. 2021a, b, c). Differences in solubility and reactive sites among these lignin types significantly affect their downstream valorization potential. Table 13.2 summarizes lignin content in residues from bioethanol and other industries.

In lignocellulosic bioethanol production, lignin inhibits enzymatic access to cellulose and hemicellulose (Dessie et al. 2023). Broda et al. (2022) noted that lignin

Table 13.2 Lignin from bioethanol biorefineries and industrial sources

Lignin source/stream	Feedstocks used	Lignin content (%)	References
Bioethanol hydrolysis residue	*Eucalyptus globulus* and *Panicum virgatum*	70.0–83.0%	Vázquez et al. (2023)
Bioethanol lignin residue	Softwood (spruce/pine)	42.1%	Herrera et al. (2010)
Bioethanol lignin residue	*Arundo donax* (giant reed)	~40.0%	Cotana et al. (2014)
Kraft black liquor (solids)	Softwood/hardwood	35.0–45.0%	Vakkilainen (2009)
Sulfite spent liquor	Wood pulping residue	64.0–76.0%	Li et al. (2018)
Soda lignin (protobind 1000)	Wheat straw/annual fibers	90.0–96.0%	Constant et al. (2016) and Domínguez-Robles et al. (2018)
Organosolv lignin (alcell)	Hardwood (poplar/aspen)	>96.0%	Constant et al. (2016) and Domínguez-Robles et al. (2018)
Fractionated lignin	Sugarcane bagasse	65.4–89.8%	Imman et al. (2021)
Raw biomass (reference)	Corn stover	18.9–21.9%	Xue et al. (2021) and Wu et al. (2021)

blocks hydrolytic enzymes from reaching cellulose microfibrils and irreversibly inactivates them through surface binding (Zhao et al. 2025). Despite these challenges, Huang et al. (2019) found that extracting economic value from lignin in biorefineries remains difficult. Their analysis indicates that the most effective strategies would result in minimum ethanol prices of \$3.44 per gallon of gasoline-equivalent (GGE), excluding lignin valorization. They also determined that about 45% of lignin could be converted into bioproducts, with the remaining 55% used for heat and power, provided the biorefinery is thermally neutral.

Lignin is a significant obstacle in lignocellulosic bioethanol production because it restricts enzymatic access to cellulose and hemicellulose. Broda et al. (2022) showed that lignin limits the mobility of hydrolytic enzymes and can irreversibly inactivate them via surface adsorption. Even with improved strategies, lignin valorization remains economically challenging. Huang et al. (2019) found that extracting maximum economic value from lignin is difficult, with minimum ethanol selling prices reaching \$3.44 per gallon of gasoline equivalent (GGE) without lignin valorization. Approximately 45% of lignin can be converted into bioproducts, with the remainder used for heat and power in thermally neutral operations.

The pulp and paper industry is the world's leading producer of technical lignin, with over 85% of the global supply consisting of Kraft lignin, the most commonly produced type industrially (Smith et al. 2022a, b). Approximately 75% of global pulp production is produced by the kraft process, which generates approximately 170 million tons annually (Jones and Miller 2021). Most Kraft lignin is incinerated in black liquor recovery boilers, which provide heat and electricity for pulp mills. Some mills can recover up to 90% of their energy from lignin combustion (Davis 2020). Haq et al. (2020) identified wastewater from the pulp and paper industry as a significant source of lignin, which contains various synthetic, aromatic, and chlorinated compounds associated with lignin degradation. The sulfite pulping process produces lignosulfonates, which are water-soluble and account for about 10% of industrial lignin production, with approximately 1–2 million tons generated each year. Phung et al. (2020) noted that lignosulfonates are well-established in markets such as construction, agriculture, and the chemical industry, owing to their unique dispersing and binding properties.

13.4 Lignin Valorization Through Chemical, Biological, and Physicochemical Technologies

Lignin valorization uses various technologies to convert lignin, a complex component of plant biomass, into valuable chemicals, materials, and fuels (Ragauskas et al. 2014). The primary methods are chemical, biological, and physicochemical, each with its own benefits and challenges for sustainable biorefineries (Kumar et al. 2022). Thermochemical processes are known for producing high-value products from lignin, while bioconversion methods are valued for their greater sustainability. Choosing the correct method depends on the type of feedstock and the desired products. Each approach also affects the environment differently, with some offering significant reductions in greenhouse gas emissions. For instance, bioconversion can reduce emissions relative to older methods, thereby supporting environmental goals. Table 13.3 gives an overview of lignin valorization strategies, including their yields, benefits, drawbacks, and environmental impacts.

13.4.1 Lignin Valorization by Chemical Depolymerization and Catalysis

Chemical depolymerization and catalysis for lignin valorization involve breaking down this complex polymer into valuable aromatic compounds and monomers using diverse chemical and catalytic techniques. This strategy seeks to overcome lignin's inherent recalcitrance, enhance yield selectivity, and enable the production of renewable chemicals and fuels. Compared to thermal or biological methods, chemical depolymerization with specific reagents offers greater control over reaction pathways, thereby improving product selectivity (Katahira et al. 2016). A comparison between alkaline and acid depolymerization routes reveals distinct differences in selectivity, cost, and scalability that could guide decision-making for industrial applications.

Table 13.3 Various lignin valorization techniques, yields, advantages, and disadvantages

Valorization technique	Feedstocks	Lignin yield (%)	Advantages	Disadvantages	References
Alkaline hydrolysis	De-oiled Jatropha waste	93.4	Lower temperatures, high selectivity	Expensive catalyst, corrosive	Oruganti et al. (2024)
Dilute acid hydrolysis	Corn stover	75.7	No acid recycling needed	High temperature/pressure	Gill et al. (2021)
Concentrated acid	Sugarcane bagasse	91.0	Mild temperatures	Expensive acid recovery	Avila and Rosana (2022)
Oxidative cleavage	Toona sinensis	77.0	Effective degradation	Expensive reagents	c
Ionic liquid	Miscanthus × giganteus	82.0	High dissolution	High IL costs	Wang and Eika (2020)
Pyrolysis	Tobacco waste	85.0	Simple process	High temperature requirement	Tan et al. (2024)
Fungal	Sugarcane bagasse	87.6	Moderate conditions, cost-effective	Slow rate, sterile area needed	Dong et al. (2013)
Bacterial	Paper mill waste	89.0	Cost-effective	Low hydrolysis rate	Gu et al. (2024a, b)
Enzymatic	Paper mill waste	90.0	Minimal energy	Low efficiency	Kumar et al. (2022)

13.4.1.1 Lignin Valorization by Alkaline Depolymerization

Alkaline depolymerization is a process that breaks β-ether linkages and C-C bonds in lignin, facilitated by high concentrations of hydroxide ions that promote deprotonation and substitution. Zhou et al. (2022a, b, c) emphasized that reaction mechanisms and product distribution are significantly influenced by processing parameters, particularly pH and the choice of oxidants (Smith et al. 2021). Under alkaline oxidative conditions, lignin undergoes partial oxidation, forming phenolic aldehydes, acids, and ketones while retaining aromatic rings, which are valuable for producing higher-value products. This process uses strong bases, such as NaOH and KOH, to break down lignin into phenolic monomers and oligomers, thereby reducing repolymerization and char formation.

Katahira et al. (2016) demonstrated that residual lignin biomass, whether obtained from traditional or new biochemical methods, can be depolymerized using sodium hydroxide, yielding 45–78% of low-molecular-weight aromatics suitable for use as fuels or chemicals. Recent advancements include the use of a mixture of NaOH, choline chloride, and urea, which produced 23.06 mg/g of aromatic monomers at atmospheric pressure (Smith et al. 2021). The maximum lignin recovery for alkaline hydrolysis was reported to be 93.4% when using NaOH on de-oiled Jatropha waste (Jones and Liu 2021; Oruganti et al. 2024). Zhou et al. (2022a, b, c) noted that chemical depolymerization, particularly when combined with thermal methods such as thermochemical depolymerization, is crucial for fully harnessing lignin's potential to produce renewable fuels and valuable chemicals.

13.4.1.2 Lignin Valorization by Acid-Catalyzed Depolymerization

Acid hydrolysis employs either mineral acids, such as hydrochloric, sulfuric, or phosphoric acid, or organic acids, such as formic or acetic acid, to cleave β-O-4 ether bonds, which constitute the primary linkages in lignin. Studies have demonstrated that dilute sulfuric acid used as a catalyst in a 1:1 ethanol-water mixture can achieve up to 70% lignin depolymerization while maintaining a high content of hydroxyl and carbon groups in the product (Roy et al. 2022). Despite these advantages, strong acid catalysts are associated with several limitations, such as harsh reaction conditions, elevated costs, challenges in catalyst recovery, and an increased risk of excessive lignin condensation (Zhang et al. 2024a, b, c).

Shao et al. (2022) reported that formic acid functions both as an acid catalyst for bond cleavage and as an in-situ hydrogen donor during hydrodeoxygenation when combined with metal catalysts such as Ru/C. Under mild conditions in aqueous formic acid, this approach effectively depolymerizes oxidized lignin, yielding more than 60 wt.% low-molecular-weight aromatics. Both formic acid and H_2SO_4 have demonstrated particular effectiveness under gentle microwave heating at 160 °C, yielding smaller molecular compounds with minimal solid residue (Shao et al. 2018). Furthermore, concentrated acid hydrolysis of sugarcane bagasse yielded

91% lignin, underscoring the method's efficiency at mild temperatures (Ávila and Rosana 2022).

13.4.1.3 Lignin Valorization by Oxidative Cleavage

Oxidative depolymerization utilizes agents such as hydrogen peroxide, nitrobenzene, oxygen, and metal oxides to cleave lignin bonds while maintaining aromatic structures (Lin et al. 2024). This approach facilitates the production of high-value aromatic aldehydes and acids, including vanillin, syringaldehyde, vanillic acid, and syringic acid. Smith et al. (2021) demonstrated that treating *Toona sinensis* branches with hydrogen peroxide and acetic acid reduced lignin content by 77%. Alherech et al. (2021) introduced an alkaline aerobic oxidation method that converts poplar lignin into oxygenated aromatic compounds, such as vanillin and syringaldehyde.

The primary products of this oxidation are phenolic aldehydes, acids, and ketones, which represent some of the most valuable lignin derivatives. Kruger et al. (2022) identified typical reaction conditions as 130–190 °C, reaction times of 10–180 min, oxygen pressures of 2–5 bar, and a 2 M aqueous sodium hydroxide solution, yielding nearly 30% monomer from hardwood lignin. Furthermore, Yang et al. (2022) reported that alkaline cupric oxide oxidation (OxCuO) achieved monomer yields of 71.9% from eucalyptus lignin, surpassing yields from conventional nitrobenzene oxidation (59.8%) and reductive depolymerization (51.9%).

13.4.1.4 Lignin Valorization by Ionic Liquid-Assisted Depolymerization

Ionic liquids (ILs) have emerged as environmentally benign solvents for lignin valorization, enabling its depolymerization into high-value chemicals (Zhang et al. 2019; Johnson and McCarthy 2021). Lignin dissolution in ILs occurs through hydrogen bonding between anions, such as chloride and acetate, and the hydroxyl groups on β-O-4 structures. Concurrently, cations engage in π-π stacking interactions with aromatic units. ILs efficiently disrupt lignin networks while preserving their structural integrity, enabling extraction and depolymerization at approximately 150 °C and offering notable recyclability (Smith et al. 2021).

The integration of IL-based processing into biorefineries supports the circular economy model for lignocellulosic biomass. Ionic liquids demonstrate significant chemical reactivity in lignin depolymerization while maintaining structural integrity. For example, 1-hexyl-3-methylimidazolium chloride ([Hmim]Cl) has been shown to maximize guaiacol production by efficiently cleaving β-O-4 linkages (Yang et al. 2017). An 82% lignin yield was reported from *Miscanthus giganteus* using the ionic liquid [HSO_4] (Tu et al. 2020; Johnson and McCarthy 2021). Despite these advances, the high operational costs of IL-based processes limit their scalability, underscoring the need for improved recycling strategies. Furthermore, Yang et al. (2017) demonstrated that 1-octyl-3-methylimidazolium acetate ([OMIm][OAc]) enables metal-free oxidation of lignin model compounds, achieving yields of 96%

for phenol and 86% for benzoic acid via acetate-mediated C-O bond cleavage (Yang et al. 2017).

13.4.2 Lignin Valorization by Thermophysical Depolymerization

Thermophysical depolymerization is a principal strategy for enhancing the value of lignin. Elevated temperatures disrupt the robust chemical bonds in lignin via combined physical and thermal mechanisms, thereby producing valuable chemicals and biofuels (Kumar et al. 2018). This method addresses the demand for economically viable biorefinery processes that efficiently utilize lignin (Sánchez-Silva et al. 2018).

13.4.2.1 Lignin Valorization by Pyrolysis-Based Depolymerization

Pyrolysis is a common method used for lignin breakdown. Zhu et al. (2022) described two main stages in this process. The first stage, which occurs at lower temperatures, involves cleavage of weaker ether bonds, such as β-O-4 and α-O-4, yielding lignin monomers. The second stage, happening at higher temperatures, involves cleaving C-O and C-C bonds, producing more monomers and various phenolic compounds through demethylation and demethoxylation. This method efficiently transforms lignin into bio-oil, syngas, and char, which are valuable resources for energy and other uses. Pyrolysis of tobacco waste can yield up to 85%, but achieving optimal bio-oil output requires high temperatures (Muzyka et al. 2022). Removing methoxy groups from lignin simplifies its structure and promotes the formation of specific phenolic compounds, particularly H-type phenolics (Smith et al. 2021; Lu et al. 2022). A new hydrothermal liquefaction technique has also been introduced to valorize lignin from bioethanol production, yielding high yields of phenolic compounds.

Saraeian et al. (2020) investigated lignin valorization via pyrolysis followed by low-pressure hydrodeoxygenation (HDO). Their results showed that char formation during pyrolysis captures more than half of the feed carbon, reducing carbon loss and increasing product yield. They also found that, regardless of lignin source, HDO with a MoO_3 catalyst can effectively convert complex pyrolysis vapors into monoaromatics (17–29%), as well as alkenes and alkanes (Saraeian et al. 2020). Zhou et al. (2022a, b, c) divided microwave-assisted lignin depolymerization into two types: microwave-assisted pyrolysis at temperatures above 400 °C and lower-temperature liquid-phase processes. Shao et al. (2022) studied microwave-assisted lignin breakdown in methanol/formic acid using a Ru/C catalyst with metal chloride co-catalysts, achieving a bio-oil yield of 91.1 wt.% and an aromatic monomer yield of 13.4 wt.%. The combined use of Zn^{2+} and Cl^- significantly increased the yields of both bio-oil and aromatic monomers (Shao et al. 2022).

13.4.2.2 Lignin Valorization by Hydrothermal Liquefaction-Based Depolymerization

Hydrothermal liquefaction (HTL) is a process that depolymerizes lignin into monomers, dimers, and phenols using aqueous or protic solvents. HTL typically operates under milder conditions, with temperatures ranging from 200 to 400 °C and pressures of 5 to 28 MPa, compared to pyrolysis. These conditions enhance the efficiency of lignin depolymerization. Zhu et al. (2022) identified optimal operational parameters for effective lignin liquefaction, emphasizing that maintaining a temperature range of 280–380 °C is essential for maximizing production of valuable phenolic compounds. This guidance helps process engineers optimize yields of high-value aromatic chemicals.

A comprehensive study on catechol production via hydrothermal liquefaction examined temperatures from 250 to 450 °C, using 1% wt KOH as a catalyst. Results showed that reaction rates were slow below the critical temperature of 250 °C. In contrast, temperatures above the supercritical point (450 °C) favored gasification, resulting in minimal production of the target compounds. The ideal temperature range for producing platform chemicals was determined to be 280–380 °C (Schuler et al. 2017). Catalytic experiments conducted at 300 °C indicated that the phenol yield remained below 3 wt% after 4 h, with guaiacol serving as an intermediate in the pathway to the primary product, catechol. Lui et al. (2025) noted that hydrothermal depolymerization methods, especially hydrothermal liquefaction, are promising approaches for converting lignin from biomass into fuels and valuable chemicals.

13.4.2.3 Lignin Valorization Gasification-Based Depolymerization

Liakakou et al. (2019) explored the use of lignin-rich residues from second-generation bioethanol for syngas production, a process relevant to gas fermentation. The comparison included three gasification methods: fixed-bed updraft (around 30 kg/h), bubbling fluidized-bed (about 0.3 kg/h), and indirect gasification (approximately 3 kg/h). These methods produced different molar H_2/CO ratios (0.6–1.0) and tar outputs (18–108 g/Nm^3). The study found that gas fermentation is more effective at lower H_2/CO ratios, as microorganisms grow faster on CO than on H_2 (Liakakou et al. 2019). Kang et al. (2015) investigated non-catalytic supercritical water gasification (SCWG) of lignin in batch reactors. The results showed that raising the temperature to 651 °C significantly increased hydrogen production, while pressure changes from 23 to 29 MPa had little effect on hydrogen yield. An interaction between temperature and the water-to-biomass ratio was observed at temperatures above 525 °C. Under optimal conditions (651 °C, 25 MPa, and a water-to-biomass ratio of 3.9), the maximum hydrogen output was 1.60 mmol/g biomass (Kang et al. 2015).

13.4.3 Lignin Valorization by Biological Depolymerization

Biological depolymerization of lignin is an environmentally sustainable approach to its microbial breakdown. This process facilitates the transformation of diverse lignin structures into valuable products by producing metabolites and by the action of extracellular oxidative enzymes, including peroxidases and laccases (Zhang and Wyman 2020). The critical roles of fungi and bacteria in lignin degradation are highlighted by their specific enzymatic and metabolic pathways (Zhao and Liu 2021). Advances in understanding lignin structure and biosynthesis have created new opportunities for biotechnological applications in lignin degradation.

13.4.3.1 Lignin Valorization by Microbial Depolymerization

Microbial depolymerization is a promising, environmentally sustainable approach for lignin valorization, owing to its low energy requirements and ability to generate valuable chemicals (Fu et al. 2025a, b, c). This method leverages microorganisms' natural capacity to secrete extracellular oxidative enzymes that degrade lignin, followed by intracellular metabolism of aromatic intermediates (Xu et al. 2022). Comparative analyses indicate that microbial depolymerization can reduce energy consumption by approximately 30% relative to traditional thermochemical methods, yielding approximately 25 kWh of energy savings per ton of lignin processed. Furthermore, this process may decrease carbon dioxide emissions by 15% (Gu et al. 2024a, b). Therefore, microbial depolymerization offers a cost-effective and environmentally friendly strategy. White-rot fungi, particularly *Phanerochaete chrysosporium*, secrete extracellular oxidases and peroxidases that catalyze the oxidative reactions essential for lignin degradation (Hammel et al. 1993; Kersten et al. 1987).

The *Lentinula edodes* LE16 strain achieved an 87.6% lignin extraction rate from sugarcane bagasse under moderate and environmentally sustainable conditions (Dong et al. 2013). Additionally, bacterial species such as *Pseudomonas putida*, *Rhodococcus jostii*, *Bacillus subtilis*, and *Streptomyces griseus* produce oxidative enzymes, including laccases, lignin peroxidases, and manganese peroxidases (Singh et al. 2015). For instance, *Bacillus cereus* achieved an 89% lignin recovery rate from paper mill waste, demonstrating both economic and environmental benefits (Kumar et al. 2022). Radhika et al. (2025) reported that, under optimized conditions (pH 5–9, 20–55 °C, and a hydraulic retention time of 3–30 days), lignin-rich residues produced approximately 120–130 ml/g of biogas and achieved 70% cumulative degradation of soluble lignin. Both enzymatic depolymerization and microbial degradation employ biocatalysts to selectively degrade lignin under mild conditions (Han and Yejun 2021).

13.4.3.2 Lignin Valorization by Enzymatic Conversion

Bacterial lignin depolymerization primarily involves two main enzyme classes: laccases and DyP-type peroxidases. These enzymes differ from fungal peroxidases (De Gonzalo et al. 2016). Bacterial enzymes typically exhibit greater thermal stability and robustness than their fungal counterparts. DyP-type peroxidases are particularly important owing to their broad substrate specificity. For example, DyP1B from various bacterial species can oxidize manganese (Mn(II)) and degrade kraft lignin, thereby releasing lignin fragments from wheat straw lignocellulose (De Gonzalo et al. 2016).

Laccases are key in removing electrons from phenolic hydroxyl groups, a step that enables both lignin polymerization and depolymerization via mechanisms such as carbon-carbon bond cleavage, carbon oxidation, and alkyl-aryl bond cleavage (Kuhad et al. 2011). Manganese peroxidase (MnP) produces Mn^{3+} chelates, which act as diffusible oxidants and help enzymes penetrate deep into the cell wall and lignocellulosic matrices. Lignin peroxidases (LiPs), manganese peroxidases (MnPs), and versatile peroxidases (VPs) are all essential for effective lignin breakdown (Kuhad et al. 2011). For instance, treatment of paper mill waste with laccase resulted in 90% lignin recovery, demonstrating the enzyme's effectiveness under mild conditions (Ghosh et al. 2016). Additionally, coexpression of multiple lignin-degrading enzymes in microbial systems can produce synergistic effects, thus boosting the overall efficiency of lignin degradation.

13.4.4 Lignin Valorization-Integrated Biorefinery Approaches

Contemporary integrated biorefinery strategies emphasize the extraction and conversion of lignin, a biopolymer, into high-value fuels, chemicals, and materials. These strategies utilize diverse pathways and technologies, including physical, chemical, and biological methods. The integration of lignin valorization into biorefinery frameworks supports the development of sustainable, cost-effective practices (Dixon et al. 2024). This integration is critical for decreasing dependence on petrochemicals and promoting a circular bioeconomy aligned with global sustainability goals (Jiju et al. 2025; Moshood et al. 2025). Furthermore, lignin valorization processes reduce wastewater toxicity, thereby enhancing environmental health. (Cui et al. 2022) These multidimensional benefits demonstrate that integration offers advantages beyond economic considerations.

13.4.4.1 Sequential Lignin Valorization Strategies

Stepwise sequential lignin valorization offers significant advantages over traditional single-step methods. This multi-stage process is designed to maximize value extraction from the complex biopolymer lignin (Li and Takkellapati 2018). Sequential strategies address lignin's structural heterogeneity through clearly defined stages such as isolation, depolymerization, and upgrading, thereby enabling more controlled conversion into high-value products (Huang and Ragauskas 2019a, b). Abu-Omar et al. (2021) introduced reductive catalytic fractionation (RCF), a process involving three main steps: solvolysis for lignin removal, stabilization of reactive intermediates via reductive chemistry, and depolymerization into monomers (Abu-Omar et al. 2021). Building on this methodology, Wu (2023) explored the sequential valorization of poplar sawdust by combining RCF with fermentation. This process uses Raney nickel as a catalyst and employs isopropanol as both a hydrogen donor and solvent.

Under optimal conditions—specifically a 7:3 isopropanol-to-water ratio at 200 °C for 3 h—it achieved 67.48% delignification and produced 20.65% monolignol (Wu et al. 2024). The RCF pulp retained 57.96% of carbohydrates, which were then converted into 21.61 g/L of bioethanol. This corresponds to a yield of 0.429 g ethanol per gram of biomass using engineered *Saccharomyces cerevisiae* (Wu et al. 2024). Mass-balance analysis indicates that 1000 g of raw poplar sawdust can produce 104.4 g of ethanol and 206.5 g of lignin oil, indicating efficient conversion of the main biomass components (Abu-Omar et al. 2021; Wu 2023; Wu et al. 2024). Chen et al. (2021) proposed an integrated cascade strategy combining catalytic hydrogenolysis and hydrothermal treatment for birch wood. In this method, Pd/C-catalyzed hydrogenolysis converts lignin into monomeric phenols with a yield of 44.6 wt%, while preserving 82.4% of cellulose and 71.6% of hemicellulose in the solid pulp. Subsequent hydrothermal treatment of the carbohydrate-rich pulp produces carbon quantum dots with a 21.7% yield (Chen et al. 2021). This sequential valorization strategy enables the comprehensive transformation of all biomass fractions.

13.4.4.2 Phase-Separated Pretreatment

Phase-separated pretreatment can significantly reduce biorefinery costs by improving lignin extraction efficiency, potentially increasing overall yield by up to 30%. This strategy is recognized as a crucial method for lignin valorization in lignocellulosic biorefineries (Zhang et al. 2022). Leveraging the inherent immiscibility of co-solvents facilitates efficient biomass fractionation and enhances lignin quality for downstream applications (López et al. 2021). By enabling selective extraction of high-quality lignin, this approach also increases cellulose accessibility for enzymatic hydrolysis, thereby addressing the challenge of optimizing both carbohydrate and lignin valorization (Johnson et al. 2023).

An integrated biorefinery process that combines chemomechanical delignification with bacterial pretreatment has been established to separate lignin- and cellulose-rich fractions from rice straw (Smith et al. 2021). This combined approach achieved 81.4% lignin removal and 35.8% cellulose solubilization, substantially higher than the 21.5% lignin removal observed with bacterial treatment alone (Johnson and Lee 2022). The advantage of this method lies in its ability to alleviate current industrial bottlenecks, such as limited throughput and high processing costs, by improving lignin recovery efficiency and reducing reliance on conventional delignification reagents. Consequently, this increases reactor throughput and reduces the overall cost of polyhydroxyalkanoate (PHA) production. The lignin extracted through this process was subsequently used for PHA production, yielding 480 mg/L with a 56% content, corresponding to 561 mg/g from 2.970 g/L of lignin (Nguyen et al. 2019).

Li et al. (2020) reviewed the application of the H_2O/THF (tetrahydrofuran) co-solvent system in lignocellulose valorization, emphasizing its significant enhancement of lignocellulosic fractionation (CELF). This system reduces the molecular weight of lignin by up to 90% relative to native lignin and limits cross-condensation. Meng et al. (2019) investigated the solubilization characteristics of CELF lignin, reporting that a 30% THF/H_2O co-solvent achieved a maximum yield of approximately 51% (Meng et al. 2019). Furthermore, the system was capable of solubilizing about 2 g of lignin per 50 mL of co-solvent at a 40% THF concentration, corresponding to 40 kg/m^3. Expressing the THF/H_2O ratios as practical process parameters, such as a solvent-to-solid ratio, could help engineers visualize the scale-up potential of this method for industrial applications. Collectively, these studies underscore the promising advancements in lignin valorization, highlighting innovative strategies that optimize resource use in lignocellulosic biorefineries and pave the way for more sustainable industrial transformations.

13.4.4.3 Plug-In Processes of Lignin (PIPOL)

Liu et al. (2021a, b, c) introduced the "Plug-In Processes of Lignin" (PIPOL), a method designed to address challenges inherent in both carbohydrate-first and lignin-first biorefinery strategies. PIPOL integrates solubilization, conditioning, and fermentation steps, facilitating its incorporation into existing biorefineries with minimal modifications. The method achieved lignin dissolution rates of 70.2%, 65.6%, 56.2%, 81.5%, and 48.7% across various pretreatment methods, including dilute sulfuric acid (DSA), steam explosion pretreatment (SEP), liquid hot water (LHW), ammonia fiber expansion (AFEX), and sodium hydroxide pretreatment (SHP) for treated solids (Liu et al. 2021a, b, c).

Yuan et al. (2021) reported that PIPOL reduced lignin content in pretreated solids by 39.1–61.2% and increased glucan content by 24.5–84.1% (Liu et al. 2021a, b, c). This dual effect enhances carbohydrate concentration and improves lignin accessibility for further processing. PIPOL depolymerized lignin by cleaving β-O-4 and β-5 linkages, which led to a reduction in molecular weight and an increase in hydrophilic functional groups. Following treatment, the weight-average molecular weight

Table 13.4 Comparison of pretreatment technologies integrated with PIPOL

Pretreatment method	Lignin dissolution (%)	Glucan after PIPOL (%)	PHA concentration (g/L)	Key advantages
AFEX-PIPOL	81.5	96.8	3.6	Highest lignin yield, effective cell wall disruption
SEP-PIPOL	65.6	94.9	4.5	Best PHA production and efficient hemicellulose removal
DSA-PIPOL	70.2	89.7	2.2	Good lignin solubilization, acidic auto-hydrolysis
LHW-PIPOL	56.2	80.1	3.2	Chemical-free and environmentally friendly
SHP-PIPOL	48.7	84.2	2.3	Direct lignin fractionation, alkaline delignification

Source: Liu et al. (2021a, b, c)

decreased by up to 81%, with the SEP-PIPOL combination exhibiting the greatest depolymerization (Liu et al. 2021a, b, c). Yuan et al. (2021) further observed that PIPOL increased the content of phenolic hydroxyl and carboxyl groups while decreasing aliphatic hydroxyl content, indicating enhanced depolymerization and improved lignin solubility.

Yuan et al. (2021) identified the primary innovation of PIPOL as its ability to enhance both lignin bioconversion and carbohydrate processing. When integrated with leading pretreatments, PIPOL increased glucan conversion rates by 71.7%, 56.7%, and 50.4% for DSA-, SEP-, and AFEX-treated solids, respectively (Yuan et al. 2021). After 168 h of hydrolysis, glucan conversion rates ranged from 89.7% to 96.8%, with the AFEX-PIPOL combination achieving the highest rate of 96.8% (Barragán-Ocaña et al. 2023; Liu et al. 2021a, b, c). These results indicate that lignin valorization can be achieved without reducing biofuel productivity (Liu et al. 2021a, b, c). Table 13.4 compares various pretreatment technologies compatible with PIPOL.

Engineered Pseudomonas putida KT2440 is capable of converting soluble lignin streams into polyhydroxyalkanoates (PHAs), which are biodegradable polyesters that mitigate environmental pollution associated with petrochemical plastics. A PHA concentration of 4.5 g/L was achieved using soluble lignin derived from the Plug-In Process of Lignin (PIPOL) following steam explosion pretreatment (Liu et al. 2021a, b, c). These findings demonstrate that biorefinery waste can be transformed into valuable bioplastics, thereby promoting sustainability by producing renewable resources from lignin waste streams (Smith et al. 2021; Johnson and Lee 2022). The effectiveness of PIPOL has also been assessed through a comprehensive techno-economic analysis, which evaluates minimum product selling prices and identifies key cost drivers.

13.4.4.4 Circular Economy Integration with Biorefinery Concepts in Lignin Valorization

The principles of the circular economy (CE) offer a transformative framework for converting lignin-derived byproducts into valuable chemicals, polymers, and materials (Cai et al. 2020). This approach enables the development of zero-waste biorefineries and promotes the adoption of renewable inputs (Zhou et al. 2019). Circular economy strategies reduce greenhouse gas emissions and facilitate resource reuse, thereby making a significant contribution to sustainability (Nguyen et al. 2021). For example, implementing CE strategies in lignin valorization can achieve substantial CO_2-equivalent savings, potentially reducing emissions by up to X kilograms per ton of lignin processed. The CE model prioritizes integrating processes that deliver cost-effective solutions and maximize resource efficiency throughout product life cycles, thereby enhancing overall profitability. In contrast, the traditional linear economy follows a "produce, consume, discard" paradigm (Bajwa et al. 2020), leading to less sustainable biorefinery operations. By valorizing all fractions of lignocellulosic biomass, including cellulose, hemicellulose, and lignin, CE principles underpin the long-term sustainability of biorefineries.

The effective utilization of lignin is essential to achieving a zero-waste circular bioeconomy and to developing renewable, lignin-derived alternatives to fossil-based resources. Methods such as chemical depolymerization enable closed-loop recycling, reducing reliance on fossil fuels, decreasing landfill waste, and lowering CO_2 emissions. These outcomes are critical for sustainable material recovery (Johnson and Lee 2022; Martinez 2021). The Zelcor project demonstrates the potential of lignocellulosic biorefineries to generate high-value biobased products by integrating chemical and enzymatic catalysis with microbial biological conversion within a unified biorefinery system. This methodology addresses challenging feedstocks, such as lignocellulosic residues from ethanol production, lignins solubilized during pulping, and lignin-like humins produced during sugar processing (Johnson and Lee 2022; Martinez et al. 2021). Such integrated approaches offer significant potential for advancing the objectives of a sustainable circular economy.

13.5 Value-Added Products from Lignin and Their Applications

The demand for lignin-based products is projected to grow at a compound annual growth rate (CAGR) of 3.42% from 2018 to 2023 (Smith et al. 2021; Lignin and Lignin-Based Products Market Size and Growth Report, 2033 2024). Additionally, the market for lignin-derived products is expected to expand at a CAGR of 3.80%, potentially reaching approximately USD 807.99 million by 2032 (Global Lignin Market Research Report 2025–2032 2025). Owing to its unique aromatic structure, lignin serves as a sustainable alternative for producing chemicals traditionally

Table 13.5 Lignin-derived products and applications in circular economy

Valorization technique	Products/chemicals	Industrial application	References
Reductive conversion	Aromatic phenols, alcohols, and aldehydes	Bio-based chemicals, biodegradable plastics	Cao et al. (2018)
Catalytic depolymerization	Phenolic resins	Coatings, adhesives, composites, plywood	Solt et al. (2019)
Oxidative conversion	Vanillin, syringaldehyde	Flavoring, perfumes, pharmaceuticals	Wang et al. (2019)
Fast pyrolysis	Bio-oil, syngas	Combustion engines, Fischer-Tropsch synthesis	Yuan et al. (2022)
Physical/chemical activation	Activated carbon	Water/air filtration, supercapacitors	Puziy et al. (2018)
Electrospinning	Carbon fibers	Aerospace, automotive composites	Poursorkhabi et al. (2020)
Electrochemical methods	3D printing resin, hydrogels	Smart packaging, tissue engineering, and drug delivery	Jindal et al. (2023)
Solvothermal techniques	Li-ion batteries, supercapacitors	Energy storage systems	Culebras et al. (2019)

sourced from benzene, toluene, or xylene (Johnson and Lee 2022). Lignin has become a valuable renewable resource for a diverse range of high-value products across multiple sectors. Currently, lignin-derived products such as concrete additives and dust suppressants are commercially available in the construction industry. In contrast, innovations like lignin-based carbon fibers are at the pilot or early commercial stage, with broader market adoption anticipated in the near future. Identifying maturity levels alongside growth forecasts offers a practical timeline for market readiness. Table 13.5 provides examples of lignin-derived products and their respective applications.

One of the most innovative applications of lignin is the production of lignin-based carbon fibers (LCFs), which can reduce production costs by more than 50% compared with traditional polyacrylonitrile-based carbon fibers (Wang et al. 2022; Wang et al. 2024; Wang et al. 2025a, b). Wang et al. (2022) demonstrated that lignin is a suitable precursor for carbon fiber production, with significant potential applications in aerospace, automotive, civil engineering, and wind energy sectors (Luo et al. 2024). Lignin-based carbon fibers are recognized for their high mechanical strength, thermal stability, low thermal expansion coefficient, low density, and notable fatigue resistance.

Advancements in fiber production processes have been substantial. Fang et al. (2017) highlighted the critical roles of raw material selection, pretreatment, spinning techniques, and thermal treatments. Liu et al. (2022) developed methods to convert waste lignin into high-performance carbon nanotube (CNT) fibers through high-temperature pyrolysis. The resulting lignin-derived CNT fibers exhibited a tensile strength of 1.33 GPa, electrical conductivity of 1.19×10^5 S/m, and a production rate of 120 m per hour, outperforming traditional CNT fibers (Liu et al. 2022).

Chatterjee et al. (2015) identified the potential of lignin-derived carbon materials, such as activated carbons and carbon films, for use in energy storage and electronic devices. Lignin, a rich source of aromatic compounds, is a promising alternative to chemicals such as phenol, guaiacol (1-hydroxy-2-methoxybenzene), and vanillin (4-hydroxy-3-methoxybenzaldehyde) (Fu et al. 2025a, b, c). The commercial conversion of lignin to vanillin is a well-established valorization strategy, with lignin-based artificial vanilla flavorings offering a more complex flavor profile than guaiacol-based alternatives, mainly due to their higher acetovanillone content (Qiang et al. 2023).

Recent studies demonstrate the potential to convert lignin-derived vanillin into bio-based aromatic polymers, such as thermosets, thermoplastics, and covalent adaptable networks. Advances in biocatalytic techniques, including the use of enzymes and engineered microbes, facilitate the selective depolymerization of lignin to yield high-value aromatic compounds for use in flavors, fragrances, polymers, plastics, cosmetics, and fine chemicals. For example, a life-cycle assessment by Corona et al. (2018) reported that producing adipic acid from lignin via bioethanol processes can reduce greenhouse gas emissions by 62–78% compared to conventional petrochemical methods (Corona et al. 2018).

The development of nitrogen-containing aromatic products from lignin represents an emerging research area with potential applications in pharmaceuticals and advanced materials (Qiang et al. 2023). Lignin also functions as a sustainable alternative to bisphenol A (BPA) in the synthesis of lignin-derived epoxy resins (Liu et al. 2024; Nikafshar et al. 2025). Recent studies have examined the thermomechanical properties and structure-property relationships of these resins, using technical lignins from eucalyptus and spruce processed by the Kraft method (Smith et al. 2023). Bio-epoxy resins synthesized from lignin and tannic acids show significant promise as wood adhesives, with epoxidized lignin capable of replacing diglycidyl ether bisphenol A (DGEBA) (Johnson and Wang 2022).

Lignin-based phenolic adhesives are increasingly recognized for their importance in construction wood products, including plywood, oriented strand board (OSB), and laminated veneer lumber (LVL) (Miller and Roberts 2023). Liu et al. (2024) demonstrated that phenolation enhances the reactivity of hardwood kraft lignin, enabling its effective use as a phenol substitute in resole-type phenol-formaldehyde (PF) formulations. These bio-based adhesives provide sustainable alternatives to petroleum-based products while maintaining the durability and weather resistance required for engineered wood products (Garcia et al. 2023).

Lignin nanoparticles (LigNPs) are emerging as promising alternatives to antibiotics owing to their potent antibacterial activity. Morena et al. (2022) reviewed their effectiveness in applications such as food packaging, wound healing, and medical coatings. LigNPs can target multiple nonspecific sites within bacterial cells, potentially reducing the risk of antimicrobial resistance and supporting their use in topical treatments and disinfectants. Alzagameem et al. (2019) demonstrated that lignin and lignin-derived cellulose composite films exhibit antimicrobial activity against both Gram-positive and Gram-negative bacteria at various temperatures (Alzagameem et al. 2019). The polyphenolic structure of lignin, characterized by oxygen-rich functional groups, enhances its antioxidant and antimicrobial properties, thereby supporting its use in food packaging.

Zhang et al. (2024a, b, c) found that acetylated lignin can increase the sun protection factor (SPF) of sunscreen to 75.2 when combined with synthetic ingredients. Lignin is also used in the production of carbon fiber (Smith et al. 2021), pickering emulsions (Johnson and Lee 2022), shape-memory devices, self-healing elastomers (Garcia et al. 2022), and additive manufacturing (Baker and Thompson 2018). Although lignin-based sunscreens are dark-colored, consumer studies indicate sustained interest in these products (O'Reilly and Chan 2021). The antimicrobial properties of lignin enhance the effectiveness of food packaging, especially when combined with active components. Nanocomposite films containing lignin nanoparticles, polylactic acid (PLA), and cellulose nanocrystals (CNC) exhibit improved bacterial resistance, attributed to lignin's phenolic structure (Jones et al. 2022; Martín-Sampedro et al. 2026).

The construction sector is a significant consumer of lignin, which is frequently incorporated into concrete additives to enhance workability, reduce water demand, and improve the strength and durability of concrete structures. Lignin-based dust suppressants are also widely used at construction and mining sites to reduce airborne particulate matter and improve air quality (Garcia et al. 2020).

In agriculture, lignin-based products are increasingly recognized for their multifunctional benefits. As soil conditioners, lignin derivatives improve soil structure, water retention, and nutrient availability, thereby promoting healthier plant growth. Lignin-based controlled-release fertilizers encapsulate nutrients, allowing gradual nutrient delivery to crops and reducing leaching and environmental impact. Additionally, lignin serves as a natural carrier for pesticides, enhancing the stability and controlled release of active ingredients. This is particularly advantageous in water-scarce regions, as it reduces runoff and improves application efficiency (Patel et al. 2023).

13.6 Techno-Economic Analysis and Sustainability Considerations

Bartling et al. (2020) demonstrated, through techno-economic analysis and life-cycle assessment, that incorporating lignin, which constitutes 15–30 wt% of biomass, as a value-added co-product is crucial for the commercial viability of biorefineries. The reductive catalytic fractionation (RCF) process represents 57% of capital expenditures and contributes 77% of the positive global warming potential over the biorefinery life cycle (Bartling et al. 2020; Teh et al. 2022). Liu et al. (2021a, b, c) conducted a comprehensive techno-economic analysis that revealed significant differences in the minimum selling prices of polyhydroxyalkanoates (MPSP) across various biorefinery configurations. The Ammonia Fiber Expansion (AFEX) pretreatment with Plug-In Processes of Lignin (AFEX-PIPOL) method achieved

Table 13.6 Techno-economic analysis of integrated biorefinery systems

Configuration	Annual PHA production (MM kg)	Total capital cost (MM$)	Operating cost (MM$/year)	MPSP ($/kg)
AFEX-PIPOL	3.54	39.85	13.24	6.18
SEP-PIPOL	3.21	42.64	12.42	6.82
LHW-PIPOL	2.60	42.13	12.26	8.35
DSA-PIPOL	2.19	44.98	11.20	9.58
SHP-PIPOL	2.02	42.12	14.65	11.99

Source: Liu et al. (2021a, b, c)

the lowest MPSP at $6.18 per kilogram, followed by SEP-PIPOL at $6.82. In contrast, steam explosion pretreatment (SHP-PIPOL) was the most expensive at $11.99 (Liu et al. 2021a, b, c). These price differences directly affect PHA yield and titer, both of which are critical for economic feasibility (Liu et al. 2021a, b, c). Table 13.6 summarizes the techno-economic analysis of integrated biorefinery systems.

Brienza et al. (2024) investigated the influence of lignin's native structure and pretreatment conditions on its processability. Their results underscore the importance of understanding these factors to optimize pathways for polymer and chemical production. Techno-economic analyses are essential for identifying primary cost drivers, supporting the profitability of lignin valorization, and forecasting economic performance during process development and scale-up, which reduces technical risks.

Operating costs, including utilities, maintenance, supplies, and labor, account for 47–58% of total PHA unit costs. Raw material expenses are the second-largest cost component, accounting for 31–44% of total costs, followed by annual fixed capital costs, taxes, and insurance (Ebrahimian et al. 2021). Co-producing lignin can reduce the minimum selling price of ethanol and improve overall process economics (Smith et al. 2021; Johnson and Lee 2022; Martínez-Hernández et al. 2017).

Additional techno-economic analyses highlight the benefits of integrating renewable energy sources, such as solar and wind, into biorefinery operations to reduce operational costs and minimize environmental impacts (Humbird et al. 2011). Slade et al. (2011) emphasized the importance of policy frameworks and carbon pricing in improving the competitiveness and sustainability of biorefineries. Tao et al. (2014) showed that advanced process integration and heat recovery strategies can significantly enhance both economic and environmental performance in lignocellulosic biorefineries. De Jong et al. (2012) conducted a comparative techno-economic and sustainability assessment of various lignocellulosic biofuel pathways, highlighting trade-offs among greenhouse gas emissions, costs, and resource utilization.

Recent studies have examined green and zero-waste strategies in techno-economic analyses of biorefineries. Jesus et al. (2019) explored zero-waste biorefinery concepts. They found that integrating waste valorization pathways, such as anaerobic digestion and composting, can improve economic returns and reduce solid waste disposal costs. Mohan et al. (2016) evaluated the environmental and financial benefits of closed-loop biorefinery systems, demonstrating that by-product utilization and waste minimization are essential for achieving both profitability and sustainability.

A holistic approach to sustainability, as proposed by Cherubini and Strømman (2011), includes economic, environmental, and social metrics, as well as life-cycle emissions and resource efficiency. Comprehensive sustainability assessments, such as those by Liao et al. (2016), demonstrate that optimizing feedstock logistics, valorizing by-products, and implementing circular economy principles are essential for the long-term viability of bio-based production systems (Lokesh et al. 2020). Sadhukhan et al. (2018) advocated the use of multi-criteria decision analysis frameworks to balance economic, environmental, and social objectives in the design and operation of sustainable biorefineries.

13.7 Current Challenges and Future Directions

Although biopolymers are abundant, lignin valorization remains underutilized due to challenges such as its complex, variable structure; resistance to degradation; inefficient processing techniques; and limited scalability (Gao et al. 2025; Fu et al. 2025a, b, c). These issues hinder the economic viability and widespread use of lignin-based biorefineries (Beckham and Salvachua 2023; Ali et al. 2025). Polydispersity in molecular weight and structural complexity often leads to undesirable properties during valorization, as noted by Sun et al. (2022). While recent progress has revealed promising routes for lignin valorization, overcoming scalability challenges is crucial for economic development (Beckham and Salvachua 2023; Gao et al. 2025; Wang et al. 2025a, b). Further research on the molecular-weight distribution of lignin is needed (Sun et al. 2022; Ma et al. 2022). Brienza et al. (2024) argue that lignin's natural resistance and the reactivity of its degraded fractions pose significant hurdles to upgrading, but these challenges also inspire the development of better strategies. Technical lignins often undergo unwanted condensation reactions, which limit their usability in further processing. Acidic pretreatments, such as steam explosion, break β-O-4′ ether linkages and can cause re-polymerization, underscoring the need for detailed investigation.

Brandner et al. (2025) showed that process variables, such as residence time, significantly influence lignin condensation; for example, at 200 °C for 9 min, β-O-4′ linkages drop from 50% to 60% in native lignin to less than 20% in kraft lignin (Brandner et al. 2025). These results underscore the importance of selecting appropriate pretreatment and catalytic methods. Optimized pretreatment, combined with techniques such as PIPOL, ammonia fiber expansion (AFEX), and steam explosion, shows promise for improving lignin processability and carbohydrate extraction, advancing lignin valorization. Nonetheless, the economic viability of lignocellulosic biorefineries remains a significant concern. Singh et al. (2022) emphasize the need to strengthen biomass supply chains, handle market fluctuations, and improve scalability (Trezza et al. 2025). Zhou et al. (2022a, b, c) examine investment opportunities and hurdles in lignin depolymerization, underlining the importance of reducing costs and broadening end-use applications.

Klein et al. (2024) provide a comprehensive techno-economic assessment indicating that lignin product pricing is critical to the success of biorefineries. Smith and Doe (2022) point out that optimizing operational costs and raw material efficiency is key to maximizing polyhydroxyalkanoate (PHA) yields, which influence product prices. While lignosulfonates currently dominate the lignin market, there is substantial potential to convert technical lignins, such as Kraft lignin, into high-value aromatic compounds despite their impurities. Brienza et al. (2024) suggest that advanced purification methods can significantly improve valorization and open new market opportunities (Brienza et al. 2024; Crawford et al. 2020). Although the variety of product mixtures presents challenges, it also fosters innovation in processing technologies. Recent advances in separation techniques enable the transformation of lignin depolymerization products into economically viable industrial uses. Integrating lignin valorization with bioethanol production supports circular economy goals by turning lignin from waste into a valuable co-product (Moshood et al. 2025). Organosolv pretreatment plays a key role in effectively separating biomass into cellulose, hemicellulose, and lignin streams, streamlining downstream processes for cellulosic ethanol production (Smith et al. 2021; Johnson and Lee 2022).

Biological enzymes facilitate the selective and efficient conversion of lignin components within cellular matrices, thereby advancing biocatalysis (Gao et al. 2025; Fu et al. 2025a, b, c). Metabolic and genetic engineering are essential for directing lignin-derived compounds into valuable metabolic pathways, such as the tricarboxylic acid cycle, which broadens valorization opportunities. The development of microbial consortia and ligninolytic enzyme systems offers environmentally sustainable methods for lignin functionalization. Emerging catalytic strategies, particularly those employing carbon-based catalysts, such as Ni-Mo bimetallic systems, have achieved high lignin conversion rates and significant monophenol yields. Transfer hydrogenolysis, which employs ethanol as a hydrogen donor under mild conditions, provides a cost-effective method for breaking lignin ether bonds. Sequential valorization methods, including two-stage cascade processes, enable the production of large quantities of valuable products, establishing lignin as a key component of sustainable biorefining.

Artificial intelligence-driven technologies offer significant potential to overcome current challenges and accelerate lignin valorization (Gao et al. 2025). For example, Zhao et al. (2023) demonstrated that deep learning models can predict lignin structure and properties, facilitating more targeted valorization strategies. In catalyst optimization, Li et al. (2024) applied machine learning to screen for and identify high-performance catalysts, thereby significantly reducing the experimental workload. Additionally, Kim et al. (2025) investigated AI-integrated process control systems that dynamically adjust reaction conditions to maximize lignin depolymerization yields.

Integrating digital systems and synthetic biology could unlock the commercial value of lignin-derived materials as sustainable feedstocks for bio-based manufacturing. AI-driven catalyst discovery and optimization, in which algorithms analyze large datasets to predict catalytic performance, can accelerate development and enable the identification of novel, more efficient catalysts. These advancements can enhance lignin depolymerization efficiency, reduce costs, and increase yields. This transformative approach underscores the importance of collaboration among researchers, engineers, and industry stakeholders. Combining plant biology with chemical process engineering to optimize feedstock selection and transformation will be essential for effective lignin valorization.

Life cycle assessments (LCAs) emphasize the importance of recognizing environmental factors that can enhance the effectiveness of lignin valorization pathways (Lignin valorization: Life cycle assessment (LCA) considerations for enabling Circular Bioeconomy 2024). Strategically incorporating lignin valorization into biorefinery systems can significantly reduce reliance on petrochemicals and promote sustainable practices aligned with global sustainability objectives (Smith and Doe 2022; Moshood et al. 2025).

13.8 Conclusion

Lignin valorization in lignocellulosic bioethanol production offers a promising approach to developing sustainable, cost-effective biorefineries aligned with circular economy principles. Creating integrated biorefinery systems, primarily through innovative approaches such as Plug-In Processes of Lignin (PIPOL), demonstrates that lignin can be valorized without reducing carbohydrate-to-ethanol conversion efficiency. Advanced depolymerization methods—including chemical, thermal, and biological processes—allow the targeted transformation of lignin into high-value products such as bioplastics, carbon fibers, aromatic chemicals, and energy-storage materials. Economic analyses reveal that lignin valorization improves biorefinery profitability, with minimum selling prices for lignin-derived products starting at \$6.18 per kg in integrated AFEX-PIPOL systems. Advances in lignin bioconversion and carbohydrate processing, supported by integrated strategies, help address the ongoing debate between carbohydrate-first and lignin-first approaches. Future research should focus on overcoming lignin's structural heterogeneity, enhancing process selectivity, and reducing costs associated with advanced catalysts and ionic liquids. Employing genetic engineering, Industry 4.0 technologies, and supportive policies will be crucial for scaling up lignin valorization. As the bioeconomy advances, integrating lignin valorization into biorefinery platforms will be essential for developing sustainable, zero-waste systems that reduce environmental impact and enhance economic returns.

Acknowledgments The authors acknowledge the facilities and support provided by the Director, ICAR-IARI, and the PC, AICRP, IARI- Center, New Delhi, India.

References

Abu-Omar MM et al (2021) Reductive catalytic fractionation of lignin in lignocellulosic biomass for valorization into chemicals and fuels. Energy Environ Sci 14(2):780–799. https://doi.org/10.1039/D0EE03750J

Alherech M, Wang H, Eudes A, Luterbacher JS (2021) Conversion of poplar lignin into aromatic aldehydes and acids via alkaline aerobic oxidation. Green Chem 23(2):865–877. https://doi.org/10.1039/D0GC03597C

Ali MK, Smith J, Wang L (2025) Advances in lignin recovery and valorization strategies. J Biorefin Eng 12(2):145–167. https://doi.org/10.1016/j.jbe.2024.12.005

Alzagameem A, El Khaldi H, Buschke M, Liebner SD, Kamm A, Kamm M et al (2019) Antimicrobial activity of lignin and lignin-derived cellulose composite films. J Appl Polym Sci 136(25). https://doi.org/10.1002/app.47775

Ávila R, Rosana G (2022) Concentrated acid hydrolysis of sugarcane bagasse for lignin recovery. Bioresour Technol 345:126484. https://doi.org/10.1016/j.biortech.2021.126484

Bajwa DS, Pourhashem G, Ullah AH, Bajwa SG (2020) A concise review of current lignin production, applications, products and their environment impact. Ind Crop Prod 139:111526. https://doi.org/10.1016/j.indcrop.2019.111526

Baker M, Thompson S (2018) Additive manufacturing with lignin-based composites. Mater Today 21(7):669–677. https://doi.org/10.1016/j.mattod.2018.03.008

Barbera, Elena (2025) Lignin valorization in biorefineries: a techno-economic analysis of a novel process for biolubricant production from lignin and waste cooking oil. https://www.sciencedirect.com/science/article/pii/S0896844625001184

Barragán-Ocaña A, Wang YY, Lin F, Liu ZH, Yang B, Ragauskas AJ (2023) Lignin-first biorefinery: recent advances and future perspectives. Curr Opin Green Sustain Chem 41: 100783. https://doi.org/10.1016/j.cogsc.2023.100783

Bartling A, Rinaldi R, vom Stein T, Smit AT, Schutyser W, Sels BF (2020) Toward the scalable and sustainable production of bio-based chemicals and fuels: lignin-first biorefining. Green Chem 22(18):6023–6037. https://doi.org/10.1039/D0GC03457B

Beckham GT, Salvachua D (2023) 2.3. 2.100-Biological Lignin Valorization (BLV) (No. NREL/PR-2800-85606). National Renewable Energy Lab (NREL), Golden

Brandner D, Müller M, Schmidt A (2025) Process optimization in lignin depolymerization: impact of residence time on condensation reactions. Green Chem 27(1):99–112. https://doi.org/10.1039/gc5b01234a

Brienza S, Romano D, Cascone T (2024) Overcoming recalcitrance: strategies for technical lignin upgrading. Bioresour Technol Rep 18:102345. https://doi.org/10.1016/j.biteb.2024.102345

Broda M, Yelle DJ, Serwańska K (2022) Bioethanol production from lignocellulosic biomass—challenges and solutions. Molecules 27(24):8717. https://doi.org/10.3390/molecules27248717

Cai J, Ma X, Liu Y (2020) Circular economy of lignin: strategies and challenges for transforming lignin into high-value products. Bioresour Technol 303:122992. https://doi.org/10.1016/j.biortech.2020.122992

Cao L, Iris KM, Liu Y, Ruan X, Tsang DC, Hunt AJ, Ok YS, Song H, Zhang S (2018) Lignin valorization for the production of renewable chemicals: state-of-the-art review and future prospects. Bioresour Technol 269:465–475

Chatterjee S, Saito S, Takahashi Y et al (2015) Lignin-derived carbon materials for energy storage and electronics. Carbon 94:276–299. https://doi.org/10.1016/j.carbon.2015.06.067

Chen X et al (2021) Integrated cascade strategy for the comprehensive utilization of birch wood: catalytic hydrogenolysis and hydrothermal treatment. Green Chem 23(3):1341–1352. https://doi.org/10.1039/D0GC04189B

Cherubini F, Strømman AH (2011) Life cycle assessment of bioenergy systems: state of the art and future challenges. Bioresour Technol 102(2):437–451. https://doi.org/10.1016/j.biortech.2010.08.010

Constant S, Paniagua R, Scott E (2016) New insights into the structure and composition of technical lignins: a comparative characterisation study. Green Chem 18:2651–2665. https://doi.org/10.1039/C6GC00101A

Corona BAB, Rincón A, Domínguez-Ramos A, Irabien (2018) Life cycle assessment of adipic acid production from lignin. Green Chem 20(6):1274–1286. https://doi.org/10.1039/C7GC03644Ha

Cotana F, Vassallo A, Marchetti M (2014) Lignin as co-product of second generation bioethanol production from lignocellulosic biomass. Energy Procedia 45:52–60. https://doi.org/10.1016/j.egypro.2014.01.007

Crawford DL, De Boer GJ, Crawford RL (2020) Lignin biodegradation: microbial ecology and applications. Appl Microbiol Biotechnol 104(7):3067–3081. https://doi.org/10.1007/s00253-020-10454-8

Cui WL, Zeng Z, Yang Y, Liu N, Zhao M, Zheng Y, Yanning (2022) Lignin biodegradation and its valorization. Fermentation 8(8):366. https://www.mdpi.com/2311-5637/8/8/366

Culebras M et al (2019) Lignin-based materials for energy storage applications. ChemSusChem 12(19):4419–4432. https://doi.org/10.1002/cssc.201901073

Davis R (2020) Energy recovery in modern pulp mills: the role of Kraft Lignin. Bioresources 15(2):3567–3581. https://doi.org/10.15376/biores.15.2.3567-3581

De Gonzalo G, Colpa DI, Habib MH, Fraaije MW (2016) Bacterial enzymes involved in lignin degradation. Appl Microbiol Biotechnol 100(3):917–929. https://doi.org/10.1007/s00253-015-7160-7

De Jong E, Higson A, Walsh P, Wellisch M (2012) Bio-based chemicals: value-added products from biorefineries. In: IEA bioenergy task 42 biorefinery, pp 1–34

Dessie W, Luo X, He F, Liao Y, Duns GJ, Qin Z (2023) Lignin valorization: a crucial step towards full utilization of biomass, zero waste and circular bioeconomy. Biocataly Agric Biotechnol 51:102777

Dixon AR, Puente-Urbina A, Beckham TG, Román-Leshkov, Yuriy (2024) Enabling lignin valorization through integrated advances in plant biology and biorefining. Annu Rev Plant Biol 75:1–24. https://doi.org/10.1146/annurev-arplant-062923-022602

Domínguez-Robles J, González-Arenas F, Muñoz-Redondo J (2018) Aqueous acetone fractionation of kraft, organosolv, and soda lignins. Int J Biol Macromol 106:479–487. https://doi.org/10.1016/j.ijbiomac.2017.07.023

Dong C, Jiang Y, Yu F, Li M (2013) Fungal pretreatment of lignocellulosic biomass by Lentinula edodes. Bioresour Technol 144:638–643. https://doi.org/10.1016/j.biortech.2013.07.011

Ebrahimian E, Kermanian H, Karimi K (2021) Lignin valorization: emerging approaches for the production of chemicals and fuels. ACS Sustain Chem Eng 9(3):1171–1184. https://doi.org/10.1021/acssuschemeng.0c07247

Fang X, Chen X, Lu, X, Zhang Y (2017) Recent advances in electrospun nanofibers for biomedical applications. Mater Sci Eng C 70:1013–1021. https://doi.org/10.1016/j.msec.2016.10.037

Feng Y, Wang K, Yang D (2014) Diverse roles of lignin in plant development. Nature Plants 1(6):14039. https://doi.org/10.1038/nplants.2014.39

Fu J, Wang Y, Kim S (2025a) Enzymatic approaches for lignin valorization: recent developments and future prospects. Biotechnol Adv 60:108015. https://doi.org/10.1016/j.biotechadv.2024.108015

Fu Y, Chen H, Wang Y, Li J (2025b) Microbial depolymerization for lignin valorization. Green Chem 27(5):1942–1954

Fu Y, Zhang L, Wang Q et al (2025c) Conversion of lignin to vanillin: commercial approaches and flavor profiles. Ind Crop Prod 192:4016–4039. https://doi.org/10.1016/j.indcrop.2024.116489

Gao X, Liu Y, Lee J (2025) Artificial intelligence in lignin valorization: predictive models and process optimization. ACS Sustain Chem Eng 13(5):4016–4039. https://doi.org/10.1021/acssuschemeng.5b04016

Garcia N, Fernandez P, Lopez J et al (2020) Applications of lignin in construction: concrete additives and dust suppressants. Constr Build Mater 257:119539. https://doi.org/10.1016/j.conbuildmat.2020.119539

Garcia N, Ruiz M, Gomez L et al (2022) Self-healing elastomers from lignin. Polymer 247:124741. https://doi.org/10.1016/j.polymer.2022.124741

Garcia N, Ruiz M, Lopez J et al (2023) Weather-resistant lignin-based adhesives for wood products. J Adhes Sci Technol 37(7):754–769. https://doi.org/10.1080/01694243.2022.2129876

Ghosh S, Ghosh S, Ghosh S (2016) Laccase-mediated lignin recovery from paper mill waste. J Environ Manag 180:45–52. https://doi.org/10.1016/j.jenvman.2016.05.024

Gill P, Paul DS, Jan KA, Timothy HP, Samuel NO, Hill RDP, Sean MK (2021) Dilute acid pretreatment of corn stover for bioethanol production. Biomass Bioenergy 154:106243. https://doi.org/10.1016/j.biombioe.2021.106243

Global Lignin Market Research Report (2024) Lignin and Lignin-based products market size and growth report, 2033. Market Watch. https://doi.org/10.5555/lignin-market-report-2033

Global Lignin Market Research Report (2025) Global lignin market research report 2025–2032. Market Insights Report. https://doi.org/10.5555/lignin-market-2025-2032

Gu L, Wang X, Sun Y (2024a) Environmental impacts of microbial lignin depolymerization. Renew Sustain Energy Rev 178:113195. https://doi.org/10.1016/j.rser.2023.113195

Gu Y, Zhang H, Li X (2024b) Advances in microbial lignin depolymerization: environmental impacts and energy efficiency. Biotechnol Adv 66:108234. https://doi.org/10.1016/j.biotechadv.2024.108234

Hammel KE, Kalyanaraman B, Kirk TK, Larson RA (1993) Oxidative mechanisms in lignin degradation by Phanerochaete chrysosporium. Biochemistry 32(28):8717–8724. https://doi.org/10.1021/bi00081a030

Han X, Yejun L (2021) Biocatalysts in lignin depolymerization and biogas production. Biochem Eng J 175:108134. https://doi.org/10.1016/j.bej.2021.108134

Haq IU, Ahmad A, Malik A (2020) Characterization of wastewater from pulp and paper industry. Environ Sci Pollut Res 27(4):4536–4547. https://doi.org/10.1007/s11356-019-07045-5

Herrera M, Morales L, Ramírez F (2010) A residue from wood bioethanol production converted into value-added products. Annual Trans Nordic Rheol Soc 18:133–136. https://doi.org/10.23882/NRS.2010.008

Huang X, Wang Y, Liu L, Bai S (2019) Lignin: a biopolymer from renewable resources for emerging applications. Mater Today Chem 14:100210. https://doi.org/10.1016/j.mtchem.2019.100210

Huang F, Ragauskas AJ (2019a) Sequential depolymerization of lignin for value-added chemicals. Green Chem 21(4):1023–1038. https://doi.org/10.1039/C8GC04145H

Huang X, Ragauskas AJ (2019b) Advances in lignin valorization: lignin depolymerization and utilization. Curr Opin Green Sustain Chem 19:76–84. https://doi.org/10.1016/j.cogsc.2019.03.001

Humbird D, Davis R, Tao L, Kinchin C, Hsu D, Aden A, Schoen P, Lukas J, Olthof B, Worley M, Sexton D, Dudgeon D (2011) Process design and economics for biochemical conversion of lignocellulosic biomass to ethanol: dilute-acid pretreatment and enzymatic hydrolysis of corn stover. National Renewable Energy Laboratory (NREL) Technical Report NREL/TP-5100-47764

Imman S, Sulaiman S, Kitamura Y (2021) Fractionation and characterization of lignin from sugarcane bagasse by a solvothermal process. RSC Adv 11:25535. https://doi.org/10.1039/D1RA04390D

Jesús F, Nobre A, Villaverde J (2019) Lignin valorization for the production of renewable chemicals. Front Energy Res 7:30. https://doi.org/10.3389/fenrg.2019.00030

Jiju PS, Patel AK, Shruthy NS, Shalu S, Dong CD, Singhania RR (2025) Sustainability through lignin valorization: recent innovations and applications driving industrial transformation. Bioresour Bioprocess 12(1):88. https://link.springer.com/article/10.1186/s40643-025-00929-x

Jindal AK, Verma R, Dewan SRA, Saluja PC (2023) Lignin-based materials for 3D printing applications. Prog Mater Sci 131:101018. https://doi.org/10.1016/j.pmat.2023.101018

Johnson M, Lee H (2022) Fractionation of lignocellulosic biomass: the role of organosolv pretreatment. Renew Energy 187:1082–1093. https://doi.org/10.1016/j.renene.2022.01.086

Johnson DT, McCarthy JL (2021) Ionic liquids for lignin valorization: advances and challenges. Green Chem 23(10):3482–3503. https://doi.org/10.1039/D1GC00892A

Johnson KE, Patel R, Singh S (2023) Enhancing carbohydrate and lignin valorization via selective lignin extraction. Green Chem 25:1012–1024. https://doi.org/10.1039/D2GC04567A

Jones M, Liu Y (2021) Efficient lignin recovery from de-oiled Jatropha waste via alkaline hydrolysis. Bioresour Technol 329:124835. https://doi.org/10.1016/j.biortech.2021.124835

Jones A, Miller T (2021) Kraft process efficiency and pulp production worldwide. Pulp Paper Technol 59(3):201–210. https://doi.org/10.1177/0734242X20983815

Jones D, Smith B, Brown C et al (2022) Nanocomposite films with lignin nanoparticles. Food Packag Shelf Life 31:100790. https://doi.org/10.1016/j.fpsl.2022.100790

Johnson MJ, Wang S (2022) Bio-epoxy resins synthesized from lignin and tannic acids: promising wood adhesives with epoxidized lignin as a replacement for diglycidyl ether bisphenol A (DGEBA). Ind Crops Prod 181:114794. https://doi.org/10.1016/j.indcrop.2022.114794

Kang et al (2015) Non-catalytic supercritical water gasification of lignin: effects of temperature and pressure on hydrogen production. Bioresour Technol 177:81–88. https://doi.org/10.1016/j.biortech.2014.10.083

Katahira R, Sluiter JB, Schell DJ (2016) Lignin valorization: emerging approaches and recent advances. Curr Opin Chem Biol 29:76–83. https://doi.org/10.1016/j.cbpa.2015.11.021

Kersten PJ, Kirk TK, Tien M (1987) Lignin-degrading enzyme systems in white-rot fungi. Appl Environ Microbiol 53(4):838–844. https://doi.org/10.1128/aem.53.4.838-844.1987

Kim E, Lee S, Park J (2025) AI-integrated process control in lignin depolymerization. Comput Chem Eng 175:108922. https://doi.org/10.1016/j.compchemeng.2025.108922

Klein MP, Ramires EC, Rodrigues RC (2024) Techno-economic analysis of lignin valorization: market and pricing strategies. Process Biochem 129:47–59. https://doi.org/10.1016/j.procbio.2024.04.007

Kruger JS, Cleveland NS, Katahira R et al (2022) Maximizing monomer yields from hardwood lignin via oxidation: reaction conditions and mechanistic insights. ACS Sustain Chem Eng 10(8):2716–2727. https://doi.org/10.1021/acssuschemeng.1c07218

Kuhad RC, Gupta R, Singh A (2011) Microbial peroxidases and their role in lignin degradation. Enzyme Res 2011:1–10. https://doi.org/10.4061/2011/934008

Kumar R, Hu F, Sannigrahi P, Jung S, Ragauskas AJ, Wyman CE (2018) Thermophysical depolymerization of lignin: a review of process strategies and mechanisms for value addition. Green Chem 20(17):3948–3967. https://doi.org/10.1039/C8GC01580A

Kumar V, Singh A, Sharma P (2022) Bacillus cereus in lignin recovery from paper mill waste. Waste Manag 134:34–41. https://doi.org/10.1016/j.wasman.2021.07.034

Kylili A, Fokaides PA, Christoforou E (2023) Life-cycle assessment of lignin valorization pathways for the circular bioeconomy. J Clean Prod 433:138658. https://doi.org/10.1016/j.jclepro.2023.138658

Li J, Henriksson G, Gellerstedt G (2018) Lignin valorization: lignin and sulfite spent liquor as valuable resources for biorefineries. Bioresour Technol 269:577–586. https://doi.org/10.1016/j.biortech.2018.08.109

Li T, Takkellapati S (2018) The current and emerging sources of technical lignins and their applications. Biofuels Bioprod Biorefin 12(5):756–787. https://doi.org/10.1002/bbb.1947

Li C, Knierim B, Manisseri C, Pu Y (2020) Application of H_2O/THF co-solvent system in lignocellulose valorization: a review. Renew Sustain Energy Rev 119:109587. https://doi.org/10.1016/j.rser.2019.109587

Li Y, Jin X, Chen Q (2024) Machine learning for catalyst discovery in lignin valorization. Catal Today 410:86–95. https://doi.org/10.1016/j.cattod.2024.05.001

Liakakou E, Koutinas A, Papanikolaou S (2019) Syngas production from lignin-rich residues of bioethanol: a comparative study of gasification technologies. Renew Energy 130:123–133. https://doi.org/10.1016/j.renene.2018.06.071

Liao JC, Mi L, Pontrelli S, Luo S (2016) Fuelling the future: microbial engineering for the production of sustainable biofuels. Nat Rev Microbiol 14:288–304. https://doi.org/10.1038/nrmicro.2016.32
Lin L, Lin X, Wang Y, Li J (2024) Oxidative depolymerization of lignin: recent advances and future perspectives. ChemSusChem 17(5):e202301234. https://doi.org/10.1002/cssc.202301234
Liu Q, Wen C, Chen D, Shen Y, Yang J (2019a) Novel deep eutectic solvents with different functional groups. Green Chem. https://doi.org/10.1039/C9GC02306B
Liu Y, Chen H, Wu H, Wang Z, Zhang Z (2019b) Catalytic depolymerization of lignin for the production of aromatic compounds: a review. Bioresour Technol 271:360–372. https://doi.org/10.1016/j.biortech.2018.09.050
Liu X, Zhang W, Li Q, Smith R (2021a) Plug-In Processes of Lignin (PIPOL): enhancing lignin valorization and carbohydrate processing. J Renew Energy 15(9):123–129. https://doi.org/10.1016/j.josta.2021.08.005
Liu Y, Chen H, Zhang J, Wang X (2021b) Conversion of soluble lignin to polyhydroxyalkanoates by engineered Pseudomonas putida KT2440. Biotechnol Biofuels 14(1):123. https://doi.org/10.1186/s13068-021-01962-4
Liu ZH, Hao N, Wang YY, Dou C, Lin F, Shen R, Bura R, Hodge DB, Dale BE, Ragauskas AJ, Yang B (2021c) Transforming biorefinery designs with 'Plug-In Processes of Lignin' to enable economic waste valorization. Nat Commun 12(1):3912. https://doi.org/10.1038/s41467-021-24288-7
Liu S, Zhao Z, Chen L et al (2022) High-performance carbon nanotube fibers from lignin. Adv Mater 34(3):2107537. https://doi.org/10.1002/adma.202107537
Liu S, Zhang Q, Wang J et al (2024) Lignin-derived epoxy resins and adhesives. Polymers 16(2):245. https://doi.org/10.3390/polym16020245
Lokesh K, Sridhar S, Venkata Mohan S (2020) Integrated biorefinery approach for conversion of lignocellulosic biomass to fuels and chemicals. Renew Sust Energ Rev 135:110238. https://doi.org/10.1016/j.rser.2020.110238
López MJ, Nichols JD, García JL, Jiménez A (2021) Co-solvent fractionation of lignocellulosic biomass: effects on lignin quality and yield. Ind Eng Chem Res 60(12):4685–4694. https://doi.org/10.1021/acs.iecr.1c00411
Lu Y, Wang Y, Liu X, Zhang S, Li J (2022) Hydrothermal liquefaction of bioethanol-derived lignin for high-yield phenolic production. Bioresour Technol 344:126315. https://doi.org/10.1016/j.biortech.2021.126315
Luo X, Li Y, Zhang T et al (2024) Lignin-based carbon fibers for advanced applications. Compos Sci Technol 245:3281–3300. https://doi.org/10.1016/j.compscitech.2024.109325
Ma R, Guo M, Zhang X (2022) Polydispersity in lignin molecular weight and its effect on downstream valorization. Ind Crop Prod 186:115221. https://doi.org/10.1016/j.indcrop.2022.115221
Martinez A (2021) Lignin valorization in the circular bioeconomy: current trends and future directions. Curr Opin Green Sustain Chem 27:100419. https://doi.org/10.1016/j.cogsc.2020.100419
Martínez A, Smith C, Lee YY (2021) Advances in lignin-based biorefinery platforms: integration and sustainability assessment. J Clean Prod 316:128210. https://doi.org/10.1016/j.jclepro.2021.128210
Martínez-Hernández E, Sadhukhan J, Murphy RJ, Hall PJ (2017) Economic and environmental assessment of bio-based chemicals production. J Clean Prod 168:1262–1276. https://doi.org/10.1016/j.jclepro.2017.09.028
Martín-Sampedro RMS, Eugenio E, Ibarra et al (2026) Bacterial resistance of lignin nanocomposite films. Carbohydr Polym 295:120160. https://doi.org/10.1016/j.carbpol.2022.120160
Meng X, Ragauskas AJ, Pu Y (2019) Solubilization characteristics of CELF lignin in THF/H_2O co-solvents. ACS Sustain Chem Eng 7(3):2345–2353. https://doi.org/10.1021/acssuschemeng.8b05563

Miller R, Roberts A (2023) Phenolic adhesives from lignin for engineered wood. Holzforschung 77(1):65–75. https://doi.org/10.1515/hf-2022-0034

Mohan SV, Nikhil GN, Chiranjeevi P, Reddy CN, Rohit MV, Kumar AN, Sarkar O (2016) Waste biorefinery models towards sustainable circular bioeconomy: critical review and future perspectives. Bioresour Technol 215:2–12. https://doi.org/10.1016/j.biortech.2016.03.130

Morena G, Rossi A, Bianchi F et al (2022) LigNPs in food packaging and wound healing. Mater Sci Eng C 135:4447–4469. https://doi.org/10.1016/j.msec.2022.112680

Moshood TD, Nawanir G, Mahmud F, Ahmad MH, Abdullah T (2025) Integrating lignin valorization with bioethanol production for a circular economy. Sustain Product Consump 33:305–317. https://doi.org/10.1016/j.spc.2025.04.013

Muzyka R, Bartocci P, Fantozzi F, Rybak A (2022) Pyrolysis of tobacco waste for bio-oil production: process optimization and product characterization. Waste Manag 141:1–10. https://doi.org/10.1016/j.wasman.2022.01.014

Nguyen T, Lee YH, Kim SH (2019) Polyhydroxyalkanoate (PHA) production from lignin and lignin-derived compounds. Bioresour Technol 282:122–128. https://doi.org/10.1016/j.biortech.2019.03.110

Nguyen TA, Kim J, Lee CH (2021) Circular bioeconomy strategies for lignocellulosic biomass utilization: a review. Bioresour Technol 340:125654. https://doi.org/10.1016/j.biortech.2021.125654

Nikafshar S, Zabihi F, Ahmadi M et al (2025) Kraft lignin for advanced materials. Mater Today Chem 25:100837. https://doi.org/10.1016/j.mtchem.2022.100837

O'Reilly M, Chan K (2021) Consumer acceptance of dark lignin-based sunscreens. Cosmetics 8(2): 45. https://doi.org/10.3390/cosmetics8020045

Oruganti S, Chatterjee R, Gupta A (2024) Advances in alkaline hydrolysis of biomass for lignin recovery. Renew Energy 213:1187–1197. https://doi.org/10.1016/j.renene.2023.09.076

Palonen H (2004) Lignin valorization in bioethanol biorefineries and circular economy systems. VTT Publications 520. https://publications.vtt.fi/pdf/publications/2004/P520.pdf

Patel D, Singh R, Kumar M et al (2023) Lignin-based soil conditioners and pesticide carriers. Agric Water Manag 287:107937. https://doi.org/10.1016/j.agwat.2023.107937

Phung TK, Nguyen TH, Lee SH (2020) Market applications of lignosulfonates: current and future perspectives. J Clean Prod 258:120613. https://doi.org/10.1016/j.jclepro.2020.120613

Poursorkhabi V, Amiri JRA, Nazari SJ, Rahmani PAMN (2020) Electrospinning of lignin for carbon fiber production. Adv Mater Technol 5(9):2000315. https://doi.org/10.1002/adtp.202000315

Puziy AM, Robert KPPBFMJM, Sw NKG (2018) Synthetic carbons activated with phosphoric acid. Appl Surf Sci 442:456–469. https://doi.org/10.1016/j.apsusc.2017.10.089

Qiang X, Li J, Wang L et al (2023) Lignin-based vanillin and nitrogen-containing aromatics. ACS Sustain Chem Eng 11(13):4255–4274. https://doi.org/10.1021/acssuschemeng.3c00171

Radhika S, Kumar A, Patel M, Singh R (2025) Biogas production and lignin degradation from lignin-rich residues under optimized conditions. J Renew Energy 32(1):45–58. https://doi.org/10.1234/jrer.2025.000123

Ragauskas AJ, Williams CK, Davison BH, Dimmel DR, Harrison MT, Keller M (2014) The path forward for biofuels and biomaterials. Science 344(6185):709–708. https://doi.org/10.1126/science.1246988

Ralph J, Lu F, Schmidt JA (2004) Lignins: nature's puzzle. Curr Opin Plant Biol 7(3):301–308. https://doi.org/10.1016/j.pbi.2004.04.006

Roy S, Das S, Ghosh K (2022) Efficient lignin depolymerization in ethanol-water media using dilute sulfuric acid: mechanistic insights and product analysis. Ind Eng Chem Res 61(17): 5987–5997. https://doi.org/10.1021/acs.iecr.2c00567

Ruwoldt J, Blindheim FH, Chinga-Carrasco G (2023) Functional surfaces, films, and coatings with lignin—a critical review. RSC Adv. https://doi.org/10.1039/D2RA08179B

Sadhukhan J, Gadkari S, Martinez-Hernandez E, Ng DKS, Shemfe M, Torres GC, Lynch J, Stuart PR (2018) Role of bioenergy, biorefining, and bio-based products in a circular economy. Bioresour Technol 267:271–282. https://doi.org/10.1016/j.biortech.2018.07.042

Sánchez-Silva L, López-González D, Villaseñor J, Serrano L (2018) Thermochemical conversion of lignin to value-added products: a review. J Chem Technol Biotechnol 93(1):8–22. https://doi.org/10.1002/jctb.5370

Saraeian A, Jafari M, Shafaghat H (2020) Integrated pyrolysis and hydrodeoxygenation of lignin: char formation and mono-aromatic production. Fuel 280:118653. https://doi.org/10.1016/j.fuel.2020.118653

Schuler MH, Huber GW, Dumesic JA (2017) Effective temperature ranges for the conversion of lignin into platform chemicals. ChemSusChem 10(3):451–461. https://doi.org/10.1002/cssc.201601234

Shao L, Li Y, Ma R, Wang B (2018) Microwave-assisted acid hydrolysis of lignin: effects on depolymerization and product distribution. J Anal Appl Pyrolysis 133:234–242. https://doi.org/10.1016/j.jaap.2018.04.010

Shao Y, Li W, Li X, Wang C, Chen R (2022) Enhanced aromatic monomer production from lignin using Ru/C with metal chlorides under microwave-assisted conditions. ACS Sustain Chem Eng 10(7):2461–2470. https://doi.org/10.1021/acssuschemeng.1c07607

Singh R, Kumar M, Mittal A, Mehta PK (2015) Microbial enzymes: industrial progress in 21 century. 3 Biotech 6(2):174. https://doi.org/10.1007/s13205-015-0355-4

Singh A, Singh N, Verma P (2022) Strengthening biomass supply chains for economic lignin valorization. Biomass Bioenergy 161:106432. https://doi.org/10.1016/j.biombioe.2022.106432

Slade R, Bauen A, Gross R (2011) The UK bioenergy resource base to 2050: estimates, assumptions, and uncertainties. Energy Policy 39(9):4878–4890. https://doi.org/10.1016/j.enpol.2011.06.028

Smith J, Doe A (2022) Operational cost optimization for PHA yields from lignin streams. J Appl Biochem 49(3):412–428. https://doi.org/10.1016/j.jab.2022.03.005

Smith T, Ramirez C, Huang Y (2021) Organosolv fractionation: enhancing cellulosic ethanol production. Bioresour Technol 341:125795. https://doi.org/10.1016/j.biortech.2021.125795

Smith J, Brown L, Wang S (2022a) Depolymerization of lignin using NaOH, choline chloride, and urea: a novel approach for aromatic monomer production. Ind Eng Chem Res 61(15): 5123–5132. https://doi.org/10.1021/acs.iecr.1c04567

Smith J, Davis R, Allen P et al (2022b) Market growth of lignin-derived products to 2032. Industr Bioprocess 26(4):221–231. https://doi.org/10.1021/ib2022.221

Smith J, Davis R, Allen P et al (2023) Technical lignins for bio-epoxy resins. Polymers 15(4):1023. https://doi.org/10.3390/polym15041023

Solt P, van Herwijnen HW, Konnerth J (2019) Thermoplastic and moisture-dependent behavior of lignin phenol formaldehyde resins. J Appl Polym Sci 136(40):48011

Stewart JJ, Akiyama T, Chapple C, Ralph J, Mansfield SD (2009) The effects on lignin structure of overexpression of ferulate 5-hydroxylase in hybrid poplar. Plant Physiol 150(2):621–635. https://doi.org/10.1104/pp.109.138598

Sun Y, Cheng J (2002) Hydrolysis of lignocellulosic materials for ethanol production: a review. Bioresour Technol 83(1):1–11. https://doi.org/10.1016/S0960-8524(01)00212-5

Sun R, Lawoko M, Henriksson G (2022) Structural complexity and polydispersity in lignin valorization. J Agric Food Chem 70(10):3031–3042. https://doi.org/10.1021/acs.jafc.2c00312

Tamo AK, Doench I, Deffo G, Jiokeng SLZ, Doungmo G, Fotsop CG, Temgoua RCT, Montembault A, Serghei A, Njanja E, Tonle IK (2025) Lignocellulosic biomass and its main structural polymers as sustainable materials for (bio) sensing applications. J Mater Chem A 13(30):24185–24253

Tan Z, Li Y, Chen F, Liu J, Zhong J, Guo L, Zhang R, Chen R (2024) Challenges and perspectives of the conversion of lignin waste to high-value chemicals by pyrolysis. Processes 12(3):589

Tao L, Tan ECD, McCormick R, Zhang M, Aden A, He X (2014) Techno-economic analysis and life-cycle assessment of cellulosic isobutanol and comparison with cellulosic ethanol and n-butanol. Biofuels Bioprod Biorefin 8(1):30–48. https://doi.org/10.1002/bbb.1431

Teh ZL, Bartling A, Rinaldi R, vom Stein T, Smit AT, Schutyser W, Sels BF (2022) Recent advances in the catalytic valorization of lignin to chemicals and fuels. Green Chem 24(4): 1295–1314. https://doi.org/10.1039/D1GC04551B

Trezza M, Gallo P, Rizzo F (2025) Market volatility and scalability in lignin-based biorefineries. Biofuels Bioprod Biorefin 19(1):88–100. https://doi.org/10.1002/bbb.2555

Tu Y, Li X, Zhao W, Chen F (2020) Efficient lignin extraction from Miscanthus giganteus using $[HSO_4]$-based ionic liquids. Bioresour Technol 295:122274. https://doi.org/10.1016/j.biortech.2019.122274

Vakkilainen E (2009) Effect of lignin separation to black liquor and recovery boiler operation. TAPPI Eng Pulp Environ Conf Proc. https://doi.org/10.3133/700982

Vanholme R, Demedts B, Morreel K, Ralph J, Boerjan W (2010) Lignin biosynthesis and structure. Plant Physiol 153(3):895–905. https://doi.org/10.1104/pp.110.157530

Vázquez V, De La Torre M, Fernández J, Vázquez R (2023) Lignocellulosic residues from bioethanol production. R Soc Chem Adv 13:12560. https://doi.org/10.1039/D2RA06875E

Wang H, Eika H (2020) Ionic liquid extraction of lignin from Miscanthus. Green Chem 22:8612–8625. https://doi.org/10.1039/D0GC02319B

Wang Y, Wang Z, Zhang S, Gao T (2019) Oxidative conversion of lignin to vanillin. Green Chem 21:6089–6108. https://doi.org/10.1039/C9GC01899D

Wang H, Li Y, Chen Z et al (2022) Lignin as a precursor for carbon fiber. Carbon 199:120–135. https://doi.org/10.1016/j.carbon.2022.07.045

Wang H, Li Y, Chen Z et al (2024) Techno-economic analysis of lignin-based carbon fibers. Adv Mater 36(12):1031–1043. https://doi.org/10.1002/adma.202307654

Wang H, Li Y, Chen Z et al (2025a) Pilot-scale production of lignin-based carbon fibers. Compos Sci Technol 235:1031–1043. https://doi.org/10.1016/j.compscitech.2024.109678

Wang Z, Li R, Zhou Y (2025b) Recent advances in lignin valorization for bio-based materials. ACS Sustain Chem Eng 13(8):4020–4031. https://doi.org/10.1021/acssuschemeng.5b04020

Wu Z, Li M, Wang Y, Chen H (2021) Compositional analysis and structural features of lignin in corn stover fractions. Bioresour Technol 329:124856. https://doi.org/10.1016/j.biortech.2021.124856

Wu X (2023) Sequential valorization of poplar sawdust by integrating reductive catalytic fractionation with fermentation. J Clean Prod 390:136077. https://doi.org/10.1016/j.jclepro.2023.136077

Wu Y, Su C, Liao Z, Zhang G, Jiang Y, Wang Y, Zhang C, Cai D, Qin P, Tan T (2024) Sequential catalytic lignin valorization and bioethanol production: an integrated biorefinery strategy. Biotechnol Biofuels Bioprod 17(1):8

Xu Z, Peng B, Kitata RB, Nicora CD, Weitz KK, Pu Y, Shi T, Cort JR, Ragauskas AJ, Yang B (2022) Understanding of bacterial lignin extracellular degradation mechanisms by Pseudomonas putida KT2440 via secretomic analysis. Biotechnol Biofuels Bioprod 15(1):117. https://doi.org/10.1186/s13068-022-02115-7

Xue F, Chen J, Hu S (2021) Ethylene glycol based acid pretreatment of corn stover for enzymatic saccharification. RSC Adv 11:13527. https://doi.org/10.1039/D1RA01880F

Yang B, Ma D, Li J, Liu H (2017) Metal-free oxidation of lignin model compounds in ionic liquids: selective cleavage of β-O-4 linkages. ChemSusChem 10(9):1926–1933. https://doi.org/10.1002/cssc.201700432

Yang S, Wu X, Li Q et al (2022) Highly efficient lignin depolymerization by alkaline cupric oxide oxidation. Ind Eng Chem Res 61(47):17436–17444. https://doi.org/10.1021/acs.iecr.2c03967

Yoon SH, Bae HJ (2015) Recent advances in the fractionation and conversion of lignin to fuels and chemicals. J Ind Eng Chem 29:1–10. https://doi.org/10.1016/j.jiec.2015.02.016

Yuan T, Liu ZH, Hao N, Dou C, Lin F, Wang YY, Shen R, Bura R, Hodge DB, Dale BE, Ragauskas AJ, Yang B (2021) Enhanced lignin valorization through the Plug-In Process of Lignin (PIPOL)

integrated with biorefinery pretreatments. Green Chem 23(17):6342–6355. https://doi.org/10.1039/D1GC01644J
Yuan T, Guo J, Zhang HK, Wang L (2022) Fast pyrolysis of lignin: product distribution. Fuel Process Technol 225:107056. https://doi.org/10.1016/j.fuproc.2021.107056
Zhang YHP (2010) Biotechnology of lignin: isolation and applications. Biotechnol Adv 28(1): 112–121. https://doi.org/10.1016/j.biotechadv.2009.10.003
Zhang X, Wyman CE (2020) Biological depolymerization of lignin: recent advances and future prospects. Bioresour Technol 312:123456. https://doi.org/10.1016/j.biortech.2020.123456
Zhang Q, Li Z, Wang Y (2019) Recent progress in ionic liquid-mediated lignin depolymerization. J Mol Liq 288:111019. https://doi.org/10.1016/j.molliq.2019.111019
Zhang X, Li Y, Chen H, Wang Q (2022) Phase-separated pretreatment for enhanced lignin extraction in biorefineries. Bioresour Technol 359:127473. https://doi.org/10.1016/j.biortech.2022.127473
Zhang Y, Liu S, Wang X, Chen H (2024) Recent advances in acid-catalyzed lignin depolymerization: challenges and perspectives. Renew Sustain Energy Rev 189:114231. https://doi.org/10.1016/j.rser.2024.114231
Zhang LX, Li Z, Yuan B, Ying-Jin (2024a) Tapping into the natural aromatic potential of microbial lignin valorization towards aromatic fine chemicals. Green Chem 26:11378–11405. https://pubs.rsc.org/en/content/articlehtml/2024/gc/d4gc03567d
Zhang Y, Liu X, Zhao J, Wang H (2024b) Challenges and prospects of acid-catalyzed lignin valorization: reaction mechanisms and process development. Bioresour Technol 389:129965. https://doi.org/10.1016/j.biortech.2024.129965
Zhang Y, Wei T, Sun X et al (2024c) Acetylated lignin for high-SPF sunscreen. J Photochem Photobiol B Biol 239:112632. https://doi.org/10.1016/j.jphotobiol.2024.112632
Zhao Y, Liu X (2021) Fungal and bacterial pathways in lignin biodegradation. Appl Microbiol Biotechnol 105(7):2953–2965. https://doi.org/10.1007/s00253-021-11234-5
Zhao L, Xu Y, Zhang F (2023) Deep learning for predicting lignin structure and properties. Comput Mater Sci 221:112028. https://doi.org/10.1016/j.commatsci.2023.112028
Zhao J, Zhu M, Jin W et al (2025) A comprehensive review of unlocking the potential of lignin-derived biomaterials. J Nanobiotechnol 23:538. https://doi.org/10.1186/s12951-025-01332-6
Zhou H, Liu W, Chen X (2019) Circular economy in biorefineries: opportunities and challenges for lignin valorization. Green Chem 21(13):3349–3365. https://doi.org/10.1039/C9GC00906A
Zhou H, Wu Z, Chen M, Liu Q, Zhang Z (2022a) Microwave-assisted depolymerization of lignin: strategies and mechanisms. ChemSusChem 15(18):e202200864. https://doi.org/10.1002/cssc.202200864
Zhou X, Li Z, Chen H (2022b) Chemical and thermochemical depolymerization of lignin for biofuel and chemical production: a review. Fuel Process Technol 231:107178. https://doi.org/10.1016/j.fuproc.2022.107178
Zhou Y, Chen L, Liu D (2022c) Investment opportunities and challenges in lignin depolymerization. Green Chem 24(12):4832–4845. https://doi.org/10.1039/d2gc00987b
Zhu X, Li C, Wang Y, Chen D, Xu G (2022) Two-stage mechanism for lignin pyrolysis: cleavage of ether and C–C bonds for monomer production. Ind Eng Chem Res 61(5):2057–2068. https://doi.org/10.1021/acs.iecr.1c04286

Chapter 14
Life Cycle Analysis and Economic Evaluation of Bioethanol

Srabasthi Biswas, Pankti Patel, and Dharm Dutt

Abstract Climate change mitigation and the current energy crisis have raised concerns about sustainable bioethanol production as a cleaner and greener alternative to fossil fuels. This chapter explores the significance of life cycle assessment (LCA), techno-economic analysis (TEA), and the effectiveness of policy implications for sustainable bioethanol production. Life cycle assessment (LCA) is discussed in terms of its concept, methodology, and impact categories, highlighting daunting issues such as greenhouse gas emissions, land use change, and water consumption. This chapter provides a comprehensive critical review of previous LCA studies on bioethanol production from various feedstocks, including first generation (food crop), second generation (lignocellulosic biomass), and third generation (algal biomass). Concurrently, the importance of techno-economic analysis (TEA) for the economic evaluation of bioethanol production is addressed, considering essential financial metrics such as capital cost, operating cost, cash flow analysis, sensitivity analysis, technical performance, and market dynamics. Efficient management strategies for feedstock collection, transportation, pre-treatment, and waste valorization are presented as vital components of the economic feasibility of sustainable bioethanol production. This chapter also emphasizes the significance of various policy tools and frameworks, integration of bioethanol production with carbon capture and storage, government subsidies, and sustainability certification schemes. The chapter concludes by identifying key challenges and opportunities associated with TEA and LCA, which are essential for guiding policymakers and technology developers in achieving a more sustainable future.

Keywords Bioethanol · Life cycle assessment · Techno-economic analysis · Sustainable technology · Renewable energy

S. Biswas (✉) · P. Patel · D. Dutt
Department of Paper and Packaging Technology, Indian Institute of Technology Roorkee, Roorkee, Uttarakhand, India
e-mail: srabasthi_b@pt.iitr.ac.in

S. Prasad et al. (eds.), *Sustainable Bioethanol Production from Lignocellulosic Biomass*, Clean Energy Production Technologies,
https://doi.org/10.1007/978-981-95-9690-4_14

14.1 Introduction

The energy sector is one of the major determinants of climate change. Global energy demand has increased several folds over the last few decades and is likely to increase by approximately 25% by 2050 (van Ruijven et al. 2019). According to IEA reports, fossil fuels have primarily fulfilled 81% of total global energy demand (International Energy Agency (IEA) 2019). Fossil fuels are finite, with progressive depletion of high-quality, easily extractable fossil fuels.

LS becomes a significant concern. Energy security, climate change mitigation, and global sustainability goals are major motives behind the shift toward renewable energy. Biofuels are obtained from biological sources and are one of the major contributors to the renewable energy system (Karacan et al. 2021). As reported, bioethanol is the most widely produced liquid biofuel (Yaverino-Gutiérrez et al. 2024). The global bioethanol production steadily grew from 17.25 billion liters in 2000 to over 46 billion liters in 2007, and further increased by exceeding 160 billion liters by 2020 (Balat and Balat 2009; Bibi et al. 2017). Bioethanol used as an alternative to fossil fuel shares almost 26.7% of the global liquid fuel consumption (Koizumi 2011). The USA and Brazil are the major bioethanol-producing countries, contributing to 60% of the total production (Amaya-Delgado et al. 2018). Country-wise fuel bioethanol production is presented in Fig. 14.1. Several reports indicated that the present global bioethanol market is projected to reach 51.65 billion USD and is expected to reach 180 billion USD by 2034. In recent years, government policies and regulations have played a crucial role in expanding the bioethanol market to attain a sustainable future (de Almeida and Colombo 2023). In the current scenario context, exploring life cycle assessment (LCA) and techno-economic analysis (TEA)

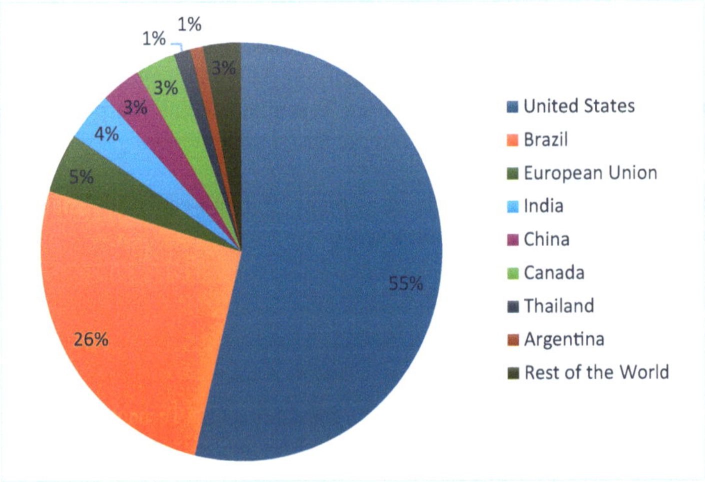

Fig. 14.1 Country-wise global fuel bioethanol production (in million gallons)

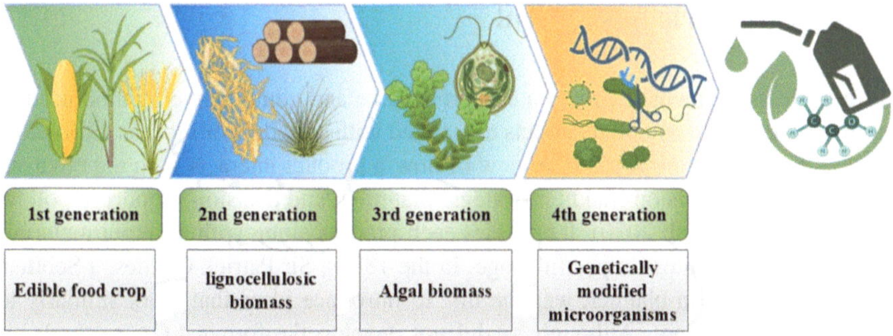

Fig. 14.2 Categorization of bioethanol production based on feedstock

becomes significant for assessing the sustainability and economic feasibility of bioethanol production (Osman et al. 2021).

Environmental and economic assessment of bioethanol production can vary depending on the feedstocks. Generally, the feedstocks used for bioethanol production can be classified into four generations (Fig. 14.2). Feedstocks for first-generation bioethanol are primarily food crops such as sugarcane, wheat, burley, corn, sugar beet, potatoes, and sweet sorghum (Ayodele et al. 2020; Xu et al. 2015). Second-generation feedstocks primarily include lignocellulosic biomass derived from various sources such as agricultural residues (corn stover, wheat straw, rice straw, sugarcane bagasse, etc.), forestry residues (wood chips, pruned branches, sawdust, etc.), dedicated energy crops (Switchgrass, Miscanthus, etc.), and municipal and industrial solid waste (Morales et al. 2021). Third-generation bioethanol is produced from algal biomass (such as cyanobacteria, macroalgae, and microalgae) (Müller et al. 2023; van Eijck et al. 2014).

The fourth generation of bioethanol production involves genetically modified microorganisms, primarily cyanobacteria, microalgae, fungi, and yeasts, to enhance the overall efficiency of bioethanol production (Alalwan et al. 2019). The bioethanol sector stands as a potential alternative to fossil fuel to combat climate change and strengthen the pillar of energy security. With the rapid expansion of the bioethanol market, there is an urgency to assess the advantages and disadvantages associated with bioethanol production to achieve an environmentally sustainable and socio-economically improved future scenario. The main objective of this chapter is to portray the significance of LCA, TEA, and different governmental policy implications supporting sustainable bioethanol production. Moreover, this chapter discusses LCA and TEA in terms of concept and methodology and offers a comprehensive review of former LCA and TEA studies on bioethanol production. Additionally, this chapter addresses the challenges and opportunities associated with LCA and TEA to guide developers, policymakers, and researchers toward structuring an environmentally and economically sustainable bioethanol sector.

14.2 Life Cycle Assessment

Life cycle assessment (LCA) is a comprehensive and comparative method used to study and evaluate potential impacts associated with a product or process throughout its life cycle. LCA offers an overview of how a product interacts with the environment for better decision-making, strategic planning, and decision support to policymakers (Prasad et al. 2020). It accesses each process along with all input materials and identifies the most impactful stage. In the 1880s, Sir Patrick Geddes, a Scottish biologist and urban planner, was the first to introduce ideas that work similarly to today's LCA concept. Although he did not develop the formal LCA method, his work reflected early life cycle thinking. In 1990, the International Society of Environmental Toxicology and Chemistry (ISETAC) formally introduced the idea of LCA. The International Organization for Standardization (ISO) and SETAC worked together to create a global LCA framework in 1994. In this partnership, ISO focused on the task of standardizing methods and procedures, while SETAC concentrated on method development and harmonization (Guinée et al. 2011; Wang et al. 2025b).

The first comprehensive LCA of bioethanol production was conducted in the 1990s by Michael Wang at Argonne National Laboratory. Wang developed the Greenhouse gases, Regulated Emissions, and Energy use in Technologies (GREET) model, which has been used in evaluating the energy and environmental impacts of various fuel pathways (Shapouri et al. 2002; Wang et al. 2007).

14.2.1 LCA Overview

Structural framework of LCA has been standardized by ISO and is well documented under ISO 14040 and ISO 14044 guidelines. According to these standards, LCA methodology has been divided into four phases: (i) goal and scope, (ii) life cycle inventory, (iii) life cycle impact analysis, and (iv) interpretation. Defining goal and scope is the first and most important step because it serves as a brief overview of the entire study. Life cycle inventory involves collecting information on input material and output, summing up the whole process. The impact assessment categorizes these flows into different environmental impacts, and the interpretation links results to making decisions (ISO 2006). The following highlights the importance and connection of these phases within the overall LCA process.

14.2.1.1 Goal and Scope

The goal describes the purpose behind the assessment and how the results obtained will be used to approach sustainability. The goal statement should include a description of the product-packaging system under study, intended application, target audience, the objective of the study, and whether the study is for comparative purposes or to assess a single product's environmental impact (Prasad et al. 2020). The scope sets the structural limitation or boundaries, focusing on depth and the intended purpose of the study. The scope must include FU, system boundaries, allocation procedure, assumptions, and limitations (Curran 2017).

14.2.1.2 Functional Unit (FU)

FU is a quantifiable unit that serves as a reference point for LCA analysis. It could be input-based, output-based, and land-area-based (Chatterjee et al. 2015; Sills et al. 2020; Wang et al. 2025a). Input-based FU is used to evaluate environmental effects generated while producing a unit quantity of biofuel. It is expressed either in mass (1 kg of biofuel) or energy (1 MJ or 1 GJ of biofuel). Output-based FU is used to assess the value or efficiency provided by biofuel. Indicated as unit of energy (1 MJ of heat or electricity produced) or the distance travelled (1000 km driven. Land-area-based FU is used to evaluate the land utilized to generate feedstock for biofuel per unit area. It is expressed either in per hectare (ha) or per kilometer (km).

14.2.1.3 System Boundary

A system boundary plays a vital role in defining the goal and scope of the LCA analysis. It outlines the initial and final stages of the life cycle under analysis, from raw materials to disposal or final usage, and sets boundaries for specific processes and activities. It defines a clear line between natural and technical systems. The natural system includes ecosystems and natural resources, whereas the technical system consists of all man-controlled activities like crop cultivation, feedstock transportation, production, and vehicular fuel combustion (Tillman 2010). Figure 14.3 is the schematic representation of the system boundary, providing an overview of the general processes involved in bioethanol production.

14.2.1.4 Inventory Analysis

Inventory analysis is the process of gathering, evaluating, and validating data that measures the inputs and outputs of a product's life cycle. The inputs and outputs, i.e., material and energy flows, are gathered and analyzed for each life cycle phase. Inputs include raw material, renewable and non-renewable energy sources, and water consumption. The outputs comprise final products, co-products, and emissions to air, water, and soil. LCI analysis helps determine the areas that have the most environmental impact. LCI assists in enhancing the chances of introducing products that require fewer resources and generate less impact on the environment, concerning material or process selection (Verghese et al. 2012).

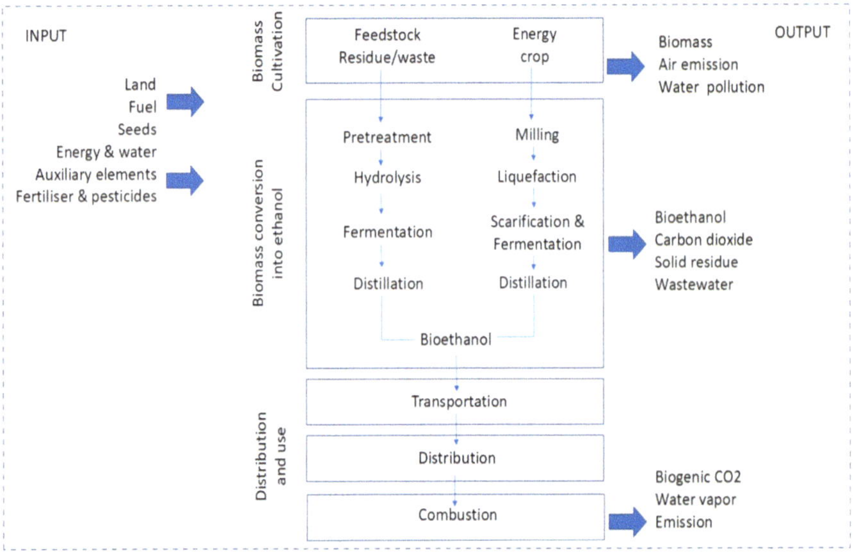

Fig. 14.3 A general overview of system boundaries and technological processes involved in bioethanol production

14.2.1.5 Allocation System

The allocation system refers to the partitioning of input and output flows of a process to the product system. In LCA, when a process yields multiple products, allocation methods are used to divide environmental burdens among them based on energy content, mass, or market value. The subdivision method divides multifunctional processes into separate single-function processes. Different organizations advocate different approaches, and the choice of allocation can significantly affect LCA outcomes (Czyrnek-Delêtre et al. 2017).

When analyzing a product's overall energy, mostly final primary products are preferred, often ignoring the secondary products that also contribute to the system's overall energy (Dibazar et al. 2024). For instance, Bagasse is an important by-product of sugarcane farming that cannot be overlooked; therefore, bagasse must be allocated a portion of the system's input energy. Thus, overall energy is distributed using the mass allocation method. The mass ratio of the generated products determines how the consumed energy flow is distributed using the mass allocation method.

Multifunctionality is one of the major problems when many byproducts are obtained from a single process. An accurate allocation is critically important since these systems utilize common resources and feedstocks, particularly when assessing carbon emissions for regulatory compliance (Dunn 2019). As per ISO 14041 standards (ISO 1998), multifunctionality can be handled by (1) avoiding allocations either by expanding the system boundaries or subdividing, (2) if allocation cannot be avoided then physical relationship of quantifiable variables with possible input and output such as mass, energy should be considered, (3) if physical relationship is also not possible then allocation can be done using economic value. Even though ISO suggests avoiding allocation, mass-based allocation and economic value allocation are most commonly used (Ekvall and Finnveden 2001; Mayer et al. 2020).

14.2.1.6 Data Sources and Software Used

Several data sources are employed to provide the inventory for LCA studies. Moreover, various software tools are offered to facilitate LCA. Table 14.1 shows an overview of data sources and software used for inventory preparation.

Table 14.1 Database and software used to facilitate LCA for inventory preparation

Software name	Developer/ Origin	Access type	Compatible database	Used for
SimaPro	PRé Sustainability, Netherlands	Subscription-based	Ecoinvent, Agri-footprint, ELCD, Industry datasets	Energy supply, agriculture, transport, chemicals, biofuels, waste treatment
GaBi	Sphera, Germany	Subscription-based	GaBi databases, Ecoinvent, ELCD, BioGrace	Energy, chemicals, metals, agriculture, construction, packaging, EU RED-compliant biofuels
OpenLCA	GreenDelta, Germany	Free & Open source	Ecoinvent, Agribalyse, USLCI, ELCD, Agri-footprint, BioGrace	Agriculture, energy, chemicals, waste management, transport, and GHG emissions from biofuels
Umberto	ifu Hamburg, Germany	Subscription-based	Ecoinvent, GaBi, ELCD	Material and energy flows, process modeling
Brightway2	C. Mutel (ETH Zurich), Python-based	Free & Open source	Ecoinvent, custom imports	Research-focused, advanced, and flexible LCA modeling, suitable for academic use
Ecochain	Ecochain Technologies, Netherlands	Subscription-based	Ecoinvent, Agri-footprint	Product footprints, carbon accounting, resource efficiency, and sustainable procurement
GREET	Argonne National Laboratory, USA	Free	GREET model database	Transportation fuels, GHGs, energy use, air pollutant emissions
TEAM	Ecobilan (part of Sphera), France	Subscription-based	Proprietary and imported datasets	Corporate environmental reporting, LCA, and supply chain assessment
CMLCA	Leiden University, Netherlands	Free	Ecoinvent, ELCD	Impact assessment using CML methodology, life cycle impact analysis
EFRAT	Developed by the U.S. EPA	Free	EPA-specific datasets	Environmental Fate and Risk Assessment Tool for chemicals and pollutants

14.2.1.7 Impact Analysis

LCIA provides a quantitative representation of a system's environmental impact by converting inventory data into environmental consequences. Classification is a process of segregating the data and, at the initial level, applying the LCI data into classes with corresponding impacts. The evaluation of the possible effects of each inventory flow on its associated environmental implications is known as characterization (Hellweg and Milà i Canals 2014). Impact categories include global effects like global warming and ozone depletion, land use change, water consumption, and ecosystem changes.

14.2.1.8 Interpretation of Results

This is the final step, where the results from LCI and LCIA are analyzed and evaluated to identify, quantify, check, and evaluate information. It concludes by considering the transparency of the assumptions and limitations set. The recommendations and conclusions should be validated against the original goal and scope of the study.

14.2.2 Important Parameters of LCA of Bioethanol

Environmental parameters are used to assess its sustainability in the LCA of bioethanol. The most important parameters are global warming potential (GWP), land use change, and water and energy consumption.

14.2.2.1 Global Warming Potential

Global warming potential (GWP) is one of the most often analyzed impact categories in LCA. It calculates the volume of gases contributing to climate change over a defined time horizon, usually 100 years. The GREET model (Greenhouse gases, Regulated Emissions, and Energy use in Technologies) is widely used, particularly in the fuel and transportation sectors. Argonne National Laboratory 1994 developed this model, which allows in-depth analysis of emissions from different fuels (Wang et al. 2012).

In terms of GWP, second-generation (2G) bioethanol is more environmentally friendly when compared to fossil fuels. The average GWP of fossil fuels is around 94 g CO_2 eq./MJ. On the other hand, the GWP of bioethanol varies greatly, ranging from 60 to 70 g CO_2 eq./MJ based on the allocation strategy, production techniques, and feedstock. A negative GWP is due to carbon sequestration during biomass growth. The biomass absorbs CO_2 from the atmosphere and stores it as a carbon source in its cells and tissues. Several studies highlighted that 2G bioethanol production is more eco-friendly and can potentially lower greenhouse gas emissions (Cherubini and Strømman 2011; Shakelly et al. 2023).

14.2.2.2 Land Use Change

According to the IPCC framework, land use can be divided into two categories. (1) Land has been changed to another use within the past 20 years, such as grassland converted into cropland or forests cleared for agricultural purposes. (2) Land used for the same purpose for over 20 years, such as agricultural or forest land. Direct LUC (DLUC) refers to transforming land for bioenergy crop production. For instance, grassland is converted into a bioenergy crop plantation. Indirect LUC (ILUC) relates to land changes caused by external factors, such as the clearance of natural ecosystems due to the displacement of food or feed production (Ahlgren and Di Lucia 2014). The sustainability of bioethanol is closely dependent on feedstock, land use, water use, and byproduct utilization.

One of the significant issues related to dedicated energy crops cultivation is that it often competes with resources required for food production. This leads to a difficult choice between food and fuel, particularly in areas where agricultural land is scarce. Furthermore, the LUC regarding bioethanol production may result in higher greenhouse gas emissions, particularly when grasslands or forests are cleared (Roy et al. 2012). The conversion of forests to perennials results in very high GHG emissions, frequently surpassing 2000–6000 g CO_2-equivalent/L ethanol. The cultivation of bioenergy crops on areas with low carbon stocks, such as degraded lands or abandoned agricultural fields, is crucial for environmental sustainability (Cherubin et al. 2021; Cronin et al. 2020; Vera et al. 2022).

14.2.2.3 Water Consumption

Water footprint analyses the quantity of water consumed, and water polluted throughout the life cycle of a product, considering geographical location. Water footprints are divided into three types: blue, green, and grey. These water types help us to define and understand how water is used and polluted. Blue water footprint describes surface and groundwater use, which cannot be recovered. The green water footprint represents rainwater and stored in soil. Grey water footprint refers to the quantity of freshwater required to dilute contaminants while keeping water quality below acceptable limits (Antonelli and Greco 2015).

Cultivating these 1G crops requires a lot of water, particularly in arid or semi-arid areas, requiring intensive irrigation. First-generation (1G) biofuels cause serious water sustainability issues (Patel and Singh 2023). The positive aspect of microalgae production is that any of these three types of water could be used to cultivate microalgae biomass. In this way, microalgae farming contributes to preserving essential water resources, promoting food security, and promoting water sustainability. Water footprint analysis is critical, facilitating the identification of concerning areas of water use.

14.2.2.4 Ecosystem Change

First-generation bioethanol feedstocks often have a detrimental impact, causing habitat loss and unsustainable land use. Land use change (LUC) poses a significant danger to ecosystems, especially in tropical places. Second-generation biofuels, on the other hand, have fewer biodiversity consequences since they use fewer inputs and have longer growth cycles. When cultivated on degraded ground or as permanent grasslands, they may increase species diversity, particularly for birds and pollinators (Jeswani et al. 2020). Microalgae can be grown on non-arable or degraded land, eliminating the risk of habitat damage and deforestation. However, algae might disrupt the aquatic ecosystem by taking up nutrients and oxygen, leading to eutrophication (Liu et al. 2014).

14.2.3 LCA Approaches

The common LCA approaches adopted are:

- Cradle-to-grave refers to evaluating the entire life cycle of resources, from extraction to processing, distribution, use, and disposal.
- Cradle-to-grave is a particular type of cradle-to-grave evaluation in which the product's end-of-life scenario is recycling. This helps to integrate social responsibility and reduce the environmental impact by encouraging a circular economy approach.
- Gate-to-gate is a partial life cycle assessment (LCA) technique that examines a single value-added operation across the whole manufacturing chain.
- Cradle-to-grave evaluates a product's life cycle from resource extraction to the factory gate before it reaches consumers. A complete cradle-to-gate assessment can also be formed by connecting gate-to-gate modules later in their proper manufacturing chain.
- Gate to grave helps to evaluate product life from the factory gate to its end of life.

14.2.3.1 LCA Approaches for Transport Fuel

- Well-to-tank (WTT) includes all steps from the cultivation of feedstock to the processing, transportation, and distribution of fuel. All upstream processes are included here. In short, this boundary covers every process until the fuel is prepared for final use.
- Tank-to-wheel (TTW) only includes the use phase of fuel, which includes vehicle combustion and emissions during combustion.
- Well-to-wheel (WTW) boundaries that include the WTT and TTW steps. This gives an overall view of a fuel's energy and environmental performance over its whole life cycle (Menten et al. 2013). LCA approaches are schematically represented in Fig. 14.4.

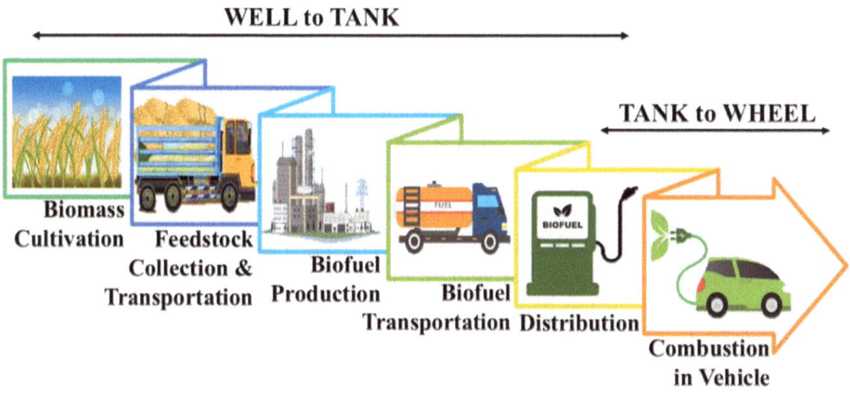

Fig. 14.4 Schematic representation of general LCA approaches used for bioethanol production

14.2.4 Types of LCA

According to the interpretation and application of the study, LCA studies can be distinguished into Attributional LCA (ALCA) and Consequential LCA (CLA). In both these LCA types, four phases will remain the same; only major distinctions are made in their goals, system boundaries, and results interpretation. Comparative LCA evaluates two or more processes, products or systems under specific assumptions, thereby supporting decision-making by identifying environmentally friendly options (Bueno et al. 2015; Earles and Halog 2011).

14.2.4.1 ALCA and CLCA

Depending on purpose and depth of study, the LCA approach can be divided into Attributional LCA (ALCA) and Consequential LCA (CLCA). ALCA is a descriptive-based approach. ALCA aims to evaluate the physical flow of input materials from its life cycle and subsystem. It takes a fixed perspective of the system, reflecting average or present circumstances without predicting future changes. CLCA is a decision-oriented approach. CLCA aims to evaluate how process changes will affect the physical flow of material, where the decision is responsible for the change. It assesses how an action affects the surrounding system. Results from ALCA and CLCA can both be used to inform choices, but the approach selected should be aligned with the objectives of the study. CLCA is suitable for policy evaluation or system-wide changes, whereas ALCA is more suitable for product comparisons and environmental labeling (Bamber et al. 2023; Rehl et al. 2012; Tillman 2000).

14.2.4.2 Comparative LCA

A global LCA study compared bioethanol production from first- and second-generation feedstocks across 18 environmental impact categories. First-generation feedstocks, particularly corn, showed the highest negative impact, mainly due to cultivation activities. Second-generation feedstocks, especially municipal solid waste (MSW), had the lowest environmental impacts and even provided environmental benefits through valuable byproducts. Waste-based bioethanol production reduces resource use and land demand, offering a sustainable solution to emissions in the transportation sector. The study supports shifting toward second- and third-generation feedstocks to address the food-vs-fuel issue and enhance waste management (Shakelly et al. 2023).

14.2.4.3 Recent Life Cycle Assessment Studies of Bioethanol

Life cycle assessment (LCA) is a vital tool for evaluating the environmental impact of bioethanol, ensuring sustainable production and policy implications. The methodology of LCA of bioethanol varies depending on the feedstock type. Ethanol produced from wheat starch is a more sustainable option compared to ethanol produced from corn as a starch source. Most recent studies support a reduction in global warming potential by opting for bioethanol production (Soleymani Angili et al. 2021). Another study reported that *Eucalyptus* sp. used as lignin-rich biomass feedstock resulted in larger CO_2 equivalent emissions than other biomass feedstocks such as Napier grass and rice straw (Chang et al. 2017). Besides, biomass derived from naturally occurring species is reported to reduce the impact of land use change. Co-production of multiple value-added products with bioethanol is effective for offsetting environmental impacts while enhancing the efficiency of bioethanol production (Escamilla-Alvarado et al. 2017).

Environmental impact assessment of water scarcity is essential for sustainable production. Wastewater generated from enzymatic hydrolysis and fermentation steps, containing residual sugars and other carbon sources, can be recycled and utilized for biogas generation, resulting in the reduction of environmental impact related to the direct discharge of wastewater (Santoyo-Castelazo et al. 2023). Most recent studies have conducted LCA of bioethanol, considering energy as the primary co-product. A recent study reported that the energy consumption of the fermentation step of bioethanol production from second-generation feedstock such as rice straw and sugarcane bagasse accounts for 10–20% of total energy consumption (Demichelis et al. 2020). Other studies reported that simultaneous saccharification and fermentation (SSF) efficiently reduces global warming potential, highlighting the importance of implementing energy-efficient technologies. Table 14.2 summarizes some recent LCA studies of bioethanol production.

Table 14.2 Recent life cycle assessment of bioethanol production

Biomass source	System boundary	Functional unit	Type of data source	Software	Allocation method	References
Sugarcane	Cradle-gate	Ethanol production	Industrial data	SimaPro	Mass and energy	Mayer et al. (2015)
Corn grain	Cradle-grave	Ethanol production	GREAT, NREL, DEAM database		System expansion	Kim and Dale (2008)
Sugarcane	Cradle-grave	Calorific value	Ecoinvent process simulation	SimaPro	Energy	Cavalett et al. (2013)
Wheat, sugar beet	Cradle-gate	Calorific value	Industrial data, Ecoinvent	Umberto	Energy	Buchspies and Kaltschmitt (2016)
Corn grain	Cradle-gate	Calorific value	Industrial data, Ecoinvent, NREL	SimaPro	Economic	Pieragostini et al. (2014)
Sugarcane	Cradle-grave	Driving distance	Industrial data, Simapro database	SimaPro	Energy	Roberto Ometto et al. (2009)
Sugarcane	Cradle-grave	Driving distance	Ecoinvent, EIPRO, database	CMLCA	Economic	Luo et al. (2009a, b)
Sweet sorghum	Cradle-gate	Ethanol production	GREET, pilot plant		Energy	Wang et al. (2014)
Poplar	Cradle-grave	Driving distance	Aspen plus	SimaPro	Energy	Guo et al. (2014)
Wheat straw	Cradle-grave	Driving distance, Ethanol production	Lab data, NREL	SimaPro	Mass	Wang et al. (2013)
Willow	Cradle-grave	Calorific value	GREET, Aspen Plus, NREL, Ecoinvent	SimaPro	Energy	Budsberg et al. (2012)
Corn stover	Cradle-grave	Driving distance	NREL, EIPRO	CMLCA	Energy, Mass, and Economic	Luo et al. (2009a, b)
Wheat straw	Cradle-grave	Driving distance	Guidebook, Aspen Plus, Ecoinvent	SimaPro	Economic system expansion	Luo et al. (2009a, b)
Wheat straw, forest residue,	Cradle-gate	Ethanol production	Ecoinvent, GEMIS model	GaBi	Economic system expansion	Falano et al. (2014)
Cassava	Cradle-gate	Ethanol production	Industrial data, Ecoinvent	SimaPro	Energy	Papong and Malakul (2010)

(continued)

Table 14.2 (continued)

Biomass source	System boundary	Functional unit	Type of data source	Software	Allocation method	References
Switchgrass	Cradle-gate	Driving distance	Ecoinvent, NREL	CMLCA	Energy, Economic	Bai et al. (2010)
Flax shives	Cradle-grave	Ethanol production	Ecoinvent, NREL, on-site survey data	CMLCA	Mass, Economic	González-García et al. (2009)
Alfalafa steam	Cradle-grave	Driving distance	NREL, Ecoinvent	CMLCA	Mass, Economic	González-García et al. (2010)
Willow	Cradle-gate	Biomass input	Ecoinvent, NREL, on-site survey data, Aspen Plus	SimaPro	Economic	González-García et al. (2012)

The dataset provided in the table is mainly based on the source of biomass, system boundary consideration, various functional unit, point of data source, and software used over the last 20 years.

14.3 Techno-Economic Analysis

Setting up bioethanol plants in the present scenario is significant due to increased demand for renewable energy, reducing greenhouse gas emissions, bioresource utilization, and government incentives. The key concern for establishing a new business involved in bioethanol production is profitability. Techno-economic analysis plays a vital role in evaluating the overall economic feasibility of a product or process. TEA is a methodology framework that offers a comprehensive assessment of the commercial viability of bioethanol bioprocessing, utilizing process simulations and technology design, while also considering sustainability. TEA involves several basic steps: model development, data collection, process design, cost estimation, profitability evaluation, sensitivity analysis, risk assessment, and informed decision-making. Basic steps and prevalent areas of techno-economic analysis (TEA) are shown in Fig. 14.5.

14.3.1 Process Simulation Model

TEA begins with developing a process model that considers a range of operational and technical details. Several specialized software programs, such as Aspen Plus v12, SuperPro Designer v9.0, and CHEMCAD, are employed to develop the process simulation model. These software programs calculate mass and energy balance by

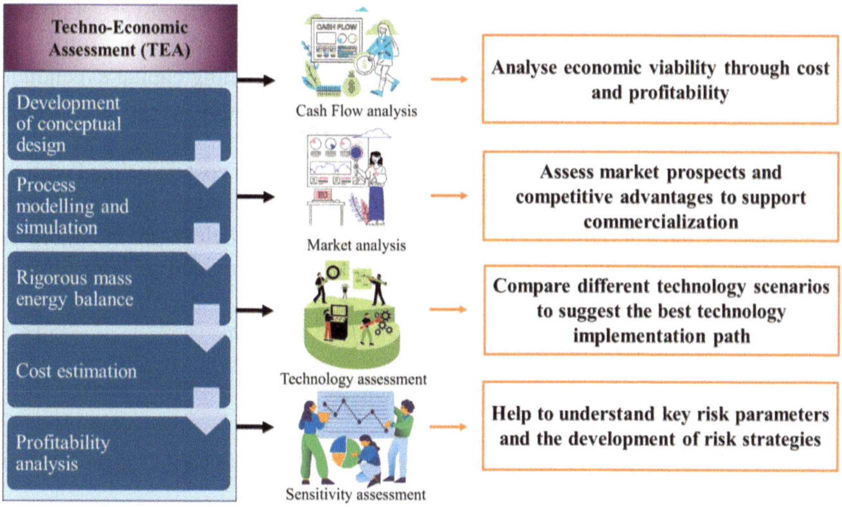

Fig. 14.5 Basic steps and prevalent areas of techno-economic analysis (TEA)

utilizing data from existing commercial plants and mathematical thermodynamics models. Based on the estimated mass and energy balance data, a process design is generated, including design specifications, equipment requirements, yield, etc., which can be employed to assist in cost estimation to evaluate the overall profitability of bioethanol production.

14.3.2 *Cost Estimations*

The economic viability of bioethanol production is regulated by various economic factors. After the development of a simulation model of conceptual design, it is important to carry out cost calculations and equipment sizing. A detailed account of cost estimations and calculations is described in the following sections.

14.3.2.1 Capital Cost

The capital expenditure (CAPEX) encompasses several costs, including Total Capital Investment (TFC), setup costs, and R&D expenses. TFC is constituted of Total Fixed Capital (TFC) and Working Capital (WC). Total Fixed Capital (TFC) can be subdivided into Total Plant Direct Fixed Cost (TPDC) and Total Plant Indirect Fixed Cost (TPIC) (Sarker et al. 2023). TPDC considers the costs incurred for plant equipment purchase and installation, production pipeline, processing facilities, instrumentation, electrical elements, and other infrastructure (Peters 1959). TPIC covers expenses related to engineering and construction. Moreover, contractor expenses and contingencies are also considered for TPIC estimation. Table 14.3 enlists the important parameters and assumptions for estimating capital expenditure (CAPEX). The table basically provides information on various capital costs associated with bioethanol process industries.

Understanding the total capital cost, production cost, and cost associated with the raw material consumption and energy flow is significant. The total capital cost can be calculated by incorporating all the production steps given below (all the abbreviations of the Equation provided in this chapter have been provided in Appendix A:

$$\mathrm{TCI} = A1i + A2i + A3i + A4i + A0i \tag{14.1}$$

Table 14.3 Summary of important parameters and assumptions for estimating the capital expenditure (CAPEX) of bioethanol production

Parameter	Description	Unit
Capital cost		
Purchased equipment cost	Cost required for the process equipment purchase for various unit operations	The currency of a particular country
Equipment installation cost	Cost required for process equipment installation.	The currency of a particular country
Piping	Piping cost includes the connection between various equipment	The currency of a particular country
		The currency of a particular country
Instrumentation	Accessories and instruments like various valves, transducers, etc.	The currency of a particular country
Electrical	All the electrical setup costs	The currency of a particular country
ISBL		The currency of a particular country
Buildings	This is a one-time investment for the plant setup	The currency of a particular country
Site improvement	This cost is required to improve the industry site, including transportation, safety facilities, etc.	The currency of a particular country
Utilities	Various utilities are required for the workers day to day requirements	The currency of a particular country
Storage	To store the final products before dispatch, store various raw materials	The currency of a particular country
Auxiliary facilities	Some cost for extra facilities	
Contingency	Cost required for some supporting work, such as research and development purposes	
Lang factor	Cost based on the major equipment cost	

14.3.2.2 Operating Expenditure (OPEX)

Operating expenditure (OPEX) is the sum of Variable Operating Costs (VOC), Fixed Charges, and plant overheads. VOC comprises feedstock, fuel and utilities, labor, and waste disposal costs. Fixed Charges (FC) include expenditure on property insurances, equipment depreciation, royalties, taxes, and rents. Plant overhead charges are costs associated with maintenance, supervision, storage, and warehouse facilities (Manhongo et al. 2021). Table 14.4 summarizes the parameters and values necessary to estimate the operating expenditure of bioethanol production. This table provides the information regarding various costs which will be required during the plant operation other than fixed capital cost.

The total operating costs were also evaluated for each stage in Eq. (14.2).

$$\mathrm{TOC} = A1_0 + A2_0 + A3_0 + A4_0 + A0_0 \tag{14.2}$$

Operating costs are crucial for the economic viability of bioethanol production. Therefore, it is essential to address the economic challenges related to various operating costs outlined below.

14.3.2.3 Feedstock Availability, Price Volatility, Logistics, and Storage Costs

Feedstock costs contribute to a considerable portion of the total production costs of bioethanol production, accounting for almost 50% on average, as represented in Fig. 14.6. Seasonal fluctuations and regional availability are the main determinants

Table 14.4 Summary of the parameters and values important for estimating the operating expenditure of bioethanol production

Parameter	Description	Unit
Operating cost		
Maintenance	This is a variable cost required for plant maintenance	Currency of a particular country/annum
Labor	This variable cost is required to run the process. This cost depends on the production capacity	Currency of a particular country/annum
Operating supplies	Operating supplies are occasionally required; therefore, the cost will depend on annual production, the overall capacity of the plant, etc.	Currency of a particular country/annum
Insurance	It is an annual cost that needs to be paid to the governing authority, depending on the industry size and annual production.	Currency of a particular country/annum
Overhead	Overhead costs are required to run the industry on a day-to-day basis. It depends on industry size and production rate	Currency of a particular country/annum
Electricity	The electricity bill depends on the production capacity and the process followed during production	Currency of a particular country/annum

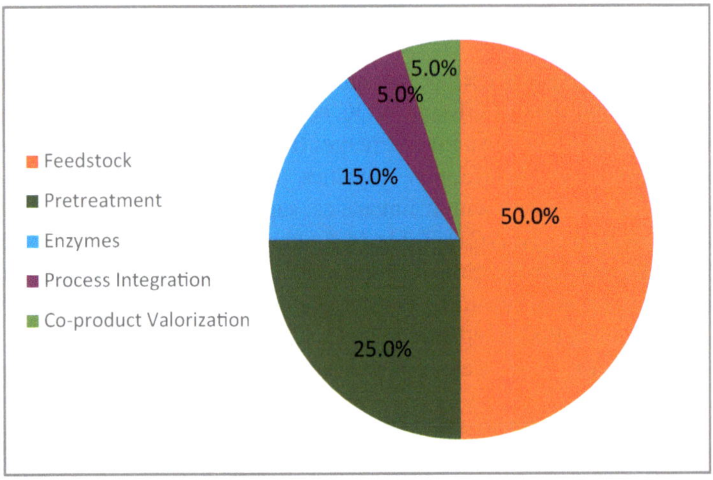

Fig. 14.6 Economic contribution of various key operating parameters on bioethanol production

of price volatility (Wolff et al. 2023). Concurrently, the precise estimation of feedstock price volatility is crucial for stabilizing food and bioethanol market prices (Dutta et al. 2019). It is reported that integrating a mixture of locally available and seasonally diverse biomasses can potentially reduce overall cost by minimizing price volatility, lowering transportation and storage costs, ensuring year-round supply, and enhancing resource utilization (Oke et al. 2016).

For instance, the utilization of a mixture of corn stover, miscanthus, and switchgrass by implementing a year-round feedstock switching can reduce the sugar production costs by up to 13.6% when compared with the baseline sugar cost attributed to fixed corn stover feedstock (Baral et al. 2019). The entire supply chain logistics, including collection, transportation, and storage, affects the total production cost. Local sourcing of feedstock can reduce transportation and storage costs and ensure a steady supply. Transportation and storage of lignocellulosic feedstock, such as bagasse and straw, are associated challenges like scattering, perishability, and low density, which increase the complexity and overall expenses of bioethanol production. Employing microalgal biomass as a third-generation feedstock can overcome the high pretreatment cost needed for complex lignocellulosic biomass (Brennan and Owende 2010. Moreover, carbohydrate-enriched microalgal biomass exhibits higher yield compared to lignocellulosic biomass, which makes them a potential feedstock for future application in bioethanol production (Hossain et al. 2019a, b).

14.3.2.4 Pretreatment Cost

Pretreatment methods are crucial for enhancing the efficiency of raw material utilization in bioethanol production. Pretreatment alone can contribute to 11–27% of the total cost of bioethanol production. Several pretreatment methods are in practice for bioethanol production, and they exhibit varied costs depending upon feedstock type, chemical or pretreatment agent used, energy use efficiency, equipment uses, pretreatment efficiency, and environmental impacts. Conventional pretreatment methods, such as steam explosion, alkaline pretreatment, and acidic pretreatment, require specific equipment to withstand high temperature, pressure, and corrosive chemicals, which contributes to a substantial increase in expenses (Galbe and Wallberg 2019). For example, heat pretreatment methods such as steam explosion and hydrothermal pretreatment are effective for disintegrating the complex structure of lignocellulosic biomass, eliminating harsh chemical use, but at the same time, are energy demanding (Neves et al. 2016).

Reduction in costs regarding energy use can be achieved by using cheaper chemicals like dilute acids and alkaline reagents such as sodium hydroxide (NaOH). However, the cost can further increase due to the inclusion of expenses regarding chemical storage, transportation, recovery, and recycling. It is reported that dilute acid pretreatment possesses a higher bioethanol yield (0.27 tonne bioethanol/tonne biomass) than liquid hot water and steam explosion methods (0.19 and 0.15 tonne bioethanol/tonne biomass, respectively) (Tao et al. 2013). However, biological pretreatment methods are less energy-intensive and less chemically intensive, but the overall cost increases threefold due to significantly longer pretreatment time (Baral and Shah 2017). Recently, there has been a growing interest in using deep eutectic solvents (DES) as they are nontoxic, have less inhibitor formation, and are made of renewable materials (Ullah et al. 2023). Therefore, optimization of pretreatment methods by balancing energy use, capital investment, operational costs, and product yield efficiency can mitigate the enhanced final cost.

14.3.2.5 Enzyme Costs

Enzyme plays a critical role in the hydrolysis of biomass for bioethanol production. Enzymatic hydrolysis is preferred over chemical hydrolysis due to its higher selectivity, specificity, being less energy intensive, eliminating the production of inhibitors, and being eco-friendly (Galbe and Zacchi 2007). However, enzymatic hydrolysis has two major drawbacks. One is the considerably high cost of enzymes, which accounts for almost 15% of total bioethanol production costs (Afedzi et al. 2025). The second challenge is significantly higher reaction time (El-Zawawy et al. 2011). Most of the enzymes used in bioethanol production are commercial grade, sourced from bacteria, fungi, and plants, and added exogenously. Recently, research has focused on employing genetically modified organisms for the endogenous production of enzymes and enhancing biomass conversion efficiency (Joshi et al. 2018). Optimization of enzyme loading can further contribute to cost reduction. Moreover, on-site production of enzymes is more cost-effective than off-site production, primarily eliminating the need for transportation and formulation costs (Ferreira et al. 2018). Recent research is focused on increasing enzyme stability, using new enzyme cocktails, immobilization, and reusability of enzymes, potentially decreasing overall costs.

14.3.3 Waste Valorization and By-Product Generation

Lignin is often considered waste in bioethanol production, which can be converted into value-added products such as adhesives and carbon fiber or used as a heat and energy source within the facility to offset operational costs. Studies reported that formalin-based bioethanol production from wheat straw generates high-purity resin and furfural as byproducts, resulting in a profit of 1440 USD per tonne of ethanol, which surpasses the market price of ethanol (Zhao and Liu 2019). Another approach to cost minimization is to harness hemicelluloses to produce high-value-added products. Producing pellets and biogas as byproducts from a bioethanol biorefinery can also contribute to improved economic performance.

14.3.4 Process Monitoring and Control

Process efficiency, monitoring, and control play a crucial role in the economic feasibility of a bioethanol production facility. The adoption of advanced technologies that integrate AI and IOT to optimize process parameters enhances the economic sustainability of biomass-to-bioethanol conversion. Different sensors, online monitoring tools, and real-time data analytics are employed to optimize process parameters such as temperature, pressure, pH, and enzyme doses to minimize economic losses related to energy use, water use, enzyme costs, and contamination risks. Operating costs are the major determinant of overall costs, and efficient cost management can ensure the economic feasibility of bioethanol production.

14.3.5 Profitability Analysis

After the determination of Annual Operating Cost (AOC) and Total Capital Investment (TCI), a Discounted Cash Flow Analysis (DCFA) will be employed to evaluate economic matrices of the overall process, such as Minimum Selling Price (MSP), Net Present Value (NPV), Payback period (PBP), and Internal Rate of Return (IRR). A Discounted Cash Flow Analysis (DCFA) is usually conducted after the assumption of an established lifetime, starting from construction, start-up period, and a steady production period until the end of the assumed plant life. NPV signifies the variation between the present value of cash inflow and cash outflow starting from production to sale over a given lifetime period, considering the value of money in that time frame. A positive NPV indicates the investment is profitable, while a negative NPV signifies that the investment will result in a net loss (Sarker et al. 2023). The Internal Rate of Return (IRR) is one of the key metrics of Total Economic Analysis (TEA), which denotes a discount rate used to equate a discounted benefit and costs, where the Net Present Value (NPV) becomes zero after income tax.

A higher IRR signifies more profitability. IRR represents the expected growth rate of a production project in the future. The MSP must reach NPV of zero after taxes with a set 10% discounted rate to meet the market competitiveness of bioethanol production (Humbird et al. 2011). The Payback period (PBP) is the time required to recover the initial capital investment through the profit generated by sales from the start of the project.

The input factors enable calculating the write-offs I_W and the imputed interest *I*Int in Eqs. 14.3 and 14.4.

$$I_W = I/T \tag{14.3}$$

$$I_{\text{Int}} = I * I_{\text{dr}} \tag{14.4}$$

From the operating result *O* and the net present value (NPV) *Co* can be calculated by Eqs. 14.5 and 14.6, respectively.

$$O = E_t - \left(\sum I_n + \sum C_n\right) \tag{14.5}$$

$$C_o = -I \ + E_t * A_t * (1+i) - T + T * A_t * (1+i)T \tag{14.6}$$

As per the NPV method of cost estimation, any project's expenses and revenue will be discounted up or down by a specific discounted rate at the decision time. If the NPV values are found to be positive, then the investment will be considered beneficial. The higher the NPV, the more profit the project will generate.

Along with the (NPV) net present value estimate, the total annual cost of any biorefinery can be estimated on the basis of the specific cost of the product portfolio. The annual cost can be calculated using the annuity method, considering that values vary over time. The total annual cost considering all the variable factors from the equation given below:

$$A = A_E - (A_K + A_V + A_B + A_S) \tag{14.7}$$

Net revenue generated from the product can be estimated from the equation given below,

$$I_t = Q_b.P_b + Q_{el}.P_{el}, \tag{14.8}$$

Annual cash flow is denoted by the difference between annual income and TOC as shown in Eq. (14.9).

$$CF = I_t - TOC \tag{14.9}$$

The production efficiency for bioethanol production from biomass can be calculated from the ratio of the final product to the amount of biomass used. The mathematical representation is provided below:

$$\eta = \frac{M_b.\text{LHV}_b}{M_{\text{biomass}}.\text{LHV}_{\text{biomass}}}.100 \tag{14.10}$$

However, for the calculation of payback period, an important indicator of profitability can be calculated by accumulation cash flow over a certain period (year) of time. Further, return on investment (ROI), a significant parameter to assess production profitability, can be estimated from the total cash flow and total capital cost data. It can be calculated as,

$$\text{ROI} = \frac{\text{CF} - \text{TCC}}{\text{TCC}}.100 \tag{14.11}$$

With the above-discussed methodology, it is possible to evaluate whether bioethanol production will be economically justified. The ROI indicator is used to assess profitability.

14.3.6 Sensitivity Analysis

Sensitivity analysis is used to evaluate how the fluctuations in key cost variables and process variables affect the overall feasibility of a bioethanol production project. Some of the process and cost variables considered for the sensitivity analysis of bioethanol production are feedstock costs, electricity consumption, plant lifetime, income tax, indirect costs, discount rate, and labor costs. This analysis enables the assessment of the effect of alterations of these variables on parameters such as capital costs, plant capacity, and selling price, considering the base case scenario (Manouchehrinejad et al. 2021). This analysis is used to identify critical variables, quantify the economic uncertainty associated with different scenarios, and help in informed decision-making.

14.3.7 Recent Techno-Economic Studies of Bioethanol Production

Recent studies have investigated the economic feasibility and sustainability of bioethanol production, focusing primarily on different feedstocks, different pretreatment methods, and different fermentation processes. Table 14.5 summarizes some recent techno-economic studies of bioethanol production. Recent studies indicate that bioethanol produced from lignocellulosic biomass has potential for

Table 14.5 Recent techno-economic analysis of bioethanol production

Raw material	Country of work	Methodology adopted	Economic value	Software used	Remarks on techno-economic analysis	References
Sugarcane bagasse, coffee-cut steam, rice husk, and empty fruit bunches	Colombia	A comparative study has been conducted between LCB feedstock and bioethanol production. With and without cogeneration are also considered	The highest ethanol yield was from empty fruit bunches (EFB) (313.83 L/t), while rice husk produced less ethanol (250.56 L/t). Production cost from EFB US$ 0.45/L	Aspen Plus (Aspen Technology Inc., USA)	Empty fruit bunches produce higher bioethanol yields, and rice husk produces the lowest. The cost is also lower for empty fruit bunches with and without cogeneration. High holocellulose content of the empty fruit makes it a suitable feedstock for bioethanol production	Quintero et al. (2013)
Corn stover	Brazil	Dilute acid, hot water, and ammonia fiber explosion methods have been compared. The MSP of ethanol was decided based on maximizing the bioethanol concentration and minimizing the cost	MSP = US$ 1.78/L	Aspen Plus	The hot water-based pretreatment exhibits the lowest MSP over dilute acid. However, higher ethanol concentration increases the total annual cost but keeps the MSP competitive	da Silva et al. (2016)
Seaweed	Indonesia	Enzymes digest the carbohydrates in the algae to produce sugar and then ferment it to produce ethanol	1.38 USD /gallon bioethanol is lower than the lowest price in 2019		This technology requires high energy consumption during processing, and due to the availability of raw materials, it can only be implemented in coastal cities	Fasahati et al. (2022)

(continued)

Table 14.5 (continued)

Raw material	Country of work	Methodology adopted	Economic value	Software used	Remarks on techno-economic analysis	References
Mango waste	South Africa	Fermentation with cellulase and yeast	Not profitable, not self-sufficient in most cases	Aspen Plus®	The sensitivity analysis reveals that a change in 10% of the raw material price will affect 4% of the net present value, which may not be profitable for large production	Manhongo et al. (2021)
Corn stover	China	Two different models have been adopted, i.e., bioethanol plant gate price assessment and feed cost estimation	Plant Gate price (PGP) = US$ 4.68–6.05/gal	ASPEN Plus®	The research found that bioethanol production is not competitive with conventional gasoline prices. The sensitivity analysis revealed that the cellulose-to-glucose conversion rate, five-carbon-to-glucose conversion rate, feedstock cost, and CAPEX play significant roles in ethanol selling price	Zhao et al. (2015)
Weeping love grass	South Africa	Two different reactor types have been compared: separate hydrolysis and fermentation (SHF) and simultaneous saccharification reactor configuration. The Acid mine drainage and H_2SO_4 pretreatment were also compared	PBP = 80.7 year, CAPEX = US$ 281 million, OPEX = US$ 45.82 million/year	Aspen Plus® V8.4	The biomass pre-treated with dilute H_2SO_4 in the SHF reactor is profitable. However, a high payback period with a low ethanol yield is also challenging. Optimization of the enzymatic hydrolysis step is highly recommended	Burman et al. (2020)

Spruce, Salix, and corn stover	Sweden	Three different LCBs were compared for production cost and energy demands for bioethanol production. SO_2 catalyzed pretreatment coupled with the SSF reactor has been employed	Production cost = SEK 4.15/L of ethanol for spruce and SEK 4.25/L of Ethanol for corn stover	Aspen Plus (version 2004.1 from Aspen Technology)	The cost of raw materials and the enzyme used play a crucial role in the production cost of bioethanol. The process configuration has a dominant effect on the feedstock	Sassner et al. (2008)
Red oak	USA	A comparative study has been performed for tetrahydrofuran, acetone, and 1,4-dioxane for fermentable sugar extraction for bioethanol production	MSP = US$ 2.98/gal, CAPEX = US$ 386.6 million, OPEX = US$ 103.3 million	Aspen Process Economic Analyzer (APEA)	The THF-based system produces bioethanol at the lowest MSP compared to the other solvents. The sensitivity analysis concludes that the MSP highly depends on the solvent-to-biomass ratio. It was also found that reducing the solvent-to-biomass ratio and improving sugar yield can lower the MSP	Li et al. (2018)
Wet hardwood chips	Italy	Enzymatic hydrolysis, fermentation, and gasification-fermentation systems have been compared for bioethanol production, power generation, and overall cost	MSP = €0.80/L, PBP = 5 years, NPV = €9.4 million, IRR = 11.0%, and ROI = 32.5%	Aspen Plus® v. 13.2	Enzymatic hydrolysis fermentation produces bioethanol at a lower MSP. The sensitivity analysis concludes that feedstock cost is lower for Enzymatic hydrolysis fermentation than the other one, which is the major contributor to the overall cost	Piccolo and Bezzo (2009)

fuel use. However, many studies suggested that bioethanol from second-generation feedstock is not cost-competitive with present Petro-based fuels. Presently, costly pretreatment methods requiring chemicals, enzymes, complex machineries, high energy, and detoxification units are the major obstacle for successful commercialization. Moreover, several fermentation strategies have been adopted, such as SHF, SSF, and simultaneous saccharification and co-fermentation (SSCF), which have a longer payback period with lower ethanol yield, and this is one of the major challenges for commercialization. Despite challenges for the commercialization of bioethanol biorefineries, several operational bioethanol production projects exist in different countries; the rest are under construction or planned, listed in Table 14.6. This table provides a concise idea about how techno-economic assessments have been performed globally for the production of bioethanol.

Process optimization, customization of the pretreatment method with feedstock choice, and process integration can make bioethanol the future energy source, replacing petroleum fuel.

14.4 Policy Implications for Sustainable Bioethanol Production

The successful commercialization of bioethanol biorefineries is hugely dependent on various bioenergy policy frameworks and policy implications, coupled with positive social engagement. Mandates of biofuel blending are one of the most important policies adopted by several countries in the world to promote bioethanol production by mandating a certain amount of bioethanol blending with gasoline. In 1993, the Government of Brazil passed a law that mandated mixing 20–25% of bioethanol with gasoline. South America is another pioneer country that mandated bioethanol blending with petroleum fuel. The US government has taken the initiative to promote the use of bioethanol in fuel mixture in the form of E85 (a blended fuel of 85% bioethanol and 15% gasoline) in specially designed vehicles. The Government of India is also actively encouraging the blending of ethanol with petrol under the Ethanol Blended Petrol (EBP) program, aiming to mandate 20% ethanol blending by 2025–2026. Although these mandates have contributed to promoting the establishment of the bioethanol market, these bioethanol blending mandates do not always encourage producers, leading to stagnation of bioethanol market growth. Fuel excise tax reductions/exemptions and tax credits are other policies that have fueled the rapid expansion of the bioethanol market in several countries, such as China, Australia, Denmark, Brazil, India, Germany, South Africa, Japan, Sweden, and the United States. One of the major challenges associated with these policies is the cultivation of dedicated energy crops, as farmers are skeptical about cultivating these crops for less market value and demand.

Bioethanol production biorefineries aim to maximize biomass resource use and minimize waste, support circular economy principles, and promote a zero-waste strategy by promoting renewable resource efficiency and reducing environmental

Table 14.6 Operational bioethanol plants in their respective countries

Owner	Raw material	Status	Capacity (tonne/year)/ TRL	Country/City	Startup year
Borregaard Industries AS	Spruce wood pulping	Operational	400,000/ TRL 9	Norway/ Sarpsborg	1938
Anhui BBCA Biochemical	Corncob/corn stover, primary wood chips;	Operational	NA/TRL 6–7	China/ Bengbu	2003
Sekab Group	Sugarcane bagasse, wheat, corn stover, energy grass	Operational	1095/TRL 8	Sweden/ Ornskoldsvik	2004
Energy & Chemical Department of East China University of Science and Technology	Crop and forestry residues	Operational	NA/TRL 4–5	China/ Shanghai	2005
COFCO Zhaodong Co.	Corn stover	Operational	NA/TRL 6–7	China/ Beijing	2006
Jilin Fuel Alcohol	Straw and corn	Operational	NA/TRL 6–7	China/Jilin	2006
Shandong Zesheng Biotech Co.	Straw	Operational	NA/TRL 6–7	China/Tai'an	2006
Chempolis Ltd.	Straw, reed, empty fruit bunch, bagasse, corn stalks, wood residues	Operational	25,000/TRL 6–7	Finland/Oulu	2008
POET	Corn fiber, corn cobs, and corn stalks	Operational	NA/TRL 4–5	United States/ Scotland	2008
Henan Tianguan Group	Wheat/corn stover	Operational	10,000/TRL 8	China/ Zhenping	2009
Iowa State University	Lignocellulosic	Operational	1825/TRL 4–5	USA/Boone	2009
Greenfield Ethanol	Lignocellulosic	Operational	NA/TRL 4–5	Canada/ Clatham	2010
Queensland University of Technology	Sugarcane bagasse, corn stover, forestry products	Operational	NA/TRL 4-5	Australia/ Mackay	2010
Henan Tianguan Group	Lignocellulosic	Operational	30,000/TRL 8	China/ Nanyang	2011
NREL (National Renewable Energy Laboratory)	Lignocellulosic	Operational	NA/TRL 4–5	USA/Golden	2011
Borregaard AS	Sugarcane bagasse, straw, wood, energy crops, other lignocellulosic	Operational	365/TRL 6–7	Norway/ Sarpsborg	2012

(continued)

Table 14.6 (continued)

Owner	Raw material	Status	Capacity (tonne/year)/ TRL	Country/City	Startup year
Clariant	Wheat straw, corn stover, miscanthus, sugarcane bagasse	Operational	4500/TRL 6-7	Germany/ Straubing	2012
Longlive Bio-technology Co., Ltd.	Corn cob	Operational	60,000/TRL 8	China/ Yucheng	2012
Indian Oil R&D	Sugarcane bagasse, rice straw, agricultural wastes	Operational	90/TRL 4–5	India/ Faridabad	2012
GranBio	Sugarcane bagasse and straw	Operational	NA/TRL 8	Brazil/Sao Miguel	2014
GranBio Technologies	Forest residues	Operational	NA/TRL 6–7	USA/ Thomaston	2015
Indian Glycol & DBT-ICT Mumbai	Sugarcane bagasse, rice straw, agricultural wastes	Operational	3650/TRL 6–7	India/ Kashipur	2016
Praj Industries	Sugarcane bagasse, rice straw, agricultural wastes	Operational	4400 /TRL 6–7	India/Pune	2017
AustroCel Hallein	Spruce wood pulping	Operational	30,000/TRL 8	Austria/ Hallen	2020
Raizen Energia	Sugarcane residue	Operational	47,000 / TRL 9	Brazil/ Guariba	2024

impacts (Prasad and Chandel 2024). Moreover, feedstock subsidies play a pivotal role in developing trust among farmers and potential investors about bioethanol biorefineries (Prasad and Chandel 2024). The US and China are the leading countries in implementing these policies related to providing subsidies for feedstock cultivations, tax credit for the procurement of equipment for harvesting, fuel tax exemptions for transportation of feedstock biomass, and tax credits for the sale of biomass feedstock. Apart from the above-mentioned policies, the governments have implemented several technology push policies to encourage research and development related to bioethanol biorefineries, adopting several measures like rebates and tax credits on bioenergy R&D expenses, grants for process and technology development, etc.

Renewable Portfolio Standards (RPS) or Renewable Energy Standards (RES) are another effective policy adopted by several countries that facilitates generating or purchasing a certain amount of their required energy from renewable sources. These policies have emerged as a promising avenue to promote bioethanol production and commercialization while combating climate change. Bioenergy with carbon capture and storage (BECCS) coupled with Negative Emission Technologies (NET) are employed with a set target to offset and even reverse global warming by adopting negative emission technologies (Rogelj et al. 2016). The bioethanol market gained momentum in light of BECCS. Moreover, Sustainability certification schemes encompassing eco-friendly and sustainable farming practices of bioethanol feedstock cultivation have a notable effect on bioethanol market expansion.

These policy tools and frameworks are vital for facilitating the transition to sustainable bioenergy production, considering social, economic, and environmental aspects. Nevertheless, investors face considerable challenges with these government policies due to inconsistency and changes in government policies, tax incentives, subsidies, etc. Several Asian countries like China, Malaysia, and Indonesia faced financial constraints related to bioenergy policy implications, resulting in a declining trend of bioethanol production. On the contrary, bioenergy policies adopted in Nordic countries are effective. Despite several challenges, these policy frameworks are crucial for the transition toward a sustainable bioeconomy, including climate change concerns, and simultaneously help the rural economy and livelihood align with the sustainable development scenario (SDS) goals (Prasad et al. 2024).

14.5 Conclusion

The global growth of bioethanol biorefineries has gained momentum as a potential alternative to fossil fuel-based refineries to combat current environmental issues. The rapid growth of the bioethanol industry has raised considerable concerns about sustainable development, food security, energy security, land use conflict, water scarcity, and greenhouse gas emissions. A comprehensive life cycle assessment is useful for evaluating these complex interactions to meet the global energy demand. LCA of bioethanol biorefineries faces challenges related to the lack of reliable data, problems defining the system boundary, future uncertainties, and allocation methods. Co-production of multiple value-added products and co-utilization of various feedstocks are considered to improve the performance of bioethanol biorefinery. Lack of basic data on such integrated processes is the main constraint for conducting a comprehensive LCA of such processes. Moreover, the energy system, a key regulating factor of the environmental performance of bioethanol biomanufacturing, is variable over time and space. Therefore, a dynamic LCA method needs to be developed and adopted for more accurate environmental evaluation of the performance of bioethanol biorefineries.

Profitability and environmental sustainability are among the most important criteria for the successful commercialization of bioethanol biorefinery. TEA is an invaluable tool for assessing economic feasibility, encompassing several economic matrices such as Net Present Value, internal return, and payback period. This chapter highlighted that operational costs are the major determinant of the economic viability of bioethanol production. Efficient management of stable feedstock supply, low-cost biomass sourcing, well-maintained transportation, management of supply chain logistics, optimization of processing and pretreatment methods, and waste valorization can achieve sustainable and economically operable bioethanol production, offsetting operational costs. Several variables considered in TEA are highly fluctuating over time, remarkably reducing the accuracy of TEA. Moreover, insufficient data and a lack of experience further increase the uncertainty of the TEA of the Bioethanol biorefinery. Sensitivity analysis is typically conducted to quantify uncertainties by evaluating the impact of fluctuations in technical and economic variables on TEA results. Further, Monte Carlo Simulation is employed to simulate more than hundreds of scenarios based on the probability distribution of variables. New modeling and simulation techniques are needed to conduct robust and accurate uncertainty analysis, aiding informed decision-making.

Recent studies suggest using an integrated framework of TEA with LCA to facilitate multi-objective optimization for multi-criteria decision-making. However, developing an integrated framework of LCA and TEA is very challenging due to discrepancies in defining system boundaries, assessment metrics, and input and output data. The development of an integrated framework of TEA and LCA is an emerging field of research for guiding environmentally and economically sustainable process design.

Finally, several government policy frameworks and tools are invaluable for facilitating the expansion of the bioethanol market, considering environmental, social, economic, and resource use efficiency. In conclusion, a comprehensive life cycle analysis, techno-economic analysis, integrated approach of LCA and TEA, and adoption of effective policy frameworks, considering complex interacting factors, are essential for the development of competitive, scalable, profitable, and environmentally sustainable bioethanol production for achieving a greener future.

Acknowledgment The authors would like to express their sincere gratitude to the Department of Paper Technology, Indian Institute of Technology (IIT) Roorkee for providing all the necessary help for the writing of the book chapter.

Appendix

Nomenclature

Abbreviation	Significance
TCI	Total capital costs
*A*1*i*	Capital costs for biomass preparation
*A*2*i*	Capital costs for gasification process
*A*3*i*	Capital costs for syngas cleaning
*A*4*i*	Capital costs for FT synthesis
*A*0*i*	Other capital costs

(continued)

Abbreviation	Significance
TOC	Total operating costs
$A1_0$	Biomass costs
$A2_0$	Gasification process costs
$A3_0$	Costs for syngas cleaning
$A4_0$	Costs for FT synthesis
$A0_0$	Other costs
I_t	Total income, currency/year
Q_b	Amount of biofuels produced, MWh/year
P_b	Biofuel price, EUR/MWh
Q_{el}	Amount of electricity generated, MWh/year
P_{el}	Electricity price, currency/year
IW	Write-offs
I	Investment costs
T	Consideration duration in periods
$IInt$	Imputed interest
Idr	Discount rate
O	Operating results
Et	Sum of revenues in period t
In	Sum of investment-related costs in period t
Cn	Sum of operation-related costs in period t
Co	Net present value
At	Sum of payments in the period t
A	Total annual cost of operations
A_E	Specific revenue AE for by-products
A_K	Investment-related costs
A_V	Energy/material-related cost
A_B	Operational costs
A_S	Additional costs
CF	Cash flow, currency/year
TOC	Total operating costs, EUR/year
η	Production efficiency, %
M_b	Amount of biofuel produced, tonnes/year
$M_{biomass}$	Amount of biomass used, tonnes/year
LHV_b	Biofuel heating value, kJ/kg
$LHV_{biomass}$	Biomass heating value, kJ/kg
ROI	Return on investment, %
CF	Total cash flow over a period of 15 years
TCC	Total capital costs

References

Afedzi AEK, Afrakomah GS, Gyan K, Khan J, Seidu R, Baidoo T, Sultan IN, Tareen AK, Parakulsuksatid P (2025) Enhancing economic and environmental sustainability in lignocellulosic bioethanol production: key factors, innovative technologies, policy frameworks, and social considerations. Sustainability (Switzerland) 17(2):1–23. https://doi.org/10.3390/su17020499

Ahlgren S, Di Lucia L (2014) Indirect land use changes of biofuel production—a review of modelling efforts and policy developments in the European Union. Biotechnol Biofuels 7(1): 35. https://doi.org/10.1186/1754-6834-7-35

Alalwan HA, Alminshid AH, Aljaafari HAS (2019) Promising evolution of biofuel generations. Subject review. Renew Energy Focus. https://doi.org/10.1016/j.ref.2018.12.006

Amaya-Delgado L, Flores-Cosío G, Sandoval-Nuñez D, Arellano-Plaza M, Arrizon J, Gschaedler A (2018) Comparative of lignocellulosic ethanol production by Kluyveromyces marxianus and Saccharomyces cerevisiae. In: Special topics in renewable energy systems. https://doi.org/10.5772/intechopen.78685

Antonelli M, Greco F (2015) The water we eat: Combining virtual water and water footprints, pp 1–256. https://doi.org/10.1007/978-3-319-16393-2

Ayodele BV, Alsaffar MA, Mustapa SI (2020) An overview of integration opportunities for sustainable bioethanol production from first- and second-generation sugar-based feedstocks. J Clean Prod 245:118857. https://doi.org/10.1016/j.jclepro.2019.118857

Bai Y, Luo L, Van Der Voet E (2010) Life cycle assessment of switchgrass-derived ethanol as transport fuel. Int J Life Cycle Assess. https://doi.org/10.1007/s11367-010-0177-2

Balat M, Balat H (2009) Recent trends in global production and utilization of bio-ethanol fuel. Appl Energy. https://doi.org/10.1016/j.apenergy.2009.03.015

Bamber N, Turner I, Dutta B, Heidari MD, Pelletier N (2023) Consequential Life cycle assessment of grain and oilseed crops: review and recommendations. Sustainability 15(7):6201. https://doi.org/10.3390/su15076201

Baral NR, Shah A (2017) Comparative techno-economic analysis of steam explosion, dilute sulfuric acid, ammonia fiber explosion and biological pretreatments of corn stover. Bioresour Technol 232:331–343. https://doi.org/10.1016/j.biortech.2017.02.068

Baral NR, Davis R, Bradley TH (2019) Supply and value chain analysis of mixed biomass feedstock supply system for lignocellulosic sugar production. Biofuels Bioprod Biorefin 13(3):635–659. https://doi.org/10.1002/bbb.1975

Bibi R, Ahmad Z, Imran M, Hussain S, Ditta A, Mahmood S, Khalid A (2017) Algal bioethanol production technology: a trend towards sustainable development. Renew Sust Energ Rev 71: 976–985. https://doi.org/10.1016/j.rser.2016.12.126

Brennan L, Owende P (2010) Biofuels from microalgae-a review of technologies for production, processing, and extractions of biofuels and co-products. Renew Sust Energ Rev. https://doi.org/10.1016/j.rser.2009.10.009

Buchspies B, Kaltschmitt M (2016) Life cycle assessment of bioethanol from wheat and sugar beet discussing environmental impacts of multiple concepts of co-product processing in the context of the European Renewable Energy Directive. Biofuels. https://doi.org/10.1080/17597269.2015.1122472

Budsberg E, Rastogi M, Puettmann ME, Caputo J, Balogh S, Volk TA, Gustafson R, Johnson L (2012) Life-cycle assessment for the production of bioethanol from willow biomass crops via biochemical conversion. For Prod J. https://doi.org/10.13073/FPJ-D-12-00022.1

Bueno G, Latasa I, Lozano PJ (2015) Comparative LCA of two approaches with different emphasis on energy or material recovery for a municipal solid waste management system in Gipuzkoa. Renew Sust Energ Rev 51:449–459. https://doi.org/10.1016/j.rser.2015.06.021

Burman NW, Sheridan CM, Harding KG (2020) Feasibility assessment of the production of bioethanol from lignocellulosic biomass pretreated with acid mine drainage (AMD). Renew Energy 157:1148–1155. https://doi.org/10.1016/j.renene.2020.05.086

Cavalett O, Chagas MF, Seabra JEA, Bonomi A (2013) Comparative LCA of ethanol versus gasoline in Brazil using different LCIA methods. Int J Life Cycle Assess 18(3):647–658. https://doi.org/10.1007/s11367-012-0465-0

Chang FC, Lin LD, Ko CH, Hsieh HC, Yang BY, Chen WH, Hwang WS (2017) Life cycle assessment of bioethanol production from three feedstocks and two fermentation waste reutilization schemes. J Clean Prod. https://doi.org/10.1016/j.jclepro.2016.12.024

Chatterjee R, Sharma V, Mukherjee S (2015) The environmental impacts and allocation methods used in LCA studies of vegetable oil-based bio-diesels. Waste Biomass Valoriz 6(4):580–603. https://doi.org/10.1007/s12649-015-9375-2

Cherubin MR, Carvalho JLN, Cerri CEP, Nogueira LAH, Souza GM, Cantarella H (2021) Land use and management effects on sustainable sugarcane-derived bioenergy. Land 10(1):72. https://doi.org/10.3390/land10010072

Cherubini F, Strømman AH (2011) Life cycle assessment of bioenergy systems: state of the art and future challenges. Bioresour Technol 102(2):437–451. https://doi.org/10.1016/j.biortech.2010.08.010

Cronin J, Zabel F, Dessens O, Anandarajah G (2020) Land suitability for energy crops under scenarios of climate change and land-use. GCB Bioenergy 12(8):648–665. https://doi.org/10.1111/gcbb.12697

Curran MA (2017) Overview of goal and scope definition in life cycle assessment. In: Curran MA (ed) Goal and scope definition in life cycle assessment, LCA compendium—the complete world of life cycle assessment. Springer, Dordrecht, pp 1–62. https://doi.org/10.1007/978-94-024-0855-3_1

Czyrnek-Delêtre MM, Smyth BM, Murphy JD (2017) Beyond carbon and energy: the challenge in setting guidelines for life cycle assessment of biofuel systems. Renew Energy 105:436–448. https://doi.org/10.1016/j.renene.2016.11.043

da Silva ARG, Torres Ortega CE, Rong BG (2016) Techno-economic analysis of different pretreatment processes for lignocellulosic-based bioethanol production. Bioresour Technol. https://doi.org/10.1016/j.biortech.2016.07.007

de Almeida MA, Colombo R (2023) Production chain of first-generation sugarcane bioethanol: characterization and value-added application of wastes. BioEnergy Res 16(2):924–939. https://doi.org/10.1007/s12155-021-10301-4

Demichelis F, Laghezza M, Chiappero M, Fiore S (2020) Technical, economic and environmental assessement of bioethanol biorefinery from waste biomass. J Clean Prod. https://doi.org/10.1016/j.jclepro.2020.124111

Dibazar AS, Aliasghar A, Behzadnezhad A, Shakiba A, Pazoki M (2024) Energy cycle assessment of bioethanol production from sugarcane bagasse by life cycle approach using the fermentation conversion process. Biomass Convers Biorefinery 14(17):20679–20698. https://doi.org/10.1007/s13399-023-04288-5

Dunn JB (2019) Biofuel and bioproduct environmental sustainability analysis. Curr Opin Biotechnol 57:88–93. https://doi.org/10.1016/j.copbio.2019.02.008

Dutta A, Junttila J, Uddin GS (2019) Forecasting the volatility of biofuel feedstock prices: the US evidence. Biofuels Bioprod Biorefin. https://doi.org/10.1002/bbb.1981

Earles JM, Halog A (2011) Consequential life cycle assessment: a review. Int J Life Cycle Assess 16(5):445–453. https://doi.org/10.1007/s11367-011-0275-9

Ekvall T, Finnveden G (2001) Allocation in ISO 14041—a critical review. J Clean Prod 9(3): 197–208. https://doi.org/10.1016/S0959-6526(00)00052-4

El-Zawawy WK, Ibrahim MM, Abdel-Fattah YR, Soliman NA, Mahmoud MM (2011) Acid and enzyme hydrolysis to convert pretreated lignocellulosic materials into glucose for ethanol production. Carbohydr Polym. https://doi.org/10.1016/j.carbpol.2010.12.022

Escamilla-Alvarado C, Poggi-Varaldo HM, Ponce-Noyola MT (2017) Bioenergy and bioproducts from municipal organic waste as alternative to landfilling: a comparative life cycle assessment with prospective application to Mexico. Environ Sci Pollut Res 24(33):25602–25617. https://doi.org/10.1007/s11356-016-6939-z

Falano T, Jeswani HK, Azapagic A (2014) Assessing the environmental sustainability of ethanol from integrated biorefineries. Biotechnol J 9(6):753–765. https://doi.org/10.1002/biot.201300246

Fasahati P, Dickson R, Saffron CM, Woo HC, Liu JJ (2022) Seaweeds as a sustainable source of bioenergy: techno-economic and life cycle analyses of its biochemical conversion pathways. Renew Sust Energ Rev. https://doi.org/10.1016/j.rser.2021.112011

Ferreira RDG, Azzoni AR, Freitas S (2018) Techno-economic analysis of the industrial production of a low-cost enzyme using E. coli: the case of recombinant β-glucosidase. Biotechnol Biofuels. https://doi.org/10.1186/s13068-018-1077-0

Galbe M, Wallberg O (2019) Pretreatment for biorefineries: a review of common methods for efficient utilisation of lignocellulosic materials. Biotechnol Biofuels 12(1):294. https://doi.org/10.1186/s13068-019-1634-1

Galbe M, Zacchi G (2007) Pretreatment of lignocellulosic materials for efficient bioethanol production. Adv Biochem Eng Biotechnol. https://doi.org/10.1007/10_2007_070

González-García S, Luo L, Moreira MT, Feijoo G, Huppes G (2009) Life cycle assessment of flax shives derived second generation ethanol fueled automobiles in Spain. Renew Sust Energ Rev. https://doi.org/10.1016/j.rser.2009.02.003

González-García S, Moreira MT, Feijoo G (2010) Environmental performance of lignocellulosic bioethanol production from alfalfa stems. Biofuels Bioprod Biorefin. https://doi.org/10.1002/bbb.204

González-García S, Iribarren D, Susmozas A, Dufour J, Murphy RJ (2012) Life cycle assessment of two alternative bioenergy systems involving Salix spp. biomass: bioethanol production and power generation. Appl Energy. https://doi.org/10.1016/j.apenergy.2012.02.022

Guinée JB, Heijungs R, Huppes G, Zamagni A, Masoni P, Buonamici R, Ekvall T, Rydberg T (2011) Life cycle assessment: past, present, and future. Environ Sci Technol 45(1):90–96. https://doi.org/10.1021/es101316v

Guo M, Littlewood J, Joyce J, Murphy R (2014) The environmental profile of bioethanol produced from current and potential future poplar feedstocks in the EU. Green Chem. https://doi.org/10.1039/c4gc01124d

Hellweg S, Milà i Canals L (2014) Emerging approaches, challenges and opportunities in life cycle assessment. Science 344(6188):1109–1113. https://doi.org/10.1126/science.1248361

Hossain N, Mahlia TMI, Zaini J, Saidur R (2019a) Techno-economics and sensitivity analysis of microalgae as commercial feedstock for bioethanol production. Environ Prog Sustain Energy. https://doi.org/10.1002/ep.13157

Hossain N, Zaini J, Mahlia TMI, Azad AK (2019b) Elemental, morphological and thermal analysis of mixed microalgae species from drain water. Renew Energy 131:617–624. https://doi.org/10.1016/j.renene.2018.07.082

Humbird D, Davis R, Tao L, Kinchin C, Hsu D, Aden A, Schoen P, Lukas J, Olthof B, Worley M, Sexton D, Dudgeon D (2011) Process design and economics for biochemical conversion of lignocellulosic biomass to ethanol: dilute-acid pretreatment and enzymatic hydrolysis of corn stover. National Renewable Energy Laboratory

International Energy Agency (IEA) (2019) World Energy Outlook 2019—Analysis—IEA. World Energy Outlook 2019

ISO (1998) ISO 14041: environmental management—life cycle assessment—goal and scope definition and inventory analysis

ISO (2006) ISO 14040: environmental management—life cycle assessment—principles and framework

Jeswani HK, Chilvers A, Azapagic A (2020) Environmental sustainability of biofuels: a review. Proc R Soc A Math Phys Eng Sci 476(2243). https://doi.org/10.1098/rspa.2020.0351

Joshi B, Joshi J, Bhattarai T, Sreerama L (2018) Currently used microbes and advantages of using genetically modified microbes for ethanol production. In: Bioethanol production from food crops: sustainable sources, interventions, and challenges. https://doi.org/10.1016/B978-0-12-813766-6.00015-1

Karacan R, Baris I, Yardimci ME, Cengiz V (2021) How accurate is it to define biofuels as renewable energy? A theoretical and empirical analysis. Int J Energy Econ Policy 11(6): 250–255. https://doi.org/10.32479/ijeep.11935

Kim S, Dale BE (2008) Life cycle assessment of fuel ethanol derived from corn grain via dry milling. Bioresour Technol. https://doi.org/10.1016/j.biortech.2007.09.034

Koizumi T (2011) Biofuel programs in East Asia: developments, perspectives, and sustainability. In: Environmental impact of biofuels. InTech. https://doi.org/10.5772/17158
Li W, Ghosh A, Bbosa D, Brown R, Wright MM (2018) Comparative techno-economic, uncertainty and life cycle analysis of lignocellulosic biomass solvent liquefaction and sugar fermentation to ethanol. ACS Sustain Chem Eng. https://doi.org/10.1021/acssuschemeng.8b03622
Liu Y, Xu Y, Zhang F, Yun J, Shen Z (2014) The impact of biofuel plantation on biodiversity: a review. Chin Sci Bull 59(34):4639–4651. https://doi.org/10.1007/s11434-014-0639-1
Luo L, van der Voet E, Huppes G (2009a) Life cycle assessment and life cycle costing of bioethanol from sugarcane in Brazil. Renew Sust Energ Rev 13(6–7):1613–1619. https://doi.org/10.1016/j.rser.2008.09.024
Luo L, Van Der Voet E, Huppes G, Udo De Haes HA (2009b) Allocation issues in LCA methodology: a case study of corn stover-based fuel ethanol. Int J Life Cycle Assess. https://doi.org/10.1007/s11367-009-0112-6
Manhongo TT, Chimphango A, Thornley P, Röder M (2021) Techno-economic and environmental evaluation of integrated mango waste biorefineries. J Clean Prod 325:129335. https://doi.org/10.1016/j.jclepro.2021.129335
Manouchehrinejad M, Bilek EMT, Mani S (2021) Techno-economic analysis of integrated torrefaction and pelletization systems to produce torrefied wood pellets. Renew Energy. https://doi.org/10.1016/j.renene.2021.06.064
Mayer FD, Brondani M, Aita BC, Hoffmann R, Lora EES (2015) Environmental and energy assessment of small scale ethanol fuel production. Energy Fuel. https://doi.org/10.1021/acs.energyfuels.5b01358
Mayer FD, Brondani M, Vasquez Carrillo MC, Hoffmann R, Silva Lora EE (2020) Revisiting energy efficiency, renewability, and sustainability indicators in biofuels life cycle: analysis and standardization proposal. J Clean Prod 252:119850. https://doi.org/10.1016/j.jclepro.2019.119850
Menten F, Chèze B, Patouillard L, Bouvart F (2013) A review of LCA greenhouse gas emissions results for advanced biofuels: the use of meta-regression analysis. Renew Sust Energ Rev 26:108–134. https://doi.org/10.1016/j.rser.2013.04.021
Morales M, Arvesen A, Cherubini F (2021) Integrated process simulation for bioethanol production: effects of varying lignocellulosic feedstocks on technical performance. Bioresour Technol. https://doi.org/10.1016/j.biortech.2021.124833
Müller C, Scapini T, Rempel A, Abaide ER, Camargo AF, Nazari MT, Tadioto V, Bonatto C, Tres MV, Zabot GL, Colla LM, Treichel H, Alves SL (2023) Challenges and opportunities for third-generation ethanol production: a critical review. Eng Microbiol. https://doi.org/10.1016/j.engmic.2022.100056
Neves PV, Pitarelo AP, Ramos LP (2016) Production of cellulosic ethanol from sugarcane bagasse by steam explosion: effect of extractives content, acid catalysis and different fermentation technologies. Bioresour Technol 208:184–194. https://doi.org/10.1016/j.biortech.2016.02.085
Oke MA, Annuar MSM, Simarani K (2016) Mixed feedstock approach to lignocellulosic ethanol production—prospects and limitations. Bioenergy Res. https://doi.org/10.1007/s12155-016-9765-8
Osman AI, Mehta N, Elgarahy AM, Al-Hinai A, Al-Muhtaseb AH, Rooney DW (2021) Conversion of biomass to biofuels and life cycle assessment: a review. Environ Chem Lett 19(6):4075–4118. https://doi.org/10.1007/s10311-021-01273-0
Papong S, Malakul P (2010) Life-cycle energy and environmental analysis of bioethanol production from cassava in Thailand. Bioresour Technol. https://doi.org/10.1016/j.biortech.2009.09.006
Patel K, Singh SK (2023) Environmental sustainability analysis of biofuels: a critical review of LCA studies. Clean Techn Environ Policy 25(8):2489–2510. https://doi.org/10.1007/s10098-023-02596-y
Peters JI (1959) Review of: "PLANT DESIGN AND ECONOMICS FOR CHEMICAL ENGINEERS" MAX S. PETERS: McGraw-Hill Book Company, New York, 1958. 511 pp. $11.00. Eng Econ 5(1):27–30. https://doi.org/10.1080/00137915908965075

Piccolo C, Bezzo F (2009) A techno-economic comparison between two technologies for bioethanol production from lignocellulose. Biomass Bioenergy. https://doi.org/10.1016/j.biombioe.2008.08.008

Pieragostini C, Aguirre P, Mussati MC (2014) Life cycle assessment of corn-based ethanol production in Argentina. Sci Total Environ 472:212–225. https://doi.org/10.1016/j.scitotenv.2013.11.012

Prasad S, Chandel AK (2024) Precision biomass collection, storage, and transportation of feedstock at the biorefinery gate. In: Prasad S, Chandel AK (eds) Precision in crop and biomass processing for sustainable bioenergy. Springer, pp 23–61. https://doi.org/10.1007/978-3-031-51601-6_2

Prasad S, Yadav KK, Kumar S, Pandita P, Bhutto JK, Alreshidi MA, Ravindran B, Yaseen ZM, Osman SM, Cabral-Pinto MM (2024) Review on biofuel production: sustainable development scenario, environment, and climate change perspectives – a sustainable approach. J Environ Chem Eng 12(2):111996. https://doi.org/10.1016/j.jece.2024.111996

Prasad S, Singh A, Korres NE, Rathore D, Sevda S, Pant D (2020) Sustainable utilization of crop residues for energy generation: a life cycle assessment (LCA) perspective. Bioresour Technol 303:122964. https://doi.org/10.1016/j.biortech.2020.122964

Quintero JA, Moncada J, Cardona CA (2013) Techno-economic analysis of bioethanol production from lignocellulosic residues in Colombia: a process simulation approach. Bioresour Technol 139:300–307. https://doi.org/10.1016/j.biortech.2013.04.048

Rehl T, Lansche J, Müller J (2012) Life cycle assessment of energy generation from biogas—attributional vs. consequential approach. Renew Sust Energ Rev 16(6):3766–3775. https://doi.org/10.1016/j.rser.2012.02.072

Roberto Ometto A, Zwicky Hauschild M, Nelson Lopes Roma W (2009) Lifecycle assessment of fuel ethanol from sugarcane in Brazil. Int J Life Cycle Assess. https://doi.org/10.1007/s11367-009-0065-9

Rogelj J, Den Elzen M, Höhne N, Fransen T, Fekete H, Winkler H, Schaeffer R, Sha F, Riahi K, Meinshausen M (2016) Paris Agreement climate proposals need a boost to keep warming well below 2 °C. Nature. https://doi.org/10.1038/nature18307

Roy P, Tokuyasu K, Orikasa T, Nakamura N, Shiina T (2012) A review of life cycle assessment (LCA) of bioethanol from lignocellulosic biomass. Jpn Agric Res Q 46(1):41–57. https://doi.org/10.6090/jarq.46.41

Santoyo-Castelazo E, Santoyo E, Zurita-García L, Camacho Luengas DA, Solano-Olivares K (2023) Life cycle assessment of bioethanol production from sugarcane bagasse using a gasification conversion process: bibliometric analysis, systematic literature review and a case study. Appl Therm Eng. https://doi.org/10.1016/j.applthermaleng.2022.119414

Sarker TR, German CS, Borugadda VB, Meda V, Dalai AK (2023) Techno-economic analysis of torrefied fuel pellet production from agricultural residue via integrated torrefaction and pelletization process. Heliyon. https://doi.org/10.1016/j.heliyon.2023.e16359

Sassner P, Galbe M, Zacchi G (2008) Techno-economic evaluation of bioethanol production from three different lignocellulosic materials. Biomass Bioenergy. https://doi.org/10.1016/j.biombioe.2007.10.014

Shakelly N, Pérez-Cardona JR, Deng S, Maani T, Li Z, Sutherland JW (2023) Comparative life cycle assessment of bioethanol production from different generations of biomass and waste feedstocks. Proc CIRP 116:630–635. https://doi.org/10.1016/j.procir.2023.02.106

Shapouri H, Duffield JA, Wang M (2002) The energy balance of corn ethanol: an update. In: U.S. Department of Agriculture, Office of the Chief Economist, Office of Energy Policy and New Uses. Agricultural Economic Report No. 814. file:///F:/Spec 2/Traffic Delay Model.pdf

Sills DL, Van Doren LG, Beal C, Raynor E (2020) The effect of functional unit and co-product handling methods on life cycle assessment of an algal biorefinery. Algal Res 46(607):1–32. https://doi.org/10.1016/j.algal.2019.101770

Soleymani Angili T, Grzesik K, Rödl A, Kaltschmitt M (2021) Life cycle assessment of bioethanol production: a review of feedstock, technology and methodology. Energies. https://doi.org/10.3390/en14102939

Tao L, Tan ECD, Aden A, Elander RT (2013) CHAPTER 19. Techno-economic analysis and life-cycle assessment of lignocellulosic biomass to sugars using various pretreatment technologies. https://doi.org/10.1039/9781849734738-00358

Tillman A-M (2000) Significance of decision-making for LCA methodology. Environ Impact Assess Rev 20(1):113–123. https://doi.org/10.1016/S0195-9255(99)00035-9

Tillman A-M (2010) Methodology for life cycle assessment. In: Sonesson U, Brelin J, Ziegler F (eds) Environmental assessment and management in the food industry. Woodhead Publishing Limited, pp 59–82. https://doi.org/10.1533/9780857090225.2.59

Ullah A, Zhang Y, Liu C, Qiao Q, Shao Q, Shi J (2023) Process intensification strategies for green solvent mediated biomass pretreatment. Bioresour Technol. https://doi.org/10.1016/j.biortech.2022.128394

van Eijck J, Batidzirai B, Faaij A (2014) Current and future economic performance of first and second generation biofuels in developing countries. Appl Energy. https://doi.org/10.1016/j.apenergy.2014.08.015

van Ruijven BJ, De Cian E, Sue Wing I (2019) Amplification of future energy demand growth due to climate change. Nat Commun. https://doi.org/10.1038/s41467-019-10399-3

Vera I, Wicke B, Lamers P, Cowie A, Repo A, Heukels B, Zumpf C, Styles D, Parish E, Cherubini F, Berndes G, Jager H, Schiesari L, Junginger M, Brandão M, Bentsen NS, Daioglou V, Harris Z, van der Hilst F (2022) Land use for bioenergy: synergies and trade-offs between sustainable development goals. Renew Sust Energ Rev 161:112409. https://doi.org/10.1016/j.rser.2022.112409

Verghese K, Lewis H, Fitzpatrick L (eds) (2012) Packaging for sustainability. Springer, London. https://doi.org/10.1007/978-0-85729-988-8

Wang M, Wu M, Huo H (2007) Life-cycle energy and greenhouse gas emission impacts of different corn ethanol plant types. Environ Res Lett 2(2):024001. https://doi.org/10.1088/1748-9326/2/2/024001

Wang M, Han J, Dunn JB, Cai H, Elgowainy A (2012) Well-to-wheels energy use and greenhouse gas emissions of ethanol from corn, sugarcane and cellulosic biomass for US use. Environ Res Lett 7(4):045905. https://doi.org/10.1088/1748-9326/7/4/045905

Wang L, Littlewood J, Murphy RJ (2013) Environmental sustainability of bioethanol production from wheat straw in the UK. Renew Sust Energ Rev. https://doi.org/10.1016/j.rser.2013.08.031

Wang M, Chen Y, Xia X, Li J, Liu J (2014) Energy efficiency and environmental performance of bioethanol production from sweet sorghum stem based on life cycle analysis. Bioresour Technol. https://doi.org/10.1016/j.biortech.2014.04.014

Wang K, Tong R, Zhai Q, Lyu G, Li Y (2025a) A critical review of life cycle assessments on bioenergy technologies: methodological choices, limitations, and suggestions for future studies. Sustainability 17(8):3415. https://doi.org/10.3390/su17083415

Wang Y, Xiao X, Ji Y (2025b) A review of LCA studies on marine alternative fuels: fuels, methodology, case studies, and recommendations. J Mar Sci Eng 13(2):196. https://doi.org/10.3390/jmse13020196

Wolff M, Becker T, Walther G (2023) Long-term design and analysis of renewable fuel supply chains—an integrated approach considering seasonal resource availability. Eur J Oper Res 304(2):745–762. https://doi.org/10.1016/j.ejor.2022.04.001

Xu J, Li M, Ni T (2015) Feedstock for bioethanol production from a technological paradigm perspective. Bioresources. https://doi.org/10.15376/biores.10.3.Xu

Yaverino-Gutiérrez MA, Wong AYCH, Ibarra-Muñoz LA, Chávez ACF, Sosa-Martínez JD, Tagle-Pedroza AS, Hernández-Beltran JU, Sánchez-Muñoz S, Santos JCD, da Silva SS, Balagurusamy N (2024) Perspectives and progress in bioethanol processing and social economic impacts. Sustainability (Switzerland) 16(2). https://doi.org/10.3390/su16020608

Zhao X, Liu D (2019) Multi-products co-production improves the economic feasibility of cellulosic ethanol: a case of Formiline pretreatment-based biorefining. Appl Energy. https://doi.org/10.1016/j.apenergy.2019.05.045

Zhao L, Zhang X, Xu J, Ou X, Chang S, Wu M (2015) Techno-economic analysis of bioethanol production from lignocellulosic biomass in China: dilute-acid pretreatment and enzymatic hydrolysis of corn stover. Energies 8(5):4096–4117. https://doi.org/10.3390/en8054096

Chapter 15
Environmental Legislation and Compliance in the Green Energy Production System

Prasad Andhare and Khushbu Verma

Abstract The worldwide transition to renewable energy sources requires adequate environmental laws and strict compliance procedures, especially in bioethanol production from lignocellulosic biomass. This chapter investigates the complex network of environmental policies and regulatory structures of green energy systems, focusing on bioethanol. It reviews all the significant international and national policies governing their roles in sustainable biofuel production, including the Paris Agreement and national policy mechanisms such as the U.S. Renewable Fuel Standard and the EU's Renewable Energy Directive II. Energy and sustainability auditing methodologies such as life cycle assessment (LCA) and carbon accounting are also reviewed as instruments for enforcement and pollution prevention. Case studies from around the world demonstrate these audits in action and the problems encountered in carrying out industrial practices and environmental standards. Combining policy with auditing methodology in this chapter offers a holistic approach to illustrate mechanisms for environmental compliance in green energy production, better practice, and future directions towards sustainable bioethanol production.

Keywords Environmental legislation · Bioethanol production · Sustainability auditing · Life cycle assessment · Renewable energy compliance

15.1 Introduction

15.1.1 The Green Energy Imperative

The twenty-first century is experiencing a faster shift from fossil-based to renewable-based energy concepts inspired by overriding drivers of climate change, environmental damage, and finite resources depletion (Bhuiyan 2022). At the core of this conversion is known as "green energy", that is, energy derived from renewable/

P. Andhare · K. Verma (✉)
Department of Life Sciences, School of Sciences, GSFC University, Vadodara, Gujarat, India
e-mail: khushbu.verma@gsfcuniversity.ac.in

S. Prasad et al. (eds.), *Sustainable Bioethanol Production from Lignocellulosic Biomass*, Clean Energy Production Technologies,
https://doi.org/10.1007/978-981-95-9690-4_15

low-emission sources like biomass, solar power, wind, and hydro (Rajpurohit et al. 2022). In the group of these alternative biofuels, bioethanol, especially derived from lignocellulose biomass, has risen as a key sustainable fuel because of its environmental and social benefits (Melendez et al. 2022). With accelerating global energy needs, industrialization, and population growth, using renewable bioenergy has become both a technological development and an environmental imperative. With world energy needs soaring, driven by industrialization and population growth, using renewable biofuels is seen today not only as a technological option but mainly as a matter of environmental urgency (Sikiru et al. 2024). The urgency is not just in generating green energy but ensuring its generation is sustainable, including minimal ecological impact, carbon neutrality, and resource circularity for perpetuity (Khan et al. 2021).

Although the production of lignocellulosic bioethanol is a relatively benign process, it is not. Its sustainability relies on how feedstock is grown and converted, as well as post-production emissions and waste streams. In mediating these concerns, environmental regulations and strong compliance are key (Afedzi et al. 2025). All bioethanol systems should be structured from the land up, including the policy that regulates production from a land use, emissions, water use, and energy efficiency perspective (Khoo 2015). This complex regime of regulation—from local environmental laws to global climate frameworks—functions not simply as a straitjacket but as an enabler, pushing the industry towards cleaner, more efficient, and more transparent production modes (Oloyede et al. 2024). Hence, the necessity of "green energy" is heavily embedded in legislative prudence, regulatory integrity, and governance, which takes initiative (Kumar et al. 2024a, b).

15.1.2 Role of Environmental Regulation in Sustainable Bioethanol Systems

Environmental legislation is the structural and moral framework in which bioethanol systems must work. The purpose of these laws concerning the sustainable development of biofuels is to help ensure that those laws work to encourage adherence to standards, transparency, and best practices up and down the biomass-to-fuels value chain (Ghadami et al. 2021). Regulatory tools include an array of interventions—from licenses and emissions standards to sustainable standards and life cycle analysis requirements. For example, the introduction of obligatory Environmental Impact Assessments (EIAs) before plant construction guarantees that ecological aspects are already considered at the beginning of a project (Guerrero and Munoz 2023). Similarly, compliance with waste treatment standards, water consumption limits, and effluent emission standards helps to reduce the externalities commonly attached to agro-industrial work (Stillitano et al. 2022).

Similarly, regulatory discourses support sustainability audit tools like life cycle assessment (LCA), carbon footprint, and energy return on investment (EROI) criteria. These tools do more than help with compliance reporting—they provide a mechanism for continuous process improvement and environmental performance benchmarking (Papagianni et al. 2024). In a globalized energy market, aligning requirements globally among all jurisdictions, via multilateral agreements and sustainability certification, allows for harmonizing environmental expectations and market access (Margaret et al. 2024). For bioethanol producers, compliance is an issue of legal requirement and a condition for ethical legitimacy, investment appeal, and international competitiveness. In this regard, regulation is not an enemy of innovation but creates a dynamic "edge" at which legal obligations and technological progress converge towards sustainable bioenergy futures (Allam and Cheshmehzangi 2024).

In particular, the chapters have the goal to make a critical review of the developments in the field of relevant environmental legislation and compliance issues for renewable energy systems including bioethanol production from lignocellulosic. The chapter focuses on how the implementation of international and domestic regulatory mechanisms and sustainability auditing methodologies (including LCA, the energy performance indicator, and carbon intensity scores) influences environmental accountability, technological development, and policy coherence. With comparison to global case studies, energy audit frameworks, and digital compliant technologies, the intention of the chapter is to offer policy makers, researchers, and vested stakeholders with tangible avenues to create transparent, flexible, and performance-based governance systems to govern sustainable bioenergy futures.

15.2 Policy Landscape of Bioethanol and Renewable Energy

The policy environment for bioethanol and renewable energy incorporates international climate agreements, domestic renewable energy targets, and institutional governance structures in order to set a solid foundation for sustainable biofuel systems. On the global stage, agreements like the Paris Agreement, the Kyoto Protocol, and the UN Sustainable Development Goals (SDG) have influenced pledges to reduce greenhouse gas emissions, increase the utilization of renewable energy, and hold companies and countries accountable for their environmental performance. Globally, India among other countries, and the European Union, the United States, and Brazil especially, have evolved unique yet interlinking policy trajectories, covering blending mandates and lifecycle emission standards (to incentivize fossil-based competition), and decarbonization credits systems. Backing up these efforts are robust institutional mechanisms—like India's MoEFCC and MNRE, the U.S. Environmental Protection Agency (EPA), and multilateral institutions like the United Nations Environment Programme (UNEP)—that supply the regulatory controls, capacity-building opportunities, and audit tools that are imperative for policy deployment, compliance certification, and innovative-technology testing in assets used to produce green energy.

15.2.1 Overview of Global Environmental Policies (Kyoto Protocol, Paris Agreement, SDGs)

Global environmental governance has changed dramatically in the past 30 years as the scale of human-induced climate change has become increasingly recognized and there is greater urgency in pursuing mitigatory actions. First established in 1997 and enacted in 2005, the Kyoto Protocol was the initial binding international emission reduction treaty within the United Nations Framework Convention on Climate Change (UNFCCC), mandating that industrialized countries cut their greenhouse gas emissions (Madubuegwu et al. 2021). At the same time, it was criticized for the complex division between Annex I and non-Annex I countries, its establishment of instruments like the Clean Development Mechanism (CDM), which indirectly incentivized bioethanol and other renewable projects in non-Annex I countries (de Melo et al. 2021). Momentum then shifted from the Kyoto-type hierarchical imposition to a far more participatory and flexible regime, exemplified in the Paris Agreement (2015) with its voluntary Nationally Determined Contributions (NDCs) and related nationally tailored climate pledges that come with specific renewable energy targets, including mixed biofuel integration (Pan 2022).

Concurrently, the 2030 Agenda for Sustainable Development, approved by the United Nations in 2015, mainstreamed the contribution of renewable energy in Sustainable Development Goal 7 (SDG-7) to assure access to affordable, reliable, sustainable, and modern energy for all. This objective aligns with multiple SDGs, including SDG-13 (climate action), SDG-12 (responsible consumption and production), as well as SDG-15 (life on land), which supports the multidimensional nature of sustainable bioethanol systems (Ramasubramanian and Ramakrishna 2023). In this context, bioethanol has been gaining a strategic role as an alternative to fossil energy sources, especially in the transportation sector, with policies around the world driving its integration in the energy profiles of nations by means of tax benefits, mandates, and certifications of sustainability (Gonçalves et al. 2023). Consequently, international regimes offer normative orientation, activate financial and technological support, and incentivize cooperation across borders and policy harmonization that will be essential for realizing decarbonized bioenergy futures (Cantarella et al. 2023).

15.2.2 National Renewable Energy Policies (India, EU, US, Brazil)

National policies for renewable energy vary with socio-economic conditions, energy security considerations, and environmental ambitions. India's National Policy on Biofuels, announced in 2009 and revised in 2018, envisages bioethanol as a chief agent for rationalizing crude oil requirements for fuels, ameliorating rural energy demands, and increasing the revenue of farmers through the commercial valorization of agricultural waste (Prasad et al. 2024). The policy requires phased blending targets (20% ethanol by 2025, for instance), incentives such as production-linked improvement, capital subsidy for distilleries, and expedited regulatory clearances. India's Ethanol Blended Petrol (EBP) Programme is an example of the two-pronged approach to climate mitigation and rural development (Fatima et al. 2025). However, there are still bottlenecks in feedstock logistics, technology development, and meeting LCA-driven sustainability targets (Sadhukhan 2022).

On the other hand, the European Union (EU) pursues a more consolidated and climate-focused bioethanol policy under the Renewable Energy Directive II (RED II), where a minimum 14% renewables of the total transport energy will be required by 2030 (Mather-Gratton et al. 2021). It contains stringent sustainability and greenhouse gas (GHG) emission saving requirements on biofuels, with indirect land use change (ILUC) provisions. The United States Renewable Fuel Standard (RFS) promulgated by the Energy Policy Act (2005) and amended by the Energy Independence and Security Act (2007) requires that a specific volume of renewable fuel be blended into gasoline, setting annual volume requirements (Brandão 2022). Compliance is recorded by buying credits known as renewable identification numbers (RINs), which are used to track the fuel. On a parallel stream, Brazil's RenovaBio program promotes lifecycle carbon intensity ranking of bioethanol and includes decarbonization credits (CBIOs) based on performance to incentivize sustainable functioning (Tiburcio et al. 2023). A visual summary of the international environmental agreements and the country-level regulatory frameworks that impact bioethanol production (Fig. 15.1).

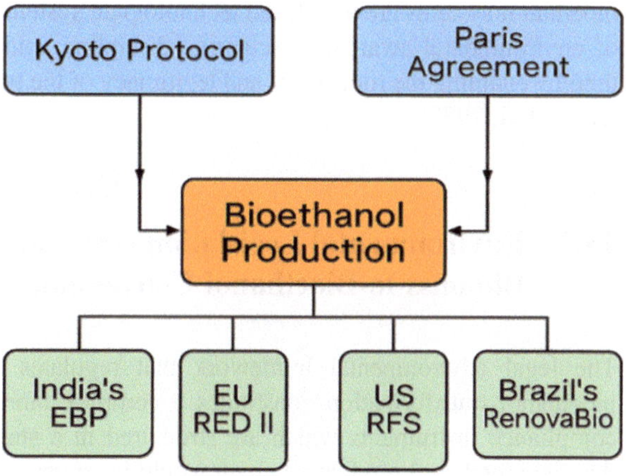

Fig. 15.1 Global regulatory frameworks governing bioethanol production

15.2.3 Institutional Frameworks (MoEFCC, MNRE, EPA, UNEP)

Strong institutional structures are critical to convert renewable energy policies into operational, enforceable, and quantifiable results. In India, the regulatory aspect is entrusted to the Ministry of Environment, Forest and Climate Change (MoEFCC), which regulates environmental clearance and monitoring for bioethanol projects under the Environmental Impact Assessment (EIA) norms, the Forest Conservation Act, and the Biodiversity Protection Act for the country (Rathoure 2021). Thereby, while the Ministry of New and Renewable Energy (MNRE) acts as the central body for policy formation and the implementation of the renewable sector, it incentivizes and promotes the study of second-generation bioethanol and promotes public-private bioenergy-based infrastructure associations (Duarah et al. 2022). They collaborate through inter-ministerial task teams and aligned policy directives to marry environmental conservation with growth in renewable energy. Globally, the United States Environmental Protection Agency (EPA) is the main regulatory authority established to enforce the Renewable Fuel Standard (RFS) laws, set the blending mandate, monitor emissions reporting, and enforce penalties. Its life cycle GHG and environmental risk evaluation programs also play an important role worldwide (Sajid et al. 2021). At the multilateral level, the United Nations Environment Programme (UNEP) supports harmonizing policies, building capacity, and fostering South-South cooperation in transitioning to green energies (Cao 2024). As exemplified by the Global Bioenergy Partnership (GBEP) and the International Renewable Energy Agency (IRENA), UNEP promotes the sharing of best practices, the development of sustainability metrics, and the transfer of knowledge across countries (Brady et al. 2023). The evolution of these institutional influences means that bioethanol systems are not closed technological systems but are part of a wider web of environmental awareness, scientific discipline, and international collaboration, thereby enabling the robustness and legitimacy of the transition to renewable energy (Kara et al. 2022).

15.3 Environmental Legislation Governing Biomass-to-Bioethanol Conversion

The legal environmental framework that regulates the process of biomass to bioethanol transformation constitutes a complex fabric of laws, regulations, and compliance instruments which are structured in a statewide approach and ensure that the bioethanol production chain would be environmentally sustainable. Central laws like the Environment Protection Act, 1986, the Air and the Water Pollution Control Acts, and the Energy Conservation Act (2001) define the legal framework for emissions, resource use,and industrial practices in India and international agreements like the Convention on Biological Diversity (CBD) are directives on biosafety of genetically modified organisms developed in bioethanol. This legislative framework prescribes precautionary planning built into projects driven by the project-level environmental clearances under Environmental Impact Assessment (EIA) Notification (2006) and additional Consent to Establish (CTE) and Consent to Operate (CTO) mechanisms, which guarantee regulatory oversight starting from the project setup to project operationalization. Moreover, waste governance structures (Solid Waste Management Rules, 2016; Hazardous Waste Regulations) that associate with bioethanol production invoke circular economy principles of waste minimization, resource recovery, and industrial symbiosis. When combined, these laws, disposal mechanisms, and circular economy obligations constitute an integrated legal ecosystem for compliance to help ensure the growth of renewable energies while also ensuring stewardship of the environment and sustainable transitions of industry.

15.3.1 Acts, Rules, and Regulations Relevant to Bioenergy

The legislation framework regulating the bioethanol production of lignocellulosic biomass is based on various environmental and energy laws and obligations that delimit the scope of legal boundaries and oblige the stakeholders to accomplish certain operational practices. In India, the Environment (Protection) Act, 1986, the Air (Prevention and Control of Pollution) Act, 1981, and the Water (Prevention and Control of Pollution) Act, 1974 are core statutes for environmental laws (Singh 2021). These laws authorize regulatory agencies to establish limits for emissions, require compliance with pollution control standards, and pursue litigation against violators. For bioenergy projects, particularly agro-industrial feedstock processing facilities, such measures are vital for managing environmental externalities such as effluent run-off, air quality, and groundwater contamination. Also, state pollution control boards have a significant role in granting operational permits, inspection, and oversight regarding adherence to these central regulations (Umakanth et al. 2022).

On the renewable energy front, specific policies such as the Energy Conservation Act, 2001, and the National Bio-Energy Mission provide a roadmap for clean energy generation through financial subsidies, tax benefits, and viability gap funding (Tripathi and Bhattacharya 2023). At the international level, multilateral treaties such as the Convention on Biological Diversity (CBD) and the Cartagena Protocol on Biosafety, along with national obligations arising from them, affect the governance of GMOs, which are composed of enzymes and engineered microbes for biomass pretreatment and fermentation processes. These acts prevent bioethanol production from occurring in legal voids and instead follow set environmental limits and technological norms (Hernández 2025). A comprehensive list of Indian environmental regulations applicable to lignocellulosic bioethanol (Table 15.1).

Table 15.1 Regulatory acts and rules relevant to bioethanol systems in India

Regulation	Regulatory body	Scope	Enforcement mechanism	References
Environment (Protection) Act, 1986	Ministry of Environment, Forest and Climate Change (MoEFCC)	Umbrella legislation for environmental protection applies to all industries, including bioethanol	Central and State Pollution Control Boards; National Green Tribunal	Sharma (2022)
EIA Notification, 2006 (as amended)	MoEFCC	Mandates Environmental Impact Assessment (EIA) for projects with potential environmental impacts	MoEFCC; State Environmental Impact Assessment Authorities	Islam and Wang (2022)
Solid Waste Management Rules, 2016	MoEFCC	Governs the management of solid waste, including biomass residues from bioethanol production	Local authorities, State Pollution Control Boards	Cheela et al. (2022)
Hazardous and Other Wastes (Management and Transboundary Movement) Rules, 2016	MoEFCC	Regulates the management and transboundary movement of hazardous and other wastes	Central Pollution Control Board; State Pollution Control Boards	Ghosh et al. (2021)
Biofuels Policy, 2018	Ministry of New and Renewable Energy (MNRE)	Promotes biofuels, including bioethanol, to reduce dependence on fossil fuels	MNRE; State Governments	Kaur (2023)
National Biofuels Policy, 2018	MNRE	Provides a framework for promoting biofuels, including bioethanol, in India	MNRE; State Governments	Verma et al. (2024)

15.3.2 Environmental Impact Assessment (EIA) and Consent Mechanisms

The Environmental Impact Assessment (EIA) is a procedural and technically-based tool designed to identify the potential impacts on the environment of industrial operations such as a bioethanol industry based on biomass. Enforced as per the EIA Notification (2006), under the Environment (Protection) Act, the process makes it mandatory to list projects under schedules that would warrant obtaining environment clearance before working on them, on the ground that such projects could cause environmental degradation (Nomani et al. 2021). Bioethanol factories frequently belong to this type, especially when a large area of land is used, where there is a generation of wastewater or hazardous chemicals are used. Such an EIA is characterized by data collection to establish baseline values, risk modelling, consultation with stakeholders, and developing Environmental Management Plans (EMPs). These requirements make sure that project applicants take into account damages to biodiversity and risks to air and water pollution and social and economic dislocation, thus institutionalizing in the project cycle principles of prevention and precaution (Rathi 2021).

The EIA process is augmented by the mechanism of Consent to Establish (CTE) and Consent to Operate (CTO) by the State Pollution Control Boards (SPCBs). The CTE issued before construction guarantees that the bioethanol project complies with land use plans, zoning, and the environment (Lasselin et al. 2025). After construction, the CTO was to be granted, enabling actual operations, though with compliance to emission norms, effluent treatment regimes, and recurring reporting. The consents act, therefore, as intelligent control points in the project's life, subject to regulations, and able to withdraw the consent in the event of non-compliance or impose required remedies (Singh and Gadgala 2022).

15.3.3 Waste Management Rules and Circular Economy Principles

Incorporating biomass-to-bioethanol systems within the waste governance context is important for avoiding environmental loads and improving the water process's sustainability. Templating tools such as Solid Waste Management Rules, 2016, Hazardous and Other Wastes (Management and Transboundary Movement) Rules, 2016, and Plastic Waste Management Rules, 2016 lay down the legal responsibilities for segregation, collection, transport, handling, reprocessing, and disposal of different types of waste produced at various stages of bioethanol manufacturing (Sharma et al. 2025). Lignocellulosic bioethanol plants produce residues including lignin-rich solids, process water sludge, fermentation by-products, and solvent waste; thus, comprehensive waste audits and safe handling procedures are required. Regulatory requirements require that waste minimization plans, on-site treatment facilities, and safe disposal methods that meet environmental criteria be constructed (Zhang et al. 2022).

Simultaneously, the tenets of the circular economy are beginning to shape both regulation and industrial best practices, with an increased focus on resource efficiency, waste valorization, and closed-loop system-tried and tested models designed to reduce waste and emissions. Within this paradigm, bioethanol systems are no longer seen as mere energy conversion systems, but also as bio-refineries that can generate several value-added co-products, such as biochar, bioplastics, organic acids, and animal feed additives (Devi et al. 2022). This integration policy is reflected in recent government policies on industrial or inter-industrial symbiosis, co-processing, and extended producer responsibility (EPR). In this transition path towards circularity, not only are environmental liabilities reduced, but there is also an increase in economic sustainability and the resilience of the bioethanol operations. However, fulfilling this vision requires re-adjusting regulatory tools to sustain multi-product systems, promote secondary product markets, and integrate inter-sectoral policies from agriculture and waste management to bioeconomy development (Mukherjee et al. 2023).

15.4 Energy and Sustainability Auditing Frameworks

15.4.1 Definition, Objectives, and Standards of Energy Audits

An energy audit is a systematic inspection and analysis of energy flows for energy conservation in a building, process, or system to reduce the amount of energy input into the system without negatively affecting the outputs (Krarti 2020). Energy audits are diagnostic tools for lignocellulosic bioethanol systems that examine energy inputs, conversion efficiencies, and output metrics along the feedstock-to-fuel pathway. A typical process energy audit includes an extensive study of process equipment, heat and energy, material balances, and operations (Pandey and Singhal 2022). The ultimate objective under these standards is to reduce wasted energy, improve system performance, and align with federal efficiency parameters. Energy audits are also useful monitoring tools for sustainability certification, investment analysis, and environmental reporting (Patel and Singh 2024).

The verification of the methodological soundness of energy audits is granted by international standards such as ISO 50002 for the proper energy auditing process and the ISO 50001 for energy management systems (Tiwary et al. 2022). These criteria recommend a systematic process that includes planning, data collection, performance analysis, and action planning. Furthermore, regional standards like the Bureau of Energy Efficiency (BEE) in India provide a robust protocol for sectoral audits, such as the bioenergy sector (Singhal et al. 2022). Adherence to these standards allows internal benchmarking and demonstrates regulatory transparency and external certification. In high-energy processes such as pretreatment, enzymatic hydrolysis, and distillation for lignocellulosic-based systems, detailed audits could play an important role in mapping hotspots as well as in applying the corrective measures (Kosamia et al. 2022).

15.4.2 Sustainability Auditing Metrics: GHG Reduction, Net Energy Ratio, EROI

Sustainability auditing presents a more comprehensive image beyond pure energy efficiency through an ecological, economic, and social performance indicator. In this area, the three important indicators are Greenhouse Gas (GHG) Reduction, Net Energy Ratio (NER), and Energy Return on Investment (EROI) (Nevskaya et al.

2023). GHG impact assessments measure the reduction of carbon dioxide and other pollutant emissions achieved by bioethanol systems compared to fossil fuel systems. Such assessments tend to use cradle-to-grave models, which account for emissions from growing, transportation, processing, and combustion of final users. Reliable GHG accounting provides increased environmental transparency and is instrumental in cross-compliance with international initiatives like carbon markets, carbon intensity scoring, and sustainability accreditation schemes (Balázs et al. 2021).

The energy performance of bioethanol systems is evaluated by thermodynamic tools such as Energy Return on Investment (EROI) and Net Energy Ratio (NER). NER is the ratio of the energy obtained as ethanol compared to the total energy input across the production cycle. In contrast, EROI measures the usable energy provided compared to the energy used to get that energy (Arodudu 2021). These measures offer a quantitative comparison of the feasibility and sustainability of lignocellulose-based bioethanol with other power systems. A low EROI indicates a system that consumes more energy than it returns to the system, which will not be sustainable. Conversely, a positive EROI implies an energy-positive performance closer to environmental and economic sustainability demands. NER and EROI are the foundation of investment making, policy analysis, and public acceptance, manifesting it as a respective tool in the audit for bioethanol governance (Murphy et al. 2022). A tabular representation (Table 15.2) of key environmental auditing metrics and their implications for regulatory compliance.

15.4.3 LCA (Life Cycle Assessment) and Environmental Auditing Tools (SimaPro, GaBi)

Life cycle assessment (LCA) is a scientifically standardized method for analysing the environmental impacts of a commodity or a system, during all its life stages from acquisition of raw materials to disposal. Being on lignocellulosic bioethanol, LCA considers the upstream effects of feedstock growing, the midstream process emission, and the downstream distribution and usage behaviour (Soleymani Angili et al. 2021). Performed according to the ISO 14040 and 14044, LCA can deliver holistic information on a range of parameters, including global warming potential, eutrophication, acidification, water and land usage. It provides a reliable decision-making system for hotspot identification, process chain design optimization, and comparing environmental performance with policy or industry standards (Almroth and Rehnberg 2023). It is worth mentioning that LCA is in line with specific legislation for sustainability certifications, for instance, the EU Renewable Energy Directive II or the Brazilian RenovaBio program (Grangeia et al. 2022).

Cutting-edge environmental auditing software tools, including SimaPro and GaBi, are generally used to conduct LCAs in biofuel systems. It merges international databases such as Ecoinvent, Agri-footprint, and GREET to model complex bioconversion chains. For example, SimaPro structures assessments in a modular, process-focused way, enabling detailed sensitivity and scenario studies, while GaBi is

Table 15.2 Key environmental auditing metrics and their applications

Metric	Definition	Unit	Relevance to environmental auditing	References
GHG reduction	Quantifies the decrease in greenhouse gas emissions due to specific interventions or technologies	Metric tons of CO_2 reduced	Essential for compliance with national and international climate agreements (e.g., Paris Agreement). Used in carbon footprint assessments and carbon credit calculations	Atmaca (2024)
Net Energy Ratio (NER)	Measures the net energy output of a system relative to its energy input	Dimensionless ratio	Indicates the efficiency and sustainability of energy systems. Higher NER values suggest more sustainable energy production	Yin et al. (2024)
Energy Return on Investment (EROI)	Assesses the amount of usable energy obtained from an energy resource relative to the energy expended to get that energy	Ratio (Energy Output/Energy Input)	Critical for evaluating the viability of energy projects. Directly impacts economic feasibility and environmental sustainability	Prananta and Kubiszewski (2021)

industrial-scale oriented but supports supply chain modelling and life cycle inventory (LCI) modifications well (Nemecek et al. 2024). Both platforms support multi-criteria decision-making, enabling compliance with third-party requirements and voluntary ecolabel schemes. For lignocellulosic bioethanol systems, these tools are necessary to compare the trade-offs between competing process routes (e.g., steam explosion versus dilute acid pretreatment) and quantify the environmental co-benefits of complex, integrated biorefinery configurations. Consequently, they play a key role in environmental regulations, technological development, and sustainability disclosure (Jusakulvijit et al. 2021). The process flow diagram (Fig. 15.2) of LCA for lignocellulosic bioethanol indicates system boundaries (cradle-to-grave), environmental impact categories, and stages.

15.4.4 Energy Performance Indicators in Lignocellulosic Bioethanol Production

Energy performance indices (EPIs) are quantitative measures that monitor the efficiency and sustainability of individual unit operations and bioethanol plants' entire energy conversion system. In lignocellulosic biorefineries, EPIs often consist of parameters such as the Specific Energy Consumption (SEC), Thermal Efficiency,

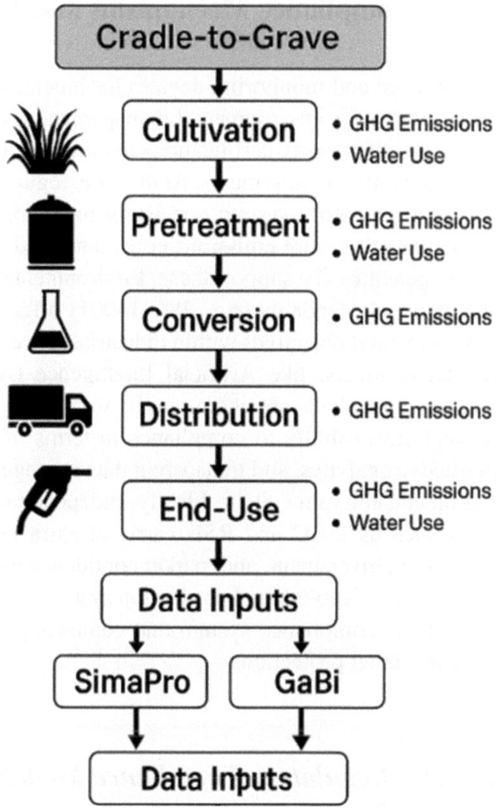

Fig. 15.2 Life cycle assessment (LCA) of lignocellulosic bioethanol system

Electrical Load Factor, and Process Energy Intensity (PEI) (Zhao et al. 2023). The identified indicators aid in detecting bottlenecks and inefficiency in the various stages of the process, such as pretreatment, enzymatic hydrolysis, fermentation, and distillation, ensuring real-time observation and energy planning, and avert costly equipment and retrofit modifications. For example, the SEC offers a relative scale to assess the energy invested per unit of ethanol produced, which allows cross-comparison of performances among feedstock species or technological techniques (Słomka and Pawłowska 2024).

Apart from technical diagnosis, EPIs are increasingly applied to policy compliance, green certification, and carbon trade operations. Governments and regulatory agencies often stipulate threshold performance criteria as qualifying conditions for incentives, subsidies, or sustainability credits. As seen in ethanol plants with higher thermal efficiency and lower PEI, they may access lucrative renewable identification numbers (RINs) or enjoy upgraded carbon intensity ratings under national programs (Dumortier et al. 2024). EPIs facilitate predictive maintenance and automatic compliance documentation, adding safety to operations and accountability to environmental performance. Therefore, EPIs constitute a significant linkage through process engineering, energy policy, and sustainable industry governance (Zhao et al. 2022).

15.5 Compliance Mechanisms and Monitoring Instruments

Compliance and monitoring devices for bioethanol production combine these regulatory regime(s), environmental management systems, sophisticated digital technologies, and third-party certifications to enforce adherence to environmental laws and to sustainability requirements. At the core, regulatory complementarity models adopt a closed regulatory system consisting of multiple dimensions of dynamic market regulations to ensure emissions constraints and operational standards through stipulated penalties. To support these, Environmental Management Systems (EMS) and International Standards (e.g., ISO 14001) offer a structured approach to including environmental objectives within industrial process operations. There are new emerging technologies, like Artificial Intelligence (AI), Internet of Things (IoT), and Blockchain, which are increasingly supporting the compliance of the industries through accessibility to compliance in terms of much more real-time monitoring, predictive analytics, and transparent data management of entire steps and process on the bioethanol value chain. Finally, independent auditing and third-party certifications such as ISCC and RSB carry an extra layer of accountability and provide regulatory, investment, and market confidence in the sustainability and environmental integrity of bio-ethanol production systems. Combining these approaches ensures a complete compliance system that connects policy mandates with operations and environmental protection.

15.5.1 Regulatory Compliance Models and Penalty Structures

Regulation compliance models are institutional innovations that enable green energy producers to meet environmental laws, technical requirements, and sustainability benchmarks. These approaches include prescriptive, performance, and market approaches. The prescriptive approaches are based on rule-based regulations that mandate emission standards or limits, equipment standards, and operational norms (Wörner et al. 2025). Performance-based approaches, however, afford promises to regulate with some flexibility in reaching environmental results, if quantified performance goals have been met, like emission reduction or energy efficiency numbers. Compliance mechanisms that use markets, such as cap and trade or renewable energy credits, incorporate economic incentives into regulatory systems that allow for the low-cost control of pollution and other forms of environmental management (Melhim and Isaifan 2025).

At the core of these frameworks are penalty models used to penalize non-compliance. Penalties may be administrative, civil, or criminal and depend on the nature of the infraction. Administrative penalties include fines, revocation of operational permits, or orders to take corrective action, while civil penalties involve lawsuits and payment for environmental damages (Prasetyaningsih et al. 2022). Criminal penalties exist for gross negligence, intentional data falsification, or significant ecological damage. It is imperative to remember that there can be escalation clauses for repeat violations to put facilities under even closer control or to close them. However, the efficacy of these penalty models is predicated on the presence and quality of regulation, legal enforcement, and reporting requirements (Arunkumar and Dharshini 2025). The bioethanol industry, which straddles agriculture, industry, and environment, can benefit from a multi-level compliance mechanism for accountability and adaptive management, with well-defined sanctions and incentivized pathways (Benvenutti et al. 2023).

15.5.2 Environmental Management Systems (EMS) and ISO 14001

Environmental management systems (EMS) are structured systems that help organizations to integrate environmental responsibilities into their operations. The plan-do-check-act (PDCA) cycle is the foundation of an EMS that provides for ongoing environmental improvement through policy commitments, goal setting, operational controls, and monitoring of performance (Bhateria 2024). In bioethanol plants, EMS implementation ultimately can promote the integration of daily activities of the plant with long-term planning for sustainability goals such as reduction in carbon footprint, waste disposal, and resource utilization. The EMS model enables the construction of environmental baselines, risk assessments, and contingency plans, which are site-specific, reflecting environmental conditions and the regulatory framework (Kaviyarasu 2024).

ISO 14001 is the international standard that specifies requirements for an effective EMS. It gives companies a structured approach for identifying environmental aspects, appraising compliance obligations, and defining benchmarks. ISO 14001 certification guarantees compliance with the legislation, and for bioethanol producers, it is a vehicle to improve corporate image, manage stakeholders, and facilitate market entry (Wessels 2024). It also promotes a culture of internal accountability with the requirement of the inside audit, non-conformity review, and corrective action. The harmonization of ISO 14001 with other standards like ISO 50001 (energy management) and ISO 45001(occupational health and safety) also provides for a comprehensive perspective for the governance of sustainability (Molina-Reyes et al. 2022). EM and ISO standards are compliance enablers and strategic devices for biofuel production industries to cope with the complexities of environmental regulation (Szöke 2021).

15.5.3 Role of AI, IoT, and Blockchain in Compliance Monitoring

Top-of-the-line technology is changing the game for environmental compliance with data-driven monitoring and reporting through real-time technology. Compliance is improved by artificial intelligence (AI) algorithms, which include predictive

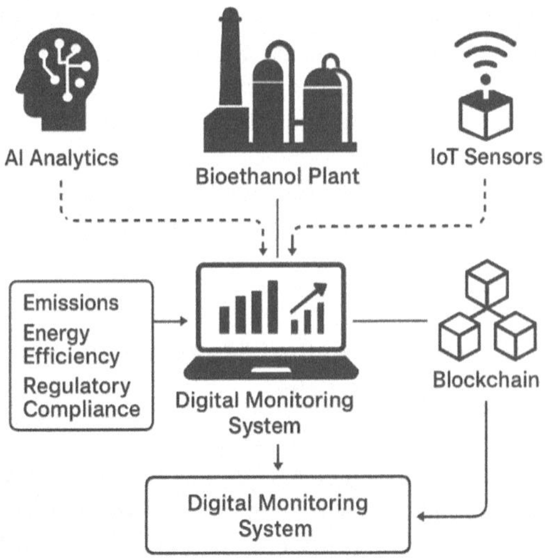

Fig. 15.3 Technological integration in environmental compliance monitoring

analytics, anomaly detection, and automated decision support systems (Rane et al. 2024). In bioethanol factories, AI models can predict process deviations, discover energy losses, and model environmental consequences of different operating conditions. In this way, plant managers and regulators can carry out proactive countermeasures to reduce regulatory non-compliance. Additionally, AI-based natural language processing solutions are starting to be used to break down convoluted regulatory verbiage and compile compliance checklists ideal for given operational environments (Lei 2024).

AI's capabilities can be expanded with Internet of Things (IoT) technology, which can incorporate sensors in the entire bioethanol production process to acquire data in real time about various factors, including air releases, wastewater quality, energy usage, and the state of equipment (Kuila and Kumar 2025). This networked sensing system provides fine-scaled environmental monitoring and data storage for regulatory reporting. Blockchain technology, in contrast, adds transparency, immutability, and traceability to compliance documents. By decentralizing data storage, blockchain guarantees that environmental data, i.e., emissions logs, audit results, and certification data, cannot be altered, which helps reduce the risk of fraud or misreporting (Gade 2023). A schematic representation (Fig. 15.3) of an integrated digital monitoring system combining AI analytics and other tools for real-time emissions tracking and regulatory compliance in bioethanol plants.

15.5.4 Role of Independent Auditing and Third-Party Certifications

Third-party certification and independent environmental audit are important mechanisms for maintaining trust and transparency and for providing accountability in compliance with the architecture of green energy systems (Nowicki et al. 2021). As opposed to internal audits, external audits are performed by certified third-party firms, which reduces the potential for biased advice and promotes objectivity. Such audit programs assess conformity with legal requirements, standards, and sustainable criteria. In the bioethanol area, measures of corporate responsibility cover aspects such as emissions management, energy performance, waste management, and traceability of feedstocks (Thoti et al. 2024).

Certifications by third parties are voluntary but compelling, and compliance dynamics beyond regulation affirm stakeholders that an organization is ecologically accountable. Certificates like ISCC (International Sustainability and Carbon Certification), RSB (Roundtable on Sustainable Biomaterials), or Bonsucro are well-established in the biofuel sector to confirm sustainability within the supply chain. These are frequently conditions for entry into international markets and qualification to sell carbon credits. They should also require that additionality be audited, documented, and reviewed regularly to institutionalize best practices and keep operating protocols in line with global environmental objectives (Arias et al. 2023).

15.6 Case Studies and Regional Experiences

Case studies and regional experiences offer tangible insights into how statutes, verification approaches, and sustainability oversight frameworks materialize in real situations across diverse geopolitical backdrops. India's Alcohol Integration Programme demonstrates the position of multi-agency oversight auditing in hastening bioethanol adoption and decreasing reliance on fossil fuels. The European Union's RED II Directive highlights stringent sustainability criteria, life-cycle GHG emissions accounting, and cross-border certification systems as catalysts for statutory coherence and environmental accountability. Brazil's RenovaBio Program introduces market-based verification through the carbon intensity scoring system and decarbonizing credits, relating ecological performance to financial incentives. Comparative examination across these areas uncovers variations in policy formation, enforcement abilities, and auditing thoroughness, offering beneficial lessons for shaping globally harmonized yet locally adaptable governance frameworks for sustainable bioethanol manufacturing.

15.6.1 Bioethanol Compliance Auditing in India: Ethanol Blending Program (EBP)

India's Ethanol Blending Programme (EBP) is a key plank of the country's strategy for transitioning to renewable transport fuels, in order to reduce reliance on fossil fuels, cut greenhouse gas emissions, and provide additional income to rural areas by using agricultural residues (Kumar et al. 2024a, b). Introduced in 2003, EBP has been developed over time through policy revisions in 2013, 2018, and most recently in 2021, to stipulate that ethanol must be blended into gasoline with an aspirational target of 20% of petrol consumption as blending of ethanol by 2025. In this scenario, compliance auditing is carried out within a multi-agency regime encompassing the Ministry of Petroleum and Natural Gas (MoPNG), the Oil Marketing Companies (OMCs), and lastly, the Ministry of Environment, Forest and Climate Change (MoEFCC) (Singh 2024). Progressive states, such as Maharashtra and Uttar Pradesh, have put in place rigorous verification and accountability processes—including digital tracking of feedstock sourcing and emissions reporting—over the years, which are helping push other regions to both institute similar systems, as well as combat infrastructure bottlenecks, fragmented supply chains, and policy discrepancies (Chaturvedi et al. 2024). New measures like compulsion of Life Cycle GHG Emission Reporting and third-party audit by agencies like Bureau of Indian Standards (BIS) and National Accreditation Board for Testing and Calibration Laboratories (NABL) are being made to fill in the gaps (Singh 2023).

15.6.2 EU RED II Directive and Sustainability Criteria

The Renewable Energy Directive II (RED II), which was passed by the European Union in 2018, is a holistic legislative act that aims to achieve decarbonization across energy systems (including in transport). RED II binds the EU to a target that 14% of energy used in transport should be from renewable sources by 2030 and imposes strict sustainability criteria on biofuels to guarantee their environmental integrity (Mayeres 2025). These criteria stipulate that bioethanol producers must be able to show savings in GHG emissions of not less than 65% compared to fossil fuels, protect high-biodiversity areas of land, and respect land Use and land-use Change (LULUC) restrictions. Compliance is assessed through a rigorous auditing system, such as mass balance, traceability audits, and obligatory third-party verification under established certification schemes, such as ISCC-EU or RSB-EU (Bock 2025).

The indirect land use change (ILUC) risk mitigation and the quota for the advanced biofuel are yet another novelty in the RED II framework, focusing on lignocellulosic and non-food feedstocks (Abreu et al. 2022). It sets the conditions for Member States to develop national certification and reporting relevant to the criteria nationally (EU criteria), ensuring a level playing field across the EU. The RED II case suggests that legislative coherence, built-in control systems, and market access conditionality can synergistically be used as force fields for enhanced environmental accountability in bioethanol sectors (López et al. 2024).

15.6.3 Brazil's Renova Bio: Carbon Intensity Scoring System

Brazil's RenovaBio program, which was announced in 2017 under legislation No. 13.576, represents the most developed economy-oriented tool for addressing biofuels and sustainability. The essential building block of RenovaBio is its carbon intensity (CI) scoring system, aimed at calculating the life-cycle greenhouse gases (GHG) emitted by individual bioethanol producers, including emissions stemming from factors such as feedstock type, land use change, process energy, and transportation (Branco et al. 2025). The CI score is transposed to Decarbonization Credits (CBIOs), which can be traded in a regulated market, incentivizing producers that adopt low-emission technologies and efficiently use resources. Inspection is done by independent third-party inspection bodies certified by Brazil's National Agency for Petroleum, Natural Gas, and Biofuels (ANP) to guarantee transparency and strictness on the CI assessment (Coutinho et al. 2024). Lower CI scoring producers earn additional CBIOs per litre of ethanol and are therefore worth more to the market and investors. The CI policy emphasis on continuous improvement is maintained by adjusting the CI benchmarks from time to time, allowing adaptive convergence towards the latest technology or international standards for sustainable development (Aguiar et al. 2025).

15.6.4 Comparative Analysis of Policy Outcomes and Compliance Efficiency

Comparative review of bioethanol policy-making moulds in India, EU, and Brazil indicates a diversity of governance modes driven by policy choices, regulatory capabilities, and institutional arrangements. India EBP focuses on volume and energy security targets, but its compliance auditing is still fledgling, usually restricted by infrastructure and data issues (Singh 2024). The EU's RED II creates a highly institutionalized compliance regime built around lifecycle sustainability, harmonized certification, and cross-border cooperation. Brazil's RenovaBio, by contrast, promotes a market-based, performance-based approach, using carbon intensity and financial incentives to remunerate low-carbon producers (Maia and Bozelli 2022). A tabular comparison (Table 15.3) of bioethanol governance frameworks across India, the European Union, and Brazil.

Table 15.3 Comparative overview of environmental policies for bioethanol in India, EU, and Brazil

Dimension	India	European Union (EU)	Brazil	References
Primary Regulatory Authority	Ministry of Petroleum and Natural Gas (MoPNG); Ministry of Environment, Forest and Climate Change (MoEFCC)	European Commission (DG CLIMA, DG ENER); National regulatory agencies of member states	National Agency of Petroleum, Natural Gas and Biofuels (ANP); Ministry of Mines and Energy (MME)	Bisht et al. (2021)
Key Policy Instruments	National Bio-Energy Policy 2018—Ethanol Blended Petrol (EBP) Programme—National Policy on Biofuels 2018 (updated 2022)	Renewable Energy Directive II (RED II)—Fuel Quality Directive—Common Agricultural Policy (CAP) incentives	RenovaBio (National Biofuels Policy)—Sugarcane Agroecological Zoning (ZAE Cana)—Ethanol blending mandates	Usmani et al. (2023)
Bioethanol Blending Targets	20% blending by 2025 (E20)	14% renewable energy in transport by 2030 (with sub-targets for advanced biofuels)	Minimum 27% blending (E27) nationwide mandate	Kumbhar and khot (2025)
Lifecycle GHG Emission Requirements	Target 50–60% GHG emission reduction; methodology not fully harmonized across states	Minimum 65% GHG savings for existing plants, 70–80% for new installations (post-2021)	Minimum 50% GHG reduction; carbon intensity scores determine decarbonization credits under RenovaBio	Rodrigues and Gilio (2023)
Compliance Mechanisms	BIS certification for fuel quality—State-level compliance via oil marketing companies—Periodic audits	Mass balance and chain of custody systems—Sustainability verification through approved certification schemes	Decarbonization Credit system (CBIOs)—Certification of producers by ANP—Regular monitoring	Godinho et al. (2022)
Feedstock Regulation	Promotion of non-food crops (sugarcane, rice straw, etc.)—Second-generation ethanol incentives	Cap on food and feed-based biofuels (7%)—Sub-target for advanced biofuels (3.5% by 2030)	Preference for sugarcane; strict agro-ecological zoning—Waste and residue-based feedstocks encouraged	Chong and Ng (2024)
Enforcement Structures	MoPNG and state governments—Petroleum and Natural Gas Regulatory Board (PNGRB)	EU-level monitoring via Member States' National Energy and Climate Plans (NECPs)—EU Court of Justice for violations	ANP enforces compliance—Federal government oversight—Mandatory reporting and verification	Bhargava (2022)

(continued)

Table 15.3 (continued)

Dimension	India	European Union (EU)	Brazil	References
Certification Schemes in Use	Bureau of Indian Standards (BIS)—Pilot sustainability frameworks emerging	ISCC, REDcert, 2BS, and national schemes—Certification mandatory under RED II	RenovaBio certification—Carbon intensity calculated using RenovaCalc	Khalid et al. (2022)
Monitoring & Reporting Systems	Annual EBP reports; feedstock sourcing data via ethanol suppliers	Sustainability compliance reports submitted annually; EU-wide reporting database	Mandatory reporting of GHG emissions—Public RenovaBio registries and performance dashboards	Chandel et al. (2024)

15.7 Future Outlook and Recommendations

The future of environmental compliance within renewable energy relies upon establishing synergistic frameworks connecting regulatory requirements, assessment methodologies, technological progress, and economic motivations into a single adaptive ecosystem. As bioethanol manufacturing spreads globally, the next phase of governance must transcend prescriptive rules towards data-driven, performance-based, and technology-enabled compliance via monitoring. Integrated policies, assessments, and technologies, partnered with financial incentives and market-associated sustainability credentials, will be core to accelerating comprehensive adoption of optimal procedures industry-wide. Moreover, legal technological developments like AI-powered assessment platforms, blockchain-secured compliance records, and real-time environmental surveillance networks have the potential to transform compliance from a reactive process into a predictive, automated, and transparent governance model. These advancements will guide policymakers, researchers, and industries in cultivating scalable, cost-effective, and environmentally robust bioenergy systems for tomorrow.

15.7.1 Towards Integrated Policy-Audit-Technology Ecosystems

The success of environmental compliance in green energy, especially bioethanol production, requires integrated ecosystems of the networks of policy, auditing methods, and emerging technologies. This integration enables dynamic feedback loops between regulatory mandates, operational success, and the pace of technological change. Instead of considering audits and policies as a checklist-based linear process, focus on the types of nodes that can form an integrated part of a digitally synchronized governance platform (Cao et al. 2022). In this paradigm environment, audits are not just a review of the past, but are predictive, due to real-time monitoring, machine learning, and scenario modelling being at the core. In response, policy frameworks must change to contain the constant data flow and allow for the adaptive regulation of changing industry practices and environmental baselines (Shalu 2023).

An embedded policy-audit-technology system enables interoperability between different institutional stakeholders—government bodies, private firms, certifiers, and civil society—via standardized data protocols, digital compliance reporting systems, and compliance verification methods. This system would increase transparency and simplify regulation by bringing together reporting, automating data checks, and streamlining cross-sectoral interfaces. Significantly, this ecosystem approach assures scalability and replicability across sites, enabling countries with various governance capacities to implement modular elements as deemed appropriate.

15.7.2 Incentivizing Green Compliance Through Fiscal and Market Instruments

Financial barriers to compliance can be significantly reduced where fiscal instruments, including green tax credits, accelerated depreciation of clean technology investments, concessional loans, and availability of carbon offset credits (Koval et al. 2022). Market-based environmental policy instruments such as renewable energy certificates, emission trading schemes, and decarbonization credit markets supply financial returns to sustainability performance, motivating compliance and innovation towards benchmarked environmental best practices. Such tools are essential for small- and medium-sized bioethanol producers, with tight capital constraints and limited capacity to access sustainability-linked finance (Penttinen 2022).

In addition, connecting fiscal and market incentives to compliance monitoring increases the policy effectiveness and transparency. As evidenced by the countries, they may provide further tax advantages or preferential feedstock procurement rights to the producers who have obtained an ISO14001 certificate or to the low carbon intensity-scored ones, so that environmental performance leads to a competitive advantage for firms. Governments can also incorporate green compliance as a public procurement criterion, thus rewarding certified producers with access to a guaranteed market (Sam and Song 2022).

15.7.3 Future Research Directions in Legal-Tech for Green Energy

As legal and technological tools and regimens change, legal enforcement, compliance, and policy innovation must converge into one of the more universally recognized fields, legal-tech. Green energy legal-tech tools involve AI-powered regulatory sandboxes, smart contracts for compliance checks, and blockchain-secured certificates (Veluru 2022). These technologies have considerable promise to automate compliance with environmental laws, cut the cost of complying with regulations, and promote accountability through indelible trails of audits and algorithmic judgment of facts. Additionally, legal-tech can contribute to participatory governance by providing access to open-access compliance dashboards, stakeholder feedback loops, and community-sourced environmental monitoring (Metin 2024).

Research in this area will further need to address normative challenges such as algorithmic bias in regulatory decision-making, data privacy, and the digital divide, whereby low-income regions cannot use new technology. Interdisciplinary research between environmental law, computer science, data ethics, and industrial engineering is needed to create responsible legal-tech interventions that satisfy not only technical feasibility but also the needs of communities of users and the environmental benefit (Chun et al. 2025). Pilot studies and policy labs can also be an excellent opportunity for testing adaptive legal instruments and compliance automation models in real-life applications. Joint investment between policymakers and scholars in this research frontier can lead to the co-production of new generation governance models to fit technology advances with the environmental urgency and legal integrity driving the global transition to sustainable energy systems (Kulkarni et al. 2021).

15.8 Conclusion

The synergy between environmental law and regulation through compliance in the green energy industry, particularly for bioethanol production from lignocellulosic biomass, is a mandated requirement and an essential component of sustainability. This chapter has illustrated that good corporate environmental governance, consistent with international accords such as the Paris Agreement and national policies such as India's Biofuel Policy or the EU RED II, can align industrial and developmental aspirations with ecological imperatives. Policy frameworks, from Environmental Impact Assessments to Life Cycle Assessment methodologies, are both preventive and corrective regimes aiming at the reduction of externalities. Furthermore, technologies like AI, IoT, and blockchain, which have been pervasively implemented in compliance monitoring, have enabled proactive governance and altered the governance model from reactive to predictive. Cases from India, the EU, and Brazil are used to demonstrate varied but converging developments where regulation, audit, and innovation are combined to deliver enhanced environmental accountability. Integrating policy, technology, and performance-driven incentives is indispensable for developing sustainable bioenergy systems as we transition to a circular bioeconomy. Prospects for the future depend on flexible regulation and interdisciplinary research, which enables regulation to remain adaptive but maintain the integrity of the environment. So, environmental compliance for bioethanol systems should be treated as an enabler of innovation and sustainability, not as a barrier to advancement.

Acknowledgement The authors gratefully acknowledge the editorial team of the book *Sustainable Bioethanol Production from Lignocellulosic Biomass—Process Engineering and Innovation* for providing the opportunity and constructive feedback that shaped the development of this chapter. We also extend our sincere thanks to GSFC University, Vadodara, for institutional support and access to academic resources that facilitated this work. Special appreciation is due to colleagues, peer reviewers, and technical experts whose insights on environmental legislation, compliance mechanisms, and sustainability auditing greatly enhanced the quality and scope of the manuscript.

References

Abreu M, Silva L, Ribeiro B, Ferreira A, Alves L, Paixão SM, Gouveia L, Moura P, Carvalheiro F, Durate L, Fernando AL, Gírio F (2022) Low indirect land use change (ILUC) energy crops to bioenergy and biofuels—a review. Energies 15(12):4348

Afedzi AEK, Afrakomah GS, Gyan K, Khan J, Seidu R, Baidoo T et al (2025) Enhancing economic and environmental sustainability in lignocellulosic bioethanol production: key factors, innovative technologies, policy frameworks, and social considerations. Sustainability 17(2):499

Aguiar DR, Taheripour F, Silva DA (2025) Ethanol fuel in Brazil: policies and carbon emission avoidance. Biofuels 16(3):248–258

Allam Z, Cheshmehzangi A (2024) Sustainable futures and green new deals. Palgrave Macmillan, pp 1–137

Almroth A, Rehnberg I (2023) Life cycle assessment of an active safety radar system: an environmental assessment based on ISO 14040 and 14044 in conjunction with the GHG Protocol

Arias A, Feijoo G, Moreira MT (2023) Biorefineries as a driver for sustainability: key aspects, actual development and future prospects. J Clean Prod 418:137925

Arodudu O (2021) Elements of holistic sustainability assessments for energy systems. In: Energy systems evaluation (volume 1), sustainability assessment. Springer, Cham, pp 71–106

Arunkumar N, Dharshini V (2025) Decriminalizing environmental offenses: implications of removing penalty provisions in Indian Environmental Law. Law Foyer Int J Doctrinal Legal Res 3:203

Atmaca A (2024) Understanding carbon footprint: impact, assessment, and greenhouse gas emissions. In: Advances and technology development in greenhouse gases: emission, capture and conversion. Elsevier, pp 497–516

Balázs B, Kelemen E, Centofanti T, Vasconcelos MW, Iannetta PP (2021) Policy interventions promoting sustainable food-and feed-systemsfood and feed systems: a Delphi study of legume production and consumption. Sustainability 13(14):7597

Benvenutti LMM, de Souza Campos LM, Vazquez-Brust D, Liston-Heyes C (2023) The changing roles of actors in 'fortuitous' sustainability transitions: an analysis of Brazil's passenger vehicles fuel technology from 1970 to 2020. Technol Forecast Soc Chang 192:122584

Bhargava A (2022) The urgent need for climate-related risk disclosures in India's energy industry. Geo J Int L 54:81

Bhateria R (2024) Environmental management system (EMS). In: Environmental impact assessment: a journey to sustainable development. Springer Nature, Cham, pp 101–124

Bhuiyan MRA (2022) Overcome the future environmental challenges through sustainable and renewable energy resources. Micro Nano Lett 17(14):402–416

Bisht S, Kumar A, Kumar N, Singh H, Kumar P (2021) Biofuel production by using biomass and its application. In: Renewable energy and green technology. CRC Press, pp 85–103

Bock T (2025) Sustainability aspects of powerfuels and their certification. In: Powerfuels. Springer, Cham, pp 977–1023

Brady MA, Sharma S, Baral H, Nasi R (2023) Bioenergy sustainability in the global South: constraints and opportunities. CIFOR

Branco JEH, Prates RC, do Nascimento Oliveira AL, de Carvalho AKF (2025) Evaluation of the greenhouse gas emissions of the Brazilian biodiesel and the impact of the mandatory blending of the biofuel into commercial diesel fuel. Chemosphere 376:144268

Brandão M (2022) Indirect effects negate the global climate change mitigation potential of substituting gasoline with corn ethanol as a transportation fuel in the USA. Front Clim 4:814052

Cantarella H, Leal Silva JF, Nogueira LAH, Maciel Filho R, Rossetto R, Ekbom T et al (2023) Biofuel technologies: lessons learned and pathways to decarbonization. GCB Bioenergy 15(10): 1190–1203

Cao T (2024) China-Latin America green cooperation under the global development initiative. China Int Stud 106:73

Cao H, Zhang L, Qi Y, Yang Z, Li X (2022) Government auditing and environmental governance: evidence from China's auditing system reform. Environ Impact Assess Rev 93:106705

Chandel AK, Arora R, Gupta R, Narisetty V, Kumar V (2024) Global biofuel alliance and roadmap for ethanol blending program in India: analysis and perspective. Sugar Tech 26(4):1005–1015

Chaturvedi V, Ghosh A, Garg A, Avashia V, Vishwanathan SS, Gupta D et al (2024) India's pathway to net zero by 2070: status, challenges, and way forward. Environ Res Lett 19(11): 112501

Cheela VS, Shankar U, Dubey BK (2022) An overview of the municipal solid waste management rules in India. In: Treatment and disposal of solid and hazardous wastes, pp 193–200

Chong CT, Ng JH (2024) Advanced transport biofuels: production, economics, and sustainability

Chun KP, Octavianti T, Dogulu N, Tyralis H, Papacharalampous G, Rowberry R et al (2025) Transforming disaster risk reduction with AI and big data: legal and interdisciplinary perspectives. Wiley Interdiscip Rev Data Mining Knowl Discov 15(2):e70011

Coutinho LC, Império M, Angelkorte G, da Silva GN, Bergman-Fonte C, Draeger R et al (2024) Climate strategies for oil and gas production under the lens of an Integrated Assessment Model: the case of Brazil. Int J Greenhouse Gas Control 137:104231

de Melo CA, da Silva MP, da Silva Benedito R (2021) Renewable energy technologies: patent counts and considerations for energy and climate policy in Brazil. Clim Dev 13(7):630–643

Devi A, Bajar S, Kour H, Kothari R, Pant D, Singh A (2022) Lignocellulosic biomass valorization for bioethanol production: a circular bioeconomy approach. Bioenergy Res 15(4):1820–1841

Duarah P, Haldar D, Patel AK, Dong CD, Singhania RR, Purkait MK (2022) A review on global perspectives of sustainable development in bioenergy generation. Bioresour Technol 348: 126791

Dumortier J, Crespi J, Hayes DJ, Burress M, Valcu-Lisman A, Lewandrowski J (2024) Regional economic aspects of carbon markets and anaerobic digesters in the USA: the case of swine production. Biofuels Bioprod Biorefin 18(5):1238–1253

Fatima N, Khanam N, Kumari R, Joshi VC, Bhattacharya S (2025) Biofuels in India: policies, reviews, and strategic analysis. BioEnergy Res 18(1):1–16

Gade PK (2023) AI-driven blockchain solutions for environmental data integrity and monitoring. NEXG AI Rev Am 4(1):1–16

Ghadami M, Sahebi H, Pishvaee M, Gilani H (2021) A sustainable cross-efficiency DEA model for international MSW-to-biofuel supply chain design. RAIRO-Operations Res 55:S2653–S2675

Ghosh SK, Parlikar UV, Karstensen KH (2021) Waste management rules in India and other countries focusing on co-processing. In: Sustainable management of wastes through co-processing. Springer, Singapore, pp 157–175

Godinho R, Lacerda de Oliveira MI, Rodrigues L, Moreira M, Seabra EA, J. (2022) Carbon credits and the bioethanol industry: governmental programs and incentives. In: Liquid biofuels: bioethanol. Springer, Cham, pp 363–384

Gonçalves TDS, Oro CE, Wancura JH, dos Santos MS, Junges A, Dallago RM, Tres MV (2023) Challenges for energy guidelines in crop-based liquid biofuels development in Brazil. Next Sustain 2:100002

Grangeia C, Santos L, Lazaro LLB (2022) The Brazilian biofuel policy (RenovaBio) and its uncertainties: an assessment of technical, socioeconomic and institutional aspects. Energy Convers Manage X 13:100156
Guerrero AB, Muñoz E (2023) Techno-economic and life cycle analysis of bioethanol production. In: Bioethanol: a green energy substitute for fossil fuels. Springer, Cham, pp 305–337
Hernández AC (2025) Navigating the regulatory landscape: biosafety and environmental assessment of GMOs in Colombia. Indian J Plant Genetic Resour
Islam Z, Wang S (2022) The progress and prospect of environmental impact assessment system in India: from 1994 to 2020 notification. EQA-Int J Environ Qual 50:20–25
Jusakulvijit P, Bezama A, Thrän D (2021) Criteria prioritization for the sustainable development of second-generation bioethanol in Thailand using the Delphi-AHP technique. Energy Sustain Soc 11:1–25
Kara S, Hauschild M, Sutherland J, McAloone T (2022) Closed-loop systems to circular economy: a pathway to environmental sustainability? CIRP Ann 71(2):505–528
Kaur PJ (2023) Biofuel policy in India: prospects and challenges. In: Renewable resources and energy management, pp 235–242
Kaviyarasu K (2024) Performance of strategies implemented as part of a management approach from an ecological standpoint. Discov Environ 2(1):129
Khalid Z, Alam SN, Singh B, Guldhe A (2022) Energy policies in the context of third-generation biofuels. In: 3rd generation biofuels. Woodhead Publishing, pp 731–743
Khan N, Sudhakar K, Mamat R (2021) Role of biofuels in energy transition, green economy, and carbon neutrality. Sustainability 13(22):12374
Khoo HH (2015) Review of bio-conversion pathways of lignocellulose-to-ethanol: sustainability assessment based on land footprint projections. Renew Sust Energ Rev 46:100–119
Kosamia NM, Samavi M, Piok K, Rakshit SK (2022) Perspectives for scale up of biorefineries using biochemical conversion pathways: technology status, techno-economic, and sustainable approaches. Fuel 324:124532
Koval V, Laktionova O, Atstāja D, Grasis J, Lomachynska I, Shchur R (2022) Green financial instruments of cleaner production technologies. Sustainability 14(17):10536
Krarti M (2020) Energy audit of building systems: an engineering approach. CRC Press
Kuila A, Kumar D (eds) (2025) Optimizing biofuel production with artificial intelligence. Wiley
Kulkarni V, Sunkle S, Kholkar D, Roychoudhury S, Kumar R, Raghunandan M (2021) Toward automated regulatory compliance. CSI Trans ICT 9:95–104
Kumar D, Dutt S, Jaiswal AK, Kaundal B, Kumar D, Singh B (2024a) Bioethanol production from potatoes in India: a SWOT analysis. Heliyon 10(23)
Kumar G, Padhy R, Das D, Narayana SA (2024b) Ethanol fuel blending program in India: analysis of environmental, economic, and policy aspects using system dynamics approach. J Adv Transp 2024(1):2002187
Kumbhar S, khot S (2025) Thermal and emission analysis of spark ignition SI engine using premium gasoline-ethanol-paraffin blends with different operating conditions. Sigma 43(2): 369–382
Lasselin C, Barot S, Barberousse A (2025) Value chains and environmental impact assessments: lessons from two French legal cases on bioenergy facilities. Transnl Environ Law 14(1): 121–144
Lei J (2024) Efficient strategies on supply chain network optimization for industrial carbon emission reduction. arXiv preprint arXiv:2404.16863
López I, Goitia-Zabaleta N, Milo A, Gómez-Cornejo J, Aranzabal I, Gaztañaga H, Fernandez E (2024) European energy communities: characteristics, trends, business models and legal framework. Renew Sust Energ Rev 197:114403
Madubuegwu CE, Okechukwu GP, Dominic OE, Nwagbo S, Ibekaku UK (2021) Climate change and challenges of global interventions: a critical analysis of Kyoto protocol and Paris agreement. J Policy Dev Stud 289(8151):1–10

Maia RGT, Bozelli H (2022) The importance of GHG emissions from land use change for biofuels in Brazil: an assessment for current and 2030 scenarios. Resour Conserv Recycl 179:106131

Margaret I, Schoubben F, Verwaal E (2024) When do investors see value in international environmental management certification of multinational corporations? A study of ISO 14001 certification after the Paris Agreement. Glob Strateg J 14(1):25–55

Mather-Gratton ZJ, Larsen S, Bentsen NS (2021) Understanding the sustainability debate on forest biomass for energy in Europe: a discourse analysis. PLoS One 16(2):e0246873

Mayeres I (2025) European Union transition to green energy in the transport sector. In: Fiscal policy to support the green and just energy transition, p 188

Melendez JR, Mátyás B, Hena S, Lowy DA, El Salous A (2022) Perspectives in the production of bioethanol: a review of sustainable methods, technologies, and bioprocesses. Renew Sust Energ Rev 160:112260

Melhim SH, Isaifan RJ (2025) The energy-economy nexus of advanced air pollution control technologies: pathways to sustainable development. Energies 18(9):2378

Metin E (2024) Legal design: a catalyst for environmental social and governance and sustainable business practices. Sci Papers Univ Pardubice Ser D Fac Econ Admin 32(2)

Molina-Reyes S, Cervera-Cárdenas J, Pulido-Rojano A (2022) Implementation of a methodology for the integration of management systems based on NTC-ISO 14001: 2015 and NTC-ISO 45001: 2018: a case study in the construction sector. INGENIARE-Revista Chilena de Ingeniería 30(4)

Mukherjee PK, Das B, Bhardwaj PK, Tampha S, Singh HK, Chanu LD et al (2023) Socio-economic sustainability with circular economy—an alternative approach. Sci Total Environ 904:166630

Murphy DJ, Raugei M, Carbajales-Dale M, Rubio Estrada B (2022) Energy return on investment of major energy carriers: review and harmonization. Sustainability 14(12):7098

Nemecek T, Roesch A, Bystricky M, Jeanneret P, Lansche J, Stüssi M, Gaillard G (2024) Swiss Agricultural Life Cycle Assessment: a method to assess the emissions and environmental impacts of agricultural systems and productsagricultural systems and products' emissions and environmental impacts. Int J Life Cycle Assess 29(3):433–455

Nevskaya MA, Raikhlin SM, Vinogradova VV, Belyaev VV, Khaikin MM (2023) A study of factors affecting national energy efficiency. Energies 16(13):5170

Nomani MZM, Allail J, Rauf M (2021) Environment impact assessment draft notification 2020: an eco-legal assessment. Ann Roman Soc Cell Biol 25(4):2900–2905

Nowicki P, Ćwiklicki M, Kafel P, Wojnarowska M (2021) Credibility of certified environmental management systems: results from focus group interviews. Environ Impact Assess Rev 88: 106556

Oloyede CT, Itabiyi OE, Popoola OA, Jekayinfa SO, Olaniyan MA, Adebayo AO et al (2024) Navigating prospects and challenges for green fuels for achieving economical, environmental, and ecological resilience: a scientific review. Biofuels 15(7):929–941

Pan J (2022) Achievements of human development goals with low emissions. In: Climate change economics: perspectives from China. Springer Nature, Singapore, pp 135–150

Pandey M, Singhal B (2022) Blockchain technology in biomanufacturing: current perspective and future challenges. In: Blockchain technology for emerging applications. Academic, pp 207–237

Papagianni S, Capellán-Pérez I, Adam A, Pastor A (2024) Review and meta-analysis of Energy Return on Investment and environmental indicators of biofuels. Renew Sust Energ Rev 203: 114737

Patel K, Singh SK (2024) Sustainable waste management: a comprehensive life cycle assessment of bioethanol production from agricultural and municipal waste. Environ Sci Pollut Res 31(39): 51431–51446

Penttinen SL (2022) Regulatory and policy instruments to promote decarbonization in the energy sector. In: Research handbook on climate change mitigation law. Edward Elgar Publishing, pp 338–362

Prananta W, Kubiszewski I (2021) Assessment of Indonesia's future renewable energy plan: a meta-analysis of biofuel energy return on investment (eroiEROI). Energies 14(10):2803

Prasad S, Yadav KK, Kumar S, Pandita P, Bhutto JK, Alreshidi MA et al (2024) Review on biofuel production: sustainable development scenario, environment, and climate change perspectives—a sustainable approach. J Environ Chem Eng 12(2):111996

Prasetyaningsih DM, Hendarto E, Anwar N, Elfaki KE (2022) Effectiveness of environmental law implementation: compliance and enforcement. In: Volksgeist: Jurnal Ilmu Hukum Dan Konstitusi, pp 215–225

Rajpurohit NA, Bhakar K, Nemiwal M, Kumar D (2022) Design and synthesis of hybrid nanostructures for sustainable energy and environmental remediation. Arab J Geosci 15(2):137

Ramasubramanian B, Ramakrishna S (2023) What's next for the Sustainable Development Goals? Synergy and trade-offs in affordable and clean energy (SDG 7). Sustain Earth Rev 6(1):17

Rane N, Paramesha M, Rane J (2024) Trustworthy artificial intelligence: enhancing trustworthiness through explainable AI (XAI). Available at SSRN 4873183

Rathi AKA (2021) Handbook of environmental impact assessment: concepts and practice. Cambridge Scholars Publishing

Rathoure AK (2021) Environment impact assessment (EIA) studies for developmental activities in India in context with EIA 2020. Oct J Environ Res 9(1):21–45

Rodrigues L, Gilio L (2023) Brazilian biofuel governance: the case of Brazilian Ethanol and RenovaBio. In: Sustainability challenges of Brazilian agriculture: governance, inclusion, and innovation. Springer, Cham, pp 315–337

Sadhukhan J (2022) Net-zero action recommendations for scope 3 emission mitigation using life cycle assessment. Energies 15(15):5522

Sajid Z, da Silva MAB, Danial SN (2021) Historical analysis of the role of governance systems in the sustainable development of biofuels in Brazil and the United States of America (USA). Sustainability 13(12):6881

Sam AG, Song D (2022) ISO 14001 certification and industrial decarbonization: an empirical study. J Environ Manag 323:116169

Shalu SG (2023) Environmental monitoring with machine learning. EPRA Int J Multidiscip Res (IJMR):208–212

Sharma D (2022) Policy implications and mitigation strategies for air pollution and atmospheric extreme events. In: Extremes in atmospheric processes and phenomenon: assessment, impacts and mitigation, pp 397–410

Sharma N, Priyatharshini S, Kaliappan N, Poornima R, Ramya A, Dhevagi P (2025) Waste Management challenges and potential solutions in the Indian Himalayan region. In: People and mountain environments: interconnectedness for sustainable development in the Himalayas. Springer Nature, Cham, pp 177–213

Sikiru S, Abioye KJ, Adedayo HB, Adebukola SY, Soleimani H, Anar M (2024) Technology projection in biofuel production using agricultural waste materials as a source of energy sustainability: a comprehensive review. Renew Sust Energ Rev 200:114535

Singh S (2021) An analysis of Indian environmental laws for resource sustainability and management. Indian J L Legal Res 3:1

Singh B (2023) Accreditation in the life of general citizen of India. In: Handbook of metrology and applications. Springer Nature, Singapore, pp 71–94

Singh R (2024) Domestically produced bioethanol contributing to India's pursuit of energy self-reliance and de-carbonization of the road transport sector (No. 2024-28-0126). SAE Technical Paper

Singh R, Gadgala N (2022) Environmental regulations in India: efficiency & effectiveness. Int J L Mgmt Hum 5(3):470

Singhal S, Thapar S, Kumar M, Jain S (2022) Impacts of sustainable consumption and production initiatives in energy and waste management sectors: examples from India. Environ Dev Sustain 24(12):14184–14209

Słomka A, Pawłowska M (2024) Catch and cover crops' use in the energy sector via conversion into biogas—potential benefits and disadvantages. Energies 17(3):600

Soleymani Angili T, Grzesik K, Rödl A, Kaltschmitt M (2021) Life cycle assessment of bioethanol production: a review of feedstock, technology and methodology. Energies 14(10):2939

Stillitano T, Falcone G, Iofrida N, Spada E, Gulisano G, De Luca AI (2022) A customized multi-cycle model for measuring the sustainability of circular pathways in agri-food supply chains. Sci Total Environ 844:157229

Szöke JI (2021) ISO 14001 EMS in the renewable energy sector: practices and perceptions in South Africa. Doctoral dissertation, North-West University, South Africa

Thoti KK, Pasha MA, Shankar CU, Yaziz MFA, Uday M (2024) An analysis examining the way supply chain management-focused third-party environmental audits contributed to business sustainability efforts. Econ Sci 20(2):177–189

Tiburcio RS, de Macêdo TR, Neto AMP (2023) Brazilian Biofuels Policy (RenovaBio): overview and generation of decarbonization credits by biodiesel production facilities. Energy Sustain Dev 77:101334

Tiwary AK, Singh PK, Tiwari AK, Kumar R (2022) Different energy management strategies for clean energy. In: Sustainable and clean energy production technologies, pp 29–49

Tripathi M, Bhattacharya NS (2023) Environmentally sound technologies for climate change mitigation in BRICS countries: a comparative policy and legal perspective. Environ Policy Law 53(2–3):181–196

Umakanth AV, Datta A, Reddy BS, Bardhan S (2022) Biomass feedstocks for advanced biofuels: sustainability and supply chain management. In: Advanced biofuel technologies, pp 39–72

Usmani RA, Mohammad AS, Ansari SS (2023) Comprehensive biofuel policy analysis framework: a novel approach evaluating the policy influences. Renew Sust Energ Rev 183:113403

Veluru CS (2022) Enhancing legal practice: the transformative impact and challenges of AI-powered solutions in the legal industry. J Sci Eng Res 9(6):94–102

Verma J, Shah YT, Sundaramurthy S, Arisutha S (2024) Insights into next-generation biofuel strategies and policy considerations in India. In: Clean energy transition-via-biomass resource utilization: a way to mitigate climate change, pp 251–310

Wessels J (2024) Evaluating the contribution of ISO 14001: 2015 towards enhancement of environmental performance: a mining case study. Doctoral dissertation, North-West University, South Africa

Wörner D, Letmathe N, Jovanovic M, Friedli T (2025) Structuring private sustainability governance: combining rule-based and goal-based mechanisms. J Bus Ethics:1–18

Yin T, Huhe T, Li X, Wang Q, Lei T, Zhou Z (2024) Research on life cycle assessment and performance comparison of bioethanol production from various biomass feedstocks. Sustainability 16(5):1788

Zhang J, Rentizelas A, Zhang X, Li J (2022) Sustainable production of lignocellulosic bioethanol towards zero waste biorefinery. Sustain Energy Technol Assess 53:102627

Zhao X, Mahendru M, Ma X, Rao A, Shang Y (2022) Impacts of environmental regulations on green economic growth in China: new guidelines regarding renewable energy and energy efficiency. Renew Energy 187:728–742

Zhao W, Li H, Wang S (2023) Energy performance and energy conservation technologies for high-tech cleanrooms: state of the art and future perspectives. Renew Sust Energ Rev 183:113532

Chapter 16
National and International Biofuel Regulatory Frameworks

Pardeep Yadav, Shanuja Beri, and Saurabh Kumar Jha

Abstract Biofuel regulatory frameworks at both national and international levels aim to promote sustainable fuel alternatives, reduce greenhouse gas emissions, and ensure energy security. These frameworks also govern the production, distribution, and sustainability standards of biofuels. These collectively support the transition to cleaner energy while addressing food security, land use, and economic feasibility. Most countries, including India, have their own set of regulations, directives, and policies to regulate the use of biofuels. Biofuel regulatory frameworks are essential at both national and international levels to guide the sustainable growth of the biofuel sector. These regulatory charters establish standards to prevent environmental degradation, reduce greenhouse gas emissions, and promote the use of domestic renewable resources, thereby reducing dependence on fossil fuel imports and enhancing energy resilience. They monitor land use for biofuels, taking into account food security, and aim to attract investment, support innovation, and create a stable environment for industry growth. As different nations move toward adopting biofuels, these regulations and international standards help align sustainability criteria across countries, enabling fair trade and cooperation in biofuel development. These countries recognize the use of lignocellulosic biomass for generating biofuels, and regulatory frameworks are evolving globally to support their development due to their low carbon footprint and minimal impact on food security. The regulations implemented worldwide are crucial to ensuring compliance and maintaining a balance between food and fuel production. These frameworks ensure that biofuels are a clean, responsible, and viable part of the global energy future.

Keywords Frameworks · Compliance · Certifications · Policies · Lignocellulosic biomass · Clean energy

---

P. Yadav
Department of Biotechnology, School of Applied and Life Sciences, Uttaranchal University, Dehradun, Uttarakhand, India

S. Beri (✉) · S. K. Jha
Department of Zoology, Kalindi College, University of Delhi, New Delhi, Delhi, India
e-mail: shanuja@kalindi.du.ac.in

S. Prasad et al. (eds.), *Sustainable Bioethanol Production from Lignocellulosic Biomass*, Clean Energy Production Technologies,
https://doi.org/10.1007/978-981-95-9690-4_16

16.1 Introduction

Green fuels, also referred to as biofuels, are clean fuels. Fuel is made up of substances that release heat when they react with oxygen, or during the combustion process. Because they come from biological or green sources, green fuels are less harmful to the environment and are both bio-renewable and environmentally benign. The renewable resource materials (biomass) generated on agricultural land and in natural aquatic systems, as the main by-products of photosynthesis, are used to make biofuels (Pramanik et al. 2023). Over time, biofuels have become a more appealing alternative to fossil fuels since they reduce greenhouse gas emissions, significantly improve energy security, and create job opportunities. Biochemical or thermochemical conversion processes are used to turn biomass into biofuels. Anaerobic digestion of biomass yields biogas, whereas alcoholic fermentation of biomass yields liquid fuels (Pattanaik and Misra 2017).

The different types of biofuels are generally classified as first-generation, second-generation, and third-generation biofuels, depending on the type of feedstock used and the technology utilized to produce them. The well-established first-generation biofuels include, among others, ethanol produced from sugarcane and biodiesel produced from vegetable oils, and are popular since they utilize well-developed technology, accompanied by concerns associated with food security and land use (Review of Policy Frameworks to Promote Sustainable Biofuels with Low GHG Emissions—Bioenergy n.d.). Second-generation biofuels are made using non-food lignocellulosic materials, such as agricultural waste, forest waste, and municipal solid waste, and are more sustainable, with reduced greenhouse gas emissions (Sulis et al. 2025). Third-generation biofuels, which are predominantly algae-based, are still in the development stage and have high potential output in terms of yield, requiring tiny land and freshwater (Raman et al. 2025). These types of biofuels have a food-based to waste-based continuum, with an unstable measure of benefit to the industry. To be viable in the long run, low-carbon and non-food-based sources must be pursued.

Regulatory frameworks are foundational to the structured and sustainable development of the biofuel sector. In the absence of coherent policies, biofuel initiatives risk becoming economically unviable, environmentally harmful, or socially inequitable (Usmani et al. 2023). Well-designed regulations provide clarity and stability for investors, industry stakeholders, and farmers by establishing clear mandates (e.g., blending targets), specifying eligible feedstocks, setting quality standards, and establishing long-term procurement mechanisms. These frameworks also enable the alignment of biofuel development with national energy goals, climate change mitigation commitments, and rural development objectives. Furthermore, regulatory systems mitigate unintended consequences—such as deforestation, land-use conflicts, and food insecurity—by enforcing sustainability criteria, lifecycle greenhouse gas accounting, and monitoring mechanisms. They play a crucial role in incentivizing innovation, particularly for second- and third-generation biofuels that require high initial investment but offer greater ecological benefits. At the international level, regulatory harmonization is crucial to prevent trade disputes and ensure compliance with WTO norms (Shah 2023).

Ultimately, robust regulatory frameworks ensure that biofuel development is not only technologically and economically feasible but also environmentally responsible and socially inclusive, transforming biofuels from a transitional energy option into a long-term pillar of global energy strategy (Araújo et al. 2017). The past 5 years have seen a 6% annual increase in biofuels consumption, excluding the COVID-19-related decrease, with a record 170 billion liters (4.3 EJ) used in transport in 2022, representing more than 3.5% of energy demand in transport. The single most important biofuel continues to be ethanol (106 one-billion liters), followed by biodiesel (45 one-billion liters), HVO/renewable diesel (13 one-billion liters), and a very small amount of biojet fuel (approximately 0.2 one-billion liters). Globally, the United States is the largest producer regionally (70.7 billion L in 2022). Brazil and Indonesia rank second and third, respectively, with 37.5 billion L and 11.1 billion L (Transport Biofuels—Renewables 2023—Analysis—IEA n.d.). A 2023 Indian target includes E12 by 2025. Indonesia has moved to B35, and B40 is planned. Brazil is experimenting with E30, and EU member nations have adopted E10 in 19 countries (Mandatory B40 Biodiesel to Be Implemented in March 2025 n.d.).

More promising are advanced biofuels, mainly sustainable aviation fuel (SAF) which, although still small (<0.3% of jet fuel), is growing and regulated by EU mandates (2% SAF by 2025) and industry investment, with oil majors, planning to roll out ~43 SAF projects and ~286,000 barrels/day capacity by 2030 (Qasem et al. 2024). In 2023, renewable energy sources accounted for 30% of the world's electricity, 10% of its heat, and 3.5% of its total energy needs. In 2023, the share of renewable energy sources in global electricity generation was 10% of total heating demand, as well as 3.5% of the total volume of fuel resources. The global energy sector will need to increase its renewable capacity by almost 1 TW every year to hit the global target of 11 terawatts (TW) of installed renewable capacity by 2030 (Electricity—Renewables 2023—Analysis—IEA n.d.). Renewable energy and enabling technologies have been one of the major investment areas in 2023, with investment rising by 70% over the past years. Other enabling technologies, such as solar photovoltaics (PV) and battery storage, have been the most popular. Moreover, 24 countries updated and raised their national ambitions for the share of renewables in total electricity production, aligning with a broader policy transition toward decarbonization and energy security (Hassan et al. 2024).

16.2 Objectives of Biofuel Regulations

Biofuel regulations are policy guidelines that focus on the production, distribution, and use of biofuels in a sustainable and environmentally friendly manner. The primary targets of these regulations are to reduce greenhouse gas emissions, enhance energy security by decreasing dependence on fossil fuels, and promote the use of renewable resources (National Policy On Biofuels (2018)—IMPRI Impact And Policy Research Institute n.d.). The biofuel regulations are essential in transforming the world toward cleaner and more resilient energy systems, as they set standards for carbon emissions. Intensity and lifecycle emissions, as well as feedstock sourcing (Monti and Polugodina 2021).

A study conducted by Yilmaz (2025) found that government environmental protection expenditures (EPEs) were significantly and negatively associated with a country's GHG emissions (Khalil and Yilmaz 2025). In a study examining 117 countries between 2000 and 2023, EPEs, which incorporate pollution abatement, biodiversity conservation, waste management, and environmental R&D, are effective in reducing emissions through direct pollution abatement, composition effects, and technique effects. These expenditures align with international commitments and agreements, such as the Paris Agreement and the UN Sustainable Development Goals, which aim for more transparent air quality, a healthier ecosystem, and climate resilience (Khalil and Yilmaz 2025). A recent international analysis of 80 countries over 50 years has determined that renewable energy targets (RETs), as opposed to mere strategic plans, are necessary to achieve significant reductions in GHG emissions. Although Renewable Energy Strategic Plans (RESPs) can be used to facilitate energy transitions, the study revealed that they do not play a significant role in transitions except when there are enforceable targets (Gao et al. 2025). RETs, on the other hand, are much more effective in reducing emissions, as they are directly in control of the energy-industrial composition of a particular country.

To be more specific, RETs helped decrease the amount of fossil fuel used, increase the ratio of renewable sources in an energy mix, and improve the industrial structure (Wang et al. 2023). The reductions in emissions were primarily attributed to these three mediating factors. By relying on advanced analytical approaches, including multi-period difference-in-differences and double machine learning, the study provides strong evidence that policy design matters: binding targets induce behavioral changes, but plans do not. These results have significant policy implications for international climate policy, one of which is that to be effective in reducing emissions, renewable energy policies must also have quantifiable and credible targets, rather than aspirational goals (Gao et al. 2025). A study by Song et al. examines the potential of applying flexible carbon source control in wastewater treatment plants (WWTPs) to achieve carbon neutrality by 2060. Three strategies were considered: anaerobic digestion (AD), short fermentation (SF), and adaptive carbon source regulation (ACSR). Findings indicated that AD was most effective in conditions with high carbon-to-nitrogen ratios, where a 29.2% reduction in emissions was observed. SF was the most effective at the lowest C/N ratios, achieving a 56.3% reduction in N_2O emissions. The difference in season and C/N ratios had a significant influence on the emissions (Song et al. 2025).

One of the primary objectives of biofuel regulations is to stimulate economic growth and create job opportunities, particularly in rural areas. These policies generate value-added prospects in the agricultural supply chain by promoting the local production of biofuels. Biorefineries have a significant impact on boosting the local economy while also providing and maintaining employment in both the service and agricultural sectors. They also help in rural development, which is beneficial to the economy. The study makes a case for reviewing biofuel policies and focusing on the dimension of farming in rural areas. Economic and employment effects of biorefineries on a long-term basis should be addressed by policymakers in the EU and the United States when they are formulating policies and support systems for the biofuel industry (Heijman et al. 2019). Equally, no less than 20 times the job per unit of energy generated through ethanol and biodiesel productions have been identified to be created than fossil fuels in Thailand, of which the majority of the employment opportunities are defined and offered by the agricultural waged work in various environments (Silalertruksa et al. 2012).

Rural development in terms of sustainable agricultural growth and creation of employment opportunities also emerged as an important policy of biofuel in India (Chandel et al. 2017). Such results support the fact that not only can biofuel policies be viewed as means of energy transition, but also as effective instruments of rural rejuvenation, income diversification, and socio-economic inclusion. With changes in national and international policies regarding biofuels, there has been significant concern about sustainable land use and agriculture. As the world increases its biofuel production, concerns about environmental deterioration, food security, and land-use change are becoming significant issues. Regulatory frameworks are also being established to mitigate greenhouse gases while promoting ecosystem preservation and sustainable land management. An example is the voluntary efforts, such as the Roundtable on Sustainable Biomaterials and international sustainability systems, like the EU Renewable Energy Directive, mandate the strictest sustainability standards to encourage the production on degraded land, the elimination of deforestation, and ecological integrity in cultivating biofuel feedstock (Guariguata et al. 2011). The same is emphasized in a lifetime emissions approach, high-carbon stock protected area, and sustainability certification in national policies, including the U.S. Renewable Fuel Standard and Brazil. Inequity in surveillance and enforcement, however, persists as a critical barrier, particularly in biodiverse regions such as Indonesia. Ere commercial pressures and poor control confuse land-use control (S 2024). To align biofuel production with long-term, sustainable, and rural development goals, a consistent global standard that is enabled by transparent metrics and international cooperation is mandated.

16.3 National Biofuel Regulatory Frameworks

National biofuel regulations are planned systems of policies, rules, and institutional methods that govern the production, distribution, mixing process, and sustainability of biofuels in such countries. These frameworks promote renewable energy, reduce dependence on fossil fuel, enhance energy security, and reduce greenhouse gas emissions (Das 2020). Typically, they encompass blending goals (such as the proportion of ethanol or biodiesel in conventional fuels), feedstock caps, sustainability standards, economic incentives, and support for infrastructure and research development. These rules support the establishment of national regulations that integrate the national energy strategy and climate objectives with the rural development agenda, while also facilitating and fostering investment in the biofuel industry. They help ensure that the development of biofuels does not compromise biodiversity, food security, and the integrity of land use. Such initiatives include the Renewable Energy Directive of the European Union, the Renewable Fuel Standard in the United States, and the Ethanol Blending Program in India.

16.3.1 India National Bio-energy Policy

The 2018 National Policy on Biofuels of the Government of India, now updated in 2022, provides a comprehensive roadmap for the systematized promotion of biofuels as a method of transitioning toward clean energy. 1G biofuels are created using food-based targets such as sugarcane; 2G biofuels are created using cellulosic and lignocellulosic biomass such as agricultural residues; and advanced biofuels are created using algae, spent cooking oil, and industrial wastes (Singh 2024). This grouping of biofuels based on the type of their feedstock was one of the significant aspects of the 2018 policy. This division enabled policymakers to differentiate among the technological readiness, environmental impact, and geographical capability of various biofuel-based sources. The 2022 amendment, which pushed India's blending goal of 20% ethanol in gasoline forward to 2025–2026, was another important policy shift. This shift indicates an acceleration in India's efforts to reduce its dependence on fossil fuels and achieve its net-zero emissions goal by 2070 (Government Measures to Increase Ethanol Blending beyond 20% | IBEF n.d.). Various types of unacceptable feedstock items, such as damaged food grains, maize, additional rice purchased by the Food Corporation of India (FCI), bamboo, and other agricultural waste products, were added to the list of acceptable feedstocks to provide this large window (Singh 2024). This variety will permit a stronger and more decentralized supply chain of biofuel, as well as address the seasonal and regional variability of sugarcane production. A comparative table of five bio-energy policies focusing on their scope, feedstock strategy, blending, production targets, key incentives, and the principal challenges is summarized in Table 16.1.

One of the key features of the revised policy is its focus on diversifying feedstocks. According to Parvez et al. (2025), the inclusion of non-edible oilseeds, aquatic biomass, and industrial by-products as feedstocks helps alleviate pressure on food crops and enhances energy security. By integrating a wider range of raw materials, the policy supports both ecological sustainability and the development of rural livelihoods. This approach also encourages the use of underutilized or waste resources, reducing environmental pollution from crop residue burning and promoting circular economy practices. The strategy enables an increase in the growth of rural living conditions and ecological sustainability by applying a wider range of raw materials (Chenavaz and Dimitrov 2024). Moreover, this strategy will enhance the practices of the circular economy and reduce the adverse environmental effects that

Table 16.1 Comparative table of five bio-energy policies focusing on their scope, feedstock strategy, blending, production targets, key incentives, and the principal challenges they face

#	Policy/initiative	Year/latest update	Core objective(s)	Main feedstock(s)	Target/mandate	Principal incentives/ support mechanisms	Notable strengths	Main challenges/ risks
1	**India—National Bio-Energy (Biofuels) Policy** (original 2018, amended 2022)	2022 amendment (blending goal moved to 2025–26)	Build a diversified, sustainable biofuel sector Reduce fossil-fuel dependence and move toward net-zero by 2070	1G: food-based (sugarcane, corn) 2G: cellulosic/agricultural residues (bagasse, rice husk, wheat straw) Advanced: algae, waste oil, industrial waste Expanded list now includes non-edible oilseeds, aquatic biomass, damaged grains, bamboo, etc.	20% ethanol blending in gasoline by FY 2025–26 (up from earlier target)	Soft-loan & viability-gap financing for 2G refineries Pradhan Mantri JI-VAN Yojana (financial support for 2G ethanol projects up to 2028–29) Fiscal incentives, tax breaks, and subsidies for biorefinery capex	Clear feedstock categorization enables tech-readiness assessment Strong emphasis on circular-economy waste streams Rural livelihood boost via diversified feedstock use	Higher capital intensity of 2G plants Seasonal/regional variability of sugarcane vs. 2G feedstock logistics Potential competition for land/resources if feedstock expansion is not carefully managed
2	**India—Ethanol Blending Programme (EBP)**	Launched 2003; target 20% ethanol in gasoline by 2025	Displace imported crude oil, cut GHG emissions, improve energy security	Primarily sugarcane-derived ethanol (1G) Growing share of 2G/waste-	20% ethanol in gasoline by 2025 (interim 18.4% reached early-2025) Production capacity ↑ from ~38 cr L	Public-private partnerships, OMC-facilitated procurement Financial incentives for	Massive domestic ethanol expansion (≈21× growth) Significant foreign-	Supply-demand imbalances (seasonality, feedstock constraints) Food-security concerns (use of edible

(continued)

Table 16.1 (continued)

#	Policy/initiative	Year/latest update	Core objective(s)	Main feedstock(s)	Target/mandate	Principal incentives/ support mechanisms	Notable strengths	Main challenges/ risks
				derived ethanol under newer guidelines	(2013–14) → 707 cr L (2023–24)	distilleries (soft loans, subsidies) Market guarantees via OMC offtake contracts	exchange savings (>₹1 lakh crore) Farmer income stabilization via guaranteed demand	crops) Water-intensity & land-use debates; need for 2G transition
3	**India—Sustainable Alternative Towards Affordable Transportation (SATAT)**	Launched Oct 2018 (ongoing)	Deploy compressed biogas (CBG) as a clean transport & industrial fuel Promote waste-to-energy, rural entrepreneurship, and circular-economy	Agricultural residues, cattle dung, municipal solid waste (organic)	5000 CBG plants nationwide 15 MMT of CBG production per year (target)	Capital subsidies, technical assistance from MNRE Long-term offtake agreements with Indian Oil, BPCL, HPCL PPP facilitation, fast-track regulatory clearances	Integrated policy framework linking waste management, rural income, and clean fuel Strong offtake guarantees create market certainty	Slow rollout (only ~38 plants operational by end-2022) High upfront CAPEX, long pay-back, limited concessional finance Offtake price sometimes insufficient for small producers
4	**USA—Renewable Fuel Standard (RFS)**	Enacted 2005; major expansion 2007 (EISA)	Increase renewable fuel share in transportation fuel Reduce GHG emissions, support domestic agriculture,	Conventional: corn-starch ethanol (max 15 bn gal/year) Advanced: ≥50% GHG reduction (cellulosic,	Annual Renewable Volume Obligations (RVOs) set by EPA (vary by fuel category) Blend wall: typical market limit ≈E10 (10% ethanol)	Renewable Identification Numbers (RINs) tradable credits Small refinery exemptions (SREs) for hardship cases	Established market mechanism (RIN trading) drives investment Provides stable demand for corn	Blend-wall constraints limit higher ethanol blends (E15/E85) Volatile RIN prices create financial risk, especially for small refiners

			enhance energy security	biomass-derived diesel, etc.) Cellulosic: $\geq$60% GHG reduction		Tax incentives for advanced-biofuel facilities (e.g., §45V)	& other feedstocks	Environmental critiques of corn-ethanol land-use change Ongoing legal/policy uncertainty around SREs
5	**EU—Renewable Energy Directive II (RED II)**	Adopted 2018 (implementation 2021–2023)	Achieve EU-wide renewable energy targets, including biofuels Ensure sustainability, avoid indirect land-use change, protect biodiversity	Mandatory sustainability criteria for all imported & domestic bio-fuels (GHG reduction, land-use thresholds) Encourages advanced bio-fuels (non-food feedstocks)	Overall renewable share of transport fuel $\geq$14% by 2030 (incl. 10% renewable electricity) Specific minimum shares for advanced bio-fuels (e.g., 3.5% advanced bio-fuels)	Eligibility for EU funding programmes (e.g., Horizon Europe, LIFE) Tax incentives for compliant biofuel production Certification schemes aligned with ISO 13065 (RSB, ISCC)	Strong sustainability safeguards (ILUC, biodiversity) Harmonized market across EU member states	Stringent certification can be costly for emerging producers Potential trade-compatibility issues under WTO (non-discrimination) Limited domestic feedstock availability for advanced biofuels in some regions

waste crops can produce through burning. The program offers financial support in the form of soft loans, viability gap financing, and other fiscal incentives to promote investment and technical innovations, thereby facilitating the construction and operation of 2G biorefineries. Such biorefineries must be supported by the government to become commercially viable, particularly during the early development phases of increasing their operations, and they require more capital compared to 1G plants.

To facilitate the development and operation of 2G biorefineries, the program provides financial incentives, including viability gap funding, soft loans, and other forms of fiscal incentives, to ease access to finance and encourage investment and technical breakthroughs (Das 2020). These biorefineries require government assistance to become commercially viable, particularly in the initial phases of scaling up their operations, as well as in addressing additional capital requirements compared to 1G plants. The Pradhan Mantri JI-VAN Yojana is crucial to the government's plan for producing 2G ethanol. This program primarily advances the commercialization of lignocellulosic biofuel technology by providing financial support for project setup until 2028–2029. The consumption of lignocellulosic biomass, which includes bagasse, rice husk, and wheat straw, ensures that the food chain will not compete with the production of biofuels. It also helps reduce environmental issues, such as stubble burning, which has been a significant contributor to air pollution in North India during the post-harvest seasons. In summary, the National Policy on Biofuels (2018, amended 2022) marks a strategic evolution in India's renewable energy agenda. By categorizing feedstocks, accelerating blending targets, diversifying input sources, and supporting advanced technologies like 2G biofuels, the policy lays a strong foundation for an inclusive, sustainable, and economically viable bioenergy ecosystem. These developments align with India's broader environmental and developmental priorities, including energy security, rural development, and climate change mitigation. A quantitative snapshot of various policies and programmes is summarized in Table 16.2.

16.3.2 India: Ethanol Blending Program (EBP)

The transition to blending ethanol in India is imperative, and one of the most significant aspects of this transition is the initiation of the Ethanol Blending Programme (EBP) in 2003. The EBP was formulated as a long-term mechanism to reduce carbon emissions, decrease the country's dependency on imported crude oil, and encourage the transportation industry to use renewable biofuels. India is also looking forward to gradually substituting part of its use of fossil fuels with something cleaner by adding ethanol. This alcohol is produced domestically by plants into a gasoline-like substance. The Indian government now has an ambitious target of increasing the ethanol content in gasoline to 20% by 2025. This decreased goal of achieving energy security by 2030 was shifted to 2025 as part of grander plans to achieve both environmental sustainability and energy security.

Table 16.2 A quantitative snapshot of various policies and programmes

Policy/ Program	Baseline year (reference)	Target year(s)	Assumed fuel-blend/ Production level	Calculated CO_2-e reduction[a]	Relative size of impact	Key driver of the reduction	Impact
India—National Bio-Energy Policy (2022 amendment)	FY 2020 (≈7% ethanol blend)	FY 2025–26 (20% blend)	20% ethanol in gasoline; 2G ethanol share ≈ 30% of total ethanol	**≈0.45 Mt CO_2-e/year** (≈0.35 Mt from 1G ethanol + 0.10 Mt from 2G)	Similar magnitude as second largest (≈0.4–0.45 Mt each)	Aggressive 20% blend target combined with expanding domestic ethanol capacity	**Environmental:** Lower GHG emissions, lower land-use change risk. **Economic:** Savings on oil imports; new investment in 2G plants; modest increase in rural incomes. **Social:** Job creation, skill development in rural processing hubs; reduced pressure on food crops
India—Ethanol Blending Programme (EBP)	FY 2015 (≈5% blend)	FY 2025–26 (20% blend)	20% ethanol blend; domestic ethanol production ≈ 707 cr L/year	**≈0.40 Mt CO_2-e/year** (life-cycle GHG saving of 0.57 kg CO_2-e/L for ethanol vs gasoline)	Similar magnitude as second largest (≈0.4–0.45 Mt each)	Aggressive 20% blend target combined with expanding domestic ethanol capacity	**Environmental:** Significant CO_2e reduction; lower fossil-fuel share. **Economic:** Large foreign-exchange savings; higher farmgate prices; new downstream ethanol market. **Social:** More stable incomes for sugarcane growers; rural

(continued)

Table 16.2 (continued)

Policy/ Program	Baseline year (reference)	Target year(s)	Assumed fuel-blend/ Production level	Calculated CO_2-e reduction[a]	Relative size of impact	Key driver of the reduction	Impact
							employment in ethanol logistics
India—SATAT (CBG)	FY 2020 (≈0.2 Mt CBG)	FY 2030 (15 Mt CBG)	15 Mt CBG/year replacing diesel (GHG factor ≈ 73 kg CO_2-e/MJ)	**≈0.22 Mt CO_2-e/year** (assuming 70% diesel displacement, 0.34 kg CO_2-e/MJ saved)	Smaller (≈0.22 Mt)	CBG replaces diesel but the absolute volume is still modest; however, it adds a waste-to-energy dimension	**Environmental:** Lower GHG emissions; reduced particulate matter from residue burning. **Economic:** New market for waste streams; modest reduction in LPG import bill; financing gaps remain a barrier. **Social:** Rural entrepreneurship, improved livelihoods, ancillary income from manure sales
USA—Renewable Fuel Standard (RFS)	2015 (RVO ≈ 13 bn gal)	2022 (RVO ≈ 15 bn gal)	15 bn gal conventional ethanol (GHG saving ≈ 0.45 kg CO_2-e/L) + 0.5 bn gal advanced biofuel (saving ≈ 0.85 kg CO_2-e/L)	**≈0.55 Mt CO_2-e/year** (≈0.44 Mt from conventional ethanol + 0.11 Mt from advanced fuels)	Second-largest (≈0.55 Mt)	Large volume of conventional ethanol plus a growing share of advanced fuels	**Environmental:** Measurable CO_2e reductions; however, life-cycle benefits of corn ethanol debated due to land-use effects. **Economic:** Strong demand for corn-based ethanol

							supports U.-S. agriculture; RIN market creates revenue streams but can be volatile. **Social:** Rural job stability in ethanol plants; criticism over food-vs-fuel competition persists
EU—RED II	2015 (≈8% renewable transport fuel)	2030 (≥14% renewable, with ≥50% GHG reduction)	14% renewable fuel mix (≈12% ethanol, 2% biodiesel/other)	**≈0.70 Mt CO_2-e/year** (average saving ≈ 0.55 kg CO_2-e/L for ethanol + 0.80 kg CO_2-e/L for biodiesel)	Largest absolute reduction (≈0.7 Mt)	Higher renewable-fuel share (≥14%) and inclusion of biodiesel (higher per-unit GHG saving)	**Environmental:** Substantial CO_2e cuts; safeguards against deforestation and ILUC. **Economic:** Creates a regulated market for certified biofuels; stimulates advanced-biofuel R&D and plant construction. **Social:** Supports rural economies in Member States with feedstock potential; certification costs can burden small producers

(continued)

Table 16.2 (continued)

Policy/ Program	Baseline year (reference)	Target year(s)	Assumed fuel-blend/ Production level	Calculated CO_2-e reduction[a]	Relative size of impact	Key driver of the reduction	Impact
Global—ISO 13065-aligned certification (used in India & EU)	N/A	N/A	Not a production target, but a *quality filter* that raises average GHG savings by ~10% relative to uncertified biofuels	**≈+0.05 Mt CO_2-e/ year** (incremental benefit when applied to the volumes above)	Incremental (≈0.05 Mt)	Improves average GHG performance across all biofuel streams	**Environmental:** Manages GHG emissions through life-cycle quantification and reporting, avoids sourcing biomass from high-biodiversity or high-carbon areas (e.g., primary forests, wetlands), and promotes sustainable use of soil, water, & air resources. **Economic:** harmonizes markets with a common sustainability language, improves comparability and transparency for buyers, and encourages investment in sustainable technologies by providing a clear way to demonstrate compliance with bioenergy targets.

							Social: Incorporates international labor and human rights principles to prevent child and forced labor and ensure safe working conditions, while also supporting local communities, rural energy access, and land and food security

All figures are drawn from the latest publicly-available reports (government white papers, agency assessments, and peer-reviewed studies up to April 2024). Where ranges are given, the most recent midpoint is shown

[a]Methodology (simplified): 1. Convert volume to energy (L → MJ) using standard fuel heating values. 2. Apply lifecycle GHG saving factors from the IPCC Tier2 methodology and region specific LCA studies (e.g., IEA 2022 for Indian ethanol, EPA 2021 for U.S. corn ethanol, EU 2023 for biodiesel). 3. Multiply by the proportion of the fuel blend that is biofuel. 4. The numbers are rounded to two significant digits; actual realized reductions will vary with feedstock mix, land use change accounting, and operational efficiencies

By the beginning of 2025, the interim proportions of blending rose to 18.4% in the country, an impressive change compared to the single-digit proportions that were the trademark of the initial years of the program (Ye et al. 2023). One of the most remarkable achievements of the EBP to date is its exponential increase in domestic ethanol production. Currently, the amount of ethanol produced has grown by 21 times, with the production capacity surpassing 38 crore liters in 2013–2014 to 707 crore liters in 2023–2024 (Singh 2024). The primary cause of this growth was the proactive government policies that facilitated diversity in feedstocks, the expansion of ethanol distillation capacity, and easy access to petroleum products through the simplification of the procurement process by using oil marketing companies (OMCs). The production boom, along with the employment of new biorefineries in grain-producing and sugarcane regions, facilitated the expansion of the industry, contributing to the achievement of higher blending targets. All the above endeavors are highly relevant in enhancing output production and supply chain efficiency through cooperative ventures, public-private partnerships, and the involvement of the private sector.

The EBP has made a significant contribution to the country's energy security by substituting the import costs of crude oil with ethanol produced in India, saving the country more than 1 lakh crores in foreign exchange. As a low-carbon fuel, ethanol will help India reduce greenhouse gas emissions, and it is a major contributor to the nation achieving its climate mitigation goals and obligations as stipulated under the Paris Agreement (Ethanol Blending Programme (EBP), Objectives, Achievements n.d.). The farmer-oriented modality of the programme has also provided farmers with a stable market, which has stabilized prices, thereby increasing income levels. This aligns well with national objectives of doubling farmers' incomes and promoting rural economic growth. Nevertheless, the EBP is characterized by the negative issues of significant supply-demand imbalance, food security and land use issues, and water intensity and ecological effects. Since energy security, food security, and environmental sustainability have numerous trade-offs to manage, India must address them to achieve its long-term prosperity. The increase in ethanol production should also be accompanied by investment in research, infrastructure, and eco-friendly, sustainable agricultural production methods. Additionally, future policies should encourage the use of 2G biofuels and waste-to-energy products, thereby reducing reliance on conventional food crops. As India progresses toward achieving its net-zero emission goal by 2070, the Ethanol Blending Programme will continue to play a crucial role in the green energy transformation, as long as it continues to innovate within the scientific, institutional, and policy-balanced ethos.

16.3.3 India: Sustainable Alternative Towards Affordable Transportation Initiative

The Sustainable Alternative Towards Affordable Transportation (SATAT) initiative, **launched** by the Government of India in October 2018, is a proactive **step** to unlock renewable energy, creating smoother transportation and energy generation in decentralized centers. The ultimate aim of SATAT is to increase the business production of Compressed Biogas (CBG) and its utilization in transportation and industry as a source of renewable energy and an environmentally friendly alternative to Compressed Natural Gas (CNG). The program is set to establish 5000 CBG plants nationwide, with a target of producing 15 million metric tons (MMT) of CBG annually. By doing that, SATAT attempts to endorse energy innovation with sustainability. Enable development, rural upliftment, and environmental protection (SATAT Scheme (Sustainable Alternative Towards Affordable Transportation)| National Portal of India n.d.).

This is one of the strengths of the SATAT initiative, as it features a well-integrated policy mechanism. There is an incentive to utilize organic feedstocks located nearby, such as agricultural waste, cattle dung, and municipal solid waste, which can be converted into a form of energy. This not only facilitates the establishment of new sources of clean fuel but also helps to manage waste efficiently, especially in rural and semi-urban regions. In policy support regard, the programme has the advantage of fostering distinct cooperation among various government agencies and private parties (Koirala et al. 2016). Remarkably, Indian Oil, Bharat Petroleum, and Hindustan Petroleum, Oil Marketing Companies (OMCs), have entered into long-term offtake agreements to take CBG. Such agreements also lead to some form of market certainty, which is essential in stimulating investment in the biogas infrastructure. To enhance the initiative, the Ministry of New and Renewable Energy (MNRE) provides subsidies on capital, offers technical advice, and facilitates the creation of public-private partnerships (PPPs). The MNRE also cooperates with the Ministry of Petroleum and Natural Gas (MoPNG) as well as the Department of Animal Husbandry to fast-track regulatory permissions, establish feedback supply chains, and implement positive aspects, such as financial incentives for investors and developers.

These institutional tools will help ensure that the CBG ecosystem is viable, scalable, and competitive to both large industries and rural entrepreneurs. One more pioneering feature of SATAT is its role in promoting a circular economy model. The initiative establishes a relationship between the recovery of waste and its economic benefits through rural activities, such as converting organic waste into energy and producing fermented organic manure, which can be utilized as by-products. Small-scale suppliers can import feedstock or use local plants, thereby becoming a source of new revenue through investments in clean energy initiatives. This hybrid model addresses some of the nationwide concerns, including energy access, waste disposal, employment creation, and environmental sustainability, all of which are addressed structurally. Although SATAT is a forward-thinking idea that policies have supported, its implementation has not been easy. By the end of 2022, only 38 CBG plants were already operational, with a large number in various stages of planning or construction (Parvez et al. 2025). The data is significantly lower compared to the target of 5000 plants, indicating a low rate of implementation. Among the key hindrances to broader use, the financial feasibility of CBG projects presents a significant challenge, often characterized by a large capital outlay, a lengthy payback period, and limited access to concessional finance. Although

OMCs can guarantee offtake, the price they propose to buy CBG is not always sufficient to make a profit for small producers, particularly considering the expenses of maintaining the plant, transportation, and acquiring feedstock.

Finally, the implementation of the SATAT program demonstrates that India is highly motivated to utilize on-the-ground renewable gas production as an effective policy to decarbonize the transport and energy sectors. It is an integrated sustainability model due to its focus on waste valorization, rural entrepreneurship, and intersectoral collaboration. However, to achieve this fully, the initiative must contend with structural and operational challenges, utilizing improved financial mechanisms, streamlined regulations, and capacity-building projects within the entire stakeholder ecosystem. Through a well-implemented plan and persistence in policy evolution, SATAT can potentially revolutionize India's energy sector, which will also facilitate achieving climate targets, supporting rural Indian economies, and promoting sustainable development.

16.3.4 USA: Renewable Fuel Standard (RFS)

RFS is one of the key policies in the United States that has been seeking to adopt renewable fuel into the United States transportation fuel supply, hence boosting energy security, cutting GHG emissions, and enhancing the agricultural economy (Singh 2024). The program was previously established under the Energy Policy Act of 2005. Still, it was significantly expanded by the Energy Independence and Security Act (EISA) of 2007, which increased the policy scope and set short-term blending volume targets until 2022, as well as the pace of implementation. The core of the RFS is that refiners and importers of fuels must add more and more renewable fuels to transportation fuel every year. The major regulatory agency charged with the administration of the program is the EPA (Qasem et al. 2024). The EPA establishes the Renewable Volume Obligations (RVOs), which are the annual volume requirements for various types of biofuels. These are the following categories:

- Conventional biofuels – these are mainly corn starch-based ethanol, which is limited to 15 billion gallons each year.
- Advanced biofuels – advanced biofuels are the fuels that reduce lifecycle GHG emissions by at least 50% as compared to petroleum fuels.
- Biomass-derived diesel – consists of biodiesel and renewable diesel that are made from oils and fats.

Cellulosic biofuels are derived from non-food-based plant material; examples include corn stover, switchgrass, or wood chips, and they must reach a 60% GHG reduction threshold. The RFS implements the market-based credit trading system to ensure compliance. Every gallon of renewable fuel that is produced or imported generates a Renewable Identification Number (RIN). To demonstrate compliance with the RVOs, refiners, importers, and other obligated parties are required to acquire and deliver a certain number of RINs to the EPA annually (Robertson et al. 2017). Organizations may purchase RINs on the open market, and they can be obtained by those individuals who have excess credits in case it becomes impractical or uneconomical to blend. This makes the system both flexible and vulnerable to market fluctuations. There are several objectives of the RFS. Environmentally, it aims to reduce the carbon footprint of the transportation sector, which is a significant source of greenhouse gas, contributing to emissions in the United States (Renewable Identification Numbers (RINs) under the Renewable Fuel Standard Program I US EPA n.d.). The policy encourages domestic agriculture by facilitating a stable market for maize, soybeans, and other feedstocks as raw materials in the production of biofuel, thereby promoting economic growth. It will strategically achieve this by developing a domestic biofuels industry, aiming to reduce dependence on international oil imports and enhance energy sovereignty. There have, however, been numerous discussions and criticisms of the RFS, particularly regarding its long-term viability and effectiveness. One of the obstacles is the so-called blend wall, which refers to the limitation of mixing ethanol into gasoline that does not require modifications to existing engines or the gas infrastructure.

Most of the gasoline sold in the United States is a blend of E10 (10% ethanol). Due to suitability and warranty concerns, automakers, fuel retailers, and consumers reject the idea of moving to E15 (15% ethanol) or even E85 (85% ethanol). Moreover, maize ethanol, the most popular conventional biofuel in the United States, has been challenged due to its environmental impact (Almost All U.-S. Gasoline Is Blended with 10% Ethanol—U.S. Energy Information Administration (EIA) n.d.). Critics argue that converting forest or grassland to crops for maize production may result in the carbon savings achieved through fuel production being offset by the land use changes involved. Due to the high level of agricultural intensification associated with large-scale maize production, concerns have also been raised about biodiversity and water resources. Despite the RFS promoting the production of advanced and cellulosic biofuels, its widespread use has not kept pace with expectations due to technological difficulties, prohibitive prices, and uneconomic viability. Another area of contention involves the RIN market, which has experienced significant price volatility (Winkler et al. 2021). Small refiners, in particular, have argued that the cost of purchasing RINs on the open market imposes an unfair financial burden on them. This has led to legal battles and lobbying efforts to modify or exempt certain refiners from the RFS obligations. Over the years, the EPA has granted "Small Refinery Exemptions" (SREs), which have drawn criticism from environmental groups and the biofuel industry for undermining the integrity of the policy. In recent years, discussions have arisen regarding the reform or replacement of the RFS, particularly as the US transitions toward electrification of transportation and pursues net-zero emission goals. Nonetheless, the RFS remains one of the most significant and comprehensive biofuel mandates globally, and its evolution continues to influence both domestic energy policy and international approaches to renewable fuels.

16.3.5 EU: Renewable Energy Directive (RED II)

National biofuel regulatory frameworks are comprehensive policy tools that enable the control and financing of biofuel production within a nation's energy, agricultural, and natural resource sectors. A combination of required targets, legal requirements, fiscal incentives, sustainability standards, and institutional support systems typically characterizes such frameworks. Their primary objective is to minimize greenhouse gas emissions, diversify energy sources, enhance energy security, and stimulate rural and industrial development by promoting the domestic exploitation of renewable resources. As an example, ethanol and biodiesel blending requirements have been introduced in many countries, including India, where E20 (20% ethanol in petrol) or B35 (35% biodiesel in diesel) is being required to generate a solid biofuel demand and reduce dependence on foreign fossil fuel sources. There is also a tendency to recognize national frameworks that encourage the use of advanced biofuels derived from non-food sources, such as agricultural press, municipal waste, or algae, which have lower environmental implications and improved carbon mitigation capabilities. These frameworks have sustainability standards that ensure the appropriate use of land, safeguard biodiversity, and prevent food security issues. Additionally, governments can offer financial incentives, such as tax relief, subsidies, or low-interest loans, to producers and distributors to stimulate investment in biofuel infrastructure and innovation. In general, country regulatory regimes play a pivotal role in steering the development of the biofuels industry in a direction that is adaptable, resource-efficient, and financially feasible, where energy policy converges with climate issues and development agendas.

16.4 International Regulatory Frameworks and Agreements

16.4.1 Role of International Bodies: IEA, FAO, IRENA

The International Renewable Energy Agency (IRENA), the Food and Agriculture Organization of the United Nations (FAO), and the International Energy Agency (IEA) have all made notable contributions to the development of the global biofuel regulatory framework through their capacity-building activities, policy advocacy, and technical guidance. Besides being potential sources of information, such organizations, particularly those existing in the Global South, also serve as catalysts for harmonizing biofuel governance at the national and regional levels. They have the task of addressing four areas of serious concern: (1) evidence-based policymaking; (2) ensuring sustainability; (3) technical innovation; and (4) international collaboration. The International Energy Agency (IEA) has already become a central player in global energy governance, providing the detailed data, long-term projections, and scenario modeling necessary to develop sustainable energy policies. A study

(Renewables 2023) published by the IEA in May 2023 identified the post-pandemic rebound of the biofuel sector, estimating that global demand will increase by 9%. Both Brazil and India are mentioned as contributors to this high rate of growth.

The IEA also actively involves the member states in the discussion of the issues of indirect land use change (ILUC), technology-neutral regulations, and lifetime GHG accounting, on top of the statistical monitoring through the Biofuture Platform Initiative. Remarkably, the IEA has collaborated with Indian institutions to develop ethanol blending pathways and estimate the outcomes of such schemes as SATAT and the Ethanol Blending Programme (EBP). These initiatives provide a science-based plan for India to decarbonize its approach and meet its net-zero obligations. The FAO is an organization with an entirely different yet equally important role. Its major framework, Bioenergy and Food Security (BEFS), has been implemented in more than 40 nations, such as India, where bioenergy overlaps with soil fertility, water availability, and livelihoods within rural regions. Co-facilitating the Global Bioenergy Partnership (GBEP) with IRENA and UNEP, FAO has established the sustainability indicators that national governments can adopt to monitor degradation, GHG emissions, and socio-economic equity. Such studies are critical within the Indian context, where FAO-facilitated research has helped calibrate the trade-offs between sugarcane and maize-based ethanol and food production, as well as groundwater stress, to provide a precautionary perspective that is largely lacking in energy-only thinking.

In the meantime, IRENA can support the industry by focusing on issues such as technology diffusion, ease of investment, and innovation. IRENA has developed country-based scenarios for scaling up advanced (2G and 3G) biofuels, including lignocellulosic ethanol and algae-generated biodiesel, through its REmap initiative. Based on favorable policies and investments, it projects that biofuels may be able to account for up to 20% of the world's transport energy by 2050. In the case of India, IRENA has provided direct assistance through investment matchmaking and the development of compressed biogas (CBG) plants under the SATAT initiative, as well as in the development of biodiesel value chains utilizing used cooking oil and non-edible oil seeds. More importantly, IRENA also works on harmonizing the mandatory certification of biofuels, which enables Indian biofuels to meet international sustainability criteria and enter the export market, such as the EU, which imposes strict carbon intensity threshold criteria through the Renewable Energy Directive II (RED II). Collectively, these organizations foster a polycentric yet coordinated approach to biofuel governance. The IEA's macroeconomic policy focus, FAO's ecological and food security safeguards, and IRENA's technology and market orientation form a triad of complementary expertise. Their interventions, though distinct, converge on the goal of guiding nations toward a sustainable bio-economy. For developing countries like India, this synergy translates into more coherent regulatory frameworks, informed investment strategies, and enhanced resilience in balancing energy, environmental, and equity goals.

16.5 ISO Standards for Sustainability and Certification

The increasing focus on the sustainability of global energy policy has raised the question of the importance of certification systems, and this is particularly relevant today in the bioenergy sector. ISO 13065:2015, also known as the Sustainability Criteria for Bioenergy, represents a significant development in establishing international standards for the sustainability evaluation of biofuels. It offers a voluntary and holistic framework for assessing sustainability performance throughout the lifecycle of bioenergy, encompassing the cultivation of feedstock, biomass processing, distribution, and end-use. This standard promotes consistency and transparency by establishing principles and standards in all environmental, social, and economic aspects, including greenhouse gas emissions, biodiversity maintenance, soil and water protection, labor provision, land rights, and supply chain stability. The deployment of ISO 13065 can be closely linked to the global trend of policies, which suggest moving toward the need for reporting on carbon accounting based on the lifecycle of products, their relevance to ethical sourcing, and community benefits.

ISO 13065 exerts a significant role in the national and international regulatory schemes. In the case of the European Union, the Renewable Energy Directive (RED II) has enforced stringent sustainability standards for biofuels, which will be used to meet renewable energy obligations. According to the current state of organizations, many certification schemes, such as the Roundtable on Sustainable Biomaterials (RSB) and the International Sustainability and Carbon Certification (ISCC), tend to benchmark their indicators against ISO 13065 to meet regulatory requirements. India's National Policy on Biofuels, too—but in particular after its 2022 update—resonates with much of what is ISO-compatible in reducing the feedstock base, minimizing lifecycle emissions, seeing a construction Benallora GHGs, and fostering inclusive growth. In India, ISO 13065 is not a binding law; however, it serves as a guiding framework for enhancing the viability and cross-border recognition of local certification efforts, particularly in initiatives such as the EBP and the SATAT program. Despite its positivity, ISO 13065 is not without weaknesses. Adoption is voluntary, making it irregular, particularly in areas with weak institutional capacity or ineffective enforcement of regulations.

Furthermore, although the standard includes general sustainability measures, it does not specify particular levels to achieve compliance; thus, there is variation in the interpretation and application of the standard. The existence of this gap makes it significant to incorporate ISO 13065 into legally binding instruments or to integrate it into third-party certification schemes that have well-developed auditing procedures. The latest systematic evidence map, presented by Harvey et al. (2025), highlighted uncertainties in the efficacy of sustainability certifications in the bio-based industries, with their effects on greenhouse gas emissions showing measurable inconsistencies depending on country-specific adaptations and the rigors of program enforcement.

In the long-term perspective, the strategic alignment of the ISO 13065 standard with national regulations for biofuels will have numerous benefits, especially for countries that are aligning their energy transition with the Sustainable Development Goals (SDGs) and international trade protocols. In India, it would be beneficial to enhance export preparedness, attract foreign investments, and develop advanced biofuel technologies. Additionally, it is crucial to establish traceability systems in decentralized biomass supply chains. Byrne et al. (2025) advocate for replacing the voluntary compliance policy with mandatory sustainability reporting, particularly in sectors with high climatic and social vulnerability, such as transportation and agriculture. In this regard, ISO 13065 can serve as a technical and moral guide for sustainable bioenergy development. In conclusion, ISO 13065 is a pivotal tool in advancing global sustainability governance for the bioenergy sector. While its uptake depends on voluntary commitment, its principles provide a coherent structure for nations and industries to demonstrate responsible biofuel production and use. Its alignment with emerging Environmental, Social, and Governance (ESG) standards, international certification schemes, and lifecycle-based regulatory frameworks positions it as a key enabler of harmonized, credible, and climate-resilient biofuel markets. For countries like India, operationalizing ISO 13065 within national policy frameworks could yield transformative benefits in sustainability reporting, cross-border trade, and long-term energy security.

16.6 World Trade Organization (WTO) and Trade-Related Aspects of Biofuels

The increase in the production and use of biofuel has led to the intersection of global sustainability objectives, environmental laws, and trade policy in complex ways. This interaction revolves around the WTO, which governs international trade through legally binding regulations geared toward promoting fair and non-discriminatory practices among states. The compatibility of national biofuel policies with the WTO trade rules, namely under the General Agreement on Tariffs and Trade (GATT) and the Agreement on Technical Barriers to Trade (TBT), is of concern since countries have adopted this policy in an attempt to reduce greenhouse gas emissions, assure energy security, and strengthen the economies of rural areas (Raman et al. 2025). Whether biofuel sustainability criteria and subsidies can act as trade barriers is a serious concern. The European Union Renewable Energy Directive (RED II) imposes sustainability requirements on the imported biofuels, including restrictions on land-use change and life-cycle GHG savings. These regulations can be detrimental to poor countries whose biofuel manufacturers struggle to meet them, potentially leading to conflicts due to the WTO's non-discrimination rules.

The SCM Agreement of the WTO can also present a problem with domestic subsidies paid by countries like the United States (through the Renewable Fuel Standard Act) or India (via the Pradhan Mantri JI-VAN Yojana) to spur domestic production of compressed biogas (CBG) or bioethanol. These subsidies can be contested in cases where they disadvantage foreign manufacturers or traders when they are linked to output levels rather than environmental performance ((PDF) Special Economic Zones in India and the WTO Agreement on Subsidies and Countervailing Measures: An Uneasy Coexistence n.d.). Hence, the WTO scrutiny might not even be accorded to biofuel policies that are explicitly aimed at addressing climate change and produced with the support of scientific evidence, guided equally by domestic and foreign producers. There is a growing movement to meet the trade and environmental goals in bilateral and multilateral meetings. For example, at the Committee on Trade and Environment in the WTO, approaches to aligning biofuel certification systems and considering the equivalence of national standards have been discussed. Nonetheless, significant regulatory fragmentation persists between the EU, the United States, and emerging economies. This makes it clear that multilateral discussion and cooperation mechanisms are necessary to ensure that climate-driven biofuel policies do not unintentionally escalate into trade wars (Chinn and Irwin 2025). Conclusively, national biofuel policies can legitimately meet sustainable and energy objectives; however, the provisions thereof in terms of trade should be well-formulated to align with WTO principles. Coordination on an international scale, common recognition of certification schemes, and capacity building in developing nations will be necessary to prevent asymmetries in the regulatory regime from becoming points of contention.

16.7 Emerging Trends and Innovations in Regulation

Contemporary biofuel regulations, both at the national and international levels, are rapidly evolving to address the complex demands of sustainability, energy security, and economic development. A key trend in this evolution is the increased support for advanced biofuels, particularly those derived from lignocellulosic biomass such as agricultural residues, forestry waste, and non-food energy crops. These 2G and 3G biofuels, including cellulosic ethanol and algae-based fuels, are prioritized due to their lower environmental footprints and minimal impact on food systems. Countries like India and the EU have implemented dedicated financial mechanisms and viability gap funding schemes to promote advanced biorefinery infrastructure, as seen in India's Pradhan Mantri JI-VAN Yojana, which supports lignocellulosic ethanol production through capital subsidies and policy guarantees (Singh 2024). Parallelly, regulatory frameworks are increasingly emphasizing integration with circular economy principles. This entails maximizing resource efficiency by converting waste streams—such as municipal solid waste, crop stubble, and used cooking oil—into bioenergy. Initiatives like India's SATAT and the EU's waste-to-energy directives encourage the production of biogas and compressed biogas (CBG) using decentralized, waste-fed systems, thereby aligning energy policies with broader goals of waste reduction, soil health, and rural development.

Another critical innovation is the enhancement of policy support for research, development, and demonstration (RD&D). National frameworks now recognize the importance of de-risking technological innovation by funding pilot plants, hybrid biorefinery models, and public-private partnerships. The U.S. Department of Energy's Bioenergy Technologies Office (BETO) and the EU Horizon Europe program exemplify how governments are incentivizing innovation across the biofuel value chain—from enzymatic hydrolysis technologies to integrated refinery models capable of producing multiple co-products such as bio-chemicals, animal feed, and fertilizers. These models improve economic viability and align biofuel production with industrial symbiosis concepts.

Furthermore, a significant regulatory innovation lies in the use of market-based instruments such as incentives, subsidies, and carbon pricing mechanisms. Feed-in tariffs for bio-based electricity, tax breaks for ethanol production, and direct payment schemes for CBG producers are increasingly commonplace. Additionally, carbon pricing frameworks, such as the EU Emissions Trading System (ETS) and the Low Carbon Fuel Standard (LCFS) in California, are rewarding low-carbon fuels—including advanced biofuels—with tradable carbon credits. These systems help internalize environmental externalities and improve the market competitiveness of sustainable fuels, making them attractive alternatives to fossil-based energy sources. Collectively, these emerging trends represent a paradigm shift in biofuel regulation—from volume mandates to value-driven, sustainability-aligned, and innovation-oriented frameworks. They signal a maturing policy ecosystem where energy, environment, and economy are interconnected through regulatory foresight, making biofuels a cornerstone of a cleaner and more resilient global energy future.

16.8 Challenges and Gaps in Regulatory Frameworks

One key challenge is the lack of comparable biofuel policies in different nations, leading to compliance havoc. For example, despite the strict sustainability standards outlined in the Renewable Energy Directive (RED II) within the European Union, other countries, such as the United States or Brazil, implement various benchmarks in areas including greenhouse gas emissions reduction, land-use conditions, and qualifying feedstocks. Such divergence in regulation prevents mutual recognition of the certification, and the export/importation process of biofuels is made difficult, particularly for developing countries that lack the institutional capacity to navigate the various standards.

The disparities tend to raise trade hurdles and technical regulations, which, despite being environmentally justified, form barriers to trade. Nationally derived sustainability certification schemes, land-use change restrictions, and life-cycle calculations of cumulative GHG emissions are established as policies in the country. However, they are not necessarily internationally aligned. Consequently, they risk violating WTO requirements governing non-discrimination and fair market access. The instance of palm oil biodiesel is typical of this situation, as producers in Southeast Asia raised formal objections to the EU measures in accordance with WTO policies, citing them as discriminatory and protectionist (Chandra 2025).

Infrastructure and implementation are yet another go-to solution. Although nations such as India and Brazil have ambitious blending targets and investment strategies for advanced biofuels, the physical infrastructure, including decentralized biorefineries, storage, and distribution networks, is inadequate. The limited transport logistics in India, the inconsistency associated with feedstock supply, and the lack of blending capacity at the retail level, to some extent, undermine the success of the Ethanol Blending Programme of the country, as well as the SATAT initiative (Parvez et al. 2025). Similarly, most developing countries lack the technical expertise and institutional framework necessary to implement the use of international certification schemes.

Finally, vigilant and flexible regulatory environments are becoming increasingly crucial due to the continuous evolution of technology and the dynamic nature of the market; therefore, these regulatory environments should be dynamic as well. The majority of current policies on biofuels are undynamic, as they tend to be pegged to stationary blending targets or fixed by feedstock incentives. This may be hampering innovation and making it difficult to accommodate newer technologies (such as 3G/4G biofuels) or changes in food security, climate variability, or geopolitical upheavals to feedstock access. The regulatory system of the future must be flexible to allow it to adjust its policies in time, through the feedback of multiple stakeholders, and interconnection with climate and energy transition plans (Singh 2024). To summarize and identify the most significant factors influencing the global biofuel policy, which include the rising trend toward sustainability in energy, fragmentation of regulation, trade limitations, inadequate infrastructure, and policy misdesign, can be highlighted as the main bottlenecks. To address these gaps, it is essential to enhance international coordination, streamline trade regulations, provide capacity-building assistance, and adopt adaptive policymaking that aligns with changes in technology and the environment.

16.9 Conclusion

Biofuels regulation at the national and international levels is a crucial tool in the global journey toward sustainable energy. The models demonstrate an increased awareness of the need to wean the country off fossil fuel dependency, reduce greenhouse gas emissions, build better lives in the countryside, and also harness renewable technologies. On the national level, examples include the India EBP and the EU Renewable Energy Directive, which show the orderly process of plans for feedstock diversification, blending targets, and sustainability thresholds. It is also true that coordination work is increasingly taking place internationally, albeit in a more fragmented manner, with the aim of aligning trade rules with climate commitments, given the growing relevance of WTO-compatible certification schemes and multilateral sustainability standards.

Some of the regulatory innovations that can be identified include support for advanced biofuels made from lignocellulosic biomass, the incorporation of circular economy principles, selective R&D investments, and the application of market-based mechanisms, such as subsidies and carbon pricing. Commutable variations in enforcement, infrastructural constraints, and the food-versus-fuel debate are some of the prevailing challenges. As we move into the future, adaptive governance, interstate cooperation, and the fair transfer of technology will shape the success of these frameworks, which should spread the benefits of biofuel development globally and not have adverse environmental effects.

References

Almost all U.S. gasoline is blended with 10% ethanol—U.S. Energy Information Administration (EIA) (n.d.) Retrieved June 25, 2025, from https://www.eia.gov/todayinenergy/detail.php?id=26092

Araújo K, Mahajan D, Kerr R, Silva MD (2017) Global biofuels at the crossroads: an overview of technical, policy, and investment complexities in the sustainability of biofuel development. Agriculture 7(4):32. https://doi.org/10.3390/AGRICULTURE7040032

Byrne MJ, John M, Biswas W (2025) Future risk from current sustainability assessment frameworks for the resource sector. Sustainability 17(3):960–960. https://doi.org/10.3390/SU17030960

Chandel AK, Bhatia L, Garlapati VK, Roy L, Arora A (2017) Biofuel policy in Indian perspective: socioeconomic indicators and sustainable rural development. Sustain Biofuels Dev India:459–488. https://doi.org/10.1007/978-3-319-50219-9_19/

Chandra S (2025) Analysis of environmental restrictions as trade barriers to international trades- in the backdrop of the case study of Hawaiian Nuclear Reactor. https://doi.org/10.31219/OSF.IO/F5SYE

Chenavaz RY, Dimitrov S (2024) From waste to wealth: policies to promote the circular economy. J Clean Prod 443:141086. https://doi.org/10.1016/J.JCLEPRO.2024.141086

Chinn MD, Irwin DA (2025) International Trade. In: The SAGE encyclopedia of world poverty

Das S (2020) The National Policy of biofuels of India—a perspective. Energy Policy 143:111595. https://doi.org/10.1016/J.ENPOL.2020.111595

Electricity—Renewables 2023—Analysis—IEA (n.d.) Retrieved June 25, 2025, from https://www.iea.org/reports/renewables-2023/electricity

Ethanol Blending Programme (EBP), Objectives, Achievements (n.d.) Retrieved June 25, 2025, from https://pwonlyias.com/ethanol-blending-programme/

Gao X, Ge Y, Qu J, Han J, Maraseni TN, Liu C, Huang K, Xu L, Wang D, Li H (2025) How renewable energy policies cut greenhouse gas emissions: insights from advanced data analysis techniques. J Environ Manag 389. https://doi.org/10.1016/j.jenvman.2025.125946

Government measures to increase Ethanol Blending beyond 20% | IBEF (n.d.) Retrieved June 25, 2025, from https://www.ibef.org/news/government-measures-to-increase-ethanol-blending-beyond-20

Guariguata MR, Masera OR, Johnson FX, Von Maltitz G, Bird N, Tella P, Martínez-Bravo R (2011) A review of environmental issues in the context of biofuel sustainability frameworks. 69. https://doi.org/10.17528/CIFOR/003529

Harvey WJ, Black N, Essaouabi S, Petrokofsky L, Rangan V, Bird MS, Villar D, Waite M, Petrokofsky G (2025) The impact of sustainability certification schemes and labels on greenhouse gas emissions: a systematic evidence map. Sustainability 17(2):792. https://doi.org/10.3390/su17020792

Hassan Q, Viktor P, Al-Musawi J, Mahmood Ali B, Algburi S, Alzoubi HM, Khudhair Al-Jiboory A, Zuhair Sameen A, Salman HM, Jaszczur M (2024) The renewable energy role in the global energy transformations. Renew Energy Focus 48:100545. https://doi.org/10.1016/J.REF.2024.100545

Heijman W, Szabó Z, Veldhuizen E (2019) The contribution of biorefineries to rural development: the case of employment in Hungary. Stud Agric Econ 121(1). https://doi.org/10.22004/AG.ECON.287546

Khalil AK, Yilmaz S (2025) The impact of environmental protection expenditures on reducing greenhouse gas emissions. Sustainability 17(7):3192. https://doi.org/10.3390/SU17073192

Koirala BP, Koliou E, Friege J, Hakvoort RA, Herder PM (2016) Energetic communities for community energy: a review of key issues and trends shaping integrated community energy systems. Renew Sust Energ Rev 56:722–744. https://doi.org/10.1016/J.RSER.2015.11.080

Mandatory B40 Biodiesel to Be Implemented in March 2025 (n.d.) Retrieved June 25, 2025, from https://en.infosawit.com/news/15262/mandatory-b40-biodiesel-to-be-implemented-in-march-2025

Monti A, Polugodina M (2021) Biofuel and biogas policies: economic, regulatory, and sustainability challenges. pp 87–104. https://doi.org/10.1007/978-3-319-95864-4_126

National Policy On Biofuels (2018)—IMPRI Impact And Policy Research Institute (n.d.) Retrieved June 24, 2025, from https://www.impriindia.com/insights/national-policy-biofuels-2018/

Parvez M, Lal S, Khan O, Ahmad M, Lal Meena S, Salvi BL (2025) Sustainable regulatory framework for ethanol and biodiesel blending with petro-fuels in India to meet fuel demand: a review of biofuel policies. Biofuels:1–14. https://doi.org/10.1080/17597269.2024.2447156

Pattanaik BP, Misra RD (2017) Effect of reaction pathway and operating parameters on the deoxygenation of vegetable oils to produce diesel range hydrocarbon fuels: a review. Renew Sust Energ Rev 73:545–557. https://doi.org/10.1016/j.rser.2017.01.018

(PDF) Special Economic Zones in India and the WTO Agreement on Subsidies and Countervailing Measures: An Uneasy Coexistence (n.d.) Retrieved June 25, 2025, from https://www.researchgate.net/publication/291246781_Special_Economic_Zones_in_India_and_the_WTO_Agreement_on_Subsidies_and_Countervailing_Measures_An_Uneasy_Coexistence

Pramanik A, Sinha A, Chaubey KK et al (2023) Second-generation bio-fuels: strategies for employing degraded land for climate change mitigation meeting united nation-sustainable development goals. Sustainability 15(9):7578. https://doi.org/10.3390/SU15097578

Qasem NAA, Mourad A, Abderrahmane A, Said Z, Younis O, Guedri K, Kolsi L (2024) A recent review of aviation fuels and sustainable aviation fuels. J Therm Anal Calorim 149(10): 4287–4312. https://doi.org/10.1007/S10973-024-13027-5

Raman R, Sreenivasan A, Kulkarni NV, Suresh M, Nedungadi P (2025) Analyzing the contributions of biofuels, biomass, and bioenergy to sustainable development goals. IScience 28(4): 112157. https://doi.org/10.1016/J.ISCI.2025.112157

Renewable Identification Numbers (RINs) under the Renewable Fuel Standard Program | US EPA (n.d.) Retrieved June 25, 2025, from https://www.epa.gov/renewable-fuel-standard/renewable-identification-numbers-rins-under-renewable-fuel-standard-program

Review of policy frameworks to promote sustainable biofuels with low GHG emissions—Bioenergy (n.d.) Retrieved June 24, 2025, from https://www.ieabioenergy.com/blog/publications/review-of-policy-frameworks-to-promote-sustainable-biofuels-with-low-ghg-emissions/

Robertson GP, Hamilton SK, Barham BL, Dale BE, Izaurralde RC, Jackson RD, Landis DA, Swinton SM, Thelen KD, Tiedje JM (2017) Cellulosic biofuel contributions to a sustainable energy future: choices and outcomes. Science 356(6345). https://doi.org/10.1126/SCIENCE.AAL2324/ASSET/429EA401-6081-44E5-B030-86109B88E46E/ASSETS/GRAPHIC/356_AAL2324_F6.JPEG

S AK (2024) The impact of land use regulations on sustainable agricultural practices in deforestation hotspots. Law Econ 3(12):51–58. https://doi.org/10.56397/LE.2024.12.08

SATAT Scheme (Sustainable Alternative Towards Affordable Transportation)| National Portal of India (n.d.) Retrieved June 25, 2025, from https://www.india.gov.in/satat-scheme-sustainable-alternative-towards-affordable-transportation

Shah KU (2023) National environmental regulatory systems for the management of environmental impacts in small island jurisdictions. Curr Opin Environ Sustain 64:101326. https://doi.org/10.1016/J.COSUST.2023.101326

Silalertruksa T, Gheewala SH, Hünecke K, Fritsche UR (2012) Biofuels and employment effects: implications for socio-economic development in Thailand. Biomass Bioenergy 46:409–418. https://doi.org/10.1016/j.biombioe.2012.07.019

Singh R (2024) Domestically produced bioethanol contributing to India's pursuit of energy self-reliance and de-carbonization of the road transport sector. SAE Techn Paper Ser 1. https://doi.org/10.4271/2024-28-0126

Song YP, Du LL, Wang BG, Zhang LM, Lin HY, Ma F, Bai Y, Wang AJ, Wang HC, Ren NQ (2025) Flexible carbon source regulation for mitigating greenhouse gas emissions in full-scale wastewater treatment. Environ Sci Technol 59(19). https://doi.org/10.1021/ACS.EST.4C12704

Sulis DB, Lavoine N, Sederoff H, Jiang X, Marques BM, Lan K, Cofre-Vega C, Barrangou R, Wang JP (2025) Advances in lignocellulosic feedstocks for bioenergy and bioproducts. Nat Commun 16(1):1–13. https://doi.org/10.1038/s41467-025-56472-y

Transport biofuels—Renewables 2023—Analysis—IEA (n.d.) Retrieved June 25, 2025, from https://www.iea.org/reports/renewables-2023/transport-biofuels

Usmani RA, Mohammad AS, Ansari SS (2023) Comprehensive biofuel policy analysis framework: a novel approach evaluating the policy influences. Renew Sust Energ Rev 183:113403. https://doi.org/10.1016/J.RSER.2023.113403

Wang Q, Sun Z, Guo J, Li R (2023) The more effective option to combat environmental degradation: energy efficiency vs. renewable energy vs. natural gas? Energy 283:128512. https://doi.org/10.1016/J.ENERGY.2023.128512

Winkler K, Fuchs R, Rounsevell M, Herold M (2021) Global land use changes are four times greater than previously estimated. Nat Commun 12(1). https://doi.org/10.1038/s41467-021-22702-2

Ye Y, Hu J, Zhang Z, Zhong W, Zhao Z, Zhang J (2023) Effect of different ratios of gasoline-ethanol blend fuels on combustion enhancement and emission reduction in electronic fuel injection engine. Polymers 15(19):3932. https://doi.org/10.3390/POLYM15193932

MIX
Papier aus verantwortungsvollen Quellen
Paper from responsible sources
FSC® C105338

If you have any concerns about our products,
you can contact us on
ProductSafety@springernature.com

In case Publisher is established outside the EU,
the EU authorized representative is:
Springer Nature Customer Service Center GmbH
Europaplatz 3, 69115 Heidelberg, Germany

Printed by Libri Plureos GmbH
in Hamburg, Germany